COMPOSITE STRUCTURES

4

VOLUME 1

Analysis and Design Studies

Proceedings of the 4th International Conference on Composite Structures, held at Paisley College of Technology, Scotland, from 27th to 29th July 1987, co-sponsored by the Scottish Development Agency, the National Engineering Laboratory, the US Air Force European Office of Aerospace Research and Development and the US Army, Research, Development and Standardisation Group, UK.

Volume 1: Analysis and Design Studies
Volume 2: Damage Assessment and Material Evaluation

Also published by Elsevier Applied Science Publishers:

COMPOSITE STRUCTURES

(*Proceedings of 1st International Conference, 1981*)

COMPOSITE STRUCTURES 2

(*Proceedings of 2nd International Conference, 1983*)

COMPOSITE STRUCTURES 3

(*Proceedings of 3rd International Conference, 1985*)

ELSEVIER APPLIED SCIENCE PUBLISHERS LTD
Crown House, Linton Road, Barking, Essex IG11 8JU, England

Sole Distributor in the USA and Canada
ELSEVIER SCIENCE PUBLISHING CO., INC.
52 Vanderbilt Avenue, New York, NY 10017, USA

WITH 92 TABLES AND 284 ILLUSTRATIONS

Softcover reprint of the hardcover 1st edition 1987

British Library Cataloguing in Publication Data

International Conference on Composite Structures
(4th: 1987: Paisley College of Technology)
Composite structures 4.
1. Composite materials 2. Composite construction
I. Title II. Marshall, I. H. III. Scottish
Development Agency
624.1'8 TA664

Library of Congress Cataloging-in-Publication Data

Composite structures 4.

Proceedings of the 4th International Conference on Composite Structures, held at Paisley College of Technology, Scotland, from July 27–29, 1987, and co-sponsored by the Scottish Development Agency and others.
Bibliography: p.
Includes index.
1. Composite construction—Congresses. 2. Composite materials—Congresses. I. Marshall, I. H. (Ian H.) II. Scottish Development Agency. III. International Conference on Composite Structures (4th: 1987: Paisley College of Technology, Scotland, UK)
IV. Title: Composite structures four.
TA664.C636 1987 620.1'18 87-13475

ISBN-13:978-94-010-8047-7 e-ISBN-13:978-94-009-3455-9
DOI:10.1007/978-94-009-3455-9

COMPOSITE STRUCTURES

4

VOLUME 1

Analysis and Design Studies

Edited by

I. H. MARSHALL

Department of Mechanical and Production Engineering,
Paisley College of Technology, Scotland, UK

ELSEVIER APPLIED SCIENCE
LONDON and NEW YORK

Preface

The papers contained herein were presented at the Fourth International Conference on Composite Structures (ICCS/4) held at Paisley College of Technology, Scotland in July 1987. The Conference was organised and sponsored by Paisley College of Technology. It was co-sponsored by the Scottish Development Agency, the National Engineering Laboratory, the US Air Force European Office of Aerospace Research and Development and the US Army Research, Development and Standardisation Group—UK. It forms a natural and ongoing progression from the highly successful First, Second and Third International Conferences on Composite Structures (ICCS/1, ICCS/2 and ICCS/3) held at Paisley in 1981, 1983 and 1985 respectively.

There is little doubt that composite materials are rightfully claiming a prominent role in structural engineering in the widest sense. Moreover, the range and variety of useful composites has expanded to a level inconceivable a decade ago. However, it is also true that this increasing utilisation has generated an enhanced awareness of the manifold factors which dictate the integrity of composite structures.

This is indeed a healthy attitude to a relatively new dimension in structural engineering which will have an increasingly dominant role as the century progresses. Both the diversity of application of composites in structural engineering and the endeavours which will ensure their fitness for purpose are reflected herein.

It is also inescapably true that traditional engineering design principles and practices are largely inappropriate when contemplating the viability of composites as alternatives to traditional engineering materials. Consequently, engineers and scientists are being confronted with a new and sometimes daunting, range of design concepts which they are, by and large, ill equipped to appreciate. Only by critically reassessing the education and training of today's engineers and scientists can proper advances in composite structures be sustained. Moreover, this also demands that the fruits of current research and development are available in a realistically digestible form. Without both of these ingredients there is little doubt that the projected future of composites in structural engineering will be impeded, or indeed compromised. To this end it is hoped that the present volume will provide both an overall appreciation of the current and future potential of composite structures and an awareness of the factors which dictate their safe usage.

Authors and delegates from in excess of twenty countries have combined to make the present conference a truly international forum of specialists in composite structures, reflecting a truly global appreciation of this expanding area of technology. Topics addressed range from the use of composites in wind turbine engineering to their usage in outer space, a truly expansive spectrum.

As always, an international conference can only take place and contribute to knowledge through the willing and enthusiastic efforts of a number of individuals. In particular, thanks are due to the following:

The International Advisory Panel

E. Anderson	Battelle Laboratories (Switzerland)
J. Anderson	Paisley College of Technology (UK)
W. M. Banks	University of Strathclyde (UK)
A. M. Brandt	Polish Academy of Sciences (Poland)
A. R. Bunsell	Ecole des Mines de Paris (France)
W. S. Carswell	National Engineering Laboratory (UK)
T. Hayashi	Japan Plastic Inspection Association (Japan)
R. M. Jones	Virginia Polytechnic & State University (USA)
L. N. Phillips, OBE	Consultant, Farnborough (UK)
J. Rhodes	University of Strathclyde (UK)
S. W. Tsai	Air Force Materials Laboratory (USA)
J. A. Wylie	Paisley College of Technology (UK)

The Local Organising Committee

S. K. Harvey
J. Kirk
G. Macaulay
J. S. Paul

The Conference Secretary

Mrs C. A. MacDonald

Grateful thanks are due to other individuals who contributed to the success of the event. A final thanks to Nan, Simon, Louise and Richard for their support during the conference.

I. H. Marshall

Contents

Buckling of Platework Structures

(*Chairman:* C. W. BERT, *University of Oklahoma, Norman, USA*)

Postbuckling of Platework Structures

(*Joint Chairmen:* J. RHODES, *University of Strathclyde, Glasgow, Scotland, UK and* G. J. SIMITSES, *Georgia Institute of Technology, Atlanta, USA*)

Aerospace Structures

(*Joint Chairmen:* R. F. MOUSLEY, *Royal Aircraft Establishment, Farnborough, UK* and E. DEMUTS, *Wright-Patterson AFB, Ohio, USA*)

Wind Turbine Design

(*Chairman:* J. G. GOREE, *Clemson University, South Carolina, USA*)

Pipes and Pressure Vessels

(*Chairman:* J. D. A. HUGHES, *National Engineering Laboratory, East Kilbride, Scotland, UK*)

Analysis of Shell-Type Structures

(*Chairman:* D. H. MORRIS, *Virginia Polytechnic, Blacksburg, USA*)

Buckling of Shell-Type Structures

(*Joint Chairmen:* C. Y. CHIA, *The University of Calgary, Alberta, Canada* and G. J. TURVEY, *University of Lancaster, Bailrigg, UK*)

Structural Sections

(*Joint Chairmen:* A. J. WOOTTON, *National Engineering Laboratory, East Kilbride, Scotland, UK* and J. G. R. HANSEN, *Air Force Office of Scientific Research, London, UK*)

Structural Optimisation

(*Chairman:* J. R. VINSON, *University of Delaware, Newark, USA*)

Thermal Loading

(*Chairman:* J. M. WHITNEY, *Wright-Patterson AFB, Ohio, USA*)

Vibration of Platework Structures

(*Chairman:* A. W. LEISSA, *Ohio State University, Columbus, USA*)

Finite Element Analysis

(*Joint Chairmen:* J. WOOD, *Paisley College of Technology, Scotland, UK* and R. S. BARSOUM, *US Army Research Office, Research Triangle Park, USA*)

Plenary Paper

An Overview of Composite Plate Buckling

ARTHUR W. LEISSA

Department of Engineering Mechanics, Ohio State University, Columbus, Ohio 43210, USA

ABSTRACT

An overview is presented of considerations involved and some of the more interesting results found in the buckling of laminated composite plates. In the case of plates which are symmetrically laminated, classical orthotropic or anisotropic plate theory applies, and numerous theoretical results are available. For unsymmetrical laminates coupling exists between bending of the plate and stretching of its midplane, and a considerably more complicated theory must be used. Additional complicating factors are considered, including: interior holes, shear deformation, sandwich construction involving other materials, local effects, nonlinear stress–strain relationships, and hygrothermal effects. Postbuckling behavior and the effects of initial geometric imperfections are also discussed. The results summarized are taken from a recent comprehensive study of the world's literature on this subject, encompassing approximately 300 references.

1. INTRODUCTION

Laminated composite plates are becoming increasingly used in structural applications. Accurate knowledge of critical buckling loads, mode shapes and postbuckling behavior is essential for reliable and lightweight structural design.

Analysis of buckling for *isotropic, homogeneous* plates began with the work of Bryan[1] a century ago, whereas serious study of laminated composites has been carried on only half as long, beginning with plywood plates (cf. Ref. 2). Development of high-strength, high-stiffness composite materials for aerospace structures gave further impetus to buckling studies approximately 25 years ago. Since that time the amount of research on composite plate buckling has continued to increase at an accelerating rate.

A reasonably comprehensive literature search has recently resulted in a monograph[3] containing 400 references relevant to the buckling of rectangular composite plates and shell panels (a shell panel being similar to a plate, except that it has cylindrical curvature instead of being flat). Stiffened composite plate and shell panels were also included in the study. The present paper deals with unstiffened plates only, which involves 300 of the 400 aforementioned references, and presents some of the more interesting highlights uncovered in these references.

Buckling analysis for plates fabricated from laminated composite materials is considerably more complicated than for homogeneous, isotropic ones, such as ordinary metals. Theoretical analysis is complicated, at least, because additional calculations must be made to account for fibre matrix material properties, fiber orientations, and stacking sequences. Moreover, if the plies are not symmetrically stacked, coupling exists between bending and midplane stretching as transverse deflection takes place, which complicates the problem further. Reliable experimental results are also more difficult to realize, particularly because of increased difficulty in simulating desired edge constraints.

In the general case of an unsymmetric laminate the inplane stress resultants N_x, N_y, N_{xy} (forces per unit length) and moment resultants M_x, M_y, M_{xy} (moments per unit length which occur *during* buckling) are related to the midplane strains ε_x, ε_y, γ_{xy} and midplane curvature changes κ_x, κ_y, κ_{xy} by six simultaneous equations, which may be written in matrix form as

$$\begin{bmatrix} N_x \\ N_y \\ N_{xy} \\ \hline M_x \\ M_y \\ M_{xy} \end{bmatrix} = \left[\begin{array}{ccc|ccc} A_{11} & A_{12} & A_{16} & B_{11} & B_{12} & B_{16} \\ A_{12} & A_{22} & A_{26} & B_{12} & B_{22} & B_{26} \\ A_{16} & A_{26} & A_{66} & B_{16} & B_{26} & B_{66} \\ \hline B_{11} & B_{12} & B_{16} & D_{11} & D_{12} & D_{16} \\ B_{12} & B_{22} & B_{26} & D_{12} & D_{22} & D_{26} \\ B_{16} & B_{26} & B_{66} & D_{16} & D_{26} & D_{66} \end{array}\right] \begin{bmatrix} \varepsilon_x \\ \varepsilon_y \\ \gamma_{xy} \\ \hline -\kappa_x \\ -\kappa_y \\ -\kappa_{xy} \end{bmatrix} \quad (1)$$

Thus, the A_{ij} coefficients represent the stretching stiffnesses of a plate, the D_{ij} represent bending stiffness, and the B_{ij} indicate bending–stretching coupling. For a symmetric laminate all B_{ij} are zero.

As in the case of isotropic, homogeneous plates, the standard procedure for determining theoretical values of buckling loads is to solve a mathematical eigenvalue problem; that is, a problem governed by differential equations and boundary conditions, all of which are homogeneous. In a relatively few cases, this can be done exactly. For most problems approximate procedures such as the Ritz, Galerkin, series (or superposition), finite element or finite difference methods must be used. The latter procedures, if properly used, will approach the exact solutions as closely as desired as sufficient terms (or degrees of freedom) are retained in the solution, although the roots of very large determinants may be required for accurate results.

Experimental difficulties in obtaining accurate buckling loads for isotropic homogeneous plates are well known. These are mainly associated with obtaining desired conditions at the plate edges in order to achieve the desired inplane loading conditions (e.g. uniform stress) and boundary conditions (e.g. simply supported or free). For composite plates additional difficulties arise. Physical discontinuities and exposed fibers at the plate edges make the desired loadings and boundary conditions even more difficult to achieve, and internal discontinuities (e.g. delamination or debonding) degrade the reliability and reproducibility of results.

This review is limited to composite plates of *rectangular planform*. Although one can find relatively few references dealing with circular, elliptical, and parallelogram (or skew) plates having rectangular orthotropy, they will not be included here. Plates having curvilinear fibers (e.g. circular plates and polar orthotropy) are also not considered. It is possible that future design optimization will include the layout of fibers in curvilinear patterns with varying spacing (i.e. inplane heterogeneity for the continuum model).[4]

2. ORTHOTROPIC PLATES

Consider first composite plates where bifurcation buckling is governed by the differential equation

$$D_{11}\frac{\partial^4 w}{\partial x^4} + 2(D_{12} + 2D_{66})\frac{\partial^4 w}{\partial x^2\,\partial y^2} + D_{22}\frac{\partial^4 w}{\partial y^4} = h\left(\sigma_x\frac{\partial^2 w}{\partial x^2} + 2\tau_{xy}\frac{\partial^2 w}{\partial x\,\partial y} + \sigma_y\frac{\partial^2 w}{\partial y^2}\right) \qquad (2)$$

where the D_{ij} are stiffness coefficients as used in eqn. (1) σ_x, σ_y, τ_{xy} are the *initial* (i.e. before buckling) inplane normal and shear stress components, and w is the transverse displacement of the plate in its buckled mode. This is the classical equation for the buckling of a plate having rectangular orthotropy. It is applicable to two important types of symmetric laminate configurations:

(a) Parallel fiber (all fibers parallel to each other and to a set of plate edges).
(b) Cross-ply (fibers of adjacent plies oriented at 90° to each other, and parallel to the plate edges).

For such plates, not only is bending behavior uncoupled from stretching (all $B_{ij} = 0$ in eqn. (1)), but twisting is also uncoupled ($D_{16} = D_{26} = 0$).

A simple, exact solution exists for the case of an orthotropic plate having all edges simply supported (i.e. $w = M_n = 0$ on each boundary) and subjected to a uniform, biaxial state of initial stress (i.e. σ_x = constant, σ_y = constant, $\tau_{xy} = 0$). The boundary conditions at $x = 0, a$ and $y = 0, b$ are exactly satisfied by assuming

$$w_{mn} = c_{mn} \sin \frac{m\pi x}{a} \sin \frac{n\pi y}{b} \qquad (m, n = 1, 2, 3, \ldots) \tag{3}$$

for the deflected shape of the mn buckling mode. Substituting eqn. (3) into eqn. (2) yields the nondimensional buckling stress

$$\frac{K_x}{\pi^2} = -\frac{\sigma_x h b^2}{\pi^2 D_{22}} = \frac{\dfrac{D_{11}}{D_{22}}\left(\dfrac{b}{a}\right)^2 m^2 + 2\left(\dfrac{D_{12}}{D_{22}} + 2\dfrac{D_{66}}{D_{22}}\right)n^2 + \left(\dfrac{a}{b}\right)^2 \dfrac{n^4}{m^2}}{1 + \left(\dfrac{\sigma_y}{\sigma_x}\right)\left(\dfrac{a}{b}\right)^2\left(\dfrac{n}{m}\right)^2} \tag{4}$$

where a negative value of σ_y/σ_x is used to denote a tensile stress acting in the y-direction, simultaneously with a compressive stress in the x-direction.

It is interesting to examine the behavior of this exact solution for various types of composite plates subjected to different loading conditions. In the special case of uniaxial loading ($\sigma_y/\sigma_x = 0$) the nondimensional buckling stress parameter K_x/π^2 given by eqn. (4) is plotted in Fig. 1 versus the plate aspect ratio (a/b) for the material stiffness ratio $(D_{12} + 2D_{66})/D_{22} = 1$. For $D_{11}/D_{22} = 1$ the curve shown is the well-known one for an isotropic plate (cf. Ref. 5, p. 353), and is the envelope of the critical (i.e. lowest) buckling loads. For $0 < a/b \leq \sqrt{2}$ the plate buckles into a mode shape having one half-sine wave ($m = 1$) in the direction of the loading, whereas for $\sqrt{2} \leq$

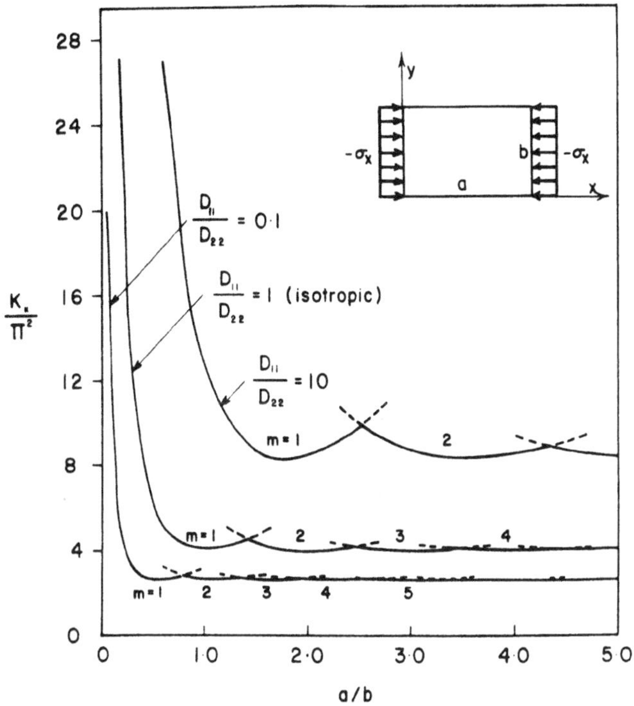

FIG. 1. Uniaxial buckling stress ($\sigma_y/\sigma_x = 0$) of SSSS plates with various D_{11}/D_{12}, for $(D_{12} + 2D_{66})/D_{22} = 1$.

$a/b \leq \sqrt{6}$ it has two half-sine waves ($m = 2$) and so on. In each case the buckled mode shape for the critical load has only one half-wave in the y-direction ($n = 1$). The minimum value of K_x/π^2 is 4, and occurs at $a/b = 1, 2, 3, \ldots$. For a composite plate which is much stiffer in the direction of loading ($D_{11}/D_{22} = 10$) Fig. 1 shows that not only higher buckling loads are achieved (minimum values of $K_x/\pi^2 = 8{\cdot}324$), but that the critical mode shapes have fewer longitudinal waves ($m = 1$ for $0 < a/b \leq 2{\cdot}515$). Conversely, if the fibers lie primarily perpendicular to the direction of loading, the buckling load is lower, and the plate is inclined to have more longitudinal waves in its critical mode. This is clear from an energy point of view, for the plate buckles into a mode shape which minimizes the total potential energy of the system, and relatively little energy is required to deform the plate in the direction perpendicular to the fibers.

Figures 2 and 3 show similar curves generated by eqn (4) for two cases of biaxial loading. In the case of hydrostatic loading ($\sigma_y/\sigma_x = 1$, Fig. 2),

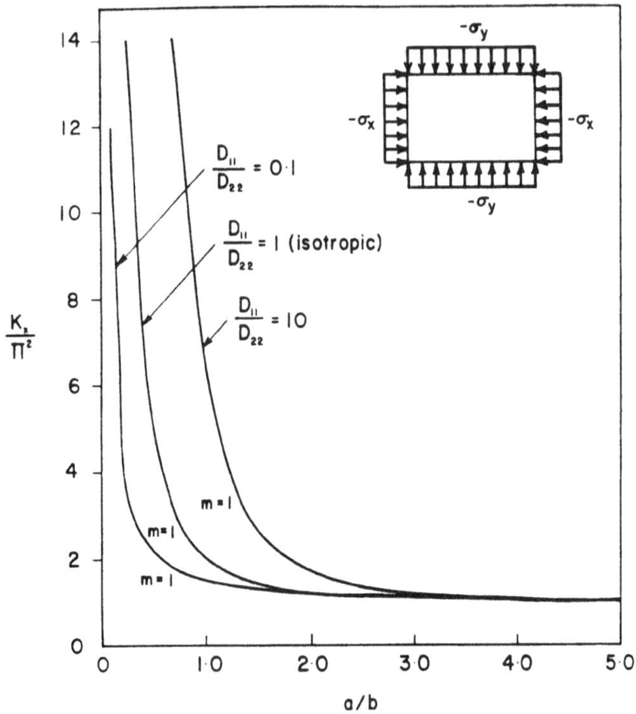

FIG. 2. Hydrostatic buckling stress ($\sigma_y/\sigma_x = 1$) of SSSS plates with various D_{11}/D_{22}, for $(D_{12} + 2D_{66})/D_{22} = 1$.

all plate configurations have a critical mode shape with only one half-wave in each direction, and the minimum value of K_x/π^2 is 1. For tension–compression buckling ($\sigma_y/\sigma_x = -1$, Fig. 3), which corresponds to pure shear on the 45° planes, more half-waves are found in the direction of the compressive loading than for uniaxial loading, although the compressive stress required for buckling is higher.

Equation (4) and Figs. 1–3 are particularly important for they demonstrate the buckling phenomenon clearly without requiring complicated and/or approximate analysis, and some forms of them may be found in numerous references.

Wittrick[6] made the interesting observation that, for small values of σ_y/σ_x: specifically, for $\sigma_y/\sigma_x < \pi^2/K_x$, eqn. (4) may be rewritten in the form

$$k = \left(\frac{m}{\lambda} + \frac{\lambda}{m}\right)^2 \tag{5}$$

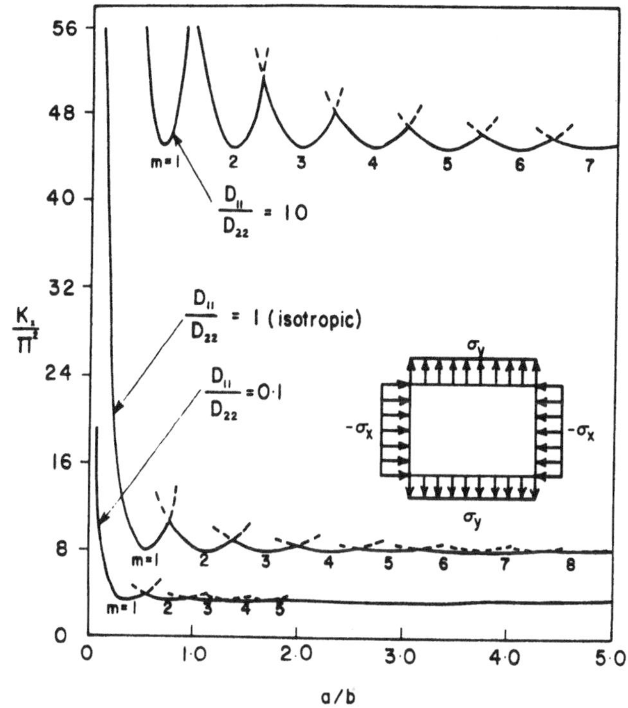

FIG. 3. Tension–compression buckling stress ($\sigma_y/\sigma_x = -1$) of SSSS plates with various D_{11}/D_{22}, for $(D_{12} + 2D_{66})/D_{22} = 1$.

where m, as before, is the number of half-waves in the direction of loading, and k and λ are two other mathematical parameters. Equation (5) has the same form as the well-known formula for the *uniaxial* buckling of an *isotropic* plate (cf. Timoshenko and Gere,[5] page 352). Thus, eqn. (5) permits the representation of all SSSS (all sides simply supported) orthotropic plate biaxial buckling solutions as a single curve, rather than requiring many families of curves. Nevertheless, Figs. 1–3 are useful in understanding the changes in buckling due to changes in the *physical* parameters of the problem.

No other inplane loading conditions for SSSS orthotropic plates of finite size permit exact solutions for the bifurcation buckling problem to be found. In the case of pure shear ($\sigma_x = \sigma_y = 0$, $\tau_{xy} = \text{constant}$) an exact solution was found by Bergmann and Reissner[7] for the case of an infinite strip ($b/a \rightarrow 0$) loaded in shear by following the same procedure used previously by Southwell and Skan[8] for an isotropic infinite strip. That is, a separation of variables is effected by choosing a solution form as

$$w(x, y) = f(y)\exp(\mathrm{i}\kappa x/b) \tag{6}$$

where $i = \sqrt{-1}$ and κ is a wave-length constant to be determined. Substituting eqn. (6) into eqn. (2) yields a fourth-order, ordinary differential equation having real and imaginary coefficients which has an exact solution. Substituting this into the four boundary conditions at $y = 0, b$ permits both κ and the critical value of τ_{xy} to be found for *arbitrary* (but continuous) conditions along those edges, including the case of simple supports.

Solutions to eqn. (2) more general than that of eqn. (3) may be found in the case of uniform, biaxial loading ($\sigma_x = \text{constant}$, $\sigma_y = \text{constant}$, $\tau_{xy} = 0$). These take the form

$$w(x, y) = Y_m(y)\sin\frac{m\pi x}{a} \qquad (m = 1, 2, \ldots) \tag{7}$$

guaranteeing that simply-supported edge conditions result at $x = 0, a$. Substituting eqn. (7) into eqn. (2) results in a fourth-order, ordinary differential equation having real coefficients which has an exact solution. This exact solution is capable of satisfying arbitrary (but continuous) boundary conditions along the remaining two edges $y = 0, b$, including the possibility of translational and/or rotational elastic edge constraints. The particular edge conditions applied yield a fourth-order characteristic determinant, the roots of which are the eigenvalues (nondimensional buckling stresses) of the problem. The characteristic determinant will be of the same form as in the case of free vibration[9] for corresponding edge conditions.

In addition to the six plate configurations having two opposite sides simply supported, it has been shown that 15 other distinct arrangements of simple edge conditions may exist.[10] These vary from all edges being clamped (CCCC) to all being free (FFFF). Of course, for each of these configurations innumerable loading conditions may exist which may cause buckling. However, none of these problems have exact solutions. A large number of results have been obtained by approximate methods such as Ritz and finite elements.

An interesting buckling analysis was carried out by Simitses and Giri[11] for uniaxially loaded, SSSS plates having rotational constraints along all four edges. These represented folding cartons made of paperboard.

3. ANISOTROPIC PLATES

Completely general anisotropic plate theory may be used to determine critical buckling loads of rectangular composite plates for two important types of symmetric laminate configurations:

(a) Parallel fiber or cross-ply, where the fibers are not parallel to the plate edges, but skewed.
(b) Angle-ply (an odd number of adjacent plies oriented at an alternating angle, $+\theta$, $-\theta$, $+\theta, \ldots$, with respect to the plate edges).

The bifurcation buckling of such plates is governed by the equation

$$D_{11}\frac{\partial^4 w}{\partial x^4}+4D_{16}\frac{\partial^4 w}{\partial x^3\,\partial y}+2(D_{12}+2D_{66})\frac{\partial^4 w}{\partial x^2\,\partial y}+4D_{26}\frac{\partial^4 w}{\partial x\,\partial y^3}+D_{22}\frac{\partial^4 w}{\partial y^4}$$
$$=h\left(\sigma_x\frac{\partial^2 w}{\partial x^2}+2\tau_{xy}\frac{\partial^2 w}{\partial x\,\partial y}+\sigma_y\frac{\partial^2 w}{\partial y^2}\right) \qquad (8)$$

That is, terms containing D_{16} and D_{26}, which involve the bending–twisting coupling are added to eqn. (2). For the alternating, angle-ply laminate (case (b) described above), D_{16} and D_{26} are largest for the smallest number of plies ($N = 3$), and become smaller relative to the other coefficients (D_{11}, D_{22}, D_{12}, D_{66}) as N is increased. Thus, for an angle-ply plate having a large (odd) number of alternating plies, D_{16} and D_{26} may be quite small. However, as Jones[12] pointed out, even small D_{16} and D_{26} may cause significantly different results from those cases in which D_{16} and D_{26} are exactly zero.

In the preceding section it was seen that orthotropic plates having two opposite sides simply supported, subjected to uniform inplane normal stresses (σ_x, σ_y), had exact solutions for the buckling loads. The addition of the terms containing D_{16} and D_{26} makes the exact solution of problems for plates of finite dimensions impossible (an exceedingly complex exact solution was achieved by Wittrick[13] for infinite strips). Moreover, results exist for only 4 of the 21 sets of simple boundary conditions (SSSS, SCSC, SFSF, CCCC).

Housner and Stein[14] used a finite difference energy method to make parametric studies for angle-ply, graphite–epoxy plates having a large number of alternating ($\pm\theta$) plies and either simply-supported (SSSS) or clamped (CCCC) edges. Variation of a nondimensional uniaxial buckling stress parameter versus fiber orientation ($\pm\theta$) and aspect ratio (a/b) is seen in Fig. 4. The buckling parameter contains E_1, which is the elastic modulus

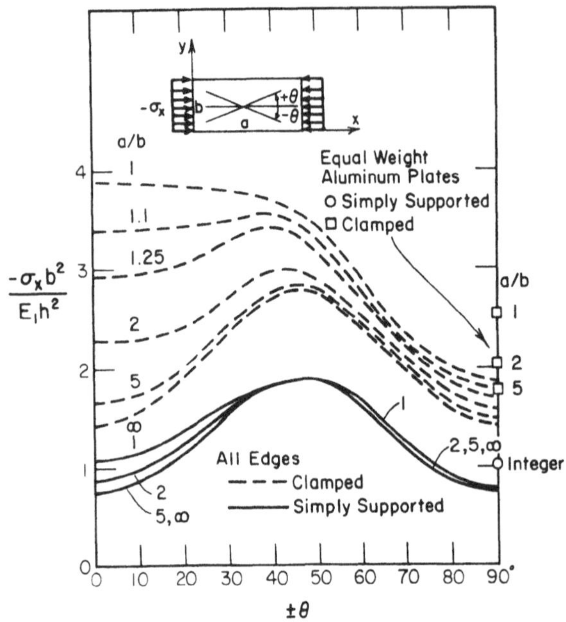

FIG. 4. Uniaxial buckling stress parameters for graphite–epoxy, angle-ply plates.

of a ply in the direction of its fibers, and which was approximately 9 times as great as the transverse modulus (E_2) in this study. Two particularly interesting conclusions may be drawn from Fig. 4:

(a) The critical buckling stress is maximized for plies having $\pm\theta$ of approximately 45°. The resulting maximum may be more than twice as great as in the case of parallel-fiber plies ($\pm\theta = 0$ or 90°).
(b) A range of fiber orientations exists for which the buckling stress of the graphite–epoxy plates exceeds that of an equal weight, aluminum plate having the same planar dimensions (*a* and *b*) and boundary conditions. The latter data are shown as points on the right-hand ordinate of Fig. 4.

Similar plots are seen in Fig. 5 for the case of shear loading. In this case the angle of optimum fiber orientation shifts to values exceeding 45° as *a*/*b* becomes greater than unity.

Another extensive set of results for compression (uniaxial) and shear buckling of SSSS, angle-ply plates were obtained by Fogg,[15] using a finite element plate analysis program. From this study it could be concluded that

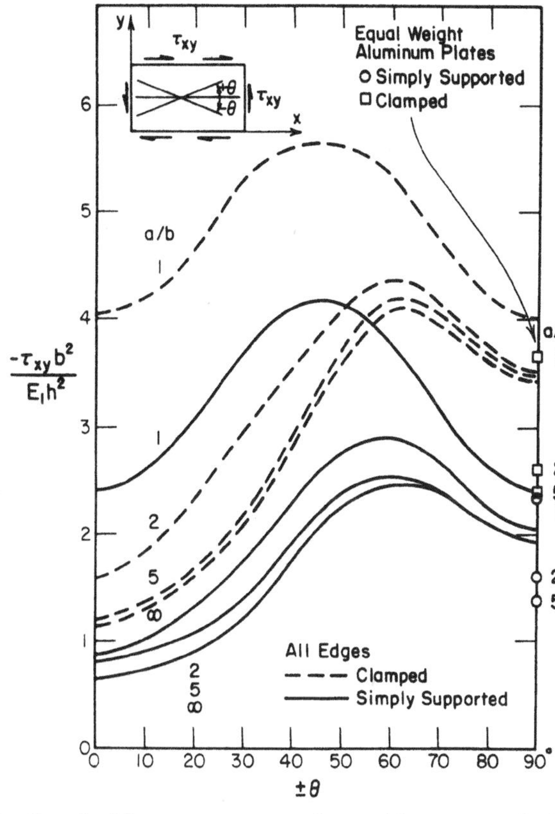

FIG. 5. Shear buckling stress parameters for graphite–epoxy, angle-ply plates.

some of the effects of neglecting the bending–twisting (D_{16} and D_{26}) terms in an analysis are:

(a) The buckling loads are always increased.
(b) In the case of uniaxial compression longer plates always have larger increases.
(c) Shear buckling stresses are more greatly increased than those due to uniaxial compression.

However, other theoretical calculations by Mandell[16,17] for SSSS angle-ply plates indicate that buckling stresses are in some cases increased and in other cases decreased by the neglect of the D_{16} and D_{26} terms. Ashton and Waddoups[18] showed that the critical buckling stress of an SSSS, $\pm 45°$

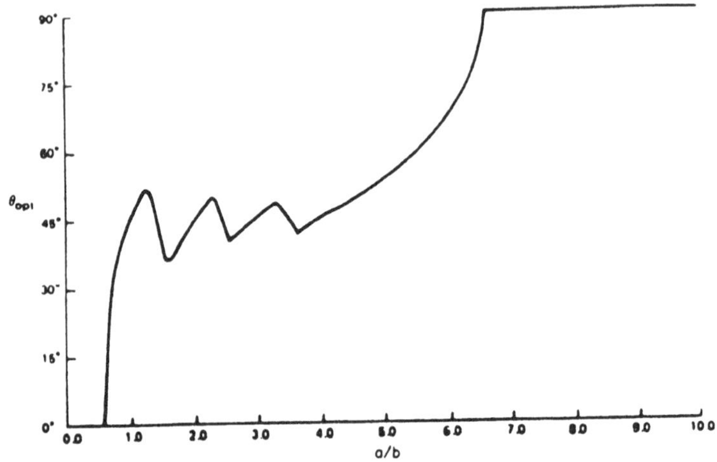

FIG. 6. Optimum material axis orientation versus aspect ratio for a uniaxially loaded SSSS plate (unidirectional, medium orthotropy laminate).

angle-ply plate loaded in shear will be different depending upon whether the shear stress is in one direction (positive) or the other (negative).

Crouzet-Pascal[19] conducted optimization studies for the buckling of parallel fiber plates having skewed fibers. Figure 6 shows the optimum skew angle (θ), measured from the direction of loading (x), for uniaxially stressed glass–epoxy plates having SSSS edges. This is considered to be a composite material of moderate orthotropy ($E_1 = 7{\cdot}8 \times 10^6$ psi, $E_2 = 2{\cdot}6 \times 10^6$ psi, $G_{12} = 0{\cdot}65 \times 10^6$ psi, $\nu_{12} = 0{\cdot}25$). The optimum value of fiber orientation is seen to vary between 30 and 55° for plates of moderate aspect ratio ($0{\cdot}7 < a/b < 4$).

4. UNSYMMETRIC LAMINATES

For an unsymmetrically laminated composite plate, bending and midplane stretching are coupled together. The equations governing bifurcation buckling may be written as

$$\begin{bmatrix} L_{11} & L_{12} & L_{13} \\ L_{21} & L_{22} & L_{23} \\ L_{31} & L_{32} & (L_{33} - F) \end{bmatrix} \begin{Bmatrix} u \\ v \\ w \end{Bmatrix} = \begin{Bmatrix} 0 \\ 0 \\ 0 \end{Bmatrix} \tag{9}$$

where u and v are inplane displacement components of the midplane *during buckling*, in the x- and y-directions, respectively, and w is the transverse displacement. The L_{ij} are differential operators given by

$$L_{11} \equiv A_{11}\frac{\partial^2}{\partial x^2} + 2A_{16}\frac{\partial^2}{\partial x\,\partial y} + A_{66}\frac{\partial^2}{\partial y^2}$$

$$L_{22} \equiv A_{22}\frac{\partial^2}{\partial y^2} + 2A_{26}\frac{\partial^2}{\partial x\,\partial y} + A_{66}\frac{\partial^2}{\partial x^2}$$

$$L_{33} \equiv D_{11}\frac{\partial^4}{\partial x^4} + 4D_{16}\frac{\partial^4}{\partial x^3\,\partial y} + 2(D_{12} + 2D_{66})\frac{\partial^4}{\partial x^2\,\partial y^2} + 4D_{26}\frac{\partial^4}{\partial x\,\partial y^3} + D_{22}\frac{\partial^4}{\partial y^4}$$

$$L_{12} = L_{21} \equiv A_{16}\frac{\partial^2}{\partial x^2} + (A_{12} + A_{66})\frac{\partial^2}{\partial x\,\partial y} + A_{26}\frac{\partial^2}{\partial y^2} \tag{10}$$

$$L_{13} = L_{31} \equiv -B_{11}\frac{\partial^3}{\partial x^3} - 3B_{16}\frac{\partial^3}{\partial x^2\,\partial y} - (B_{12} + 2B_{66})\frac{\partial^3}{\partial x\,\partial y^2} - B_{26}\frac{\partial^3}{\partial y^3}$$

$$L_{23} = L_{32} \equiv -B_{16}\frac{\partial^3}{\partial x^3} - (B_{12} + 2B_{66})\frac{\partial^3}{\partial x^2\,\partial y} - 3B_{26}\frac{\partial^3}{\partial x\,\partial y^2} - B_{22}\frac{\partial^3}{\partial y^3}$$

and F is a differential operator representing the inplane loading,

$$F \equiv h\left(\sigma_x\frac{\partial^2}{\partial x^2} + 2\tau_{xy}\frac{\partial^2}{\partial x\,\partial y} + \sigma_y\frac{\partial^2}{\partial y^2}\right) \tag{11}$$

The A_{ij}, B_{ij}, D_{ij} are the plate stiffness coefficients previously used in eqn. (1).

It is important to note that u and v are not the inplane displacements which occur with increasing initial inplane stresses, but rather the additional displacements which arise when the buckling load is reached and the plate is deformed in a buckled mode shape of infinitesimal amplitude. These additional inplane displacements characterize the bending–stretching coupling which exists in the deformation of an unsymmetrical laminate. In eqn. (9) the bending–stretching coupling is induced by the operators $L_{13}(=L_{31})$ and $L_{23}(=L_{32})$, which vanish only when the B_{ij} bending–stretching stiffness coefficients are all zero. Then the plate buckling problem reduces to $(L_{33} - F)w = 0$, which is eqn. (8) previously given for the anisotropic plate.

Equations (9) are an eighth-order set of differential equations which closely resemble the form of shell buckling equations, which are also of eighth order. (Indeed, the coupling between bending and stretching is what links together the two sets of fourth-order equations which would otherwise exist for inextensional and membrane shell theories.) Since the equations are of eighth order, four boundary conditions must be specified along each edge to define the problem physically, and to generate a proper mathematical eigenvalue problem.

The first satisfactory theory incorporating bending–stretching effects into the deformations of laminated plates was developed by Reissner and Stavsky.[20,21] This was done for the special case of an antisymmetrical, angle-ply plate, for which $B_{11} = B_{12} = B_{22} = B_{66} = 0$, which leaves B_{16} and B_{26} terms to cause the coupling. Governing equations for the buckling equilibrium position were expressed in terms of w and an Airy stress function (ϕ). Transverse equilibrium and inplane compatibility conditions were then used to generate an eighth order set of governing equations.

Existence of the bending–stretching coupling in unsymmetrical laminates was questioned (cf. Refs 22, 23) at the time of the development of the new theory more than two decades ago. But, since then, the effects have been demonstrated and quantified, both theoretically and experimentally. The primary effect is to decrease the stiffness of a plate; therefore, in the case of buckling, critical loads are reduced. As will be subsequently seen, the effect is strongest when only a small number of plies is used, and decreases as the number increases.

On the other hand, some researchers have questioned whether, because of the bending–stretching coupling, meaningful bifurcation buckling problems can exist for an unsymmetrical laminate—that is, whether the plate will not remain flat but undergo finite transverse displacements as finite inplane stresses are applied. In this case an eigenvalue problem of bifurcation buckling would not arise but, rather, an equilibrium problem similar in nature to those arising due to eccentric loadings or geometric imperfections. Conditions of inplane loading and edge support for unsymmetrical laminates which permit buckling to occur have been discussed in a recent paper.[24]

An approximate theory was suggested by Chamis[25] and Ashton[26] for simplifying problems involving unsymmetrical laminates. It replaces the eighth-order set of eqns (9) by that of anisotropic plate theory, eqn. (8), where the bending stiffness coefficients D_{ij} are replaced by the 'reduced bending stiffness' (RBS) coefficients. The boundary conditions are also reduced in number from four to two, and the resulting solutions are independent of the degree of inplane constraint. Recent work indicates that the RBS Method yields good results for cross-ply plates.[27] However, for angle-ply plates errors of up to 28% in the buckling loads were found, especially for unsymmetric laminates having a small number of plies.

For an unsymmetrically laminated plate, the meaning of a 'simply-supported' edge is not clear. Assuming that, as in classical plate theory, the edge must have zero transverse displacement and bending moment, there remain yet four possible combinations of 'simple' (i.e. not elastically restrained) boundary conditions, depending upon the inplane constraints, viz.

$$
\begin{aligned}
S1&: \quad w = M_n = u_n = u_t = 0 \\
S2&: \quad w = M_n = N_n = u_t = 0 \\
S3&: \quad w = M_n = u_n = N_{nt} = 0 \\
S4&: \quad w = M_n = N_n = N_{nt} = 0
\end{aligned}
\tag{12}
$$

where n and t are used to designate directions normal and tangent to a boundary, respectively. The bending moment (M_n), normal stress (N_n), and shear stress (N_{nt}) resultants at edges $x =$ constant, $y =$ constant are related to the three displacement components by equations, wherein bending–stretching coupling is generated by the B_{ij}.

Two closed form, exact solutions were found for unsymmetrically laminated plates subjected to uniform, biaxial stresses ($\sigma_x =$ constant, $\sigma_y =$ constant, $\tau_{xy} = 0$).[28,29] One is for cross-ply plates having $S2$ boundary conditions along all edges; the other is for angle-ply plates having $S3$ edges.

A representative plot of the nondimensional, uniaxial buckling stress arising from the exact solution for cross-ply plates is shown in Fig. 7 (taken from Ref. 30). The curves plotted are for a plate made of very high modulus fibers embedded in an epoxy resin matrix material. Four curves are shown, corresponding to antisymmetric laminates having $N = 2, 4, 6$ and an infinite number of plies. For $N = \infty$ the bending–stretching coupling disappears, and the plate behaves as if it were orthotropic and symmetrically laminated. The curves are seen to be similar in shape to those of Fig. 1. However, most importantly, Fig. 7 shows how the critical buckling stress is drastically reduced when a laminate is laid up unsymmetrically and is composed of only a few layers. For a square plate ($a/b = 1$) having only two plies, Fig. 7 indicates that the plate will buckle at a stress which is only about one third as much as if it had a large number of plies.

A similar plot is shown in Fig. 8 for the other exact solution—the antisymmetric angle-ply laminate with $S3$ edge supports (from Ref. 31). Here the uniaxial buckling stress is plotted versus fiber orientation angle

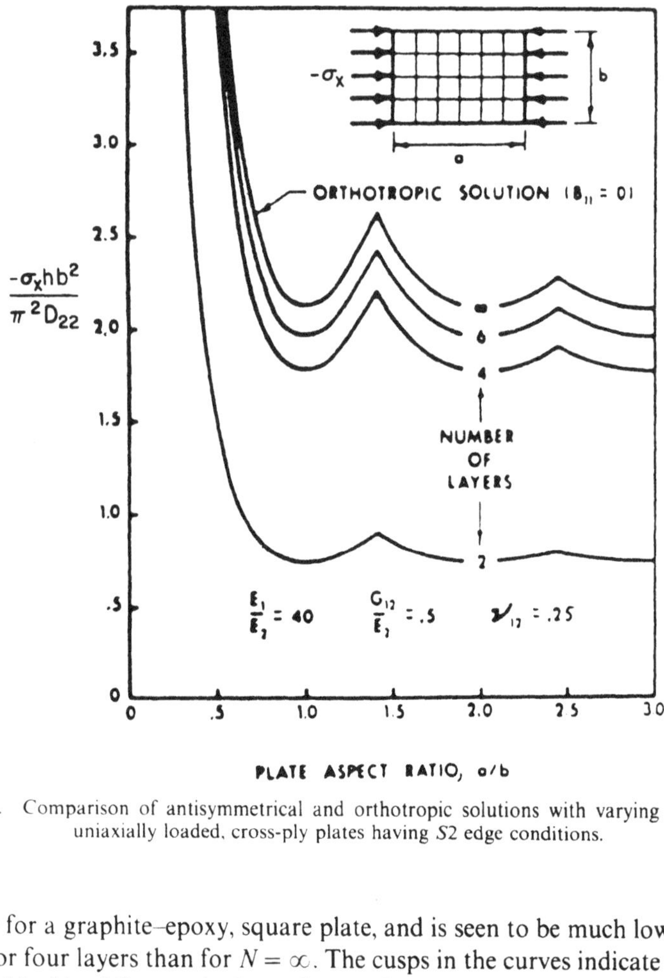

FIG. 7. Comparison of antisymmetrical and orthotropic solutions with varying a/b for uniaxially loaded, cross-ply plates having $S2$ edge conditions.

$(\pm\theta)$ for a graphite–epoxy, square plate, and is seen to be much lower for two or four layers than for $N = \infty$. The cusps in the curves indicate where the critical buckling mode shape changes from one to two half-waves in the direction of loading. The optimum value of θ was found to be $\pm 45°$.

Unsymmetrically laminated plates subjected to *shear* loading ($\sigma_x = \sigma_y = 0$, $\tau_{xy} = \text{constant}$) have been analysed by a few researchers by approximate methods. It may be shown that, similar to isotropic plates, the buckling modes of cross-ply plates separate into symmetric and unsymmetric classes, with the critical (i.e. lowest) buckling mode changing from one class to the other as a/b is varied. Representative curves of this type may be seen in Fig. 9, which is for antisymmetric, cross-ply, SSSS ($S2$) boron–epoxy plates (taken from the work of Hui[32]).

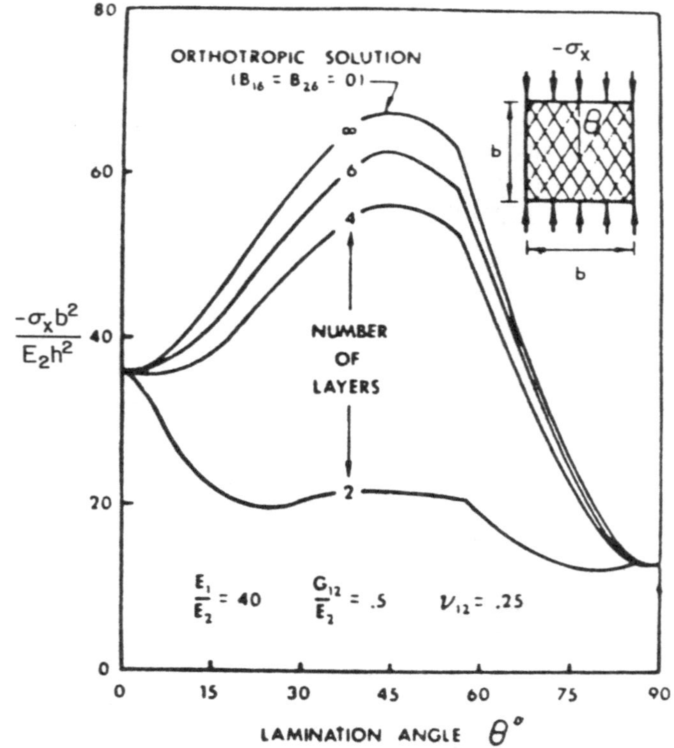

FIG. 8. Comparison of antisymmetrical and orthotropic solutions with varying lamination angle for uniaxially loaded, angle-ply plates having $S3$ edge conditions ($a/b = 1$).

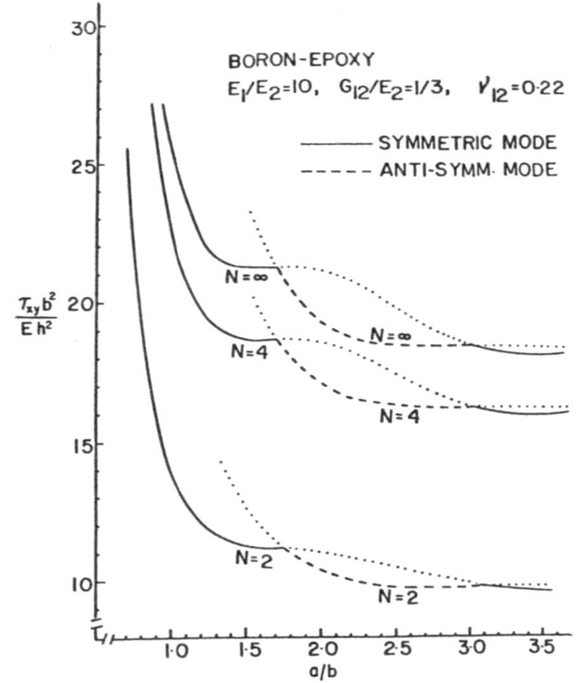

FIG. 9. Shear buckling stress parameters for antisymmetric, cross-ply SSSS ($S2$), boron–epoxy plates (including higher modes).

5. COMPLICATING FACTORS

In this section brief discussions will be presented for some of the complicating effects which enter into the buckling analysis of laminated composite plates. These complications are:

(a) interior holes;
(b) shear deformation;
(c) sandwich plates having laminated composite layers combined with other materials;
(d) local effects;
(e) nonlinear stress–strain relationships;
(f) hygrothermal effects.

The complicating factors emanate either from geometric complications to the relatively simple shapes described in the previous sections, or from generalizations of the previously used theory in order to accommodate broader classes of problems.

The problem of buckling of a composite plate containing an interior hole has received little attention.[33–35] Part of the difficulty lies in the fact that the inplane stresses caused by simple exterior loadings are typically quite complicated, and require an approximate solution before the buckling problem may be addressed. The latter then requires another approximate solution.

Consideration of transverse shear deformation effects results in added flexibility and, consequently, lower buckling stresses. Thick plate theory which is developed to include shear formation effects typically involves three dependent variables w, ψ_x, ψ_y, each a function of x and y, where now the bending rotations of a normal to the middle surface (ψ_x, ψ_y) are added to the shear rotations to obtain the total slopes $\partial w/\partial x$ and $\partial w/\partial y$. The resulting set of governing differential equations is of sixth order, which requires the specification of three boundary conditions per edge. Further generalization to unsymmetrical laminates, with the ensuing coupling between bending and stretching, would result in a tenth-order set of equations, as one finds for thick, isotropic shells. A major contribution to the analysis of thick, laminated composite plates is the book by Ambartsumyan.[36] Bert[37] made an excellent analysis of various thick plate theories as applied to laminated composite materials.

Shear deformation effects may be considerably more significant for laminates than for isotropic plates. An example of this may be seen in Fig. 10 which shows results obtained by Whitney[38] for uniaxially loaded,

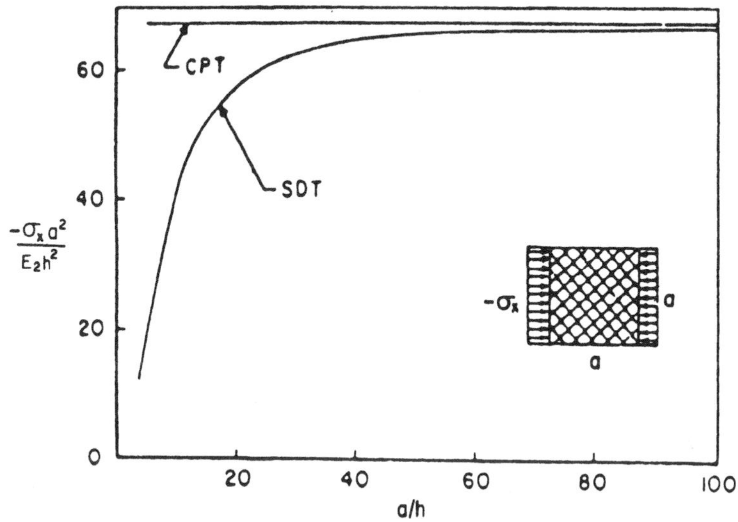

FIG. 10. Buckling of a uniaxially loaded, SSSS, $\pm 45^\circ$ angle-ply, square plate having an infinite number of layers, with and without shear deformation.

SSSS, $\pm 45^\circ$ angle-ply plates. Nondimensional buckling stress is plotted versus length-to-thickness ratio (a/h) for square plates. The two curves shown are for classical plate theory (CPT) and shear deformation theory (SDT). The decrease in buckling stress for the laminate for representative values of a/h (say, $a/h = 20$) is much greater than that predicted by isotropic theory. These types of differences have been found by many researchers. Indeed, incorporation of shear deformation effects into composite plate buckling analysis has received considerable attention in addition to that already mentioned, as may be seen by the 17 relevant publications summarized in Ref. 3.

The use of sandwich plates consisting of orthotropic or anisotropic face sheets separated by core material has become widespread, for such plates are capable of providing lightweight construction. For decades plywood was used for the face sheets. More recently the fiber composites have been used, especially for aerospace applications. Core materials may be considered as either homogeneous (e.g. foam) or heterogeneous (e.g. hexagonal or honeycomb cells, or corrugated). Numerous buckling studies, both theoretical and experimental have been conducted for these configurations (21 publications are summarized in Ref. 3).

One important local effect in buckling of composite plates which has recently become quite important is delamination.[39–41] Delamination is a breakage of the shear bonds, either between fibers and matrix, or between adjacent layers, and may be due to malfunctioning deficiencies or developments during the life of the laminate, such as impact by foreign objects. Delaminations cause a reduction in the plate stiffness, thereby reducing its buckling levels. One related study considered the buckling of laminates having discontinuous bonding.[42]

Harris *et al.*[43–45] considered localized buckling in sandwich plates having fibrous composite face sheets. The types of local instability considered were: shear crimping, dimpling or intercellular buckling, and face wrinkling. These failure modes are illustrated in Fig. 11, where they may be compared with an overall buckling mode.

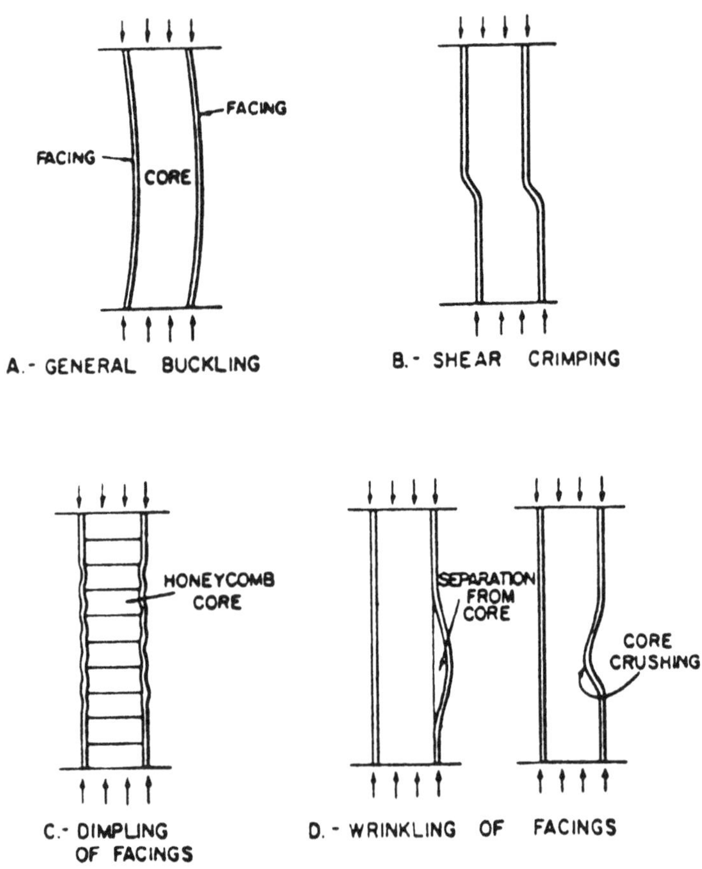

FIG. 11. Types of buckling failure for soft-core sandwich plates.

Buckling of cross-ply plates with non-linear stress–strain variation was taken up by Morgan and Jones.[46] The nonlinear behavior was characterized by assuming

$$E_{ij} = A - BU^C \tag{13}$$

for each of the material moduli, where A, B and C are constants determined experimentally for each material and U is the strain energy density of the material. The equations of equilibrium for the buckled configuration are satisfied incrementally by an iterative numerical procedure as the plate undergoes buckling deformation. Significant effects were uncovered for boron–aluminum laminates. A few other researchers have also considered nonlinear stress–strain relationships.[47–49] One paper used flow theory to describe the plastic properties of orthotropic plates undergoing buckling.[50]

Two effects of the external environment may cause significant internal strains, thereby affecting buckling loads—hygroscopic (i.e. water absorption) and thermal. Both moisture absorption and temperature serve to expand a plate, causing internal compressive stresses due to edge restraints or differential expansion. These effects in the laminated composite plate buckling problem apparently were first jointly taken up by Whitney and Ashton.[51] An interesting situation was found to develop in plates using graphite fibers, for those fibers have negative values of the coefficient of thermal expansion coefficient in the axial direction. It was found that certain orientations of angle-ply layups for graphite–epoxy plates having certain inplane boundary constraints can be buckled by *lowering* the temperature rather than raising it. Extensive studies of hygrothermal effects were also made by Flaggs and Vinson.[52–54]

6. POSTBUCKLING AND IMPERFECTIONS

As discussed previously, a linear, bifurcation buckling analysis establishes the critical value of loading for a particular plate. However, plates are typically capable of carrying considerable additional loads before the collapse (or crippling, or ultimate) load is reached. In some cases this is even several times as much as the critical load, although this capability has been found to be less pronounced for laminated composite plates than for isotropic, homogeneous ones. Theoretical analysis of postbuckling behavior of plates is nonlinear, even though the transverse displacements considered may be only 'moderately large' (i.e. on the order of a few times the plate thickness). The initial nonlinearity is due to additional inplane strains (and stresses) generated by the transverse displacements. Additional geometrical or material nonlinearities may arise during larger transverse displacements after buckling, but these are typically not considered in theoretical postbuckling analyses.

Imperfections considered are typically geometrical in nature. In the case of a plate, they are usually measures of deviation from flatness. Thus, for example, an SSSS plate may have an initial bow which may be represented by a half-sine-wave in each direction, where w_0 is the imperfection amplitude. The application of compressive loads in the flat reference plane defined by the four plate edges causes a change in the transverse displacement, no matter how small the load. Two other types of situations where this type of behavior occur are: (1) eccentricity due to loads not applied in the elastic midplane, and (2) transverse loads acting simultaneously with inplane loads. These problems all are equilibrium problems, in contrast with classical bifurcation buckling problems, which are eigenvalue problems.

The behavior of plates with geometrical imperfections may be studied with either linear or nonlinear analysis, depending upon the relative magnitude of the transverse displacement which will be permitted. For example, small displacement analysis of plates with imperfection in flatness may be carried out by linear, shallow shell theory.[55]

An excellent book by Chia[56] is available which provides a complete theory for dealing with the large displacement behavior of laminated composite plates, including orthotropic, anisotropic, and unsymmetrical laminates. Numerous example problems are also solved therein for plates undergoing transverse loading, postbuckling, imperfection behavior, and free vibrations.

For postbuckling analysis eqn. (9) is generalized by adding rather complicated, nonlinear terms to each of the three linear equations shown. These terms involve the products of u, v, w and their derivatives with respect to x and y (see Ref. 56). Although considerable simplification and uncoupling of these equations occur when a laminate is symmetric (as discussed in Section 4), the equations remain coupled because of the nonlinear terms, and a system of eight-order, nonlinear differential equations must be dealt with. These equations are homogeneous; however, the boundary conditions are not. For the postbuckling problem either the inplane stresses and/or displacements are prescribed along the edges, and the resulting equilibrium problem typically entails solving for $w(x, y)$ in terms of these edge values. Alternatively, the problem may be formulated in terms of w and an Airy stress function, in place of u, v and w.

TABLE 1
Elastic constants and uniaxial buckling stresses for the plates referred to in Fig. 12

Material	$\frac{E_1}{E_2}$	$\frac{G_{12}}{E_1}$	ν_{12}	$-\frac{\sigma_x b^2}{\pi^2 E_1 h^2}$
Isotropic	—	—	0·316	0·3704
Glass–epoxy	3	0·1667	0·25	0·1943
Boron–epoxy	10	0·0333	0·22	0·1069
Graphite–epoxy	40	0·0150	0·25	0·0916

A clear example of how the postbuckling curves for laminated composites differ from those of isotropic plates may be seen in the work of Prabhakara and Chia.[56,57] A set of *orthotropic* plates was analyzed which represented parallel-fiber configurations made of glass, boron, or graphite fibers embedded in epoxy resin. Numerical values of the moduli ratio and ν_{12} for the plates as well as for an isotropic comparison plate are listed in Table 1. Critical values of uniaxial buckling stress for $a/b = 1$ are also given. The Galerkin method was used to obtain numerical results for SSSS configurations. The variation of σ_x/σ_{cr} with w_c/h for uniaxially loaded, square plates made of the four materials is shown in Fig. 12, where σ_{cr} is the

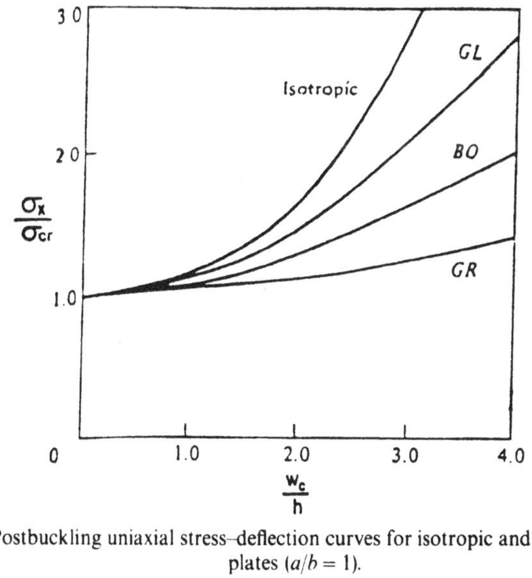

FIG. 12. Postbuckling uniaxial stress–deflection curves for isotropic and orthotropic, SSSS plates ($a/b = 1$).

critical uniaxial buckling stress, w_c is the plate deflection at its center, and where GL, BO and GR identify the glass, boron and graphite fiber plates, respectively. It may be observed that, for a given percentage increase in axial compressive stress beyond the buckling stress, the composite material plates all require greater deflection than the isotropic one, with graphite–epoxy needing the greatest increase in w_c/h. Similar curves are also available for $a/b = 1{\cdot}5$ and 2 (Ref. 57), and their slopes are found to increase with increasing a/b.

In spite of the mathematical complexity of the problem, a considerable number of other postbuckling studies have been made which are applicable to laminated composite plates (48 publications are available in Ref. 3). In addition to curves of the type shown in Fig. 12, results are often presented as plots of inplane stress versus the end-shortening displacement.

Typical of the plots exemplifying the effects of initial imperfections is Fig. 13, taken from the work of Bhattacharya[58] on symmetrically laminated, cross-ply plates subjected to uniaxial loading ($\sigma_y = \text{constant}$, $\sigma_x = \tau_{xy} = 0$). The plate edges were taken to be simply supported with elastic rotational constraint. The Galerkin method was used to solve the problem. Figure 13 is a plot of the nondimensional uniaxial loading parameter

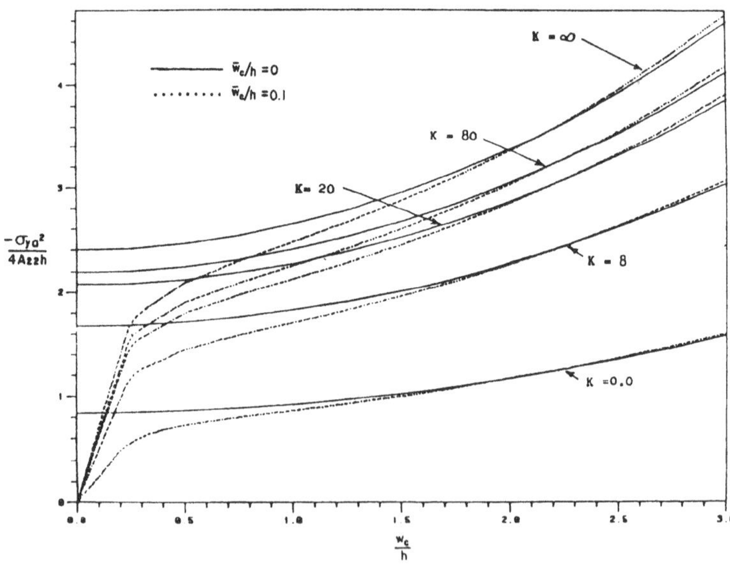

FIG. 13. Load–deflection curves for cross-ply, CFRP plates having rotational edge constraints ($a/b = 1$).

$-\sigma_y a^2/4A_{22}h$ versus $w_c h$, where w_c is the transverse displacement at the center resulting from σ_y, for carbon fiber reinforced plastic (CFRP) plates. Two sets of curves are shown, corresponding to zero and small ($\bar{w}_c/h = 0{\cdot}1$) amplitudes of the imperfection. For each set, curves are drawn corresponding to various rotational stiffness parameters K. For $K = 0$, the plate is SSSS; for $K = \infty$, it is CCCC.

The effect of sign in the imperfection (i.e. positive or negative) upon the initial postbuckling behavior was examined by Hui.[59] It was shown that, in certain circumstances, the *initial* postbuckling behavior will be unstable (i.e. negative slope in the load–displacement curve) rather than the customary stable form.

7. CONCLUDING REMARKS

The subject of buckling has a relatively short history for laminated composite plates in comparison with isotropic, homogeneous plates. Nevertheless, during this short time considerable research has taken place and an extensive repository of relevant literature has been created. This is primarily due to the fact that many laminated composites have been developed which are very efficient when compared with other materials used structurally, and are being adopted increasingly in design.

It is fortunate that symmetrically laminated composites may be represented by orthotropic and more generally anisotropic plate theory, for those theories existed in adequate form long before the need for buckling information on fibrous composites. Thus, well established and widely understood theories may be used with a considerable degree of surety. Moreover, laminated composites are most often laid up in symmetric ply configurations, so that the long existent theories are usually applicable. Nevertheless, unsymmetric laminates are needed in specific applications, and reliable results for such configurations are therefore also needed.

It is expected that research in the buckling of laminated composites will continue at least at the current rate for the foreseeable future for, in spite of all that has been done, the need for information remains great. Because of the large number of composite material combinations and ply orientations possible, parameter studies far more extensive than for ordinary plates are needed. Furthermore many investigations are needed for nonuniform loading conditions (e.g. inplane bending, effects of holes), for which more information is available for ordinary plates. Increasingly, it is also expected that more serious attention will be given to the optimization problem in design, including the use of curvilinear fiber arrangements.

REFERENCES

1. BRYAN, G. H., On the stability of a plane plate under thrusts in its own plane, with applications to the 'buckling' of the sides of a ship, *Proc. London Mathematical Society*, **22** (1891), 54–67.
2. MARCH, H., Buckling of flat plywood plates in compression, shear, or combined compression and shear, Forest Products Lab., Madison, Wis., Rept. No. 1316, 1942.
3. LEISSA, A. W., Buckling of laminated composite plates and shell panels, Flight Dynamics Laboratory, Wright Patterson Air Force Base, Dayton, Ohio, Report AFWAL-TR-85-3069, 1985, 439 pp.
4. LEISSA, A. W. and VAGINS, M., The design of orthotropic materials for stress optimization, *International Journal of Solids and Structures*, **14** (1978), 517–526.
5. TIMOSHENKO, S. P. and GERE, J. M., *Theory of Elastic Stability*, 2nd edn, New York, McGraw-Hill, 1961, 541 pp.
6. WITTRICK, W. H., Correlation betweeen some stability problems for orthotropic and isotropic plates under bi-axial and uni-axial direct stress, *Aeronautical Quarterly*, **4** (Part 1) (1952), 83–92.
7. BERGMANN, S. and REISSNER, H., Neuere Probleme aus der Flugzeugstatik. Ueber die Knickung von Wellblechstreifen bei Schubbeanspruchung, *Z.F.M.*, **20**, 1929.
8. SOUTHWELL, R. V. and SKAN, S. W., On the stability under shearing forces of a flat elastic strip, *Proc. Royal Society of London, Series A*, **105** (1924), 582–607.
9. LEISSA, A. W., The Free Vibration of Rectangular Plates, *Journal of Sound and Vibration*, **31** (1973), 257–293.
10. LEISSA, A. W., Vibration of Plates, NASA SP-160, U.S. Govt. Printing Office, Washington, D.C., 1969, 353 pp.
11. SIMITSES, G. J. and GIRI, J., Buckling of rotationally restrained orthotropic plates under uniaxial compression, *Journal of Composite Materials*, **11** (1977), 345–364.
12. Jones, R. M., *Mechanics of Composite Materials*, Washington, D.C., Scripta Book Co., 1975, 355 pp.
13. WITTRICK, W. H., Rationalization of anisotropic buckling problems, Contributions to the Theory of Aircraft Structures, Van der Neut Anniversary Vol., Delft University Press, 1972, pp. 359–377.
14. HOUSNER, J. M. and STEIN, M., Numerical analysis and parametric studies of the buckling of composite orthotropic compression and shear panels, Technical Note, NASA TN D-7996, Oct. 1975, 103 pp.
15. FOGG, L., Stability analysis of laminated materials, State of the Art Design and Analysis of Advanced Composite Materials, Lockheed California Company, Sessions I and II, 1981, 162 pp.
16. MANDELL, J. F., An experimental study of the buckling of anisotropic plates, M.S. thesis (and Report No. 23), Case Western Reserve University, Division of Solid Mechanics, Structures and Mechanical Design, June 1968, 315 pp.
17. MANDELL, J. F., An experimental investigation of anisotropic fiber reinforced plastic plates, Technical Report, AFML TR 68-281, 1968.
18. ASHTON, J. E. and WADDOUPS, M. E., Analysis of anisotropic plates, *Journal of Composite Materials*, **3** (1969), 148–165.

19. CROUZET-PASCAL, J., Buckling analysis of laminated composite plates, *Fibre Science and Technology*, **11** (1978), 413–446.
20. REISSNER, E. and STAVSKY, Y., Bending and stretching of certain types of heterogeneous aeolotropic elastic plates, *Trans. ASME, Journal of Applied Mechanics*, **9** (1961), 402–408.
21. STAVSKY, Y., On the theory of heterogeneous anisotropic plates, Doctoral thesis, Cambridge, Mass., Massachusetts Institute of Technology, 1959.
22. SMITH, C. B. (Forest Products Laboratory), Letter to E. Reissner, 1961.
23. BROUTMAN, L. J., HASLETT, W. H., KROKOSKY, E. M., ROY, J. R. and MCGARRY, F. J., Investigation of mechanics of reinforced plastics, Aeronautical Systems Div., WPAFB, WADD-TR-60-746, Pt. II, 1962.
24. LEISSA, A. W., Conditions for laminated plates to remain flat under inplane loading, *Composite Structures*, **6** (1986), 261–270.
25. CHAMIS, C. C., Buckling of anisotropic composite plates, *Proc. ASCE, Journal of the Structural Division*, **95** (ST 10) (1969), 2119–2139.
26. ASHTON, J. E., Approximate solutions for unsymmetrical laminated plates, *Journal of Composite Materials*, **3** (1969), 189–191.
27. EWING, M. S., HINGER, R. J. and LEISSA, A. W., On the validity of the reduced bending stiffness method for laminated composite plate analysis, to appear in *Composite Structures*.
28. WHITNEY, J. M., A study of the effects of coupling between bending and stretching on the mechanical behavior of layered anisotropic composite materials, Ph.D. dissertation, Ohio State University 1968, also Tech. Rept. AFML-TR-68-330, Apr. 1969, 80 pp.
29. WHITNEY, J. M. and LEISSA, A. W., Analysis of heterogeneous anisotropic plates, *Trans. ASME, Journal of Applied Mechanics*, **36**, No. 2 (1969), 261–266.
30. JONES, R. M., Buckling and vibration of unsymmetrically laminated cross-ply rectangular plates, *AIAA Journal*, **11**, No. 12 (1973), 1626–1632.
31. JONES, R. M., MORGAN, H. S. and WHITNEY, J. M., 'Buckling and vibration of antisymmetrically laminated angle-ply rectangular plates, *Trans. ASME, Journal of Applied Mechanics*, **12** (1973), 1143–1144.
32. HUI, D., Shear buckling of anti-symmetric cross-ply rectangular plates, *Fibre Science and Technology*, **21** (1984), 327–340.
33. MARSHALL, I. H., LITTLE, W. and EL TAYEBY, M. M., The stability of composite panels with holes, *Proc. Reinforced Plastics Congress*, Brighton, UK, 1984, pp. 139–141.
34. MARTIN, J., Buckling and postbuckling of laminated composite square plates with reinforced central circular holes, Ph.D. thesis, Case Western Reserve University, Solid Mechanics Division, Jan. 1972.
35. NARA, H. R., Interface and mechanics research in fiber reinforced composites, Case Western Reserve University, Cleveland, Ohio, Technical Rept. AFML-TR-71-260, March 1972, 278 pp.
36. AMBARTSUMYAN, S. A., *Theory of Anisotropic Plates*, Stamford, Conn., Technomic Publishing Co., 1970.
37. BERT, C. W., A critical evaluation of new plate theories applied to laminated composites, *Proc. Symp. on Mechanics of Composite Materials*, ASME 1983, Also University of Oklahoma, School of Aerospace, Mech. and Nuclear Engng, Rept. No. OU-AMNE-83-3, 1983, 24 pp.

38. Whitney, J. M., The effect of transverse shear deformation on the bending of laminated plates, *Journal of Composite Materials*, **3** (1969), 534–547.
39. Konishi, D. Y. and Johnson, W. R., Fatigue effects of delamination and strength degradation in graphite/epoxy laminates, in: *Composite Materials: Testing and Design* (S. W. Tsai ed.), American Society for Testing and Materials, ASTM STP 674, 1979, pp. 597–619.
40. Simitses, G. J. and Sallam, S., Delamination buckling and growth of flat composite structural elements, AFOSR Technical Report, 1984, 74 pp.
41. Shivakumar, K. N. and Whitcomb, J. D., Buckling of a sublaminate in a quasi-isotropic composite laminate, Technical Memorandum, NASA TM-85755, 1984.
42. Clark, L. G., General small deflection theory of elastic bending and buckling of laminated plates, *Proc. 4th Midwest. Conf. on Solid Mechanics*, Austin, Texas, Sept. 1959, Austin, Texas University Press, 1959, pp. 127–151.
43. Harris, B. J., Strength properties and relationships associated with various types of fiberglass reinforced facing sandwich structures, Norman, Okla., University of Oklahoma Research Institute, Rept. No. 1386-16, 1964.
44. Harris, B. J. and Crisman, W. C., Face-wrinkling mode of buckling of sandwich panels, *Proc. ASCE, Journal of Engineering Mechanics Division*, **91**, EM 3 (1965), 93–111.
45. Harris, B. J. and Nordby, G. M., Local failure of plastic-foam core sandwich panels, *Proc. ASCE, Journal of the Structural Division* (1969), 585–610.
46. Morgan, H. S. and Jones, R. M., Buckling of rectangular cross-ply laminated plates with nonlinear stress–strain behavior, *Trans. ASME, Journal of Applied Mechanics*, **46**, No. 3 (1979), 637–643.
47. Hahn, H. T., Nonlinear behavior of laminated composites, *Journal of Composite Materials*, **7**, No. 2 (1973), 257–271.
48. Durocher, L. L. and Palazotto, A. N., Elastic plastic buckling of anisotropic plates, ASCE National Structural Engineering Meeting, Preprint 1997, 1973, pp. 1–23.
49. Durocher, L. L. and Palazotto, A. N., Instability of plastically anisotropic rectangular plates, Development in Theoretical and Applied Mechanics, *Proc. 8th Southeast. Conf. on Theoretical and Applied Mechanics*, Blacksburg, VA., April 29–30, 1976, Vol. 8, pp. 123–137.
50. Teters, G. A., The effect of transverse shear upon the buckling stability of inelastic orthotropic plates, *Polymer Mechanics*, **1**, No. 2 (1965), 100–102.
51. Whitney, J. M. and Ashton, J. E., Effect of environment on the elastic response of layered composite plates, *AIAA Journal*, **9** (1971), 1708–1713.
52. Flaggs, D. L., Elastic stability of generally laminated composite plates including hygrothermal effects, M.S. thesis, Dept. of Mechanical and Aerospace Engineering, University of Delaware, 1978.
53. Flaggs, D. L. and Vinson, J. R., Elastic stability of generally laminated composite plates including hygrothermal effects, Technical Report, AFOSR TR 78-1349, July 1977, 68 pp.
54. Flaggs, D. and Vinson, J. R., Hygrothermal effects on the buckling of laminated composite plates, *Fibre Science and Technology*, **11** (1978), 353–365.
55. Hui, D. and Leissa, A. W., Effects of geometric imperfections on vibration of

biaxially compressed rectangular flat plates, *Journal of Applied Mechanics*, **50**, No. 4 (1983), 750–756.

56. Chia, C.-Y., *Nonlinear Analysis of Plates*, New York, McGraw-Hill, 1980, 422 pp.
57. Prabhakara, M. K. and Chia, C.-Y., Post-buckling behaviour of rectangular orthotropic plates, *Journal of Mechanical Engineering Science*, **15**, No. 1 (1973), 25–33.
58. Bhattacharya, A. P., Note on the postbuckling analysis of cross-ply laminated plates with elastically restrained edges and initial curvatures, *Journal of Structural Mechanics*, **10**, No. 3 (1982–83), 359–372.
59. Hui, D., Imperfection sensitivity of axially compressed laminated flat plates due to bending-stretching coupling, *International Journal of Solids and Structures*, **21** (1985).

1

A Composite Structural System for a Large Collapsible Space Antenna

L. HOLLAWAY and A. THORNE

Department of Civil Engineering, University of Surrey, Guildford, Surrey GU2 5XH, UK

ABSTRACT

Satellite systems of the future will have to cope with an ever-increasing number of small ground terminals of reduced cost. The use of these small and low-powered ground terminals will demand improved space craft radio frequency performance which will lead in turn to the use of large antennas.

A number of space missions have been proposed incorporating reflectors of very large diameter. These applications include mobile satellite systems, broadcast satellites, electronic transmission of mail and data sharing. Reflectors up to 50 m diameter may be used in the next decade.

The role of the structural engineer in the design of large space antennas is to ensure that the structure is capable of withstanding all possible load conditions without failure, to minimise deviations from the perfect geometry and so ensure high-quality performance, to produce a minimum mass and minimum package volume for low cost transportation using the space shuttle and orbiting transfer vehicle, and to devise very reliable methods for the deployment or construction of the satellite in space.

This chapter illustrates an erectable reflector, based upon the General Dynamics Corporation's extendable truss antenna, which is deployed at low earth orbit and then transported to geostationary orbit. A suitable composite material of construction for the manufacture of the reflector is suggested and the manufacturing process of this carbon fibre polyethersulphone tube is given.

Finally, compressive and buckling tests on, and mechanical properties of, the composite are discussed when it has been exposed to normal atmospheric conditions and when it has been degraded by temperature cycling in high vacuum. The temperature limits are at present between +50°C and −95°C. The initial indications are that, within the limits of the test parameters, the composite material is structurally satisfactory. However, because the environment of space is very hostile and the temperatures are extreme many more tests will have to be undertaken.

1. INTRODUCTION

The demand for both terrestrial and space services is growing rapidly. Consequently, development of satellites for use in land mobile communications systems is likely to accelerate in the near future. Ground-based mobile radio telephone systems are only cost-effective in urban areas. The use of such services in large rural areas would not be economical. If mobile communication facilities via satellites are not developed it is likely that by the turn of the century, even in the developed Western nations, a large proportion of the population will have inadequate mobile communication facilities. Reflectors over 50 m in diameter are being proposed for future projects in both the United States of America and Europe.

To enable large reflectors to be placed in space it is necessary to package them to a smaller size for launch. This may be achieved by various methods and Annandale[1] has listed the different techniques into the following classes:

(a) synchronously deployable;
(b) sequentially deployable;
(c) deployable modules with part assembly;
(d) erectable;
(e) space fabricated and assembled.

The characteristics of each of these classes of large reflectors has been given in Table 1. In addition, Ref. 1 has listed the individual company designs; these are summarised in Table 2.

Garrett and Ferebec[2] and Anderson *et al.*[3] have compared the attributes of several large antenna concepts. The most promising are:

(a) base truss (Martin Marietta Deriver Aerospace);
(b) tetrahedral truss (General Dynamics Convair);
(c) hoop-column antenna (Harris Corporation).

The tetrahedral truss reflector has been chosen as the structural system to

TABLE 1
Classes of reflectors
(based on Ref. 1)

Classes	Release from stowed configuration: Method	Release from stowed configuration: Orbit	Launch vehicle	Orbit transfer vehicle
Synchronously deployable	Simultaneous deployment of all structural elements	GEO	Space shuttle Arianne	IUS
		LEO	Space shuttle	Low-thrust
Sequentially deployable	Joined elements or modules unfolded in sequence by robotic devices	LEO	Space shuttle	Low-thrust
Deployable modules with part assembly	Modules are deployed and then joined together by EVA or RMS	LEO	Space shuttle	Low-thrust
Erectable	Elements of structure are assembled by EVA or AEM to form structure	LEO	Space shuttle	Low-thrust
Space fabricated and assembled	Members fabricated in orbit from earth's raw materials	LEO	Space shuttle	Low-thrust

be analysed in the Structural Composites Research Unit of the Department of Civil Engineering because it has many advantages. These include:

(a) high structural strength and stiffness due to the truss depth;
(b) the number of nodal joints which provide ideal locations for mounting subsystems;
(c) easy division of the truss into modules if more than one shuttle flight is needed;
(d) high degree of redundancy minimising the effect of single member failure.

The tetrahedral truss reflector has two further important considerations. Firstly, it is estimated that, within the next two decades, large manned space stations in excess of 300 m length will be placed in orbit. It is certain that these structural systems will be of a skeletal form and the technology required to manufacture and to put them into orbit will be allied to that of

TABLE 2
Company reflector designs
(based on Ref. 1)

Classes	Concepts	Company designs
Synchronously deployable	Multiple continuum petal or panel structures	TRW advanced sunflower precision deployable antenna
		Lockheed high-frequency radial-rib antenna
	Structures using frames, membranes and stringers	LMSC wrap-rib antenna
		JPL radial-rib antenna
		Harris Corporation radial-rib antenna
		Harris Corporation hoop and column antenna
	Truss structures supporting a woven metallic mesh reflector	GDC paraboidal extendable truss antenna (PETA)
	Inflatable structures	Contraves inflatable space-rigidised structures (ISRC)
	Electrostatically figured membrane reflectors (EFMR)	Massachusetts Institute of Technology EFMR
	Spin-stiffened membrane antennas	LMSC spin-stiffened membrane reflectors
Sequentially deployable	Sequentially deploying precision reflectors	ARC sequentially deployed precision reflector
	Sequentially deploying mesh reflectors	ARC sequentially deploying truss with expandable mesh RF reflector
		Martin Marietta expandable box truss antenna
Deployable modules with part assembly		LMSC deployable module concept
		GDC deployable cell module (DCM)
		GDC MOD–PETA concept
		GDC modular extendable truss antenna (META)
Erectable		NASA tapered nestable column concept
Space fabricated and assembled		Marshall space flight centre—Grumman beam builder

the truss reflector. Secondly, the truss may be constructed on earth and deployed at low earth orbit (LEO) in an automatic synchronous manner and thence transferred to the required orbit station.

Alternatively, if the Space Shuttle is used, deployable modules may be transferred to LEO and assembled by astronauts or robotic devices before being transferred to station.

It is likely that composite materials will be the leading candidate for the manufacture of space systems including reflectors and space stations because of the unique combination of high specific strength and stiffness, low weight, good dimensional stability and high specific damping capacity. However, there are three further important property requirements and these are that the material should possess high thermal and electrical conductivity, acceptable outgassing value in high vacuum and acceptable UV degradation.

Resin matrix composites which incorporate graphite fibres do satisfy these requirements also after certain additives have been incorporated into the polymer.

The most common thermosetting resin matrices which are used with graphite fibres are the epoxies. During long missions the brittle nature of this resin may be a problem but it is anticipated that graphite/epoxy composites will continue to be used for many space applications during the development of the advanced thermoplastic composites. These latter composites have several advantages over the thermosetting ones. They have shorter cycle times and handle easily. Polyethersulphone is a relatively low viscosity resin which adheres well to carbon fibres and will withstand high strains before failure.

2. THE UNFURLABLE STRUCTURE

Deployable structures have to be capable of efficient compaction and be able to unfurl to a given structural accuracy upon deployment. The release of energy during deployment must be such that no structural damage or deformation occurs to the system. The ease of deployment is dependent upon the structure geometry, joint friction manufacturing tolerances and method of deployment. Deployable structures, of necessity, do contain a large number of joints distributed throughout the configuration and this tends to make reliable deployment more difficult. However, it does provide a greater control of the truss dynamic behaviour throughout, providing greater rigidity at some joints and incorporating desirable damping characteristics in others.

3. THE UOS TRUSS

The particular structural configuration chosen for analyses by the Structural Composites Research Unit is the tetrahedral truss, based upon that of the General Dynamics Convair configuration.

Figures 1a and 1b show a perspex model of a unit of the tetrahedral flat truss in the process of being deployed and at full deployment respectively.

The truss is being developed using carbon fibres in a polyethersulphone (PES) matrix. There are various grades of PES polymers, all with their own specific properties and uses. During the current investigation PES compound grade 4100G was used. The energy loaded central joints and node joints are being developed using compatible composite materials such that the completed structure will be manufactured entirely from carbon fibre/PES composite materials.

Dubel[4] has discussed McDonnell Douglas Astronautic Company's composite tetratruss model for future space stations. In the paper he describes the design fabrication and assembly process used to construct the model. The material used in the investigation is a graphite fibre in a thermosetting resin.

4. THE MATERIALS AND METHOD OF CONSTRUCTION OF TRUSS MEMBERS

The prepregs of carbon fibres in PES matrix have been fabricated into tubes of 25 mm diameter and a nominal 2·0 mm wall thickness and 1 m in length using the film stacking[5] hot press moulding technique. Ninety per cent by weight of fibre is arranged longitudinally and 10% is in the transverse direction. The moulding temperature at the time of manufacture of the composite was 285 ± 2°C and the pressure was 7·25 N/mm^2. In addition to undertaking tests on the tubes, tests were performed on tensile coupons of dimensions $25 \times 1{\cdot}3$ mm thick and on compression coupons of $50 \times 1{\cdot}3$ mm thick. These specimens were manufactured from the film stacking hot press moulding technique with the same number of prepregs, films and stacking sequence and the same moulding pressure as the tubes. However, because of the different geometric shapes of the composites, the plates from which the coupons were cut were compressed to a thickness of 1·3 mm compared with 2·0 mm for the tube.

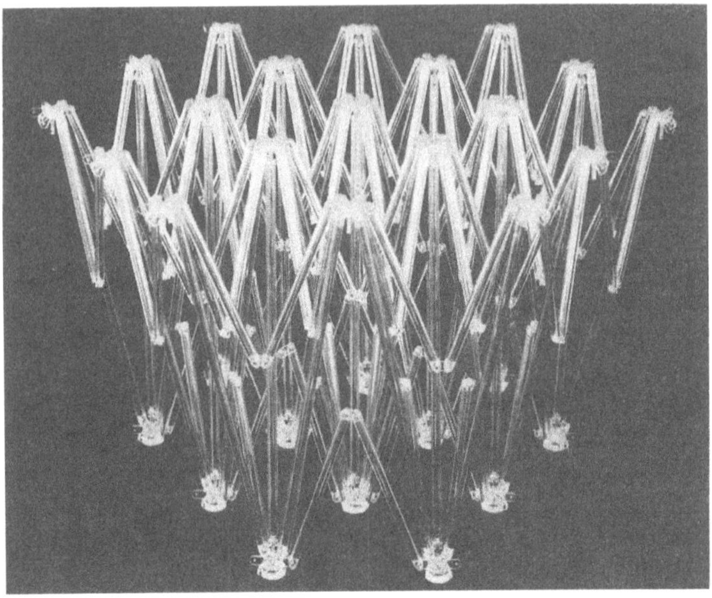

(a)

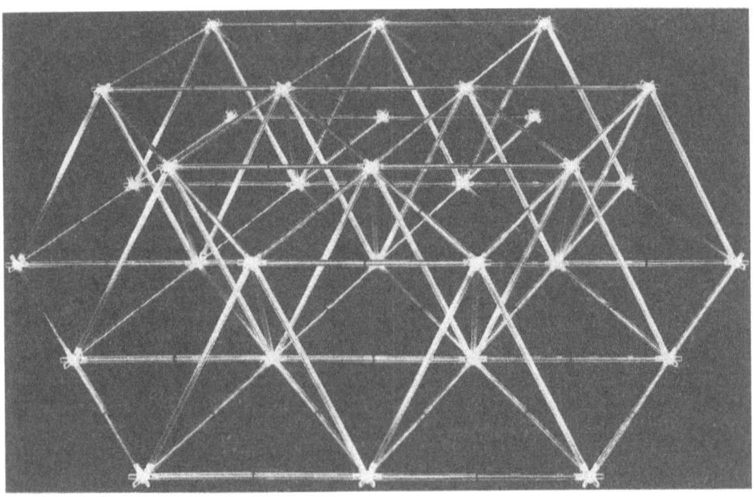

(b)

FIG. 1. (a) Perspex model of a unit of the tetrahedral flat truss in the process of being deployed. (b) Tetrahedral flat truss at full deployment.

5. TEST CONDITION

The space environment is very hostile and has many problems unrelated to those which are associated with earth conditions. Two of these problems are being investigated concurrently in the present work, namely:

(a) the structural elements exposed to high vacuum;
(b) the temperature variations.

Atmospheric density decreases with altitude and at geostationary orbit (GEO) a vacuum is very nearly reached; in conjunction with temperature cycling this may cause a number of harmful effects, one of these is a change in material property.

The magnitude and distribution of temperature within the truss will be influenced by the shadowing of the truss members by other members, the reflecting mesh and other spacecraft components such as the spacecraft bus, the feed and solar panels.

Solar eclipses of satellites in GEO occur during the autumnal and vernal equinoxes and last for a maximum of 72 min,[1] during which time the satellite is exposed to the deep cold of space (a cold sink of 4 K). During the first 12 min the temperature falls from 70°C to −120°C with a further 30°C drop to −150°C during this 72-min period. On emerging from the umbra region of the earth, the temperature of the structure will rise again to about 70°C in 4 min.

6. TEST PROCEDURE

During the current test programme the composite tube specimens were exposed to temperatures of −95°C to 50°C in high vacuum of 10^{-7} torr; some tubes were placed under a four-point bending situation whilst others were unloaded. Other specimens remained in the laboratory under constant temperature (23° ± 2°C) and atmospheric pressure. After temperature cycling under vacuum all tube specimens were taken to the laboratory and tested under compressive and tensile loads. In addition, coupon specimens were tested under these loads; the compression specimens were supported in an anti-buckling device, based upon that given in the Royal Aircraft Establishment Technical Report 85099 (Crag Test Method).[6]

The tests were carried out under constant rate of strain conditions in an Instron testing machine. To enable changes in parameters to be measured, tests were conducted at strain rates lower than those recommended in

Ref. 6. For tension and compression tests on the tubular specimens a strain rate of 0·02 cm/min was used. A strain rate of 0·01 cm/min was employed for the compressive test undertaken on coupons mounted in the anti-buckling device.

The tube specimens tested in compression were 80 and 1000 mm in length. The ends were cut perfectly square to the axes and were encapsulated into a tightly fitting inner steel plug and an end cap, as shown in Fig. 2. The inner plug and end cap were used to protect the edge of the specimens and to prevent local splitting. A ball bearing was placed in series with the end cap and the platten of the Instron machine to ensure uniaxial load application through a pin joint.

To enable tensile specimens to be held in the Instron machine crimp-bonded aluminium tubes were secured on to each end of the specimen. The crimp-bonded technique has been described by Green and Phillips.[7] The preparation for the crimp-bonded procedure involved thorough cleaning of the ends of the specimen by means of sandpaper and dust remover. The aluminium sleeves of internal diameter of 27 mm and 70 mm long were then

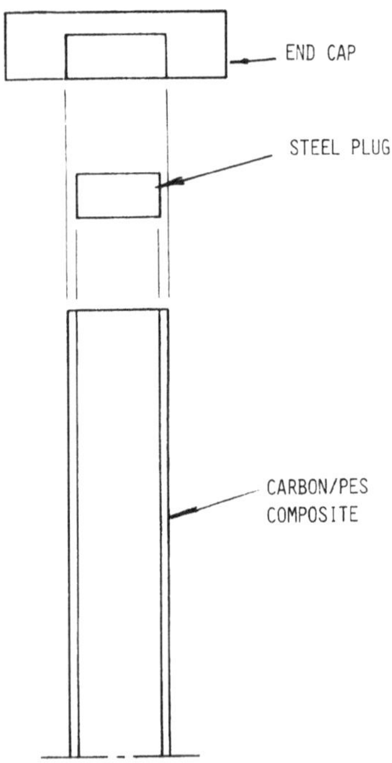

FIG. 2. End arrangement for compressive testing of carbon fibre/PES tube specimens.

FIG. 3. Test set-up for buckling tests.

crimped on to the outer surface of the carbon/PES tube. An epoxy adhesive was applied to the inner surface of the sleeves and the outer surface of the composite tube. After crimping, the aluminium sleeve was threaded and was screwed into the tensile jaws of the Instron machine.

All tests were controlled by microcomputer which enabled load strain monitoring to be undertaken during the test using screen graphics facilities. A plotting routine was used to provide a hard copy of the test; the routine used a least squares analysis on all relevant data to give a value of modulus of elasticity of the material and an indication of the accuracy of the curve fitting. At each point of interest three strain gauges were bonded on to the surface of the tube and subtended angles of 120° at the centre. Figure 3 shows the test set-up for a buckling test.

7. TEST RESULTS

7.1. Tension Tests

Tests were performed firstly to assess the joint capacity between carbon/PES composite tube and the aluminium, and secondly to determine the elastic modulus of the composite. The system failed in tension in the glue line of the joint at a load in excess of 3·5 tonnes. However, subsequent tests to determine the modulus of elasticity were undertaken up to 3·5 tonnes, which was the limit of the test facility shown in Fig. 3.

It can be seen from Fig. 4 that as the load increases to 3·5 tonnes the relationship between the load and averaged strain tends to an upward concave curve at the higher load values. This apparent increase in stiffness could be due to one of two causes, either the woven fibres had large curvatures in them, due to the nature of the weave, and were straightening out as the load increased, or the test set-up was at fault. To eliminate the latter, a steel specimen was loaded to 3·5 tonnes and the relationship between load and strain was plotted; this was shown to be a straight line. In addition, the strains were monitored in the composites by both electrical resistance strain gauges and a demountable strain recorder in which the extensometer gauge points were embedded well into the composite tube. The two methods of measurement showed close agreement, which proved that the PES-rich surface layer did not affect the true overall composite behaviour and that the resistance strain gauges were recording the true composite strains.

Although the relationship between the load and strain for the tensile coupon specimens did show a slight concave upwards curve, it was not as

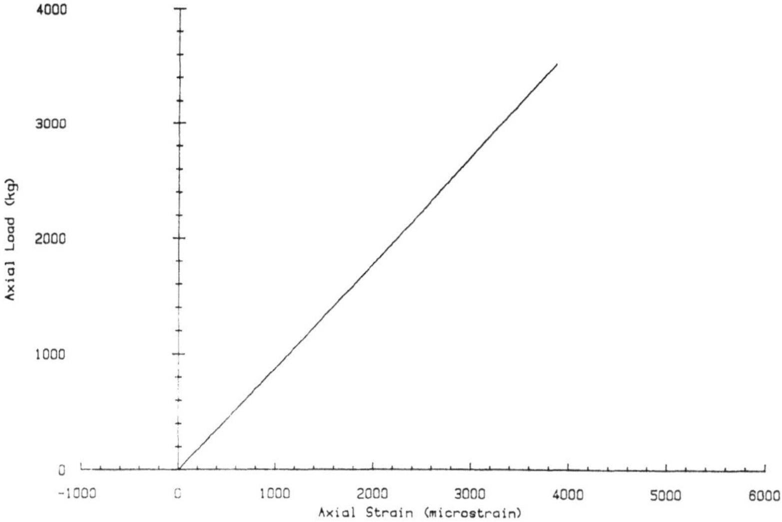

FIG. 4. Relationship between tensile load and average strain.

severe as that for the tubes. This indicated that a better compaction was achieved for the plate sample compared with that for the tubes and that the curvature in the fibres, as a result of the weave, had essentially been straightened during composite manufacture.

The secant values of the modulus of elasticity between stress values of 0 and 35 N/mm^2 was 60·5 GN/m^2, and between stress values of 0 and 245 N/mm^2 was 64·0 GN/m^2. The average value of the secant modulus between stress values of 210 and 245 N/mm^2 was 104 GN/m^2.

7.2. Compressive Tests

To determine a compressive failure stress for the composite, two lengths of tubes were used; one was 80 mm long and the other was 1000 mm long. However, the short specimen did not prove satisfactory as it tended to fail at the end cap support by hoop failure in the transverse fibres, followed by longitudinal fibre buckling and catastrophic failure of the composite. The compressive failure value varied between 87·5 and 240 N/mm^2.

The 1-m length specimens were strain gauged, as described earlier, to determine the compressive modulus of elasticity and to monitor strains at the point of buckling.

The variability in the values of the buckling stresses was thought to be a function of inaccuracies in the manufacture of the tubes and hence upon the degree to which the load was axially applied. Fifteen tubes were tested and their histories fell into three categories depending upon whether the specimens were:

(i) unloaded and stored under normal laboratory conditions;
(ii) unloaded and temperature cycled 14 times under high vacuum;
(iii) loaded and temperature cycled 14 times under high vacuum.

Three tubes were tested for category (i) and six each for categories (ii) and (iii).

The specimens in category (iii) were loaded in four-point bending at stress levels in excess of those predicted for the satellite.

Load/strain tests were not undertaken on the loaded tubes that were stored under environmental conditions defined in category (i) because the changes in performance of specimens in the loaded and unloaded states, stored in environmental conditions defined by categories (iii) and (ii) respectively, were within the tolerances exhibited by the unloaded tubes stored under laboratory conditions.

The values of the modulus of elasticity for category (i) varies between 61 and 65 GN/m^2 and the buckling stress varies between 45 and 81 N/mm^2. The equivalent values for category (ii) were 56·3 and 62·8 GN/m^2 and 50 and 89 N/mm^2 respectively, and for category (iii) were 57 and 61·4 GN/m^2 and 51 and 90 N/mm^2 respectively. A typical stress–strain behaviour for a compressive tubular specimen undergoing buckling is shown in Fig. 5.

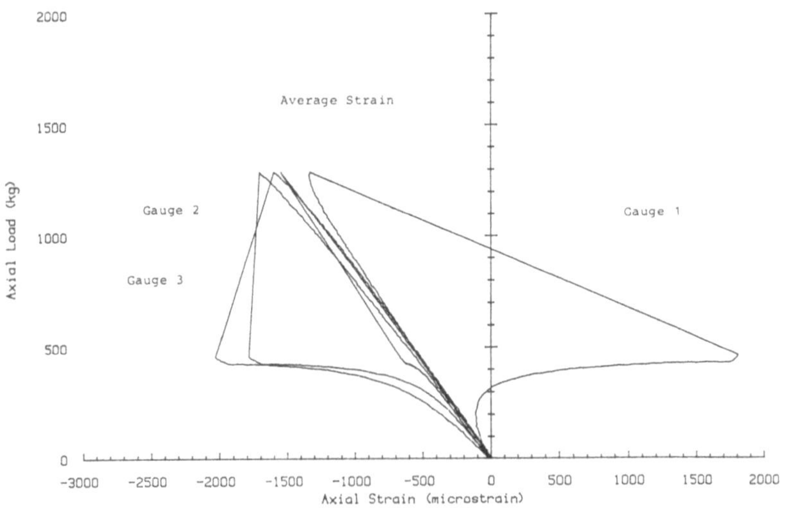

FIG. 5. Relationship between compressive load and strain.

Because of the apparent variability in the manufacture of the tubes it is not possible to say categorically that the modulus of elasticity has dropped due to the temperature cycling under high vacuum. However, the repeatability of the value of modulus of elasticity of any one tube is good and in subsequent tests the values of modulus of elasticity of specimens will be determined before and after temperature cycling under high vacuum. It would then be possible to identify to what extent, if any, the modulus value changes for a given tube.

8. OBSERVATIONS

It has been shown that there are many types of large reflectors which are available to the engineer for use in space, all of which must be packaged to a smaller size for launch. It has been agreed that the tetrahedral truss has many advantages over other types of systems. The present investigation has developed a method of film and prepreg stacking of carbon fibres in PES to form tubes and these latter have been tested in compression and tension after being exposed to temperature cycling in high vacuum; some specimens were stressed whilst others were not.

It has been shown that the technique used for the manufacture of the tubes by the film stacking hot press method does not compress the composite as efficiently as that which is used in the manufacture of the flat plates. It appears that temperature cycling 14 times between $-95°$ and $+50°$ in vacuum does not drastically impare the mechanical properties of the tubular composite in either the stressed or unstressed condition. The elastic modulus is the most easily measurable consistent property.

ACKNOWLEDGEMENTS

The work was sponsored by the Ministry of Defence and the authors would like to acknowledge the assistance of the Space Department, Royal Aircraft Department, particularly for allowing them the use of the vacuum chamber.

REFERENCES

1. Annandale, R. W., Thermal and structural analysis of large space antenna reflectors, Ph.D. thesis, University of Surrey, 1986.

2. GARRETT, L. B. and FEREBEC, M. J., Systems design and comparative analysis of large antenna concepts, Large space antenna systems technology, NASA CP2269, Part 1, 1982.
3. ANDERSON, G. C., GARRETT, L. B. and CALLESON, R. E., Comparative analysis of on-orbit dynamic performance of several large antenna concepts, *Proceedings of the AIAA/ASME/ASCE/AHS 26th Structures, Structural Dynamics and Materials Conference*, 1985, pp. 707–721.
4. DUBEL, J. F., Fabrication of a space station composite tetratruss model, *Proceedings of 31st International SAMPE Symposium, April 7–10*, 1986, pp. 1469–1475.
5. PHILLIPS, L. N., Fabrication of reinforced thermoplastics by means of the film stacking technique, Symposium, Fabrication Techniques for Advanced Reinforced Plastics, University of Salford, 22–23 April 1980, pp. 101.
6. CURTIS, P. T. (ed.), Royal Aircraft Establishment Farnborough, Technical Report 85099, Crag test methods for the measurement of the engineering properties of fibre reinforced plastics.
7. GREEN, A. K. and PHILLIPS, L. N., Crimp-bonded end fittings for use on pultruded composite sections, *Composites*, **15**, No. 3 (1982), 219–224.

2

Outgassing of Spacecraft Composites

ROBERT D. KARAM

Fairchild Space Company, Germantown, Maryland 20874-1181, USA

ABSTRACT

Diffusion and outgassing of contaminants in spacecraft composite structures are evaluated. The generalized mass and heat transfer equations are simplified for application to thin platforms, and a mathematical model is constructed for predicting outgassing rates as a function of orbital temperatures. It is found that small changes in temperature can have a major influence on the time needed for a structure to reach a specified level of residual contamination. Numerical examples are presented to illustrate application of the theory, and recommendations are given for testing and for monitoring orbital temperatures.

INTRODUCTION

One of the difficulties in using composites in spacecraft is their outgassing and contamination of surrounding elements. Preflight procedures for driving out contaminants are by no means perfect, and subsequent exposure to various environments, including what may be an extended period on the launch pad, introduces moisture and impurities which eventually diffuse outward when in space vacuum. The problem becomes crucial when composite structures are used as dimensionally stable platforms for precision instruments and experiments. Special coatings and sealants, which can reduce outgassing, cannot be used liberally in this case since they produce the counter effect of hygrothermal deformation caused by entrapped moisture.[1] A solution would be to allow outgassing to continue for a specific period in orbit while the instruments remain dormant behind special protective covers. The covers are later removed by ground command when contamination reaches tolerable levels. This procedure is being planned for Extreme Ultraviolet Explorer,[2] which is a NASA earth-orbiting spacecraft carrying instruments with highly sensitive reaction to the electromagnetic spectrum between 100 and 1000 Å. After EUVE has been inserted into orbit, a checkout phase lasting for more than a month will be used to allow outgassing of volatiles.

Knowledge of outgassing rates as function of time and environment is therefore paramount to predicting timelines for instruments operation and for assessing extraneous effects on measured data. In this chapter, the problem of diffusion of contaminants through composites is analyzed in view of the temperature profiles usually experienced by orbiting spacecraft structures. The one-dimensional case representing thin mounting platforms and antenna dishes with large surface areas is considered in some detail. Mathematical expressions for temperature distribution are derived, and a time-dependent integrated average is obtained and justified as the value on which diffusion coefficients depend. The coefficients are transformed into a function of time and substituted into the diffusion equation which is then solved for history of residual impurity.

Numerical examples are given based on an Arrhenius equation commonly used for relating diffusion coefficients to temperature. The results indicate that the duration in orbit for which a specified fraction of impurity escapes depends very strongly on temperature. The sensitivity to diffusion constants is also noted. It is concluded that accurate predictions of outgassing times require obtaining data from vacuum testing of samples constructed simultaneously with flight hardware. It is also recommended that temperatures of contaminating surfaces be monitored during a spacecraft mission for continuous assessment and updating of the outgassing problem.

THE MATHEMATICAL MODEL

Mass and Heat Transfer

Outgassing rates of volatiles in solids are obtained from the solution of the diffusion equation which relates mass change to its transfer due to a concentration gradient. In a non-reacting solid,[3]

$$\frac{\partial c}{\partial t} = -\operatorname{div} \mathbf{J} \tag{1}$$

where **J** is mass flux, t is time, and c may be defined as mass of impurity per unit volume of uncontaminated material divided by its value at zero time. The advantage of this definition is that residual impurity can then be estimated as fraction of original quantity. This serves the purpose of predicting the time at which a specified percentage has escaped by the process of outgassing.

For most practical situations Fick's law with constant density applies and

$$J_i = -D_{ij}c_{i,j} \tag{2}$$

with D_{ij} ($= D_{ji}$) representing the tensor form of the coefficient of diffusivity which is characteristic of composites.

Generally, the components of diffusivity depend on material porosity, the thermofluid properties of the contaminants, concentration gradient, and temperature and pressure.[3] It will be assumed in this discussion that the influence of material and flow parameters are inherently included in available data on diffusivity, and that the concentration gradient will not be sufficiently steep to appear in the functional relationship. The dependence on pressure will also be eliminated in view of the nearly permanent and constant pressure levels which occur immediately as space vacuum (or test chamber vacuum) is realized. Variation with temperature will be represented by the Arrhenius equation

$$D_{ij} = a_{ij}\exp(-f_{ij}/T) \tag{3}$$

where T is absolute temperature and a_{ij} and f_{ij} are predetermined constants of the contaminated composite which are assumed not to change in time.

The use of Arrhenius' law can be justified by the fact that measurements of diffusion constants for gas flows through metals, alloys and glasses show the equation to be very satisfactory.[4] Data on the diffusion of moisture in composites also show similar trends.[5] In addition, there is theoretical basis for eqn. (3) when f_{ij} are interpreted as energies of activation needed to move gas molecules through pores along the i, j planes.

The temperature is determined by solving the energy equation[7]

$$\rho c_p \frac{\partial T}{\partial t} = -\operatorname{div}\mathbf{q} \tag{4}$$

with appropriate boundary and initial conditions. Here ρ is density, c_p is specific heat, and $\mathbf{q}$ is conduction heat flux given by Fourier's law

$$q_i = -k_{ij}T_{i,j} \tag{5}$$

with k_{ij} $(=k_{ji})$ being the components of conductivity which will also be assumed not to change in time. This may be justified by recognizing that contribution of vaporized contaminants to bulk conductance is usually minor, regardless of concentration, due to the low values of conductivities of gases.

Application to Spacecraft Structures

In application to spacecraft antennas and mounting platforms, it is noted that these are usually 'thin' structures which invariably feature sealants and insulation on the edges and the inactive surface (see, for example, Fig. 3 in Ref. 8). The net heat flow is across the thermally coated surface which is exposed to the environment. The same surface can be utilized as the interface for mass diffusion into space. If the thermal treatment has the potential for interacting with the surface to seal off transpiration pores, then a perforation procedure might become necessary to insure rapid relief from impurities.

A schematic of the thermal-diffusion mathematical model is given in Fig. 1. The components of conductivity and diffusivity of the assembled stacked

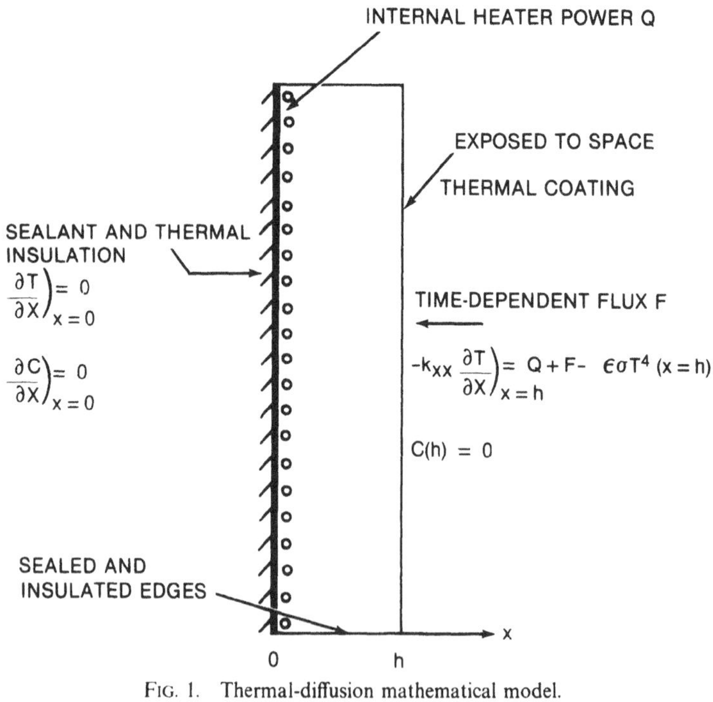

FIG. 1. Thermal-diffusion mathematical model.

laminate shall be considered known either from tests or from theoretical predictions.[5,9,10] Heating and outgassing will be uniform, and symmetry of the temperature and concentration profiles will exist in at least most of the panel away from the edges. Hence there will be no variation of T and c with y or z. In the case of temperature,

$$\frac{\partial T}{\partial y} = \frac{\partial T}{\partial z} = 0 \tag{6}$$

and eqns (5) give

$$q_x = -k_{xx}\frac{\partial T}{\partial x} \qquad q_y = -k_{yx}\frac{\partial T}{\partial x} \qquad q_z = -k_{zx}\frac{\partial T}{\partial x} \tag{7}$$

Substituting eqns (6) and (7) into eqn. (4) yields

$$\frac{\partial^2 T}{\partial x^2} = \frac{\rho c_{\mathrm{p}}}{k_{xx}}\frac{\partial T}{\partial t} \tag{8}$$

The initial condition is known and the boundary conditions are

$$\left.\frac{\partial T}{\partial x}\right)_{x=0} = 0 \qquad -k_{xx}\left.\frac{\partial T}{\partial x}\right)_{x=h} = Q(t) + F(t) - \varepsilon\sigma T_{\mathrm{h}}^4 \tag{9}$$

$Q(t)$ is internal heat (heater power), $F(t)$ is absorbed energy from environment heating sources (solar, planetary and adjacent emitters), and $\varepsilon\sigma T_{\mathrm{h}}^4$, where ε is emittance and $\sigma = 5{\cdot}668 \times 10^{-8}\ \mathrm{W/m^2K^4}$, is radiated energy.

Equation (8) with boundary conditions (9) can be solved analytically if the surface temperature to the fourth power is replaced by the linear expression[11]

$$T_{\mathrm{h}}^4 \approx T_{\mathrm{m}}^4 + 4T_{\mathrm{m}}^3(T_{\mathrm{h}} - T_{\mathrm{m}}) \tag{10}$$

where T_{m} is a mean temperature whose value is selected in view of optimum accuracy. Hence the boundary condition at $x = h$ becomes

$$-k_{xx}\left.\frac{\partial T}{\partial x}\right)_{x=h} = Q + F + 3\varepsilon\sigma T_{\mathrm{m}}^4 - 4\varepsilon\sigma T_{\mathrm{m}}^3 T \tag{11}$$

Solutions of eqn. (8) with linear boundary conditions are detailed in Ref. 7.

Similar considerations of the concentration profile give

$$\frac{\partial c}{\partial y} = \frac{\partial c}{\partial z} = 0 \tag{12}$$

and from eqn. (2),

$$J_x = -D_{xx}\frac{\partial c}{\partial x} \qquad J_y = -D_{yx}\frac{\partial c}{\partial x} \qquad J_z = -D_{zx}\frac{\partial c}{\partial x} \tag{13}$$

Substituting into eqn. (1),

$$\frac{\partial^2 c}{\partial x^2} = \frac{1}{D_{xx}}\frac{\partial c}{\partial t} \tag{14}$$

The diffusion coefficient is dependent on temperature in accordance with eqn. (3). Since temperature is a function of space and time, the dependence of D_{xx} on x and t must be incorporated into the solution of eqn. (14). The variation in the diffusion coefficient is

$$\Delta D_{xx} = \frac{-a_{xx}f_{xx}}{T^2}\exp\left(-\frac{f_{xx}}{T}\right)\left(\frac{\partial T}{\partial x}\Delta x + \frac{\partial T}{\partial t}\Delta t\right) \tag{15}$$

But for thin structures the change in temperature along the thickness is usually small, even for stacked laminates with low normal conductivity.[8] The change with time, however, can be considerably large. For example, the average temperature of a solar array orbiting part time in shadow and part time in sunlight can vary by as much as 200°C during an orbit, whereas the temperature difference across the panel at any time remains within 5°C.[12] Therefore D_{xx} is very nearly a function of time only. The inaccuracy becomes insignificant when the instantaneous temperature is taken as the integrated average across the thickness.

Equation (14) and the conditions in accordance with the definition of c and Fig. 1 are now written

$$\frac{\partial^2 c}{\partial x^2} = \frac{1}{D_{xx}(t)}\frac{\partial c}{\partial t} \qquad c(x,0) = 1 \qquad \left.\frac{\partial c}{\partial x}\right)_{x=0} = 0 \qquad c(h) = 0 \tag{16}$$

The vanishing of concentration at the surface conforms to the notion that particles exposed to space vacuum escape almost instantly.

Solutions for eqn. (16) can be found by the method of the separation of variables. By assuming

$$c = f(x)g(t) \tag{17}$$

and substituting into eqn. (16), the concentration is found to be

$$c = \sum_{n=1}^{\infty}\frac{2(-1)^{n-1}}{(2n-1)(\pi/2)}\cos(2n-1)\frac{\pi}{2}\frac{x}{h}\exp\left(-\left[(2n-1)\frac{\pi}{2h}\right]^2\int_0^t D_{xx}(t)\,dt\right) \tag{18}$$

The total concentration is obtained by integrating from $x=0$ to $x=h$:

$$c_{\text{total}} = \sum_{n=1}^{\infty} \frac{2}{[(2n-1)(\pi/2)]^2} \exp\left(-\left[(2n-1)\frac{\pi}{2h}\right]^2 \int_0^t D_{xx}(t)\,\mathrm{d}t\right) \tag{19}$$

OUTGASSING IN AN ORBITAL FLUX ENVIRONMENT

Orbiting structures undergo steady periodic changes in temperature caused by periodic environment heating. A typical heating input is solar energy impinging on a sun-oriented flat panel in a near-earth noon orbit. The absorbed energy is constant during the sunlit portion of flight, then disappears when the panel is shadowed by earth. The profile is repeated every orbit. Another example is solar heating of a cylindrical spacecraft spinning about its axis. The flux input to each point on the surface is represented by a rectified trigonometric wave.

Temperature predictions in linear problems are obtained by representing the anticipated periodic temperature by a Fourier series of unknown amplitudes but of the same base frequency as the heating function. Substituting into eqn. (8) leads to ordinary second-order differential equations which can be solved for the amplitudes. The method of the Laplace transforms is sometimes preferred because it permits rapid calculations where the Fourier series converges slowly.

Simple Harmonic Flux

Consider a simple harmonic flux impinging on the platform shown in Fig. 1:

$$F(t) = F_0 \cos \omega t \tag{20}$$

F_0 being amplitude and ω frequency. Let the structure heater power be a constant $Q_0 > F_0 > 0$ so that only positive temperatures are obtained as demanded by physical reality. Using complex variables, with meaning attached only to the real part, let

$$F = F_0 \mathrm{e}^{\mathrm{i}\omega t} \qquad \mathrm{i} \equiv \sqrt{-1} \tag{21}$$

$$\frac{T}{T_\mathrm{m}} = u_0 + u(\xi)\,\mathrm{e}^{\mathrm{i}\omega t} \qquad \xi \equiv \frac{x}{h} \tag{22}$$

$$\varepsilon\sigma T_\mathrm{m}^4 \equiv \frac{1}{2\pi} \int_0^{2\pi} (Q_0 + F_0 \cos \omega t)\,\mathrm{d}(\omega t) = Q_0 \tag{23}$$

Substituting into eqn. (8) and the linearized boundary conditions:

$$\frac{d^2u}{d\xi^2} = \frac{h^2 i\omega}{\alpha} u \qquad u'(0) = 0$$
$$u'(1) = \frac{4hQ_0}{k_{xx}T_m} u(1) - \frac{hF_0}{k_{xx}T_m} \qquad u_0 = 1 \tag{24}$$

α is thermal diffusivity which equals $k_{xx}/\rho c_p$. The solution of eqn. (24) gives the steady periodic temperature

$$\frac{T}{T_m} = 1 + \frac{F_0}{4Q_0} \operatorname{Re} \frac{\cosh \beta\xi \, e^{i\omega t}}{\cosh \beta - (k_{xx}T_m/4hQ_0)\beta \sinh \beta} \tag{25}$$

Here Re stands for 'the real part of' and $\beta = \sqrt{h^2 i\omega/\alpha}$. The integrated average temperature across the thickness is

$$\frac{T_{av}}{T_m} = 1 + \frac{F_0}{4Q_0} \operatorname{Re} \frac{\sinh \beta \, e^{i\omega t}}{\beta \cosh \beta - (k_{xx}T_m/4hQ_0)\beta^2 \sinh \beta} \tag{26}$$

This expression is of the form

$$T_{av}(\omega t) = T_m[1 + A\cos(\omega t + \phi)] \tag{27}$$

in which T_m, A and ϕ depend on system and environment properties. Table 1 lists some values based on the following data:

Flux amplitude, $F_0 = 135\ \mathrm{W/m^2}$ (one tenth a solar constant)
Flux frequency, $\omega = 2\pi/1{\cdot}5/\mathrm{h}$ (near earth orbit)
Panel thickness, $h = 0{\cdot}02\ \mathrm{m}$
Surface emittance, $\varepsilon = 0{\cdot}80$ (white coating)
Normal conductivity, $k_{xx} = 1{\cdot}35\ \mathrm{W/mK}$ (stacked GFRP laminate)
Density, $\rho = 1605\ \mathrm{kg/m^3}$ (GFRP)
Specific heat, $c_p = 864\ \mathrm{J/kgK}$ (GFRP)
$\beta = 0{\cdot}6839 i^{1/2}$

In all cases given in Table 1, the temperature difference across the thickness is less than 1°C. The history of the average temperature for 300 W heater power is shown in Fig. 2.

Total concentration over a cycle is obtained from eqn. (19) by substituting

$$D_{xx} = a_{xx} \exp\{-f_{xx}/T_m[1 + A\cos(\omega t + \phi)]\} \tag{28}$$

TABLE 1
Amplitude and phase angle in eqn. (27)

Q_0 (W)	T_{ml} (K)	A	ϕ (rad)
500	324·05	0·0130	1·3754
400	306·47	0·0138	1·4057
300	285·20	0·0148	1·4378
200	257·71	0·0164	1·4730

and performing the integration (numerically) from $t = 0$ to $t = 2\pi/\omega$. For the profile in Fig. 2, the time needed to expel 90% of original concentration ($c = 10\%$) is described in Fig. 3 as a function of a_{xx} for two values of f_{xx}. Comparison is made with diffusion at the constant temperatures that correspond to the maximum and minimum values during the cycle. It is evident that an underestimate or overestimate in the values of temperature by only a few degrees can result in a considerable variance in time calculations. When the diffusion coefficient is 2×10^{-4} m^2/s, the time for outgassing 90% original concentration is 480 h at 8·6°C versus 785 h at 16·1°C. A difference of 12·5% in the values of f_{xx} can change the time for reaching a specified concentration by a factor of 5.

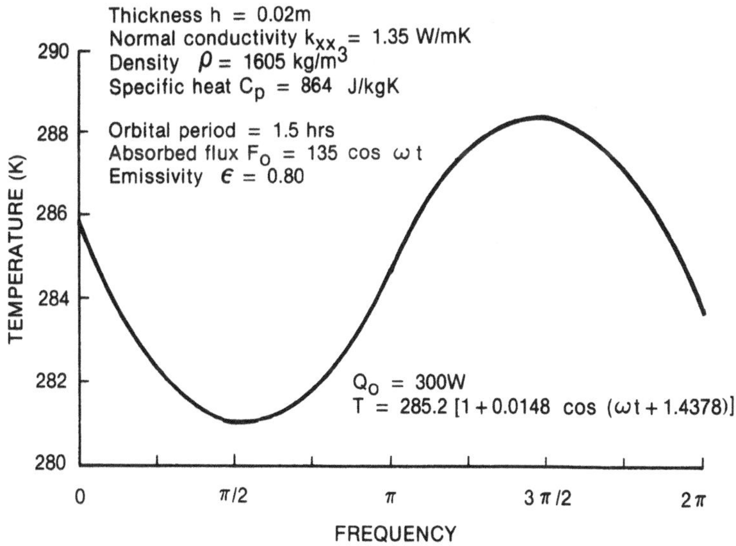

FIG. 2. Panel temperature in a simple harmonic flux environment.

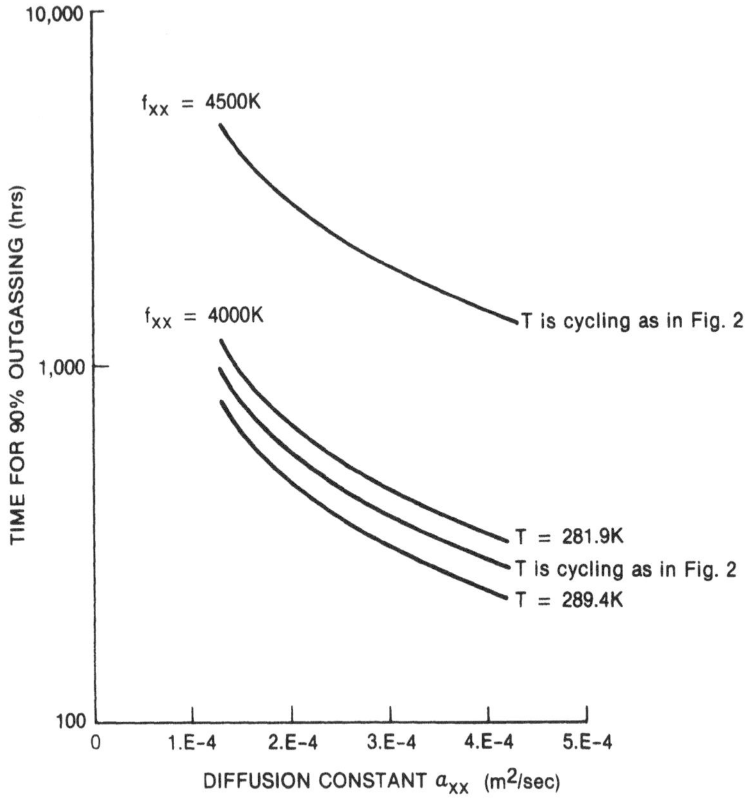

FIG. 3. Time for 90% outgassing under simple harmonic flux.

Rectangular Wave Flux

Another example which is more in line with actual orbital behavior is outgassing of the panel described above when exposed to a rectangular-shaped flux. This simulates the heating input to a sun-facing platform in geosynchronous orbit during equinox. The orbital period here is 24 h, of which 1·2 h are spent in shadow. Because of small gradients ($\Delta T < 1°C$ across the thickness) and the long orbital period, the temperature profile can be represented as follows:

(A) In sunlight, just before entering shadow:

$$T = (F/\varepsilon\sigma)^{1/4} \quad \text{(constant)} \tag{29}$$

where F is absorbed solar flux.

(B) Cool-down during shadow:

$$T = T_i\left(1 + \frac{3\varepsilon\sigma T_i^3 t}{\rho c_p h}\right)^{-1/3} \tag{30}$$

where T_i is initial temperature, which is the temperature just before entering shadow.

(C) Heat-up, as panel leaves shadow:

$$\frac{\varepsilon\sigma T_i^3 t}{\rho c_p h} = \tfrac{1}{2}\gamma^3\left[\ln\left|\frac{(1+\gamma\tau)(1-\gamma)}{(1-\gamma\tau)(1+\gamma)}\right|^{1/2} + \tan^{-1}\gamma\tau - \tan^{-1}\gamma\right] \tag{31}$$

where $\gamma = (\varepsilon\sigma T_i^4/F)^{1/4}$, $\tau = T/T_i$, and T_i is the temperature when the panel emerges from shadow.

Figure 4 is the temperature history when the solar absorptance is 0·20. The corresponding residual concentration is obtained by the numerical integration of eqn. (19) and is indicated in Fig. 5 as a function of f_{xx}. With a heater operating at 270 W/m^2 during shadow, the temperature remains constant at 4·5°C throughout the orbit. The reduction in time for the release of 90% original concentration is shown by the dotted line in Fig. 5.

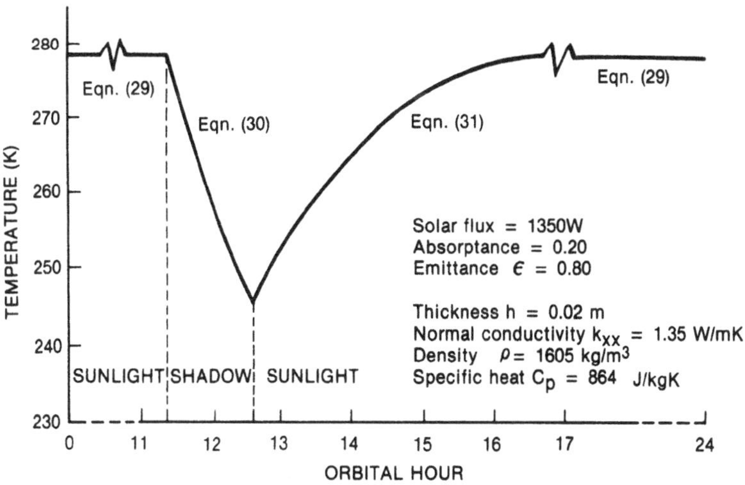

FIG. 4. Panel temperature in geosynchronous orbit.

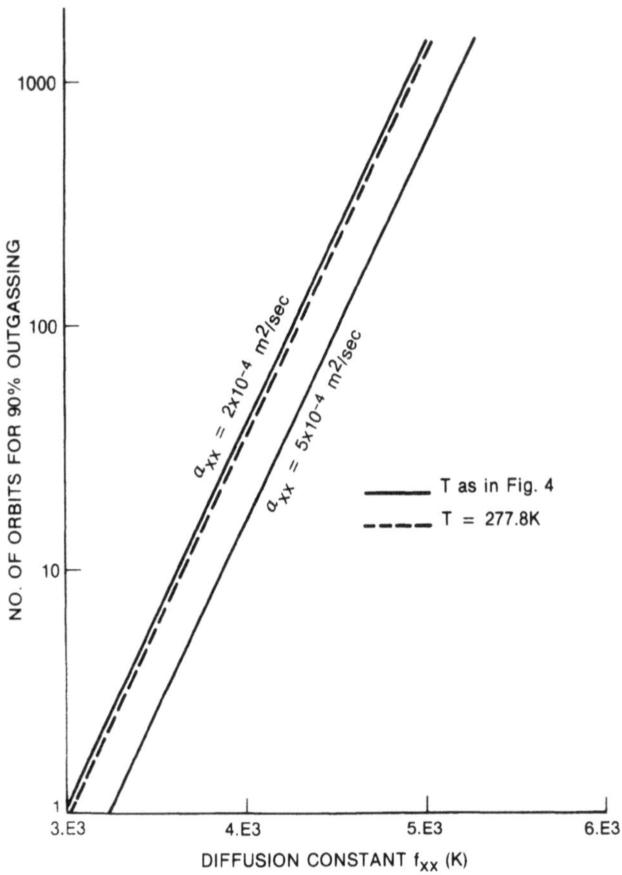

FIG. 5. Number of orbits for 90% outgassing at geosynchronous altitude.

SUMMARY AND DISCUSSION

Diffusion and the eventual outgassing of impurities in spacecraft composites are analyzed by using the conventional differential equations of heat and mass transfer. The study is prompted by the necessity to permit outgassing in order to avoid deformation due to the expansion of suppressed volatiles, and to assess potential contamination of neighboring instruments and experiments.

It is found that the temperature and diffusion constants have a strong influence on outgassing rates and must therefore be known within good accuracy for a meaningful assessment. The thermal analysis presented in this work has consistently been proven valid when the thermal characteristics of the system and environment are known. But in a contaminated environment there will be no assurance against erratic change in properties. One major consideration is the increase in temperature due to a degradation of surface absorptivity. The uncertainty can be partially resolved by monitoring the temperature through flight thermistors located in key areas of the spacecraft and mounting platform.

Correct estimates of the diffusion constants a_{xx} and f_{xx} are also essential. The values used in the numerical examples above lie within data points obtained during ground testing in an atmospheric environment. It is questionable, however, that the same data can be used for analysis in space surroundings, especially if the system is treated to achieve relatively rapid release of impurities. The problem is further complicated by the long-term deterioration and expulsion of the resin.

Testing in vacuum is therefore recommended. The complexity and extent of the test will depend on the criticality of thermal stability and the sensitivity of surrounding equipment to contamination. But determination of diffusion constants within tight tolerances should be a major goal. To this end, test specimens fabricated simultaneously with flight hardware should be considered. Understanding of the results would be enhanced considerably if the test chamber were equipped with product analyzers for identifying the outgassed materials and classifying their influence.

REFERENCES

1. Tsai, S. W. and Hahn, H. T., *Introduction to Composite Materials*, Westport, CT, Technomic Publishing Company, 1980, pp. 342–344.
2. Extreme Ultraviolet Explorer (EUVE) to Explorer Platform Interface Control Document, NASA ICD no. 408-EUVE-ICD-001, GSFC, Beltsville, MD, Jan. 1986.
3. Bird, R. B., Stewart, W. E. and Lightfoot, E. N., *Transport Phenomena*, New York, John Wiley, 1960, pp. 554–560.
4. Sherwood, J. N. *et al.* (eds), Diffusion processes, *Proceedings of the Thomas Graham Memorial Symposium*, Paris, Gordon and Breach, 1971, pp. 7–9.
5. Loos, A. C. and Springer, G. S., Moisture absorption of graphite–epoxy composites immersed in liquids and in humid air, *J. Comp. Mater.*, **13** (April 1979), 131–147.
6. Barber, R. M., *Diffusion In and Through Solids*, Cambridge, The Macmillan Company, 1941.
7. Carslaw, H. S. and Jaeger, J. C., *Conduction of Heat in Solids*, 2nd edn, London, Oxford University Press, 1959.

8. KARAM, R. D., Thermal engineering of spacecraft composite structures, in: *Composite Structures—3* (Marshall, I. H. ed.), London, Elsevier Applied Science, 1985.
9. THORNBURGH, J. D. and PEARS, C. D., Prediction of the thermal conductivity of filled and reinforced plastics, ASME publication no. 65-WA/HT4, 1965.
10. CHI-HUNG SHEN and SPRINGER, G. S., Moisture absorption and desorption of composite materials, *J. Comp. Mater.*, **10** (Jan. 1976), 2–20.
11. KARAM, R. D., Optimum solution of linearized radiation equations, in: *Numerical Methods in Thermal Problems* (Lewis, R. W. and Morgan, K. eds), Swansea, Redwood Burn Limited, 1979.
12. DUCAS, W., Thermal analysis of a conceptual Solarex/FSC solar array, Fairchild Space Company IOC Ref., ME-83-09, April 12, 1983.

3

On the Design of Prestressed and Non-prestressed Bolted Joints in Glass Fibre Reinforced UP-Laminates

ALFRED RÜBBEN
Fachgebiet Kunststoffbau, Konstruktion und Berechnung, Rhein.-Westf. Technische Hochschule Aachen, Mies-van-der-Rohe-Straße 1, D-5100 Aachen, West Germany

ABSTRACT

The correct design of joints appropriate to the specific stress–strain behaviour of the material is of eminent importance for all load-bearing structures. The usual dimensioning rules, developed for traditional materials cannot be applied when using new composites.

The present study deals with the tensile strength and failure behaviour of bolted joints in glass fibre reinforced unsaturated polyester mat laminates (GFR-UP-M) prestressed and non-prestressed. The influence of geometric conditions like the bolt diameter (hole diameter), the thickness of the material, the pitches, the type and quantity of reinforcement, and the degree of prestressing are taken into account. The different failure modes of the various joint-types as well as the decisive factors for the design, such as strength, hole and rim distances are examined in dependence on the different parameters. Short-term and long-term loadings are considered.

The investigations undertaken have the aim of providing a contribution to the development of a method of calculation, appropriate to the material, for bolted joints with different degrees of prestressing.

1. INTRODUCTION

Plastics evidence a fundamentally different material behaviour from the classic materials. Time-dependent changes of the material properties due to stress, temperature and media; long-term deformations and loss of strength under long-term load and—with regard to reinforced plastics—the absence of flow before rupture, as well as the influence of anisotropy have to be considered.

Since the above mentioned factors influence not only the load-bearing behaviour of plastics in general, but that of the areas where loads are introduced in particular, the methods of calculation for joints developed for the classic materials and construction have to be modified when applied to reinforced plastics.

Although a lot of research has been done to examine the strength behaviour of bolted joints in GFRP, for example Refs 1–4, many questions are still not answered.

2. TEST PROGRAM

The tests with glass fibre reinforced unsaturated polyester mat laminates (GFR-UP-M) were carried out to examine the tensile strength behaviour of three different types of bolted joints:

—bolted joints without inhibition of transverse deformation and prestressing;
—bolted joints with inhibition of transverse deformation without prestressing;
—bolted joints with prestressing.

The joints were loaded to the point of fracture with a constant deformation velocity of 2 mm/min at a room temperature of 20°C (short-term load). The mode of experimental procedure and specimen geometry are shown in Fig. 1.

The mean short-term tensile strengths σ_{f0} (according to DIN 53453) for GFR-UP-M laminates were 65–69 N/mm^2, depending on material thickness, with $w_f = 27\%$ (fibre weight fraction) and 79 N/mm^2 with $w_f = 37\%$.

2.1. Bolted Joints without Inhibition of Transverse Deformation and Prestressing

As presented in Refs 1–3, short-term tensile strength tests were carried out with bolted joints without inhibition of transverse deformation and prestressing (see Fig. 1a). Special emphasis was laid on a centred and momentum-free arrangement of the samples. Hardened steel was used for the bolts. All bolts were fitting.

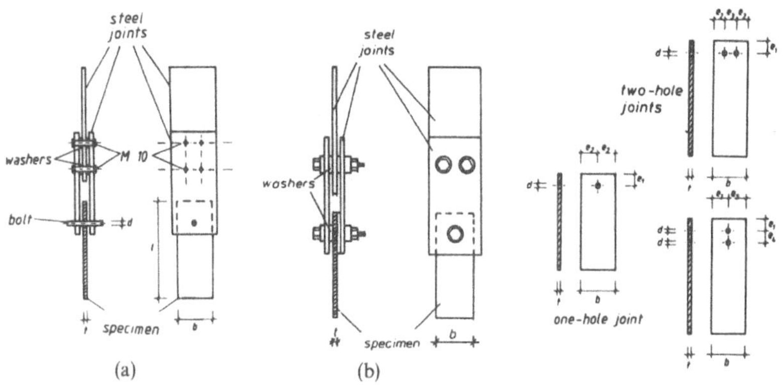

FIG. 1. Mode of experimental procedure and specimen geometry. (a) Without inhibition of transverse deformation. (b) With inhibition of transverse deformation.

Single-hole and multi-hole joints were tested in accordance with the test program shown in Table 1.

2.2. Bolted Joints with Inhibition of Transverse Deformation without and with Prestressing

In addition to this program, joints in GFR-UP-M with $w_f = 27\%$ and with inhibition of transverse deformation were tested. In the case of the non-prestressed joints washers were placed very close to the laminates in order to avoid transverse deformation. The prestressed joints were given additional prestressing of 9 kN per bolt. In the samples with a hole diameter

TABLE 1
Test program for bolted joints without inhibition of transverse deformation

Laminate thickness t (mm)	Distances from edge and pitch of holes				Hole diameter d fibre weight fractions (mm)			
	e_1/d	e_2/d	e_3/d	e_4/d	25%	27%	37%	38%
4	2, 3, 4, 5	2, 3. 4, 5	—	—	8	4, 6, 8, 10	6	8
5	2, 3, 4, 5	2, 3. 4, 5	—	—	—	—	4, 6, 8, 10	—
6	2, 3, 4, 5	2, 3. 4, 5	—	—	—	6, 8, 10	6, 8, 10	—
8	2, 3, 4, 5	2, 3. 4, 5	—	—	—	6, 8, 10	6, 8, 10	—
6	2. 3, 4, 5	2, 3. 4, 5	2, 3, 4, 5	—	—	6	—	—
6	5	5, 6. 7, 8	—	2, 3, 4, 5		6	—	

TABLE 2
Test program for bolted joints with inhibition of transverse deformation

Laminate thickness t (mm)	Distance from edge e_1/d	e_2/d	Hole diameter d (mm) fibre weight fractions 27%	Prestressing (kN)
4	2; 3; 4; 5; 6	2; 3; 4; 5; 6	6; 10	0; 9
6	3; 4; 5; 6	3; 4; 5; 6	6; 10	0; 9
8	3; 4; 5; 6	3; 4; 5; 6	6; 10	0; 9

of 6 mm, the washers had a surface area of approximately 280 mm^2, and in the joints with a 10 mm hole diameter, the surface area of the washers was approximately 280 mm^2. Figure 1b and Table 2 show the test program.

3. FAILURE MODES AND SHORT-TERM STRENGTH OF BOLTED JOINTS IN GFRP

The failure modes of bolted joints in GFRP depend upon geometry, prestressing, and matrix material, as well as type, quantity and orientation of fibres.

Three main failure modes are of importance for the materials and single-hole joints examined here (see Fig. 2):

- —Tensile strength failure in the hole cross section (failure mode 1), cracks perpendicular to the strain direction.
- —Modified tensile strength failure (failure mode 2), final failure with cracks in the strain direction.
- —Failure because of specific pressure on the hole (failure mode 3).

Depending on the geometric conditions, the following failure modes of different single-hole bolted joints occurred during short-term loading (Fig. 3).

3.1. Bolted Joints without Inhibition of Transverse Deformation and Prestressing

Depending on the geometric conditions the failure modes 1, 2 and 3 are possible, depending on thickness of material t and hole diameter d, e.g.

Form 1 at $e_1 \geq e_2$ and $e_2 \leq 4d$
Form 2 at $e_1 \leq e_2$
Form 3 at $e_1 \geq 4d$ and $e_2 \geq 4d$

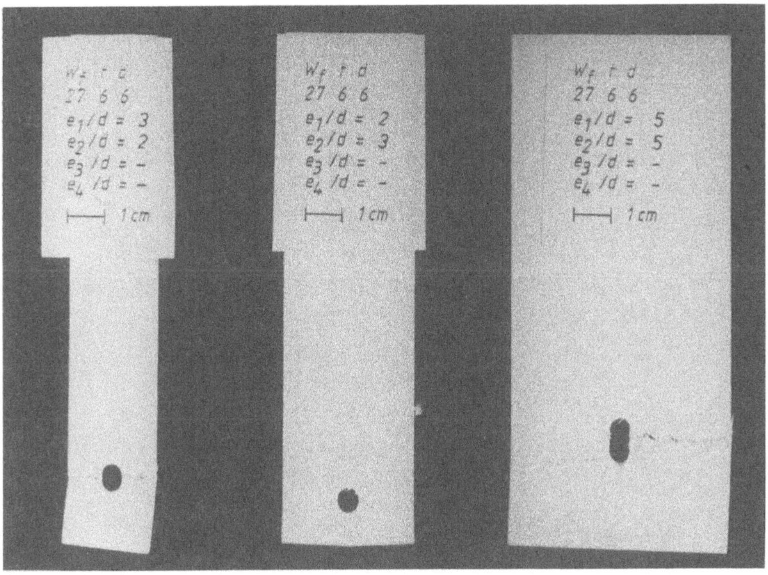

FIG. 2. Failure modes of bolted joints in GFR-UP-M.

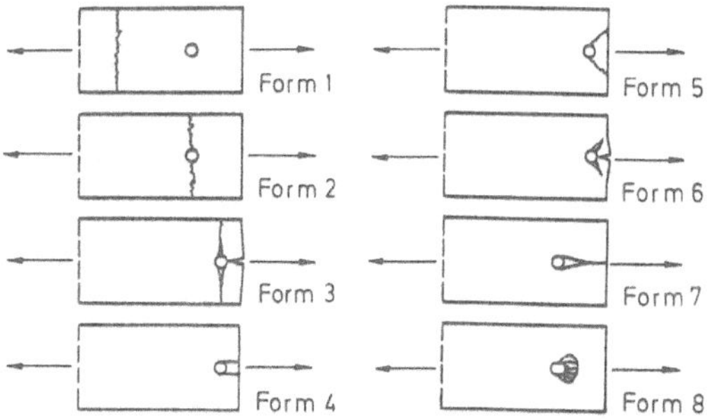

FIG. 3. Failure modes of single-hole bolted joints.

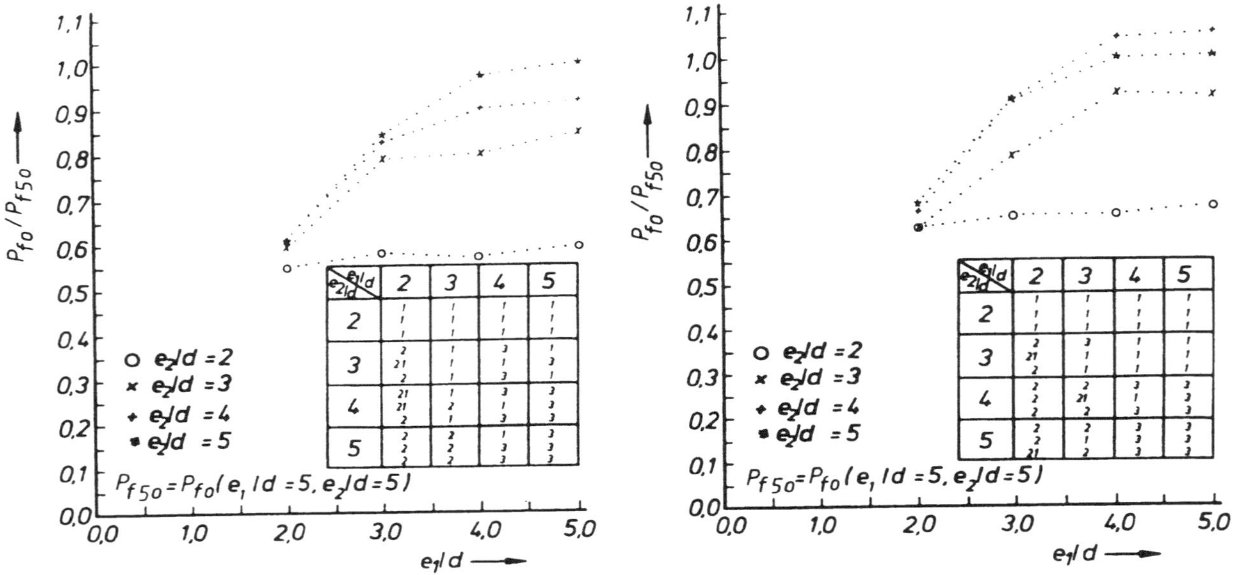

(a) (b)

FIG. 4. Short-term failure load ratio P_{f0}/P_{f50} for single-hole bolted joints without inhibition of transverse deformation. (a) $t = 6$ mm, $d = 8$ mm, $w_f = 27\%$. (b) $t = 6$ mm, $d = 8$ mm, $w_f = 37\%$.

Figure 4 shows the short-term failure loads P_{f0} of single-hole bolted joints dependent on e_1/d and e_2/d. Failure because of specific pressure on the hole P_{f50} for $e_1/d = e_2/d = 5$ shows the highest load-bearing capacity. Beyond that, no load increase could be observed. For further results, especially many-hole bolted joints, see Refs 1–3.

3.2. Bolted Joints with Inhibition of Transverse Deformation without Prestressing

The same failure modes 1, 2 and 3 occurred with non-prestressed single-hole bolted joints. The most frequent failure mode was tensile strength failure in the hole cross section. Modified tensile strength failure could be observed in approximately 30% of the cases, for $e_1/d < e_2/d$, depending on material thickness.

Failure because of specific pressure on the hole was very rare with $e_1/d \geq 5$ and $e_2/d \geq 5$. Compared to the joints without inhibition of transverse deformation, a strain increase beyond $e_1/d = e_2/d = 5$ could be observed (see Fig. 5).

3.3. Bolted Joints with Prestressing

Because of prestressing a more advantageous strain distribution within the joints is possible (see Fig. 6).

The tests do not show failure mode 2. Approximately 75% of all samples failed in consequence of tensile strength failure in the hole cross section.

Modified tensile strength failure occurred with $e_1/d \leq 4$; $e_2/d \geq 4$–5, depending on material thickness.

4. COMPARISON OF THE SHORT-TERM STRENGTH OF DIFFERENT TYPES OF BOLTED JOINTS

Figures 7 and 8 show a comparison of the short-term strength behaviour of the examined types of bolted joints.

Prestressed joints normally are more effective than non-prestressed joints. Compared to bolted joints with inhibition of transverse deformation a strength increase of approximately 10% was achieved before final failure (compared to first material damages even 30–70%).

The strength of bolted joints without inhibition of transverse deformation was beyond that of the other types, when thin laminates

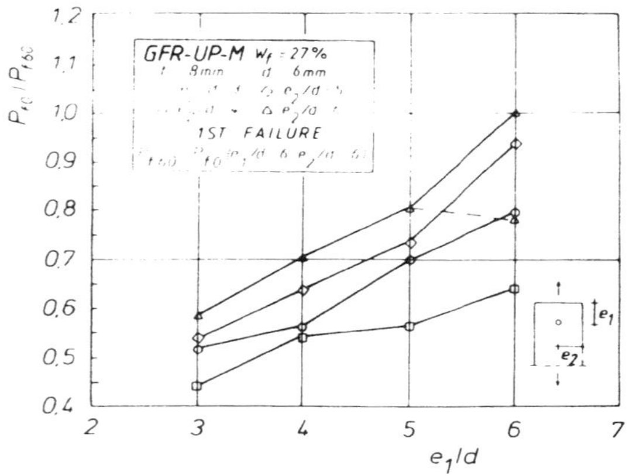

e_{2}/d \ e_{1}/d	3	4	5	6
3	1	1	1	1
4	2	1	1	1
5	2	1	1	1
6	2	2	2	(3) 1

FAILURE MODES 1, 2, 3

1. Tensile strength failure in the hole cross-section
2. Modified tensile strength failure
3. Failure because of specific pressure on the hole

(a)

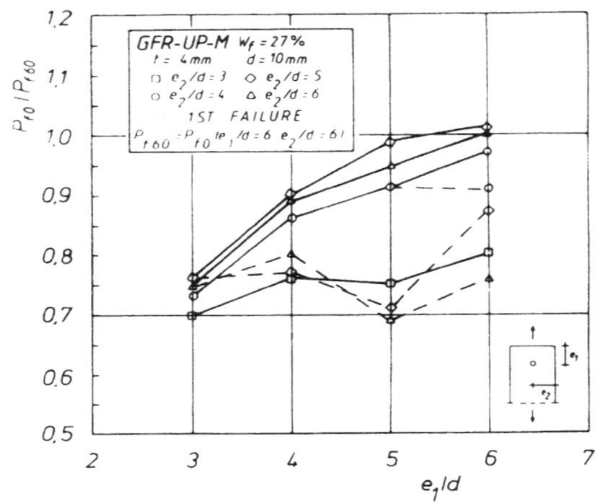

e_{2}/d \ e_{1}/d	3	4	5	6
3	1	1	1	1
4	1	1	1	(3) 1
5	2	(3) 1	(3) 1	(3) 1
6	2	(3) 2	(3) 1	(3) 1

FAILURE MODES 1, 2, 3

1. Tensile strength failure in the hole cross-section
2. Modified tensile strength failure
3. Failure because of specific pressure on the hole

(b)

FIG. 5. Short-term failure load ratio P_{f0}/P_{f60} for single-hole bolted joints with inhibition of transverse deformation but without prestressing. (a) $t = 8$ mm, $d = 6$ mm. (b) $t = 4$ mm, $d = 10$ mm.

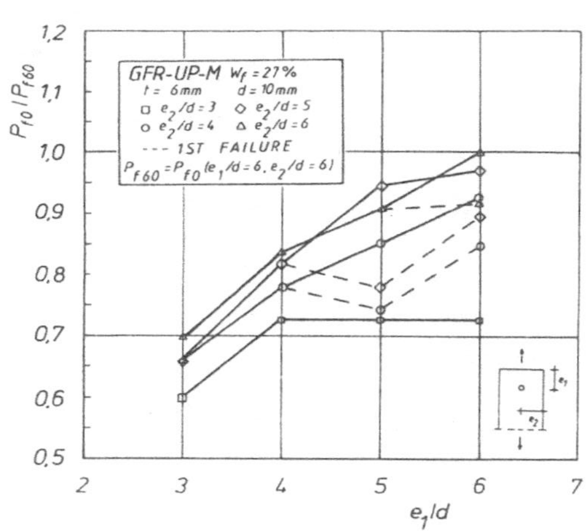

e_2/d \ e_1/d	3	4	5	6
3	1	1	1	1
4	1	1	(3) 1	(3) 1
5	2	1	(3) 1	(3) 1
6	2	1	1	(3) 1

FAILURE MODES 1, 2, 3

1. Tensile strength failure in the hole cross-section
2. Modified tensile strength failure
3. Failure because of specific pressure on the hole

(c)

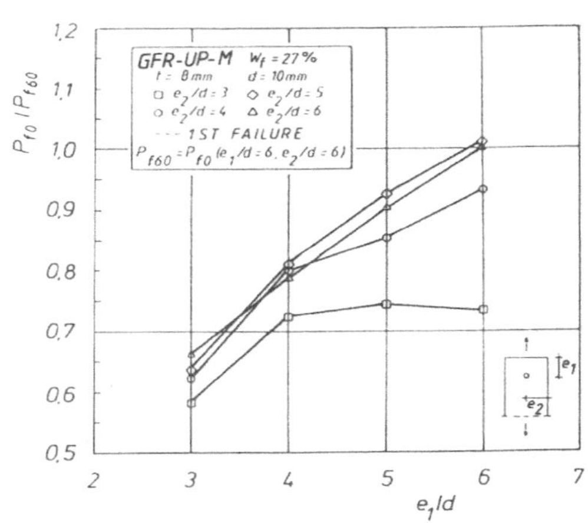

e_2/d \ e_1/d	3	4	5	6
3	1	1	1	1
4	1	1	1	1
5	1	1	1	1
6	2	1	1	1

FAILURE MODES 1, 2, 3

1. Tensile strength failure in the hole cross-section
2. Modified tensile strength failure
3. Failure because of specific pressure on the hole

(d)

FIG. 5.—*contd.* (c) $t = 6$ mm, $d = 10$ mm. (d) $t = 8$ mm, $d = 10$ mm.

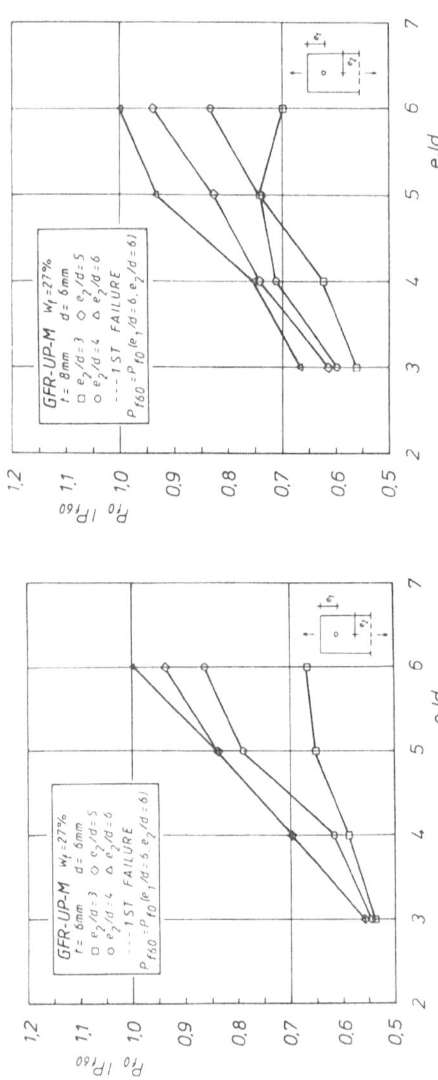

e_2/d \ e_1/d	3	4	5	6
3	1	1	1	1
4	2	1	1	1
5	2	1	1	1
6	2	2	2	1

FAILURE MODES 1, 2, 3

1 Tensile strength failure in the hole cross-section
2 Modified tensile strength failure
3 Failure because of specific pressure on the hole

(a)

e_2/d \ e_1/d	3	4	5	6
3	1	1	1	1
4	2	1	1	1
5	2	1	1	1
6	2	2	2	1

FAILURE MODES 1, 2, 3

1 Tensile strength failure in the hole cross-section
2 Modified tensile strength failure
3 Failure because of specific pressure on the hole

(b)

FIG. 6. Short-term failure load ratio P_{f0}/P_{f60} for prestressed single-hole bolted joints. (a) $t = 6$ mm, $d = 6$ mm. (b) $t = 8$ mm, $d = 6$ mm.

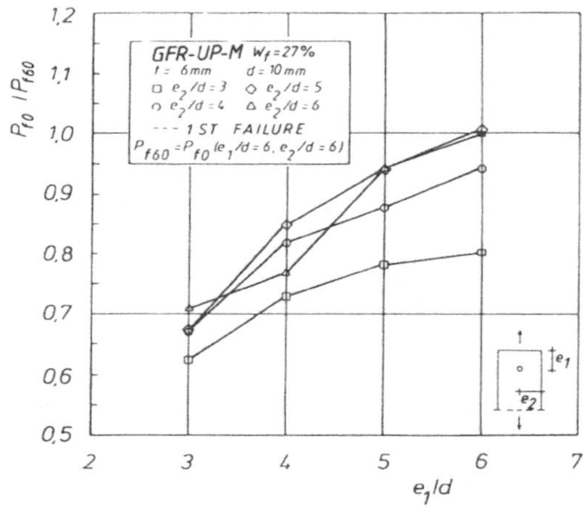

e_2/d \ e_1/d	3	4	5	6
3	1	1	1	1
4	1	1	1	1
5	2	2	1	1
6	2	1	1	1

FAILURE MODES 1, 2, 3

1. Tensile strength failure in the hole cross-section
2. Modified tensile strength failure
3. Failure because of specific pressure on the hole

(c)

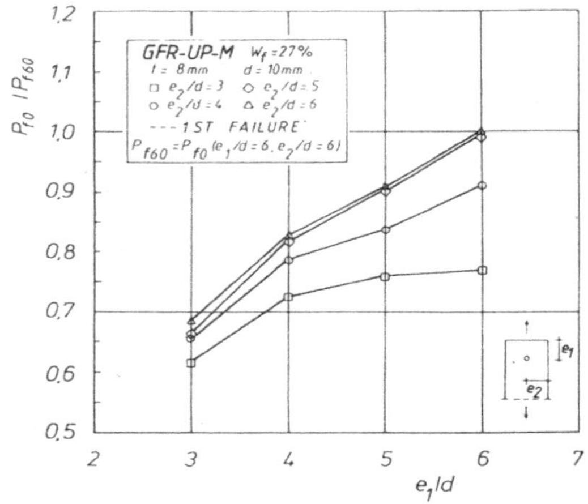

e_2/d \ e_1/d	3	4	5	6
3	1	1	1	1
4	1	1	1	1
5	2	1	1	1
6	2	1	1	1

FAILURE MODES 1, 2, 3

1. Tensile strength failure in the hole cross-section
2. Modified tensile strength failure
3. Failure because of specific pressure on the hole

(d)

FIG. 6.—*contd.* (c) $t = 6$ mm, $d = 10$ mm. (d) $t = 8$ mm, $d = 10$ mm.

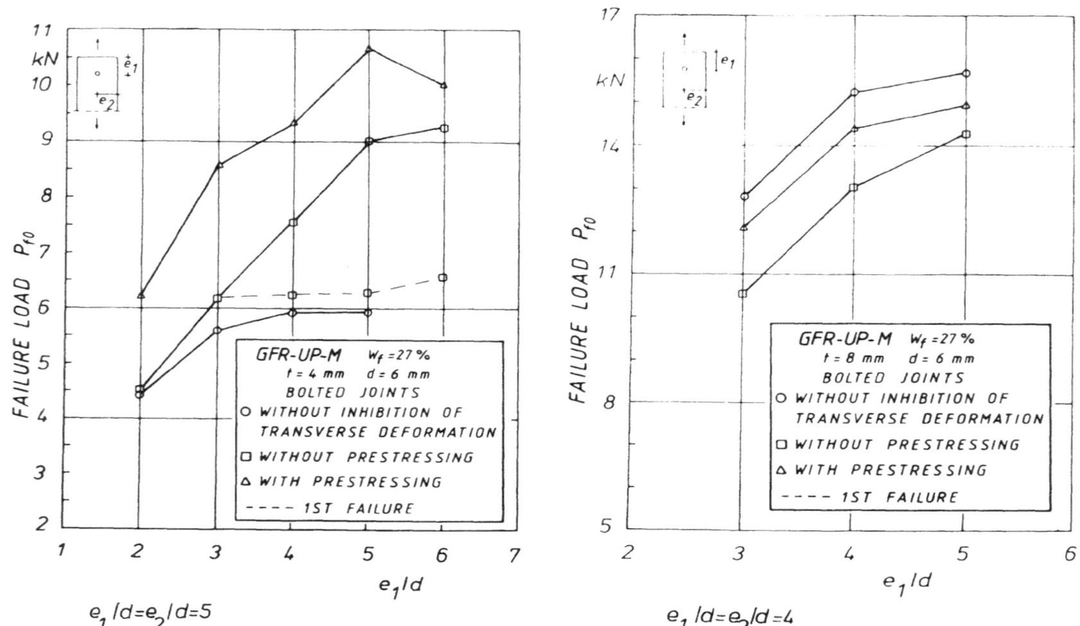

(a) (b)

FIG. 7. Short-term failure loads of single-hole bolted joints. (a) $t = 4$ mm, $d = 6$ mm. (b) $t = 6$ mm, $d = 10$ mm.

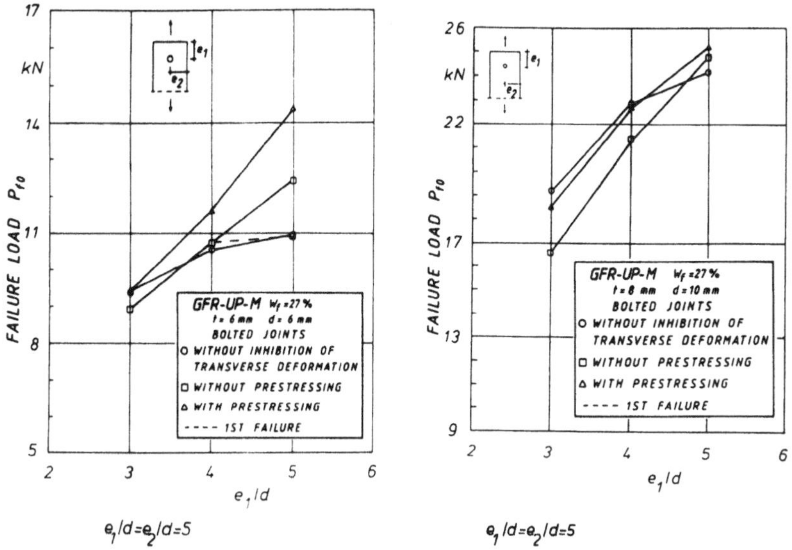

(c) (d)

FIG. 7.—*contd.* (c) $t = 6$ mm, $d = 6$ mm. (d) $t = 8$ mm, $d = 10$ mm.

($t = 4$ mm and 6 mm) were used. But with thicker laminates it was also possible to achieve even higher failure loads than with prestressed bolted joints.

5. LONG-TERM STRENGTH BEHAVIOUR OF BOLTED JOINTS

Figure 9 shows the strength decrease of bolted joints without inhibition of transverse deformation under long-term loading.

6. SUMMARY

As the tests have shown, the load-bearing behaviour of bolted joints is influenced by many factors. A correct design depends on adequate hole and rim distances. Prestressing does not always permit higher strain as compared to bolted joints without prestressing. Decrease in strength due to long-term loading, temperature, and media has to be considered.

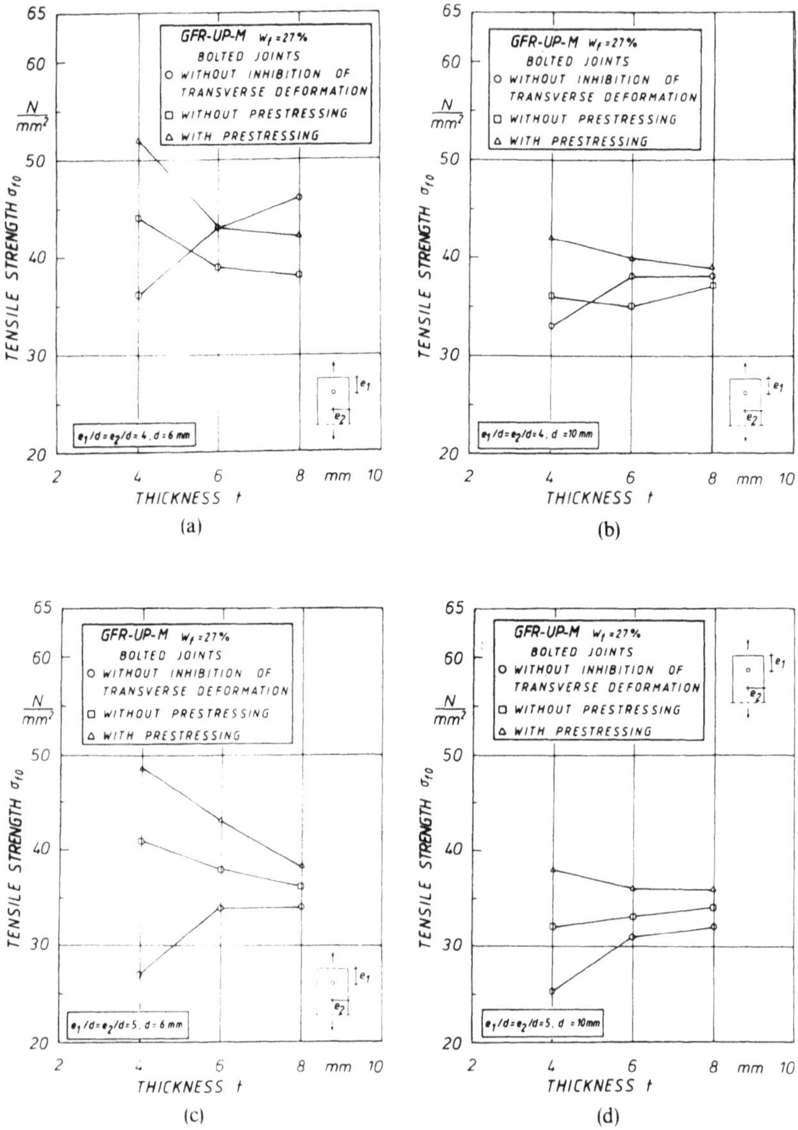

FIG. 8. Mean short-term tensile strength in the hole cross section of single-hole bolted joints. (a) $e_1/d = e_2/d = 4$, $d = 6$ mm. (b) $e_1/d = e_2/d = 4$, $d = 10$ mm. (c) $e_1/d = e_2/d = 5$, $d = 6$ mm. (d) $e_1/d = e_2/d = 5$, $d = 10$ mm.

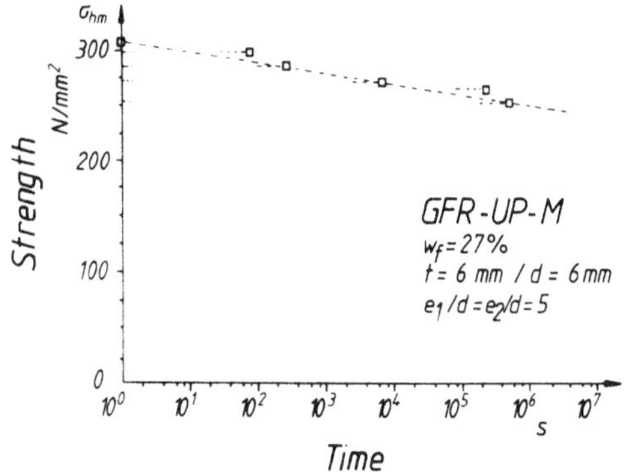

FIG. 9. Long-term behaviour of bolted joints without inhibition of transverse deformation.

REFERENCES

1. DOMKE, H. and RÜBBEN, A., Beitrag zur Bemessung von nicht vorgespannten Schraub- und Bolzenverbindungen bei Kunstoffbauteilen, *Fortschr.-Ber. VDI-Z*, Series 4, no. 63, 1983.
2. RÜBBEN, A., On the calculation of bolted joints for GRP plastics, *Composite Structures—1*, Elsevier Applied Science, London, 1983, pp. 175–187.
3. RÜBBEN, A., Fracture behaviour of non-prestressed bolted joints of GRP plastics, in: *Advances in Fracture Research*, Vol. **4**, (S. R. Valluri *et al.*, eds), New Delhi, 1984, pp. 2997–3004.
4. KRETSIS, G. and MATTHEWS, F. L., The strength of bolted joints in glass fibre/epoxy laminates, *Composites*, **16**, no. 2 (1985), 92–102.

4

Study on Static and Long-Term Strength of Mechanically Fastened GFRP and CFRP

H. HAMADA, Z. MAEKAWA, T. HORINO

Faculty of Textile Science, Kyoto Institute of Technology, Matsugasaki, Sakyou-Ku, Kyoto 606, Japan

A. KAJI

Osaka Municipal Technical Research Institute, Morinomiya, Jyouto-Ku, Osaka 536, Japan

and

Y. SHIINA

Production and Engineering Department, Sekisui Chemical Co. Ltd, Kamitoba, Minami-Ku, Kyoto 601, Japan

ABSTRACT

The deformation behaviour of mechanically fastened composite joints under creep load is examined. The notion of creep life is proposed taking practical applications of joints into consideration. In bearing failure modes the creep life is shorter than the time of the final failure, which is confirmed in the low and high cycle fatigue tests. The creep life of hybrid constitutions can be predicted by applying the law of mixtures.

INTRODUCTION

Composite materials are frequently used being connected with other materials in structural applications by means of mechanical fasteners. However the unskilful design of joints often causes reduction of the bearing capacity of composite structures, even if each member has sufficient strength and stiffness.

A number of investigators have studied the mechanically fastened joint by experimental approaches[1–4] and also by analytical approaches.[5–9] In our previous paper, the mechanical behaviour of the bolted joints in glass fibre mat reinforced plastics was experimentally evaluated, and the failure modes were predicted by two-dimensional finite element methods taking the stiffness of the bolt into consideration.[10] The numerical procedure was modified to incremental analysis in order to simulate the propagation of the damage zone in Mat-FRP.[11] The strength of hybrid constitutions (Mat/Cloth) in bolted joint were evaluated, and the designing process was proposed based on the experimental and numerical results.[12]

These investigations are, however, limited to the static behaviour of composite joints. Little attention has been given to the problem of the long-term strength of mechanically fastened composite materials.

In this chapter the creep strengths of mechanical joints are evaluated in GFRP, hybrid GFRP and CFRP. Firstly, the creep displacement is measured and the static fatigue life (creep life) is defined by the time when the creep displacement is abruptly increased. The creep life can be estimated in terms of the results of static tests. From these results the design procedure of mechanical joints can be improved.

TEST MATERIALS AND EXPERIMENTAL PROCEDURE

GFRP flat plate specimens are made from a chopped strand mat (REM-450-g1, Nippon Glass Fiber Co.) and a roving cloth (REW-55, Nippon Glass Fiber Co.) with unsaturated polyester resin (G-753-PTW, Japan Catalytic Chemical Industry Co.) by the hand layup method. The catalyst is methyl ethyl ketone peroxide in 0·5% solution. All specimens are aftercured at 80°C for 5 h. Four types of laminate constitution are chosen. Mat-FRP consists of four layers with mat and Cloth-FRP consists of four layers with cloth. The hybrid laminations are shown in Fig. 1.

The geometry of the specimen is shown in Fig. 2. The hole diameter is

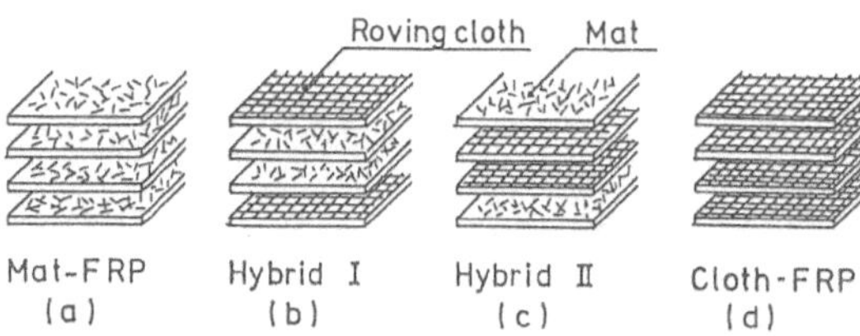

FIG. 1. Ply configuration of GFRP laminates.

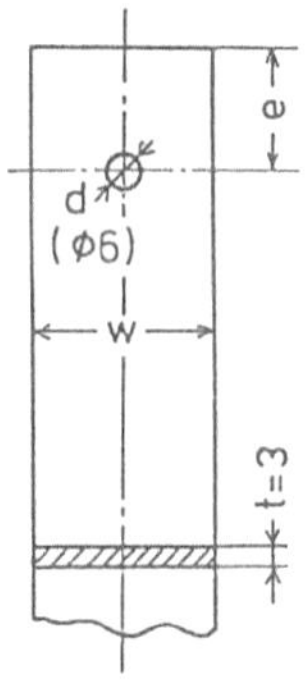

FIG. 2. Dimension of specimen.

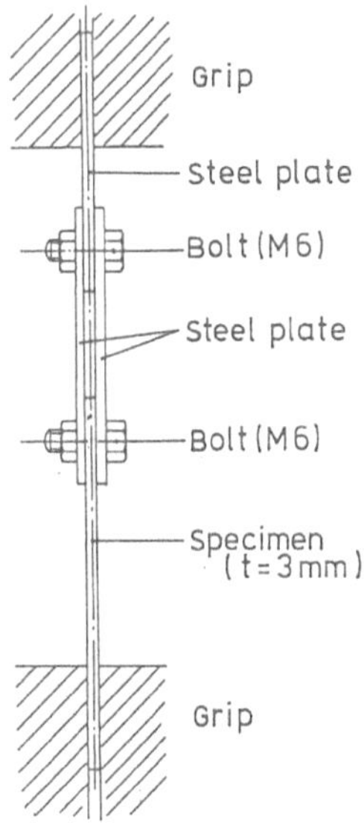

FIG. 3. Set-up of bolted joint tests.

6 mm. The joint ratio of w/d (specimen width/hole diameter) and e/d (centre of hole to edge distance/hole diameter) are varied from 5 to 9 and 2 to 9 respectively. The testing configuration of the double-lap joint is shown in Fig. 3. The specimen is placed between two flat steel plates, and the bolts are tightened by finger. The statical tensile test is conducted by an Instron type testing machine at a crosshead speed of 1 mm/min. The creep test is conducted by a single lever type creep tester. These tests are carried out under an environmental temperature of 20°C.

The low cycle fatigue test and high cycle fatigue test are also performed in order to specify the difference between creep and fatigue behaviours. The procedures of these tests are described later.

In the case of CFRP one type of angle-ply laminate with quasi-isotropic property is selected.

EXPERIMENTAL RESULTS AND CONSIDERATIONS

FRP Laminates (Mat-FRP)

Static bearing strength

In order to obtain the basic data of long-term strength the static bearing test is performed. Figure 4 shows the relationship between joint geometry

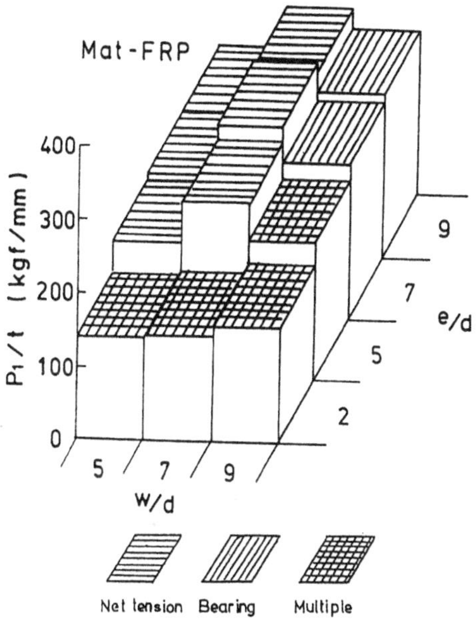

FIG. 4. Relation between normalized load and joint geometry (Mat-FRP).

(w/d and e/d) and the maximum normalized load. The final failure modes observed in this experiment are classified into three groups. These are bearing mode, net tension mode and multiple mode as shown in Fig. 5. In Fig. 4 the failure modes are illustrated as well. The normalized loads increase with increase of w/d and e/d. The failure modes change from the multiple mode to the net tension mode or the bearing mode as w/d and e/d increase. Three types of specimens are chosen for the purpose of obtaining the bolted joints with each failure mode. In the multiple mode a joint geometry of $w/d = 5$ and $e/d = 2$ (5*w*2*e* for short) is selected. The net tension mode is observed for a joint geometry of $w/d = 7$ and $e/d = 5$ (7*w*5*e*) and the bearing mode for $w/d = 9$ and $e/d = 7$ (9*w*7*e*). Figure 6 shows the load–displacement curves in 5*w*2*e*, 7*w*5*e* and 9*w*7*e*. It is clear from this figure that a rapid decrease after the maximum load is observed in the case of the multiple and net tension modes, while the several load reductions occur in the bearing mode. It is considered that the bearing failure progresses at these drop points.

Creep curves

Figure 7 shows a schematic diagram of a typical creep curve. The total displacement (δ) is divided into two parts, i.e. the static displacement (δ_s)

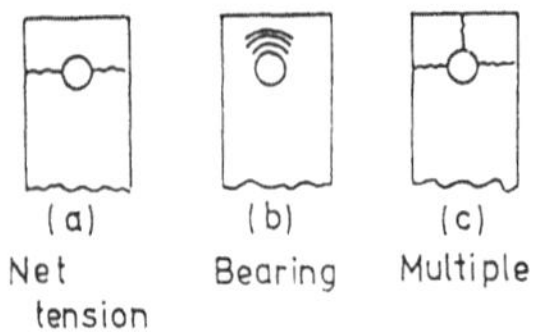

FIG. 5. Failure modes.

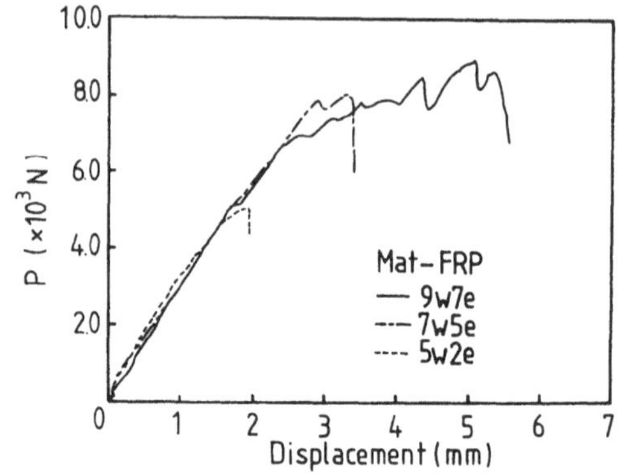

FIG. 6. Load–displacement curves in Mat-FRP.

and the creep displacement (δ_c). In our experiment the first measurement of the total displacement is performed 30 s (t_0) after the creep test starts. Therefore the value of creep displacement at t_0 is set to be zero. The creep curves shown in this experiment are the relation between the creep displacement and the time.

The creep curves of 5*w*2*e*, 7*w*5*e* and 9*w*7*e* are shown in Fig. 8. In this figure K_w is the ratio of the applied load to the maximum load and the creep displacement (δ_c) is normalized by the hole diameter (d). The final creep failure modes are the multiple mode in 5*w*2*e* and the bearing mode in 9*w*7*e*. In 7*w*5*e* the net tension mode is observed after the bearing failure. These failure modes are similar to that obtained in the static test mentioned above. From Fig. 8 in the case of 5*w*2*e* the creep displacement increases continuously until the final failure ($K_w = 0{\cdot}9$). Similar tendencies are obtained in other load levels. However the creep curve of 9*w*7*e* shows the abrupt increase of the creep displacement at an early stage ($K_w = 0{\cdot}9$). The

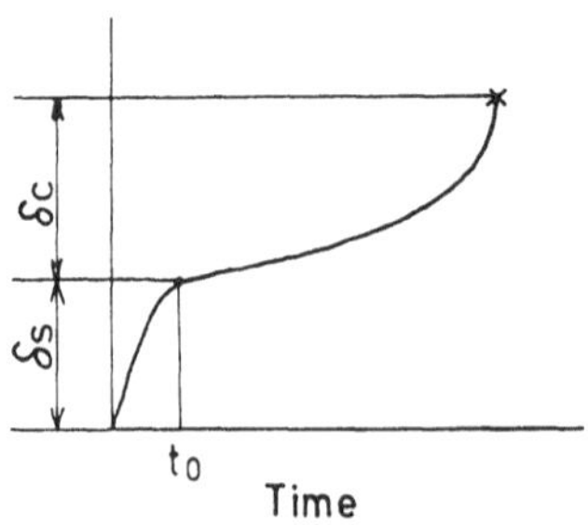

FIG. 7. Schematic diagram of creep curve.

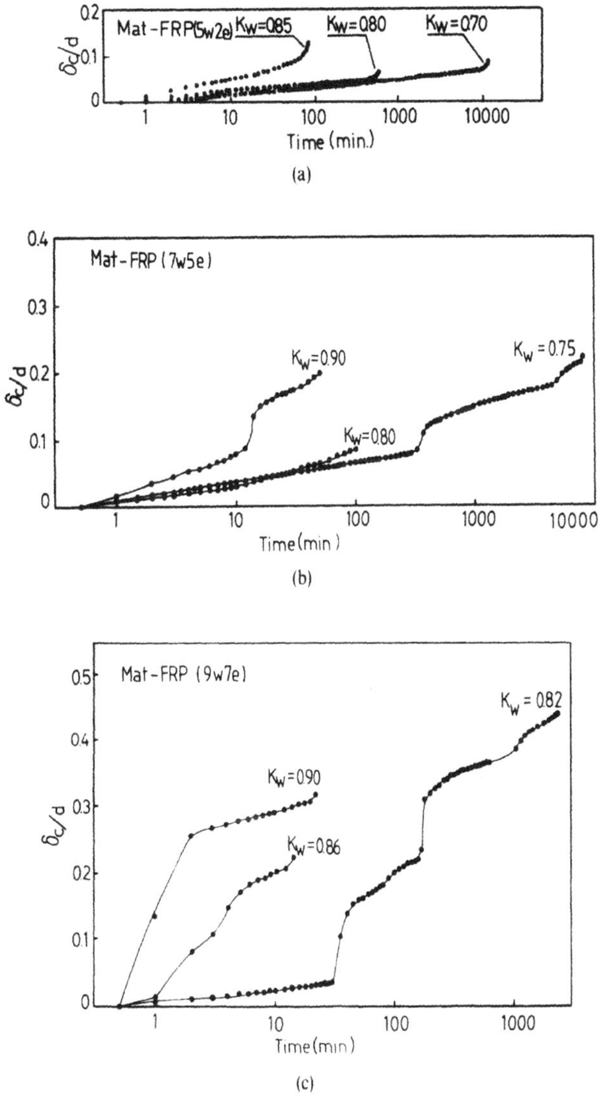

FIG. 8. Typical creep curves of: (a), 5*w*2*e*; (b), 7*w*5*e* and; (c), 9*w*7*e* in GFRP.

creep curves involve some discontinuities in the case of the lower load levels. At these points it is considered that the bearing failure is taking place. The value of the creep displacement in 9*w*7*e* is larger than that in 5*w*2*e*. The creep curves in 7*w*5*e* show the discontinuous points, but the value of the creep displacement is smaller than that in 9*w*7*e*. These phenomena are inferable from the creep failure modes.

From the results mentioned above two types of creep curves are obtained, that is, continuously increasing curve and discontinuously increasing curve.

The difference between the two creep curves is related to the static behaviour of joint. In the case of the continuous creep curve, the static load–displacement curve is also continuous as shown in Fig. 9a. On the other hand in the case of discontinuous creep the curve shows some reduction in the load in the static load–displacement curve (cf. Fig. 9b). The difference of applied load in the same joint geometry is considered as follows.

For large applied loads the drop point of the load in the static load–displacement curve does not exist between the applied load and the maximum load. At the beginning of the creep test the specimen is considerably deformed.

In lower applied loads the deformation process should pass through the drop point. The bearing failure is remarkably developed at the drop point, so that the increase in the displacement is abrupt in the creep test.

Creep life

The schematic creep curves are shown in Fig. 10. The displacement at the beginning of the creep test may be calculated by incremental analysis using FEM (for example at Point A), in other words the designer permits this deformation. However the abrupt increase of the displacement shown in

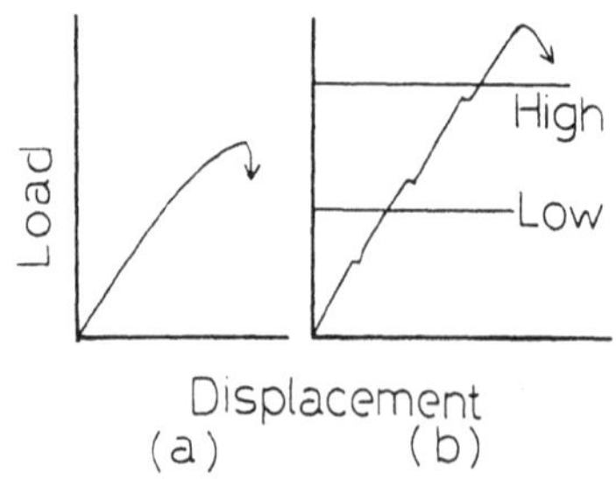

FIG. 9. Schematic diagram of static load–displacement curves.

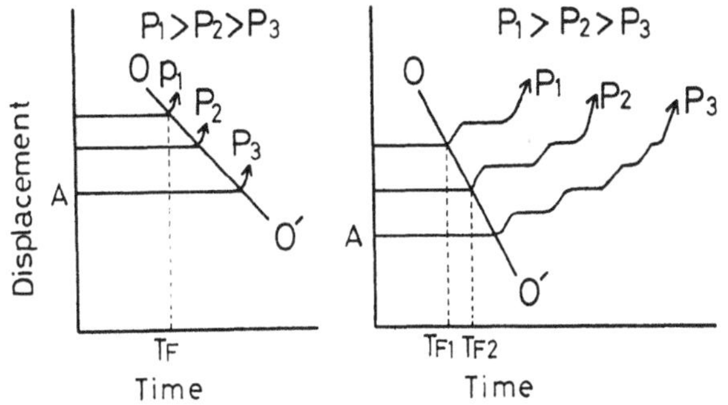

FIG. 10. Two types of creep curves.

the discontinuous creep curve can not be ignored in the design. Therefore 'The First Significant Damage Time' (T_F shown in Fig. 10) is adopted as the creep life. In the continuous creep curve T_F is the time when final failure occurs. The line connected with 'The First Significant Damage Time' (O–O') shown in Fig. 10 is presented as a function of the applied load.

Low cycle fatigue test and high cycle fatigue test

The low cycle fatigue test and high cycle fatigue test are carried out in order to confirm the existence of 'The First Significant Damage' as observed in the creep test. In low cycle fatigue tests the Instron type tensile machine is used at a crosshead speed of 1·5 mm/min. Figure 11 shows the relation between the maximum and minimum displacement and numbers of cycles. In this case the ratio of the upper applied load to the maximum load is 0·8 and the minimum applied load is nearly equal to zero. The maximum displacement increases parallel to the minimum displacement, and an abrupt increase of these displacements is observed.

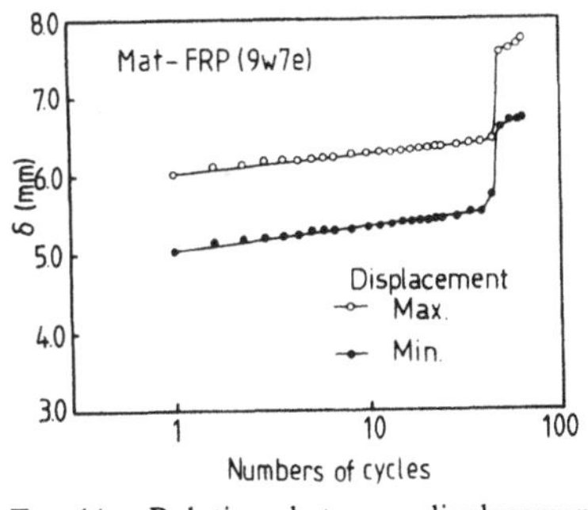

FIG. 11. Relation between displacement and numbers of cycles in low cycle fatigue test.

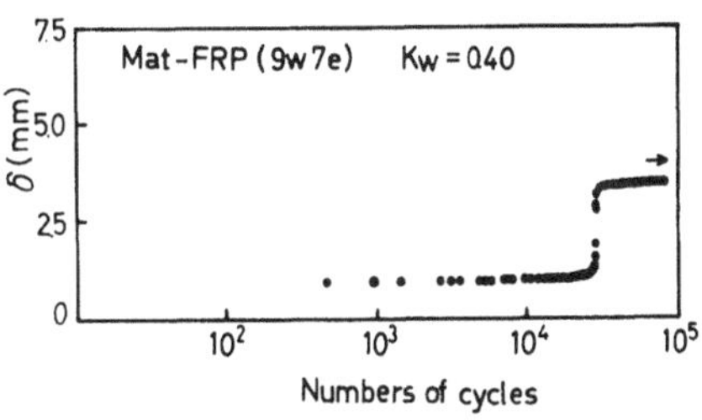

FIG. 12. Relation between displacement and numbers of cycles in high cycle fatigue test.

The high cycle fatigue test is conducted by a hydronic pressure type fatigue machine. The frequency is 10 Hz and K_w is 0·4. Figure 12 shows the relation between the maximum displacement and the number of cycles. Also the existence of 'The First Significant Damage Time' is confirmed in this test.

Therefore it must be noted that the creep life and the fatigue life are 'The First Significant Damage Times', and in particular these life times are considerably smaller than the time of final failure in the bearing failure mode.

GFRP Laminates (Cloth-GFRP, Hybrid Constitution)

In this section the effects of laminate constitution on creep are described. Figure 13 shows the creep curves of 5*w*2*e* in the Cloth-FRP, Hybrid I and II. In the case of Cloth-FRP the creep displacement is larger than that in Mat-FRP. From the results of static tests the failure modes and the deformation states of Hybrid I and II are similar to those of Mat and Cloth-FRP respectively. Also in creep tests, the same tendencies are recognized, i.e. the creep displacement of Hybrid II is larger than that of Hybrid I. The creep failure modes are the same as in the static test.

CFRP Laminates

Figure 14 presents the creep curves of CFRP laminates. The stacking sequence is $[(0/45/90/-45)_2]_s$ with quasi-isotropic properties. This specimen is made from T-300 carbon fibre/epoxy prepreg by autoclave processing.

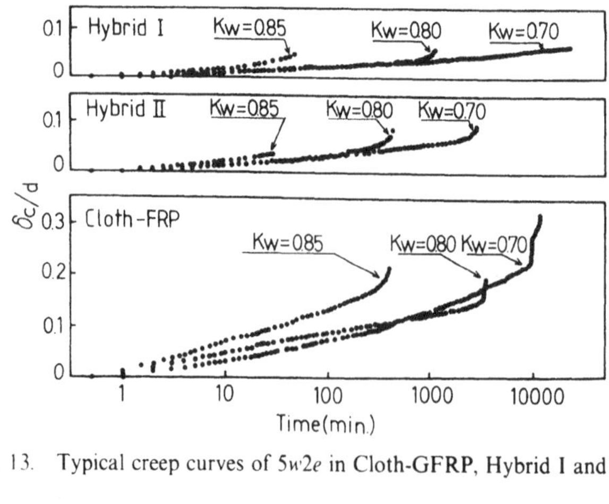

FIG. 13. Typical creep curves of 5*w*2*e* in Cloth-GFRP, Hybrid I and II.

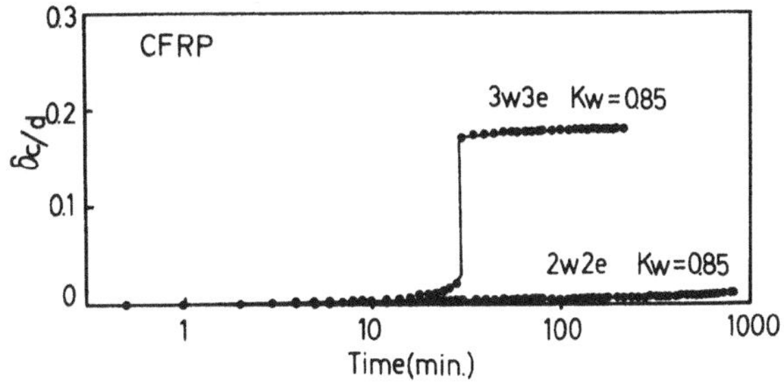

FIG. 14. Typical creep curves of 2*w*2*e* and 3*w*3*e* in CFRP.

In 2*w*2*e* the failure mode of the static test is the net tension mode and in 3*w*3*e* the failure mode is the bearing mode. From Fig. 14 the creep displacement is nearly equal to zero and the final failure suddenly occurs in 2*w*2*e*. On the other hand, 'The First Significant Damage' is observed in 3*w*3*e*. These phenomena are the same in GFRP mentioned above.

Prediction of the Creep Life of Hybrid Constitution

The relation between the creep life and the ratio of the applied load to the maximum bearing load (K_w) is discussed. In the bearing failure mode 'The First Significant Damage Time' can be adopted as the creep life. Figure 15 shows the relation between the creep life and W_w in the case of MAT-FRP. In each type of joint geometry straight lines are obtained. The creep life can be easily predicted from this relation.

Moreover the creep life of hybrid constitution is predicted by the data of

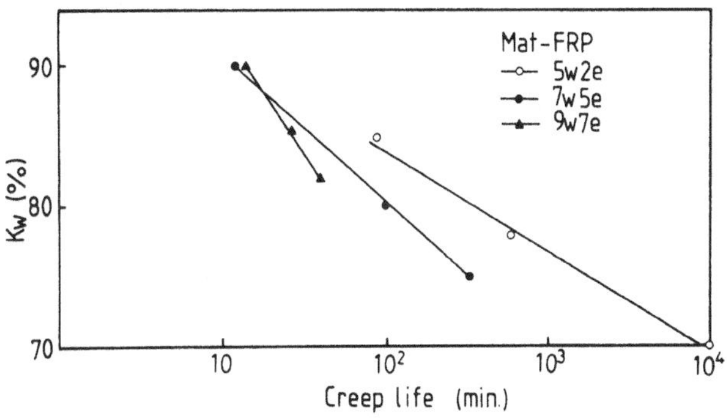

FIG. 15. Relation between creep life and K_w.

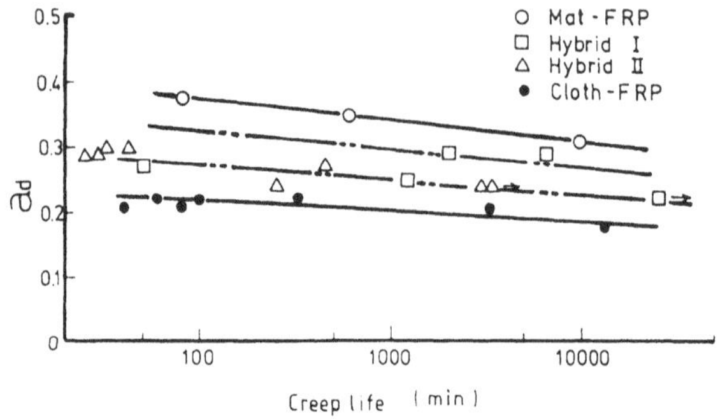

FIG. 16. Predicted results of creep life in hybrid lamination.

Mat and Cloth-FRP. The basic failure load of a mechanical joint, P_0, is assumed to be given by

$$P_0 = \sigma(w - d)^* t$$

where σ is the tensile strength of flat specimen. The creep life coefficient a_d is represented in terms of the static bearing load P_{max}, basic failure load P_0 and the ratio of the applied load to P_{max} as follows;

$$a_d = K_w^* P_{max}/P_0$$

We shall assume that the creep life coefficient a_{dh} is given in the following form analogous to the law of mixtures.

$$a_{dh} = k_1^* a_{dm} + k_2^* a_{dc} \qquad (k_1 + k_2 = 1)$$

where d_{dm} and a_{dc} are the creep life coefficients in Mat-FRP and Cloth-FRP respectively. k_1 and k_2 are effective coefficients. In hybrid I k_1 is larger than k_2. The value of k_1 is 0·684 and the value of k_2 is 0·316. In this case the estimated value of Hybrid I is fitted to the static experimental data. In Hybrid II the reverse values of k_1 and k_2 in Hybrid I are adopted. These results are shown in Fig. 16 in the case of 5*w*2*e*. From these results the creep life of Hybrid constitution can be predicted.

CONCLUSION

This chapter discusses the creep life of mechanically fastened composite joints. The method of the prediction of creep life is proposed. The main results obtained are summarized as follows;

(1) In GFRP laminates the failure modes of creep test are similar to those of static test.

(2) 'The First Significant Damage Time' should be considered as creep life, and this time is smaller than the time of final failure in the bearing mode.
(3) The occurrence of 'The First Significant Damage' is confirmed in low and high cycle fatigue tests.
(4) The method of prediction of the creep life of hybrid constitution is proposed.

ACKNOWLEDGEMENT

The authors would like to thank Dr Sinnya Motogi (Osaka Municipal Technical Research Institute) for helpful discussion.

REFERENCES

1. STOCKDALE, J. H. and MATTHEWS, F. L., The effect of clamping pressure on bolt bearing loads in glass fiber-reinforced plastics, *Composites*, January (1976), 34–38.
2. COLLINGS, T. A. and BEAUCHAMP, M. J., Bearing neglection behaviour of a loaded hole in CFRP, *Composites*, January (1984), 33–38.
3. GODWIN, E. W. and MATTHEWS, F. L., A review of strength of joints in fiber-reinforced plastics, *Composites*, July (1980), 155–160.
4. CHEN, D. K. and DHARAN, C. K. H., Bolt bearing strength of woven carbon-aramid hybrid laminates, *29th National SAMPE Symposium*, 1984, pp. 130–137.
5. CHANG, F. K., SCOTT, R. A. and SPRINGER, G. S., Design of composite laminates containing pin loaded hole, *J. Comp. Mater.*, **18** (1984), 279–289.
6. AGARWAL, B. L., Static strength prediction of bolted joint in composite materials, *AIAA Journal*, 18 (1980), 1371–1375.
7. WONG, C. M. S. and MATTHEWS, F. L., A finite element analysis of single and two hole bolted joints in fiber reinforced plastics, *J. Comp. Mater.*, **15** (1981), 481–491.
8. JONG, T. D., Stress around pin-loaded holes in elastically orthotropic or isotropic plates, *J. Comp. Mater.*, **11** (1977), 313–331.
9. ERIKSSON, L. I., Contact stresses in bolted joints of composite laminates, *Composite Structures*, **6** (1986), 57–75.
10. MAEKAWA, Z., KAJI, A., HAMADA, H. and NAGAMORI, M., Failure mode and strength predictions of mechanically fastened composite joint, *Proc. ICCM-V*, 1985, pp. 99–109.
11. HAMADA, H., MAEKAWA, Z., HORINO, T., KAJI, A. and NAGAMORI, M., Influence of laminate constitution on the failure mode and strength of mechanically fastened composite joint, *Engineering Applications of New Composites (Comp '86)* (in press).
12. HAMADA, H., MAEKAWA, Z., KAJI, A. and NAGAMORI, M., Study of failure mode and strength of mechanically fastened composite joint, *Composites '86 Recent Advances in Japan and the United States*, 1986, pp. 371–378.

5

Inserts for Fastening Sheet Molding Compounds

S. V. HOA, A. DI MARIA and D. FELDMAN

Department of Mechanical Engineering, Concordia University, Montreal, Quebec H3G 1M8, Canada

ABSTRACT

Stainless steel inserts used for fastening sheet molding compound (SMC) are being developed. Using these molded-in inserts, SMC components can be assembled without protrusions. The inserts consist of two mating surfaces which interlock through the existence of a shoulder male boss and a shoulder female counterbore. Repeated assembly and disassembly can take place without damaging the material being fastened. The strength of the joint, when using these interlocking inserts, is comparable to that when using bolts. The joint strength of the whole assembly when immersed in water for a period of time is also investigated. The difference between the dry and wet strengths for the inserted samples is less than that for the bolted samples. Load versus extension plots show that these particular inserts provide a more rigid joint than a bolted connection.

INTRODUCTION

Composite materials introduce particular problems when components must be fastened together. A successful joint must be obtained in order to minimize the possibility of component failure. Stainless steel inserts allow a convenient means of fastening Sheet Molding Compound (SMC). SMC is a brittle composite material which is made of polyester resin with glass-fiber reinforcement and calcium carbonate filler. It is widely used in the automotive industry for applications such as hoods, radiator supports, as well as subway seat benches.

Work done by previous researchers on the joining of SMC has placed emphasis on bonding and bolting.[1] It was found that epoxy adhesives give good bonding strength and that bolting gives stronger joints than bonding. Also a previous design for the inserts has been investigated and reported in Ref. 2. In that work, round inserts made of aluminum were used. Even though the strengths of the inserted samples were comparable to the strengths of the bolted samples (steel bolts), the aluminum inserts bend at large loads. In this chapter, a modified design for the inserts is made. This design consists of a hexagonal head insert and the material is changed to stainless steel. These modifications produce significant improvements on the joint strength of the SMC. The advantage of no protrusions and ease in repeated assembly and disassembly is retained.

STAINLESS STEEL INSERTS

The molded-in inserts consist of a pair of interlocking disks which have a hexagonal head and a flaired body as shown in Fig. 1. The flaired body protrudes from the large clamping head and allows the insert to be held rigidly within the SMC. The hexagonal head prevents rotation of the insert when being torqued as compared to a plain round insert.[2] Interlocking is obtained through the existence of a shoulder male boss and a shoulder female counterbore. The two SMC components are joined by interlocking the mating inserts and using a special flat head Allen screw to bolt them together. A flush assembly with no surface discontinuity is obtained.

SPECIMEN PREPARATION

The Sheet Molding Compound used was Prepreg SMC R-30 and it was obtained from Jet Molding Compound Inc. (Ontario, Canada). The prepreg sheets were cut into 10 in square sections. Holes of 0·350 in diameter were punched into the prepreg and the inserts were placed in their appropriate positions. The sheets were then compression-molded under heat (350°F) and pressure (450 psi) for 20 min in a matched metal mold.

Once the cured plate was obtained, specimens were cut and shaped into four different configurations for five different tests. The first configuration consisted of two rectangular specimens, one containing a male insert and the other, a female insert. An effective hole diameter was established as 0·469 in by calculating an average diameter for the flaired boss. Ratios of

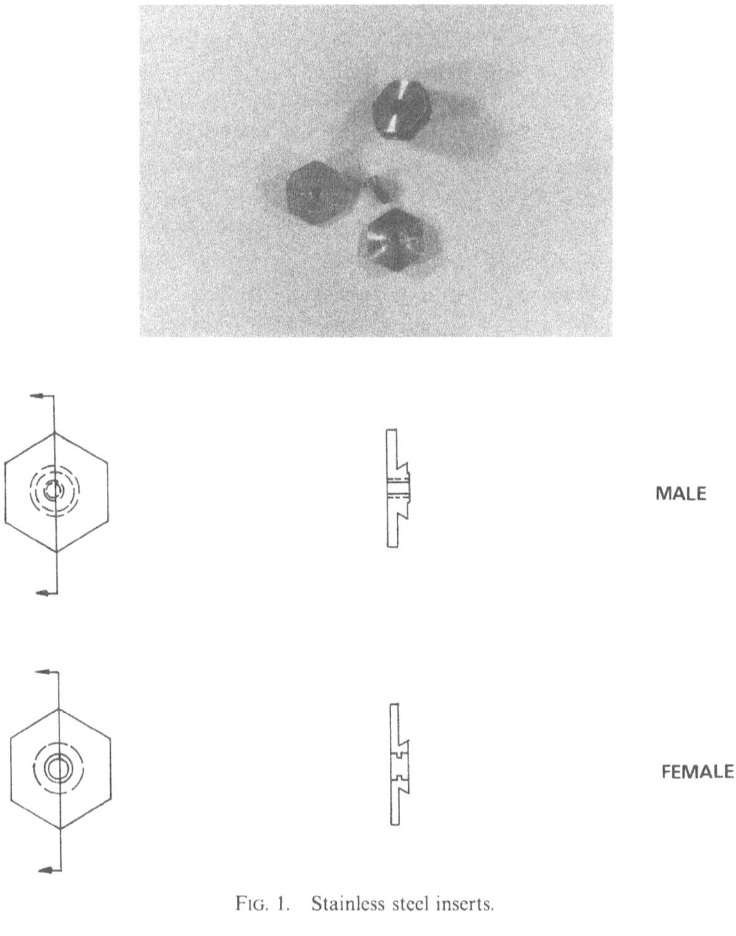

FIG. 1. Stainless steel inserts.

edge distance (e) to hole diameter (d) and width (w) to hole diameter were chosen as $e/d = 2$ and $w/d = 4$ respectively. These ratios remained constant throughout the experiment. Both these specimens were 5 in long, 1·875 in wide and 0·095 in thick. SMC tabs were bonded to the ends of the specimens in order to be able to place them within the grips of the testing machine. Two part epoxy (Ciba Geigy) was used to attach the tabs to the specimens. The samples were later assembled in a single lap joint configuration and pulled in tension (see Fig. 2a).

The second configuration consisted of a pair of specimens which were joined through a bolted connection. The individual specimens were 1 in

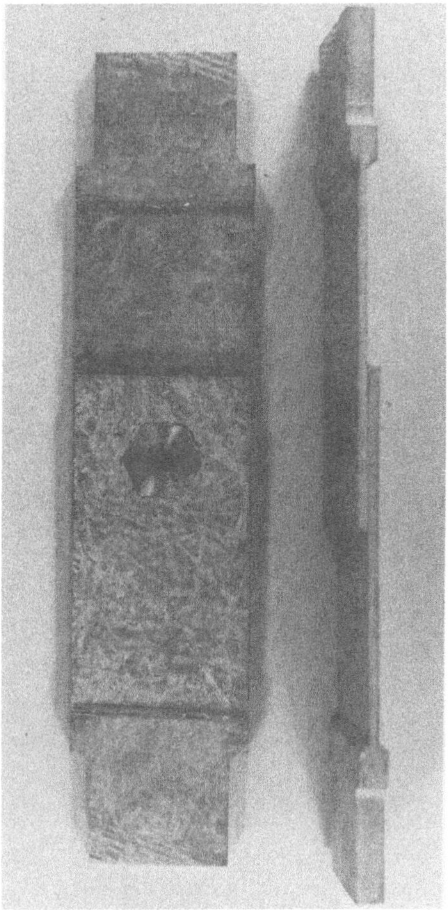

(a) (b)

FIG. 2 Jointed samples. (a) Inserted; (b) Bolted.

wide, 5 in long and 0·095 in thick. The difference in width between the bolted and inserted samples is to ensure the same $w/d = 4$ ratio. Samples with width = 1·875 in were also tested. SMC tabs were bonded to these samples as well. A hole (0·250 in diameter) was drilled into each of the specimens at an equal distance of 0·5 in from each edge. These specimens were assembled in a single lap joint configuration using a washer and a 0·250 in diameter bolt (see Fig. 2b).

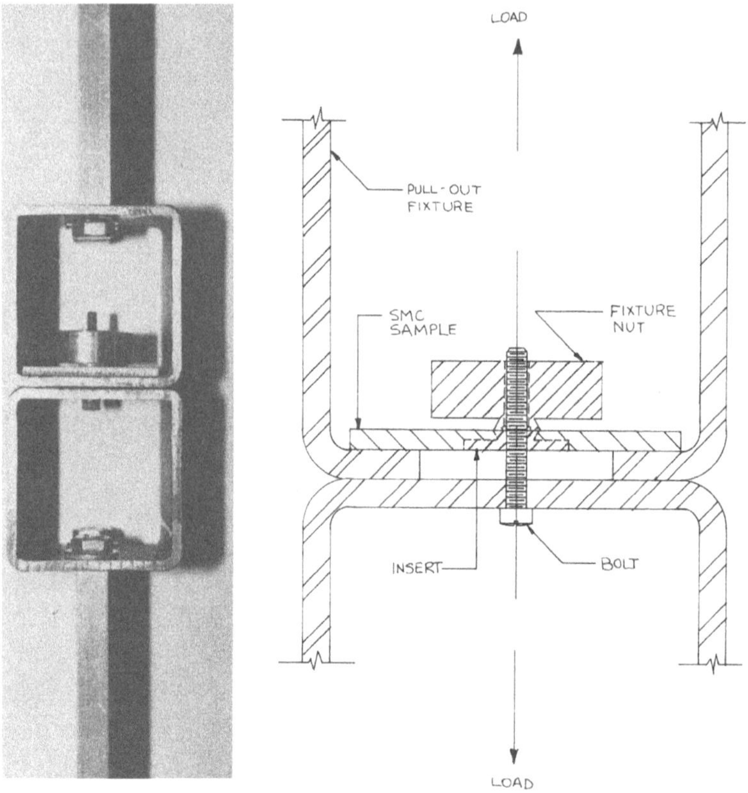

FIG. 3. Pull-out test.

The third configuration consisted only of SMC material which was cut 1 in wide, 10 in long and 0·095 in thick. This was done in order to obtain the strength of the material to be used in calculating joint efficiencies. Joint efficiency can be defined as the strength of the joint connection divided by the strength of the material.

The fourth configuration was to be used for a pull-out test as demonstrated in Fig. 3. These specimens are 1 in × 1 in and contain an insert in the center. This test was done in order to examine how rigid the insert was held in the SMC (with the aid of the flaired boss which resists the pull-out load). The same configuration of specimen was used in a torsion test in order to determine the insert's ability to resist torque.

Specimen dimensions are summarized in Table 1.

TABLE 1
Specimen dimensions

	Width	Length	Thickness
Bolted			
(smaller)	1·000	5	0·095
(larger)	1·875	5	0·095
Insert	1·875	5	0·095
Control	1·000	10	0·095

Note: All dimensions are in inches.

EXPERIMENTS

1. Water Absorption

Five samples of each configuration were placed in water at room temperature for 13 days. The water absorption specimens were weighed daily and the results documented. Water absorption by weight was found to be 1·30% for the insert connections, 2·65% for the bolted connections and 2·50% for the pull-out specimens. A graph describing the water absorption is shown in Fig. 4.

Comparing the amount of water uptake to previous work[3] shows that the joint samples absorbed more water than the regular material. It is also noted that the inserted samples absorbed less water than the bolted samples. This is due to the fact that edges were cut during hole drilling for the bolting. For the inserted samples, the edges were sealed during molding.

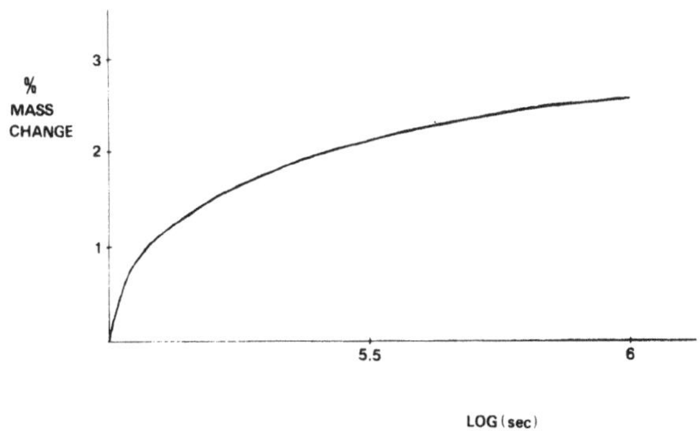

FIG. 4. Water uptake of SMC joint samples.

TABLE 2
Strength of insert connection

Specimen No.	Load (lb)	Average load (lb)	Type of failure
1W	668·5		IV
2W	642·4		IV
3W	682·6	667·3	IV
4W	688·3		IV
5W	654·9		IV
1D	710·0		IV
2D	668·0		IV
3D	680·0	675·4	IV
4D	654·0		IV
5D	665·0		IV

D = dry specimen; W = wet specimen.

TABLE 3
Strength of bolted connection

Specimen No.	Load (lb)	Average load (lb)	Type of failure
1W	584·3		I
2W	522·5		I
3W	505·6	547·2	I
4W	589·9		II
5W	533·7		I
1DS	674·2		I
2DS	589·9	625·0	III
3DS	662·9		II
4DS	573·0		III
1D	652·7		IV
2D	684·2	646·0	IV
3D	610·4		IV
4D	636·6		IV

DS = dry specimen (width = 1 in); D = dry specimen (width = 1·875 in); W = wet specimen.

TABLE 4
Strength of control specimens

Specimen No.	Load (lb)	Average load (lb)	Stress (psi)	Average stress (psi)
1	674·2		7 096·4	
2	719·1		7 569·5	
3	662·9	730·3	6 978·1	7 687·8
4	865·2		9 107·0	
5	730·3		7 687·8	

2. Mechanical Tests

Testing of all specimens was done on a MTS 810 Testing Machine of 100 kN capacity with crosshead speed of 5 mm/min. Load versus displacement plots were obtained for all tests and the results were tabulated in Tables 2–5. A torque of 60 in-lb was applied to all bolted connections prior to testing. All assemblies consisted of a single lap joint.

Tension tests were initially performed for the control specimens which were used as a basis to obtain the joint efficiencies. The loads sustained by these specimens are recorded in Table 4. An average stress of 7688 psi was obtained for these SMC control specimens. This is slightly lower than the results obtained from another batch of SMC (8400 psi) in a previous investigation.[4] Tension tests were also done for the insert and the bolted connection assemblies. Five sets of the water absorption specimens as well as five sets of the dry specimens were used for each type of connection. The

TABLE 5
Strength of inserts in pull-out mode

Specimen No.	Load (lb)	Average load (lb)
1W	59·6	
2W	67·4	
3W	69·7	70·4
4W	70·8	
5W	84·3	
1D	196·6	
2D	182·0	
3D	176·4	170·5
4D	168·5	
5D	129·2	

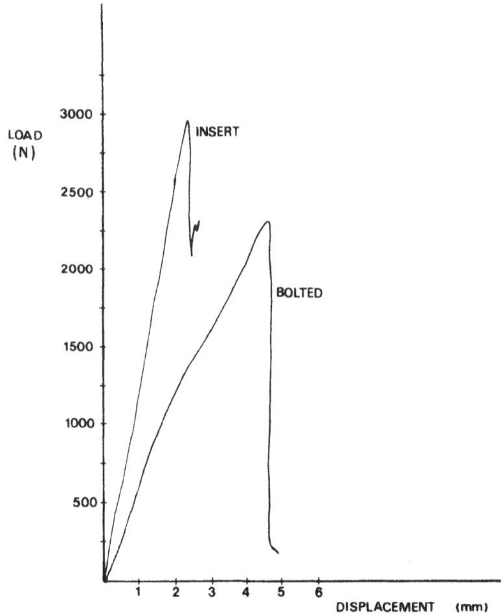

FIG. 5. Plot for insert and bolted connection (wet).

plot of load versus displacement for the insert and bolted joint is shown in Fig. 5. The plot illustrates the more rigid joint of the insert connection. For the insert specimens, and average load of 675·4 lb was obtained when dry, and 667·3 lb when wet. The bolted specimens with the same width (1·875 in) sustained an average load of 646·0 lb when dry. A load of 625·0 lb for the dry and 547·2 lb for the wet specimens of width = 1 in was obtained. A schematic diagram showing the types of failure which occurred in the tests is shown in Fig. 6. The type of fracture sustained by each of these specimens is indicated in their respective tables.

Pull-out tests were performed on five dry and five wet SMC specimens. They were tested in such a way that the inserts assumed the total applied load, which was perpendicular to the sample's surface as shown in Fig. 3. The maximum load needed to pull out the insert, which was embedded into the specimen, is depicted on the plot of Fig. 7 and the values recorded in Table 5. An average pull-load of 170·5 lb was obtained for the dry specimens and 70·4 lb for the wet specimens.

Torque retention was investigated by performing torsion tests. This was done by using a torque wrench and 6-32 NC Allen bolt. The bolt was initially threaded into the insert finger tight. Torque was then applied at a

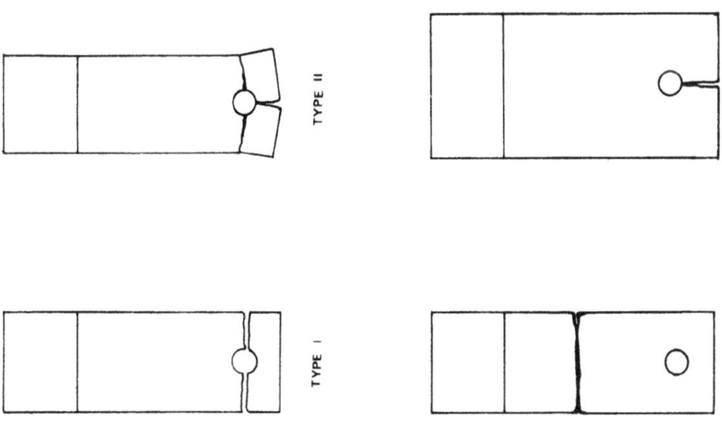

FIG. 6. Schematic of fractured specimens.

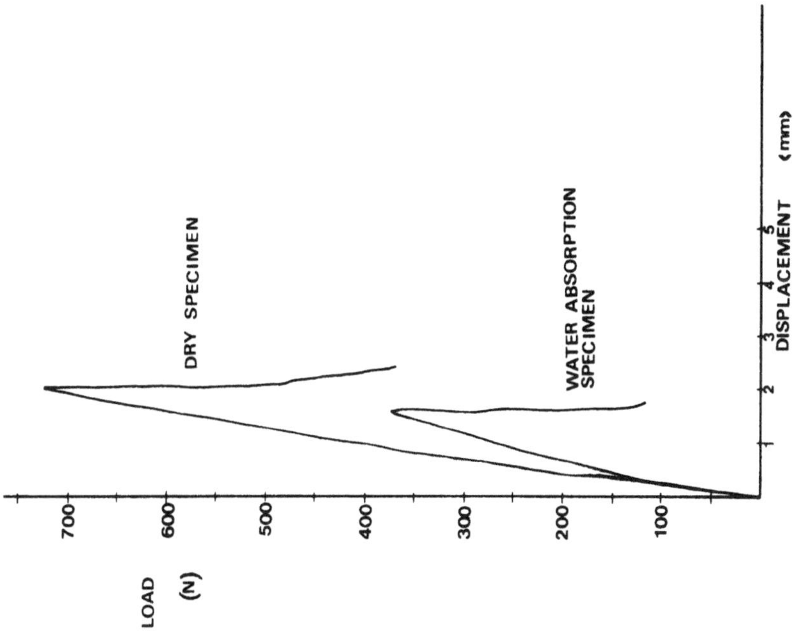

FIG. 7. Plot for pull-out test.

constant rate until the insert would begin to rotate within the SMC specimen. It was found that the Grade 8 cap screw with UTS = 180 000 psi broke without twisting the insert. Therefore, unlike the preliminary design of round inserts, the hexagonal head did not allow any rotation of the insert.

DISCUSSION

Due to the setup on the testing machine and the eccentricity of the applied load, bending of the single lap joint occurs when pulled in tension. This bending at the joint consequently reduces the strength of the joint. Less relative movement was detected at the joint with the stainless steel inserts compared to the bolted connections. The preliminary design using aluminum inserts also showed a greater amount of rigidity in the joint, however, a slight bending of the insert flange was observed. The hexagonal head on the inserts was found to possess a greater advantage in torque retention when compared to the simple, round inserts. None of the inserts were found to twist under the applied torque, instead the bolt yielded.

Different failure modes were obtained with the bolted connections of width = 1 in. However, the specimens with inserts and the bolted connections of width = 1·875 in failed with type IV mode only. These results are shown in Figs. 9 and 10. Figure 8 shows the inserted specimens after the pull-out test.

FIG. 8. Specimens after pull-out test.

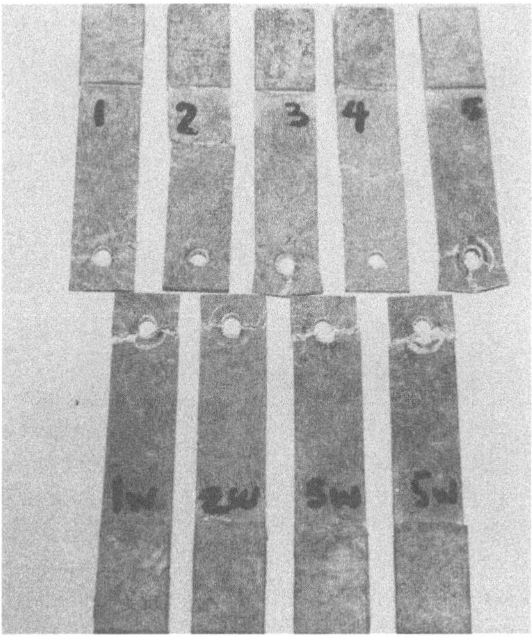

FIG. 9. Bolted samples after testing.

The specimens that absorbed water were found to be less efficient than the dry samples. A greater amount of noise, resulting from the cracking of fibers and matrix, was heard when the dry specimens were being loaded in tension compared to the wet specimens. Figure 7 demonstrates a plot of a dry and wet specimen which were undergoing the pull-out test. Overall, the dry specimens were 58·8% stronger.

When tested in tension, little difference was obtained between the dry and wet specimens. The dry insert connections were found to be 1·2% stronger than the wet specimens. However, a significant difference of 12·4% can be observed between the dry and wet specimens for the bolted connections. This can be explained by the fact that the inserts seal the edges in the joint area which leads to less water absorption. The bolted connection, however, is exposed to the environment to a greater extent. The insert connections were found to be 49·3% efficient and the bolted connections were found to be 47·2% efficient in the tensile test.

FIG. 10. Inserted sample after testing.

CONCLUSION

The objective in using these stainless steel inserts is to increase fastening performance compared to the previous design of the aluminum inserts in terms of joint rigidity, tensile and pull-out strength as well as torque retention. Water absorbed directly from the environment decreases the joint efficiency. These inserts provide a more rigid joint compared to the bolted joint. A better wet performance of these molded-in inserts versus the bolted connection is obvious. Therefore, the stainless steel inserts provide a flush assembly as well as a convenient means of fastening and can be assembled repeatedly without damage to the SMC material.

REFERENCES

1. HOA, S. V. and FELDMAN, D., Joining strength of sheet molding components, *Polymer Composites*, **3**, No. 1, Jan. 1982.
2. HOA, S. V., LULHAM, I. and SANKAR, T. S., Aluminum inserts for fastening sheet molding compounds, *Composite Structures—3* (Marshall, I. H. ed.), Elsevier Applied Science Publishers, London, 1985.
3. HOA, S. V. and OUELLETTE, P., Liquid absorption of a sheet molding compound, *Polymer Composites*, **2**, No. 4, Oct. 1981.
4. HOA, S. V., Notched strength of sheet molding compounds, *Polymer Composites*, **2**, No. 4, Oct. 1981.

6

Stability of an Orthotropic Plate Subjected to a Follower Force

LIEN-WEN CHEN and C. Y. YU

Department of Mechanical Engineering, National Cheng Kung University, Tainan, Taiwan 70101

ABSTRACT

The vibration and dynamic stability problems of an orthotropic rectangular plate subjected to a non-conservative follower force are studied. The governing differential equation derived by Brunelle and Oyibs in affine space is used. The Levy-type solution method is used to solve the vibration problems. The optimization package ADS is employed to find the dynamic stability loading. The effect of generalized rigidity ratio D on the stability boundaries is presented.*

1. INTRODUCTION

The lightweight design of structures has been used in many industrial fields. Investigations of stability problems of elastic systems subjected to non-conservative loads have become an important and interesting topic in the theory of elastic stability.

An outline of the theory of the stability of structures subjected to non-conservative loads has been given by Bolotin.[1] Herrmann and Bungay[2] studied the stability of elastic systems subjected to various non-conservative forces. Sugiyama and Kawagoe[3] studied the vibration and stability problems of elastic columns under the combined action of uniformly distributed vertical and tangential forces. The effect of a single crack on the non-conservative stability of columns subjected to vertical and follower loads was studied by Anifantis and Dimarogonas.[4]

The problem of elastic stability of plates subjected to conservative in-plane loadings is an important subject and has been studied extensively. However, information on the stability of plates subjected to non-conservative loading appears to be lacking and needs to be further studied. Culkowski and Reismann[5] studied the dynamic stability problems of a rectangular plate due to follower edge forces. The plate was simply supported at two opposite edges and clamped free at the other two opposite edges. The free edge is subjected to a follower force. Celep[6] investigated axially symmetric stability of a completely free circular plate subjected to a non-conservative edge load. Farshad[7] solved the stability problems of cantilever plates subjected to biaxial substantial loading. The Galerkin method is employed and the stability of the plate in terms of subtangency and load parameters is analysed. Irie *et al.*[8] did the analysis of the vibration and stability of non-uniform annular plates subjected to a follower force radially distributed at the outer edges in which the transfer matrix method is used. Chonan[9] dealt with the vibration and stability of annular plates elastically restrained against translation and rotation at the edges. Both axisymmetric and non-axisymmetric vibrations are considered. The asymmetric vibration and stability problems of a free circular plate subjected to non-conservative loading were presented by Celep.[10] The flutter and divergence instabilities of a rectangular plate which is subjected to the combined action of a tangential follower force and a unidirectional axial force along one edge were studied by Adali.[11]

The object of the present studies are to investigate the instabilities of an orthotropic plate subjected to a follower edge force. The governing differential equation which was derived in affine space by Brunelle and Oyibo[12] will be used. The effect of generalized rigidity ratio D^* on the stabilities is presented.

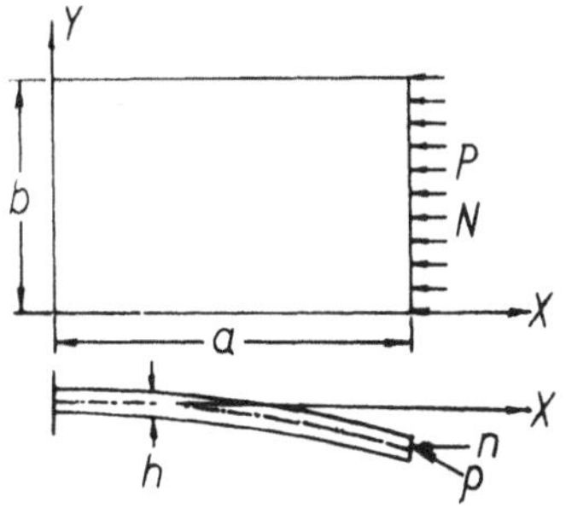

FIG. 1. Geometry and loading condition of plate.

2. FORMULATION AND ANALYSIS

The specially orthotropic rectangular plates are considered in the present report. The plate is subjected to a tangential follower force P and an axial force N at the free edge $x = a$. P and N are both uniformly distributed along the free edge, as shown in Fig. 1. The governing differential equation for an orthotropic plate in the affine space[12] is written as

$$\frac{\partial^4 w}{\partial x_0^4} + 2D^* C_0 \frac{\partial^4 w}{\partial x_0^2 \partial y_0^2} + C_0^4 \frac{\partial^4 w}{\partial y_0^4} + (n_0 + k_0) \frac{\partial^4 w}{\partial x_0^4} + \rho_0 C_0 \frac{\partial^2 w}{\partial t^2} = 0 \quad (1)$$

where $x = (D_{11})^{1/4} a_0 x_0 \qquad y = (D_{22})^{1/4} b_0 y_0$

$$D^* = \frac{D_{12} + 2D_{66}}{(D_{11} D_{22})^{1/2}} \quad \text{is the generalized rigidity ratio}$$

$$k_0 = \frac{Pa^2}{D_{11}} \qquad n_0 = \frac{Na^2}{D_{11}} \qquad C_0 = \frac{a_0}{b_0} = \left(\frac{D_{22}}{D_{11}}\right)^{1/4} \frac{a}{b} \quad (2)$$

and $\rho_0 = \rho b_0^4 = \dfrac{\rho b^4}{D_{22}}$

Considering a specially orthotropic rectangular plate which is simply supported at $y_0 = 0, 1$, clamped at $x_0 = 0$ and free at $x_0 = 1$, a steady-state free vibration mode is considered as

$$w(n_0, y_0, t) = w_0(x_0, y_0)\, e^{i\Omega t} \quad (3)$$

Substituting eqn. (3) into eqn. (1), eqn. (1) becomes

$$\frac{\partial^4 w_0}{\partial x_0^4} + 2D^* C_0^2 \frac{\partial^4 w_0}{\partial x_0^2 \partial y_0^2} - C_0^4 \omega^2 w_0 + C_0^4 \frac{\partial^4 w_0}{\partial y_0^4} + (k_0 + n_0) \frac{\partial^2 w_0}{\partial x_0^2} = 0 \quad (4)$$

where $\omega^2 = \rho h \Omega^2 b^4 / D_{22}$ (5)

The corresponding boundary conditions are

$$w_0 = 0 \quad \text{and} \quad \frac{\partial w_0}{\partial x_0} = 0 \quad \text{at } x_0 = 0 \quad (6)$$

$$-\frac{\partial^3 w_0}{\partial x_0^3} - (2 - \varepsilon) D^* \frac{\partial^3 w_0}{\partial x_0 \partial y_0^2} - n_0 \frac{\partial w_0}{\partial x_0} = 0$$

and

$$-\frac{\partial^2 w_0}{\partial x_0^2} - \varepsilon D^* \frac{\partial^2 w_0}{\partial y_0^2} = 0 \quad \text{at } x_0 = 1 \quad (7)$$

$$w_0 = 0 \quad \text{and} \quad -\varepsilon D^* \frac{\partial^2 w_0}{\partial x_0^2} - \frac{\partial^2 w_0}{\partial y_0^2} = 0 \quad \text{at } y_0 = 0, 1 \quad (8)$$

where the generalized Poisson ratio ε is defined by

$$\varepsilon D^* = D_{12}/(D_{11}D_{12})^{1/2} \tag{9}$$

To solve eqn. (4) with boundary conditions (6–7), a Levy solution method is used and the displacement function is assumed to be

$$w_0(x_0, y_0) = X(x_0) \sin m\pi y_0 \qquad m = 1, 2, 3, \ldots \tag{10}$$

Equation (10) automatically satisfies the boundary conditions at $y =$ constant edges. Substituting eqn. (10) into eqns (4), (6) and (7), we obtain an ordinary differential equation of fourth order:

$$\frac{d^4X}{dx_0^4} + (k_0 + n_0 - 2D^*C_0^2m^2\pi^2)\frac{d^2X}{dx_0^2} + C_0^4(m^4\pi^4 - \omega^2)X = 0 \tag{11}$$

with the boundary conditions

$$X(0) = 0 \qquad \frac{dX(0)}{dx_0} = 0$$

$$\frac{d^2X(1)}{dx_0^2} - \varepsilon D^*C_0^2m^2\pi^2X(1) = 0 \tag{12}$$

$$\frac{d^3X(1)}{dx_0^3} - (2-\varepsilon)D^*C_0^2m^2\pi^2\frac{dX(1)}{dx_0} + n_0\frac{dX}{dx_0} = 0$$

The solution of eqn. (11) has the general form

$$X(x_0) = e^{rx_0} \tag{13}$$

Substituting eqn. (13) into eqn. (11), we obtain the characteristic equation:

$$r^4 + (k_0 + n_0 - 2D^*C_0^2m^2\pi^2)r^2 + C_0^4(m^4\pi^4 - \omega^2) = 0 \tag{14}$$

The roots of eqn. (14) can be written as

$$r^2 = \tfrac{1}{2}(-\delta_1 \pm \sqrt{\delta_1^2 - 4\delta_2}) \tag{15}$$

where $\quad \delta_1 = (k_0 + n_0 - 2D^*C_0^2m^2\pi^2) \qquad \delta_2 = C_0^4(m^4\pi^4 - \omega^2) \qquad (16)$

Depending on the relative magnitudes of δ_1 and δ_2, the general solution $X(x_0)$ can be written as the following different expressions.

(I) $\delta_1^2 - 4\delta_2 > 0$

(a) if $\delta_1 > 0$ and $\delta_2 > 0$, then

$$X = C_1 \cos\alpha_1 x_0 + C_2 \sin\alpha_1 x_0 + C_3 \cos\alpha_2 x_0 + C_4 \sin\alpha_2 x_0 \tag{17}$$

(b) if $\delta_1 > 0$ and $\delta_2 < 0$ or $\delta_1 < 0$ and $\delta_2 < 0$, then

$$\begin{aligned} X = C_1 \cos \alpha_2 x_0 + C_2 \sin \alpha_2 x_0 \\ + C_3 \cosh \alpha_3 x_0 + C_4 \sinh \alpha_3 x_0 \end{aligned} \tag{18}$$

(c) if $\delta_1 < 0$ and $\delta_2 > 0$, then

$$\begin{aligned} X = C_1 \cosh \alpha_3 x_0 + C_2 \sinh \alpha_3 x_0 \\ + C_3 \cosh \alpha_4 x_0 + C_4 \sinh \alpha_4 x_0 \end{aligned} \tag{19}$$

where

$$\alpha_1 = \left(\frac{\delta_1 - \sqrt{\delta_1^2 - 4\delta_2}}{2} \right)^{1/2} \qquad \alpha_2 = \left(\frac{\delta_1 + \sqrt{\delta_1^2 - 4\delta_2}}{2} \right)^{1/2}$$

$$\alpha_3 = \left(\frac{-\delta_1 + \sqrt{\delta_1^2 - 4\delta_2}}{2} \right)^{1/2} \qquad \alpha_4 = \left(\frac{-\delta_1 - \sqrt{\delta_1^2 - 4\delta_2}}{2} \right)^{1/2} \tag{20}$$

(II) if $\delta_1^2 - 4\delta_2 < 0$, then

$$\begin{aligned} X = (C_1 \cosh \beta_1 x_0 + C_2 \sinh \beta_1 x_0) \cos \beta_2 x_0 \\ + (C_3 \cosh \beta_1 x_0 + C_2 \sinh \beta_1 x_0) \sin \beta_2 x_0 \end{aligned} \tag{21}$$

where

$$\beta_1 = (2\delta_2^{1/2} - \delta_1)^{1/2}/2 \qquad \beta_2 = (2\delta_2^{1/2} + \delta_1)^{1/2}/2$$

Using the boundary conditions of eqn. (12), a linear system of four homogeneous algebraic equations in the unknowns C_1, C_2, C_3 and C_4 are obtained. The determinant of this system should be equal to zero for a non-trivial solution. The resulting non-linear, transcendental characteristic equation is expressed as

$$f(k_0, n_0, \omega; D^*, \varepsilon, C_0, m) = 0 \tag{22}$$

This is a double eigenvalue problem for flutter stability; the flutter load k_0 and flutter frequency ω are to be found. The only one (eqn. (22)) cannot be used to solve the two unknowns. From the phenomenon of the flutter, we observe that the condition $\partial k_0/\partial \omega$ always exists when the flutter occurs. The condition can be used to derive the other condition, $\partial f/\partial \omega = 0$. Subsequently, the two equations are simultaneously solved for the unknowns k_0 and ω. Because of the complicated form of $f(k_0, n_0, \omega; D^*, \varepsilon, C_0, m)$, the method of computation will be very cumbersome. Instead, the problem is changed to an optimization problem as follows:

Minimize $$-k_0 \tag{23}$$

with constraints

$$f(k_0, n_0, \omega; D^*, \varepsilon, C_0, m) = 0 \tag{24}$$

$$\omega_1 \leq \omega \leq \omega_2 \tag{25}$$

$$0 \leq k_0 \tag{26}$$

The optimization package ADS[13] is used to find the flutter load k_0 and flutter frequency ω_0.

3. EXAMPLE

Now we consider the cases when the generalized rigidity ratio D^* equals 1, and the generalized Poisson ratio $\varepsilon = 0{\cdot}3$. We can see in Table 1 that the present $D^* = 1$ solutions are very close to the isotropic solution of Adali.[11] It is proved that $D^* = 1$ is equivalent to the isotropic case as is proposed by Brunelle.[12] The flutter loads and frequencies for different D^* are shown in Table 2. It can be seen that the flutter loads and frequencies increase with increasing D^*. It means that the isotropic case $D^* = 1$ is the most stable condition. The flutter boundaries are shown in Fig. 2 with various affine aspect ratios. We can see that the flutter load will decrease when the normal dead load n_0 increases from negative to positive. The increase of affine

TABLE 1
Present solutions with $D^* = 1$ and compared with the isotropic solution of Ref. 11

n_0	a_0/b_0		S. Adali's solution[11]	Present solutions
−10	0·5	k_0	44·78	44·78
		ω	50·90	50·71
0	0·5	k_0	27·11	27·11
		ω	49·58	49·52
10	0·5	k_0	10·30	10·30
		ω	36·65	36·58
−10	1·0	k_0	68·01	67·99
		ω	15·99	15·79
0	1·0	k_0	51·65	51·64
		ω	16·67	16·48
10	1·0	k_0	30·65	30·64
		ω	16·91	16·77

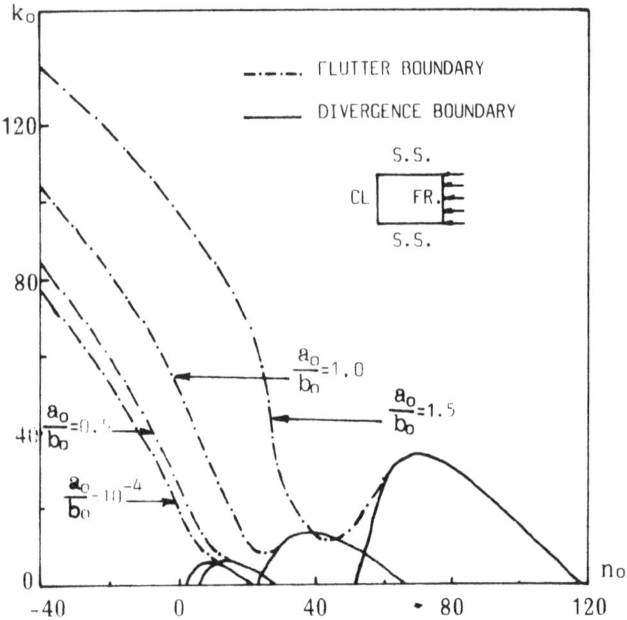

FIG. 2. Flutter boundaries ($D^* = 1{\cdot}0$, $\varepsilon = 0{\cdot}3$).

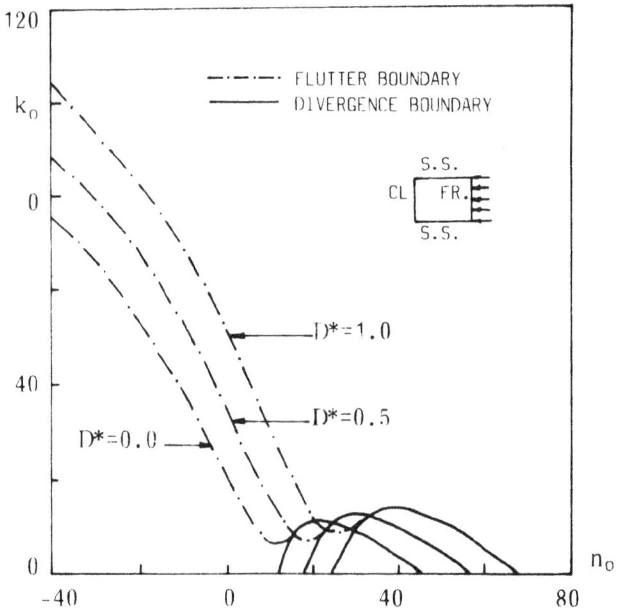

FIG. 3. Flutter boundaries with different D^* ($a_0/b_0 = 1{\cdot}0$, $\varepsilon = 0{\cdot}3$).

TABLE 2
Flutter load and frequency under different D^*
($n_0 = 0$, $a_0/b_0 = 0{\cdot}5$, $\varepsilon = 0{\cdot}3$)

D^*	0·0	0·2	0·4	0·6	0·8	1·0
K_0	20·051	21·416	22·804	24·216	25·651	27·109
ω	45·002	46·027	46·977	47·848	48·774	49·517

aspect ratio a_0/b_0 is seen to improve the flutter loading. The effect of D^* on the stability boundaries can be seen in Fig. 3. We can see that the flutter loads will be raised when D^* increases from zero to one.

4. CONCLUSIONS

The dynamic stability of an orthotropic plate subjected to a combination of in-plane follower forces and in-plane normal forces is considered in the present chapter. The governing differential equation in affine space is used to solve the dynamic stability problems. It is easy to discuss the effect of orthotropic property on the flutter load by using the generalized rigidity ratio D^*. The isotropic case $D^* = 1$ is found to be the most stable condition.

The vibrations of circular orthotropic plates and the laminated composite plates subjected to a follower force are also interesting research topics and need to be further studied.

REFERENCES

1. BOLOTIN, V. V., *Nonconservative Problems of the Theory of Elastic Stability*, Oxford, Pergamon Press, 1963.
2. HERRMANN, G. and BUNGAY, R. W., On the stability of elastic systems subjected to nonconservative forces, *ASME Trans., J. applied Mech.*, **31** (1964), 435–440.
3. SUGIYAMA, Y. and KAWAGOE, H., Vibration and stability of elastic columns under the combined action of uniformly distributed vertical and tangential forces, *J. Sound Vib.*, **38** (1975), 341–355.
4. ANIFANTIS, N. and DIMAROGONAS, A., Stability of columns with a single crack subjected to follower and vertical loads, *Int. J. Solids Struct.*, **19** (1983), 281–291.
5. CULKOWSKI, P. M. and REISMANN, H., Plate buckling due to follower edge forces, *ASME Trans., J. applied Mech.*, **44** (1977), 768–769.
6. CELEP, Z., Axially symmetric stability of a completely free circular plate subjected to a non-conservative edge load, *J. Sound Vib.*, **65** (1979), 549–556.

7. FARSHAD, M., Stability of cantilever plates subjected to biaxial subtangential loading, *J. Sound Vib.*, **58** (1978), 555–561.
8. IRIE, T., YAMADA, G. and KANEKO, Y., Vibration and stability of a non-uniform annular plate subjected to a follower force, *J. Sound Vib.*, **73** (1980), 261–269.
9. CHONAN, S., Vibration and stability of annular plates under conservative and non-conservative loads, *J. Sound Vib.*, **80** (1982), 413–420.
10. CELEP, Z., Vibration and stability of a free circular plate subjected to non-conservative loading, *J. Sound Vib.*, **80** (1982), 421–432.
11. ADALI, S., Stability of a rectangular plate under non-conservative and conservative forces, *Int. J. Solids Struct.*, **18** (1982), 1043–1052.
12. BRUNELLE, E. J. and OYIBO, G. A., Generic buckling curves for specially orthotropic plates, *AIAA J.*, **21** (1983), 1150–1156.
13. VANDERPLATES, G. N., *ADS—A Fortran Program for Automated Design Synthesis, Version 1.00*, Monterey, California, Naval Postgraduate School, 1984.

7

The Effect of Shear Deformation on the Bending and Buckling of Anisotropic Laminated Plates

James M. Whitney*

Air Force Wright Aeronautical Laboratories, AFWAL/MLBM, Wright-Patterson Air Force Base, Ohio 45433-6533, USA

ABSTRACT

A Mindlin-type shear deformation theory is applied to the bending and buckling analysis of anisotropic, rectangular plates. Approximate solutions to the governing partial differential equations are obtained utilizing the Galerkin method. Three sets of boundary conditions are considered. Numerical results are obtained in conjunction with symmetrically-laminated composite plates. Accuracy of the numerical method is investigated along with the effects of bending-twisting coupling on maximum deflection under transverse load and on critical buckling loads. Arbitrary combinations of inplane normal and shear forces are included in the buckling analysis. Numerical results indicate that the effect of bending-twisting shear coupling depends on plate thickness.

1. INTRODUCTION

Numerous solutions have been published for laminated plates based on classical Kirchhoff assumptions. Basic equations and numerical results for buckling and bending under transverse load can be found in a number of textbooks.[1-3] It is well recognized, however, that shear deformation can be more significant in laminated, anisotropic plates than in isotropic,

* Also Adjunct Associate Professor of Materials Engineering, University of Dayton, Dayton, Ohio, USA.

homogeneous plates. This is due to the large ratios of inplane tensile moduli to through-the-thickness shear moduli.

Shear deformation theories based on Mindlin's assumptions[4] have been developed for laminated, anisotropic plates.[5,6] Although analytical solutions to these theories have been obtained for specially orthotropic laminates, few solutions can be found for laminated, anisotropic plates (nonzero bending-twisting shear coupling terms). Flaggs and Vinson,[7] and more recently Wilson and Vinson,[8] used an energy approach to analyze the buckling of laminated, anisotropic plates, including the effect of shear deformation. In their solutions, however, the plate rotations were not carried as independent variables, i.e. they were assumed to be a function of the midplane displacements, thus reducing the total number of degrees of freedom of the plate. Craig and Dawe,[9] and Dawe and Craig[10,11] used the Rayleigh–Ritz and finite strip methods to investigate the effect of transverse shear deformation on the buckling and flexural vibrations of symmetrically-laminated plates. Numerical results were presented for both specially orthotropic plates and anisotropic plates.

In the present paper, the Galerkin method[12] is utilized for obtaining analytical solutions to buckling and bending problems associated with laminated, anisotropic plates, including the effect of transverse shear deformation. Three distinct sets of boundary conditions are considered. The analysis is limited to symmetric laminates, although the approach is applicable to general laminates. This limitation simplifies the solution procedure. From a practical standpoint, symmetric laminates are of more interest than unsymmetric laminates.

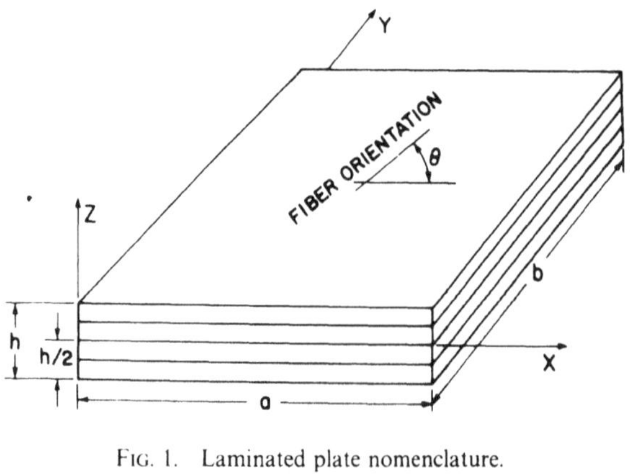

FIG. 1. Laminated plate nomenclature.

2. GOVERNING EQUATIONS

A standard x, y, z cartesian coordinate system is located at the center of the plate, as illustrated in Fig. 1. The plate thickness and inplane dimensions are denoted by h, a and b, respectively. In addition the plate is composed of an arbitrary number of layers with arbitrary fiber orientation in each layer. The Mindlin-type laminated plate equations[5,6] for bending and buckling of symmetric, anisotropic laminates are based on the assumed displacement field

$$u = z\psi_x(x, y), \qquad v = z\psi_y(x, y), \qquad w = w(x, y) \tag{1}$$

where u, v and w are the displacements in the x, y and z directions, respectively. Denoting partial differentiation by a comma, the plate constitutive relations are of the form[6]

$$\begin{bmatrix} M_x \\ M_y \\ M_{xy} \end{bmatrix} = \begin{bmatrix} D_{11} & D_{12} & D_{16} \\ D_{12} & D_{22} & D_{26} \\ D_{16} & D_{26} & D_{66} \end{bmatrix} \begin{bmatrix} \psi_{x,x} \\ \psi_{y,y} \\ \psi_{y,x} + \psi_{x,y} \end{bmatrix}$$

$$\begin{bmatrix} Q_y \\ Q_x \end{bmatrix} = k \begin{bmatrix} A_{44} & A_{45} \\ A_{45} & A_{55} \end{bmatrix} \begin{bmatrix} \psi_y + w_{,y} \\ \psi_x + w_{,x} \end{bmatrix} \tag{2}$$

where k is a shear correction factor as introduced by Mindlin.[4] Denoting normal stresses by σ_i and shear stresses by τ_{ij}, then

$$(M_x, M_y, M_{xy}) = \int_{-h/2}^{h/2} (\sigma_x, \sigma_y, \tau_{xy})\,\mathrm{d}z$$

$$(Q_x, Q_y) = \int_{-h/2}^{h/2} (\tau_{xz}, \tau_{yz})\,\mathrm{d}z \tag{3}$$

and

$$D_{ij} = 2\int_0^{h/2} Q_{ij} z^2\,\mathrm{d}z \qquad (i, j = 1, 2, 6)$$

$$A_{ij} = \int_{-h/2}^{h/2} C_{ij}\,\mathrm{d}z \qquad (i, j = 4, 5) \tag{4}$$

where Q_{ij} are the anisotropic reduced stiffnesses for plane stress and C_{ij} are anisotropic stiffnesses associated with interlaminar shear.

The equilibrium equations in terms of moment and shear resultants are those of classical plate theory, i.e.

$$M_{x,x} + M_{xy,y} - Q_x = 0$$
$$M_{xy,y} + M_{y,y} - Q_y = 0 \qquad (5)$$
$$Q_{x,x} + Q_{y,y} + N_x w_{,xx} + 2N_{xy} w_{,xy} + N_y w_{,yy} = q$$

where N_i are prebuckling inplane force resultants and q is the lateral pressure loading defined by the relationship

$$q(x, y) = \sigma_z(x, y, h/2) - \sigma_z(x, y, -h/2) \qquad (6)$$

Combining eqns (2) and (5), we obtain the governing equations in operator form

$$L_{11}\psi_x + L_{12}\psi_y - L_{13}w = 0$$
$$L_{12}\psi_x + L_{22}\psi_y - L_{23}w = 0 \qquad (7)$$
$$L_{13}\psi_x + L_{23}\psi_y + L_{33}w = q$$

The linear operators L_{ij} are defined as follows:

$$L_{11} = D_{11}(\ )_{,xx} + 2D_{16}(\ )_{,xy} + D_{66}(\ )_{,yy} - kA_{55}$$
$$L_{12} = D_{16}(\ )_{,xx} + (D_{12} + D_{66})(\ )_{,xy} + D_{26}(\ )_{,yy} - kA_{45}(\ )$$
$$L_{13} = k[A_{55}(\ )_{,x} + A_{45}(\ )_{,y}] \qquad (8)$$
$$L_{22} = D_{66}(\ )_{,xx} + 2D_{26}(\ )_{,xy} + D_{22}(\ )_{,yy} - kA_{44}(\ )$$
$$L_{23} = k[A_{45}(\ )_{,x} + A_{44}(\ )_{,y}]$$
$$L_{33} = (N_x + kA_{55})(\ )_{,xx} + 2(N_{xy} + kA_{45})(\ )_{,xy} + (N_y + kA_{44})(\ )_{,yy}$$

3. THE GALERKIN METHOD

Three sets of boundary conditions are considered in the present paper. We first consider the simply-supported conditions, BC-1

at $x = 0, a$:

$$w = \psi_y = M_x = 0 \qquad (9)$$

at $y = 0, b$:

$$w = \psi_x = M_y = 0 \qquad (10)$$

The second set of boundary conditions, BC-2, is the fully clamped case

at $x=0, a$; $y=0, b$:

$$w=\psi_x=\psi_y=0 \tag{11}$$

The third set of boundary conditions, BC-3, corresponds to the edges $x=0$, a, clamped, eqn. (11), and the edges $y=0, b$, simply-supported, eqn. (10).

A solution to these problems is sought in the form

$$\begin{aligned}\psi_x &= \sum_{m=1}^{\infty}\sum_{n=1}^{\infty} A_{mn}\Phi_{mn}(x,y)\\ \psi_y &= \sum_{m=1}^{\infty}\sum_{n=1}^{\infty} B_{mn}\chi_{mn}(x,y)\\ w &= \sum_{m=1}^{\infty}\sum_{n=1}^{\infty} C_{mn}W_{mn}(x,y)\end{aligned} \tag{12}$$

The functions Φ_{mn}, χ_{mn}, and W_{mn} must satisfy the displacement boundary conditions. The Galerkin method[9] leads to the following equations

$$\int_0^b\int_0^a (L_{11}\psi_x + L_{12}\psi_y - L_{13}w)\Phi_{mn}\,\mathrm{d}x\,\mathrm{d}y + \int_0^a (\bar{M}_{xy} - M_{xy})\Phi_{mn}\Big|_0^b\,\mathrm{d}x + \int_0^b (\bar{M}_x - M_x)\Phi_{mn}\Big|_0^a\,\mathrm{d}y = 0 \tag{13}$$

$$\int_0^b\int_0^a (L_{12}\psi_x + L_{22}\psi_y - L_{23}w)\chi_{mn}\,\mathrm{d}x\,\mathrm{d}y + \int_0^a (\bar{M}_y - M_y)\Big|_0^b\,\chi_{mn}\,\mathrm{d}x + \int_0^b (\bar{M}_{xy} - M_{xy})\chi_{mn}\Big|_0^a\,\mathrm{d}y = 0 \tag{14}$$

$$\int_0^b\int_0^a (L_{13}\psi_x + L_{23}\psi_y + L_{33}w - q)W_{mn}\,\mathrm{d}x\,\mathrm{d}y + \int_0^a (\bar{Q}_y - Q_y)W_{mn}\Big|_0^b\,\mathrm{d}x + \int_0^b (\bar{Q}_x - Q_x)W_{mn}\Big|_0^a\,\mathrm{d}y = 0 \tag{15}$$

where bars denote prescribed values on the boundary.

Equations (13)–(15) yield a set of linear algebraic equations. For the case of lateral loading, these equations are solved for the unknown coefficients

A_{mn}, B_{mn} and C_{mn}. In the case of buckling, the equations are in the form of classical eigenvalue problems from which critical buckling loads can be determined. For the case of fully clamped edges, BC-2, the boundary integrals in eqns (13)–(15) will vanish exactly, due to the requirement that Φ_{mn}, χ_{mn} and W_{mn} vanish on the boundary.

For BC-1 the following functions are utilized:

$$\Phi_{mn} = \cos\frac{m\pi x}{a}\sin\frac{n\pi y}{b} \tag{16}$$

$$\chi_{mn} = \sin\frac{m\pi x}{a}\cos\frac{n\pi y}{b} \tag{17}$$

$$W_{mn} = \sin\frac{m\pi x}{a}\sin\frac{m\pi y}{b} \tag{18}$$

For BC-2 eqns (17) and (18) are replaced by eqn. (19), i.e.

$$\Phi_{mn} = \chi_{mn} = W_{mn} = \sin\frac{m\pi x}{a}\sin\frac{n\pi y}{b} \tag{19}$$

In the case of BC-3, W_{mn} is represented by eqn. (19), while

$$\Phi_{mn} = \sin\frac{m\pi x}{a}\sin\frac{m\pi y}{b} \tag{20}$$

$$\chi_{mn} = \sin\frac{m\pi x}{a}\cos\frac{n\pi y}{b} \tag{21}$$

Equations (16)–(21) satisfy all of the displacement boundary conditions, but not necessarily the natural boundary conditions. The line integrals in eqns (13)–(15) must be considered when the natural boundary conditions are not satisfied.

It should be noted that eqns (16)–(18) represent an exact solution to eqns (7) for BC-1, eqns (9) and (10), when certain loading conditions are considered in conjunction with laminates in which the shear coupling terms vanish, i.e.

$$A_{45} = D_{16} = D_{26} = 0 \tag{22}$$

In this case eqns (16)–(18) exactly satisfy the simply-supported boundary conditions, eqns (9) and (10), and the governing eqns (7) if either critical buckling loads under biaxial inplane loading ($N_{xy} = 0$) or bending deformations under any lateral load which can be represented by the double Fourier series

$$q = \sum_{m=1}^{\infty} \sum_{n=1}^{\infty} q_{mn} \sin \frac{m\pi x}{a} \sin \frac{n\pi y}{b} \tag{23}$$

are sought.

For the general case, eqns (13) and (14) are solved for A_{mn} and B_{mn} in terms of C_{mn} and substituted into eqn. (15). If this infinite series is truncated at $m = M$ and $n = N$, then eqn. (15) produces an $M \times N$ set of algebraic equations which are solved directly in the case of transverse bending. In the case of buckling, these equations are in classical eigenvalue form.

4. NUMERICAL RESULTS

We consider laminates constructed from unidirectional material with the following elastic properties:

$$\begin{gathered} E_1/E_2 = 14,\ G_{12}/E_2 = 0{\cdot}533,\ G_{23}/E_2 = 0{\cdot}323 \\ E_3 = E_2,\ G_{13} = G_{12},\ \nu_{13} = \nu_{12} = 0{\cdot}30,\ \nu_{23} = 0{\cdot}55 \end{gathered} \tag{24}$$

where the subscripts 1, 2 and 3 denote properties parallel to the fiber direction, transverse to the fiber direction, and through-the-thickness, respectively. These properties are typical of contemporary graphite/epoxy composites.

Convergence is illustrated in Table 1 for critical buckling loads associated with uniaxial compression loading of a $[\pm 45]_s$ square plate with

TABLE 1
Convergence for uniaxial compression buckling

$[\pm 45]_s$ graphite/epoxy, $a/b = 1$, $a/h = 20$				
M	N	$N_{cr}b^2/E_2h^3$		
		BC-1	BC-2	BC-3
2	2	20·77	86·26	53·32
4	4	19·79	31·03	24·82
6	6	19·42	29·68	24·26
8	8	19·23	29·35	24·05

TABLE 2
Convergence for uniform transverse pressure loading

$[\pm 45]_s$ graphite/epoxy, $a/b = 1$, $a/h = 20$					
M	N	$W_c E_2 h^3 \times 10^3/q_0 a^4$		$M_c a^2 \times 10^2/q_0 h^2$	
		BC-1	BC-2	BC-1	BC-2
2	2	7·803	2·573	3·817	1·903
4	4	7·918	4·605	4·229	3·387
6	6	7·990	4·713	3·771	2·640
8	8	8·090	4·897	4·175	3·133

$a/h = 20$. Note that convergence is rapid for all three sets of boundary conditions under consideration. Convergence will be slower for $a/h > 100$. Bending results under uniform lateral pressure ($q = q_0 = \text{constant}$) for the same laminate are shown in Table 2 where

$$w_c = w(a/2,\ b/2),\ M_c = M_x(a/2,\ b/2) \tag{25}$$

Note that the convergence for BC-2 is slower for the center deflection than for the critical buckling load shown in Table 1. The bending moments in Table 2 show a slow oscillating type of convergence. This is not unanticipated, as bending moment generated from Galerkin or Rayleigh–Ritz type solutions often display erratic convergence!

We now consider the case of uniaxial compression in the y-direction, i.e. $N_x = N_{xy} = 0$, $N_y = -N = \text{constant}$. As the aspect ratio a/b of the plate increases, the critical buckling load should approach the solution for cylindrical bending (CB) of an infinite strip. For the case of cylindrical bending the demension a becomes unbounded and

$$\psi_x = \psi_x(y),\ \psi_y = \psi_y(y),\ w = w(y) \tag{26}$$

Closed form solutions[13] can be obtained for cylindrical bending in conjunction with simply-supported boundary conditions along the edges $y = 0, b$. Numerical results are shown in Fig. 2 for a $[\pm 45]_s$ laminate in conjunction with BC-1 and BC-3. Since both of these boundary conditions involve simple supports along the edges $y = 0, b$, the critical buckling loads should approach the cylindrical bending solution with increasing values of a/b. The anticipated behavior is observed in Fig. 2.

The effect of the bending-twisting shear coupling terms D_{16} and D_{26} depends on the length-to-thickness ratio a/h of a laminated plate. This is

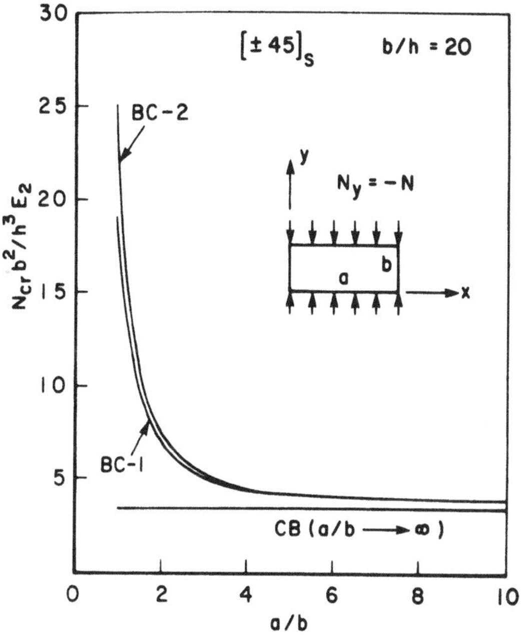

FIG. 2. Effect of plate aspect ratio on uniaxial compression buckling.

seen in Figs. 3 and 4 where critical buckling loads are shown as a function of a/b for a $[\pm 45]_s$ laminate with simply supported boundary conditions (BC-1). In both of these figures results are compared for the case where the D_{16} and D_{26} bending-twisting shear coupling terms are retained and for the case where these terms are neglected. Results are shown in Fig. 3 for uniform shear loading ($N_x = N_y = 0, N_{xy} = \pm S =$ constant). Because of the presence of the D_{16} and D_{26} bending-twisting shear coupling terms, the critical buckling load depends on the sign of the shear stress. It can easily be observed that the difference between critical buckling loads (S_{cr}) associated with positive and negative shear increases with increasing values of a/h. As a result, departure from the orthotropic solution ($D_{16} = D_{26} = 0$) must also increase for both positive and negative shear with increasing a/h. In Fig. 4 uniform compression loading ($N_y = N_{xy} = 0$, $N_x = -N =$ constant) is considered. Departure of critical buckling loads (N_{cr}) from the orthotropic solution ($D_{16} = D_{26} = 0$) with increasing values of a/h is also observed in this case. The effect of bending-twisting shear coupling appears to be less severe in this case than in the case of shear loading. It should be noted that the orthotropic solution represents an exact solution to eqn. (7).

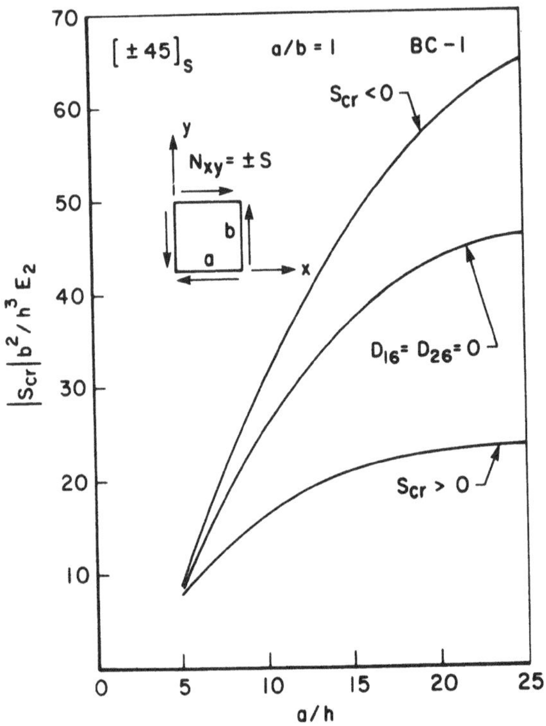

FIG. 3. Effect of plate thickness on critical shear buckling loads.

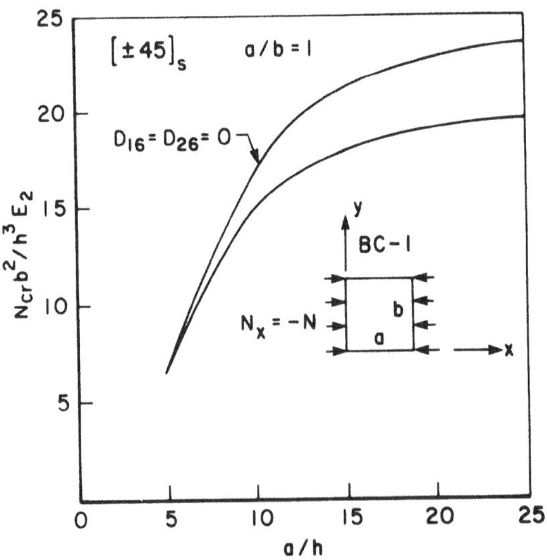

FIG. 4. Effect of plate thickness on uniaxial compression buckling.

The effect of stacking sequence on the response of symmetrical laminates is illustrated in Figs. 5 and 6 for $[0/\pm 45]_s$ and $[\pm 45/0]_s$ laminates. Increased bending stiffness in the x-direction (D_{11}) is obtained by placing the 0-degree plies on the outer surfaces, while increased twisting stiffness (D_{66}) is obtained by placing the ± 45-degree plies on the outside of the laminate. In addition, placing the ± 45-degree plies on the outside leads to increased bending-twisting shear coupling (D_{16}, D_{26}). Results are shown in Fig. 5 for shear buckling of a rectangular plate ($a/b = 2$) with clamped boundaries. The increased shear stiffness associated with $[\pm 45/0]_s$ laminates leads to increased shear buckling strength for both positive and negative shear loads compared to the $[0/\pm 45]_s$ laminate stacking sequence. It is also interesting to note that the bending-twisting shear coupling terms have a significant effect on both stacking sequences as observed by the increasing departure in critical buckling load between positive and negative shear with increasing values of a/b. In Fig. 6, the center deflection for laminates under uniform lateral pressure

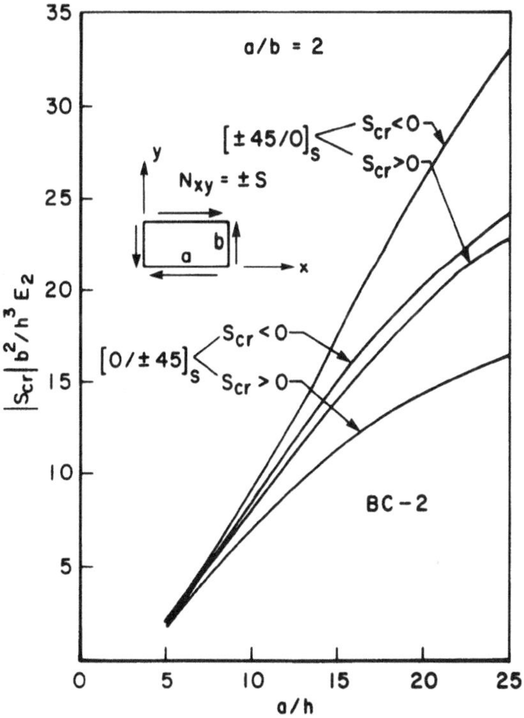

FIG. 5. Effect of laminate stacking sequence on critical shear buckling loads.

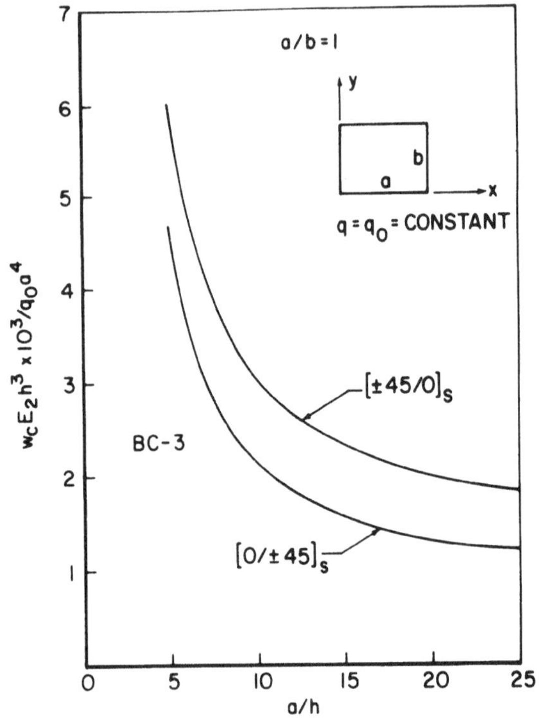

FIG. 6. Effect of laminate stacking sequence on center deflection for plates subjected to uniform transverse pressure.

($q = q_o$ = constant) is illustrated as a function of a/h for BC-3. We note that in this case the axial bending stiffness (D_{11}) is dominant. In particular, the $[0/\pm 45]_s$ laminate yields a smaller center deflection compared to the $[\pm 45/0]_s$ stacking sequence.

5. SUMMARY AND CONCLUSIONS

A numerical solution based on the Galerkin method has been presented for bending and buckling of symmetrically laminated plates with bending-twisting coupling present. Four separate sets of boundary conditions have been considered. Based on numerical results presented, the following conclusions are made:

1. The Galerkin method provides a viable approach for obtaining solutions to anisotropic laminated plate problems.

2. The effect of the bending-twisting shear coupling terms increase with increasing values of a/b.
3. The effect of the bending-twisting shear coupling terms are more severe for shear buckling than for compression buckling or bending under lateral pressure.

REFERENCES

1. WHITNEY, J. M., *Structural Analysis of Laminated Plates*, Lancaster, Pennsylvania, Technomic Publishing Company, 1987.
2. JONES, R. M., *Mechanics of Composite Materials*, Washington D.C., Scripta Book Company, 1975.
3. VINSON, J. R. and CHOU, T. W., *Composite Materials and Their Use in Structures*, New York, John Wiley, 1975.
4. MINDLIN, R. D., Influence of rotatory inertia and shear deformation on flexural motions of isotropic, elastic plates, *ASME Trans., J. Appl. Mech.*, **18** (1951), 31–38.
5. YANG, P. C., NORRIS, C. H. and STAVSKY, Y., Elastic wave propagation in heterogeneous plates, *Int. J. Solids Struct.*, **2** (1966), 665–671.
6. WHITNEY, J. M. and PAGANO, N. J., Shear deformation in heterogeneous anisotropic plates, *ASME Trans. J. Appl. Mech.*, **37** (1970), 1031–1036.
7. FLAGGS, D. L. and VINSON, J. R., Hygrothermal effects on the buckling of laminated composite plates, *Fibre Sci. Technol.*, **11** (1978), 353–365.
8. WILSON, D. W. and VINSON, J. R., Viscoelastic analysis of laminated plate buckling, *AIAA J.*, **22** (1984), 982–988.
9. CRAIG, T. J. and DAWE, D. J., Flexural vibration of symmetrically laminated, composite, rectangular plates including transverse shear effects, *Int. J. Solids Struct.*, **22** (1986), 155–169.
10. DAWE, D. J. and CRAIG, T. J., The influence of shear deformation on the natural frequencies of laminated rectangular plates, in *Composite Structures*—3 (Marshall, I. H. ed.), London, Elsevier Applied Science, 1985, pp. 660–676.
11. DAWE, D. J. and CRAIG, T. J., The vibration and stability of symmetrically-laminated composite rectangular plates subjected to in-plane stresses, *Composite Structures*, **5** (1986), 281–307.
12. KANTOROVICH, L. V. and KRYLOV, V. I., *Approximate Methods of Higher Analysis* (translated by C. D. Benster), New York, Interscience, 1958.
13. WHITNEY, J. M., Curvature effects in the buckling of symmetrically-laminated rectangular plates with transverse shear deformation, submitted to *Composite Structures*.

8

Interactive Buckling Effects in Stiffened FRP Panels

C. S. Smith and R. S. Dow

Admiralty Research Establishment, St Leonard's Hill, Dunfermline, Fife KY11 5PW, Scotland, UK

ABSTRACT

Results are described of a numerical study of the initial compressive buckling and post-buckling behaviour of GRP panels reinforced by longitudinal hat-section stiffeners. Particular reference is made to the effects of interaction between local buckling of the panel laminate and overall, column-like buckling involving bending of the stiffeners. Conclusions are reached regarding the influence of interactive effects on collapse strength and recommendations are made on allowance for these effects in design.

INTRODUCTION

Stiffened GRP panels, employing hat-section stiffeners aligned longitudinally in the direction of dominant load, have been adopted in the deck and bottom-shell structure of new minehunters currently under development for the Royal Navy.[1] A theoretical and experimental investigation[2] has identified various forms of instability which may occur in panels of this type under longitudinal compression, and has examined the relationship between initial buckling modes and post-buckling and collapse behaviour. It appeared that significant interactions could occur in some cases between local and overall buckling modes, influencing both initial buckling and post-buckling behaviour and hence possibly requiring special treatment in design. The purpose of the present paper is to describe a further study in which such interactive effects were examined by numerical analysis of some panels in which the relationship between local and overall, column-like buckling was varied systematically by adjusting the panel geometry.

METHODS OF ANALYSIS

Methods of evaluating the initial buckling behaviour of stiffened FRP panels have been described previously[2] but will be summarized below for completeness.

A typical GRP deck or shell panel, reinforced by longitudinal stiffeners of trapezoidal, top-hat section, is shown in Fig. 1. Strips of laminate forming the panel cross-section may be related to x- and y-axes parallel to their longitudinal and transverse edges. For most marine-type GRP laminates based on chopped-strand, woven-roving or unidirectional reinforcement it is possible to assume that each strip of laminate is orthotropic with principal directions of elasticity parallel to its edges and that plies of reinforcement are stacked symmetrically about mid-thickness so that no direct coupling occurs between membrane and bending deformations. In-plane deformations of each strip then satisfy the orthotropic plane stress–strain relationship:

$$\begin{bmatrix} \varepsilon_x \\ \varepsilon_y \\ \gamma_{xy} \end{bmatrix} = \begin{bmatrix} 1/E_x & -\mu_x/E_y & 0 \\ -\mu_y/E_x & 1/E_y & 0 \\ 0 & 0 & 1/G_{xy} \end{bmatrix} \begin{bmatrix} \sigma_x \\ \sigma_y \\ \tau_{xy} \end{bmatrix} \quad (1)$$

where E_x, E_y, G_{xy}, μ_x and μ_y are Young's moduli, shear modulus and Poisson ratios referring to the principal directions. The moment–curvature relationship for the strip is

$$\begin{bmatrix} m_x \\ m_y \\ m_{xy} \end{bmatrix} = -\begin{bmatrix} D_x & \mu_x D_x & 0 \\ \mu_y D_y & D_y & 0 \\ 0 & 0 & D_{xy} \end{bmatrix} \begin{bmatrix} \dfrac{\partial^2 w}{\partial x^2} \\ \dfrac{\partial^2 w}{\partial y^2} \\ 2\dfrac{\partial^2 w}{\partial x \, \partial y} \end{bmatrix} \quad (2)$$

where

$$D_x = \frac{E_x h^3}{12(1-\mu_x\mu_y)} \quad \text{and} \quad D_y = \frac{E_y h^3}{12(1-\mu_x\mu_y)}$$

are flexural rigidities per unit width in x- and y-directions. $D_{xy} = G_{xy}h^3/12$ is the twisting rigidity and h is the laminate thickness.

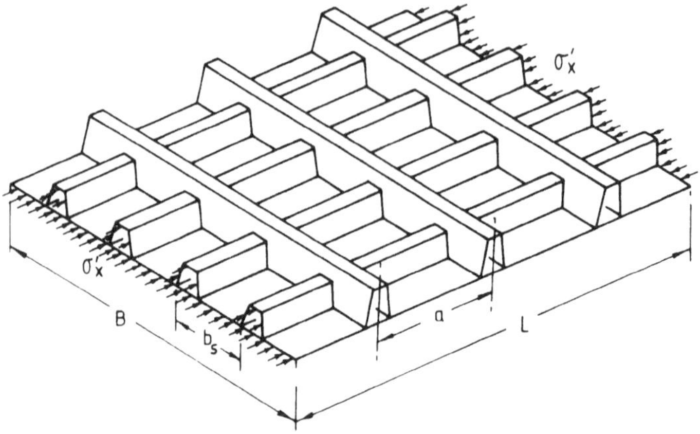

FIG. 1. Longitudinally compressed GRP panel.

In the presence of a uniform destabilizing stress σ'_x as shown in Fig. 1, the equation governing small bending displacements w of a laminate strip is

$$D_x \frac{\partial^4 w}{\partial x^4} + 2H \frac{\partial^4 w}{\partial x^2 \partial y^2} + D_y \frac{\partial^4 w}{\partial y^4} + N_x \frac{\partial^2 w}{\partial x^2} = 0 \tag{3}$$

where $H = 2D_{xy} + \mu_x D_x$ and $N_x = \sigma'_x h$. A conservative estimate of local buckling stress for a stiffened panel may be obtained by applying to each element of the cross-section the formula

$$N_{x\,cr} = \frac{2\pi^2}{b^2}(H + \sqrt{D_x D_y}) \tag{4}$$

which corresponds to a solution of eqn. (3) for a long orthotropic strip of width b with simply-supported edges: local buckling will tend to occur predominantly in the most slender element with a number of half-waves over the length (a) of the panel given approximately by

$$n = \frac{a}{b}\left(\frac{D_y}{D_x}\right)^{1/4} \tag{5}$$

A corresponding upper-bound estimate of initial buckling stress is provided by the formula for a long orthotropic strip with clamped edges:[3]

$$N_{x\,cr} = \frac{\pi^2}{b^2}(2{\cdot}4H + 4{\cdot}6\sqrt{D_x D_y}) \tag{6}$$

in which case the preferred number of buckling half-waves is approximately

$$n = 1{\cdot}5\frac{a}{b}\left(\frac{D_y}{D_x}\right)^{1/4} \tag{7}$$

Initial buckling may also take the form of column-like instability of longitudinal stiffeners and attached shell laminate between heavy transverse frames or bulkheads. Ignoring rotational restraint provided by the transverse members, this buckling stress may be estimated using the Euler formula:

$$\sigma'_{x\,\mathrm{cr}} = \frac{\pi^2 EI}{Aa^2}\Bigg/\left(1 + \frac{\pi^2 EI}{a^2 GA_s}\right) \tag{8}$$

in which EI is the flexural rigidity of a stiffener with assumed effective breadth of shell, A is the total cross-section of the stiffener with attached strip of shell, a is the spacing of transverse frames or bulkheads and GA_s is the shear rigidity in which A_s may be taken as the area of the stiffener webs.

More accurate (theoretically exact) evaluation of initial buckling stresses may be achieved using folded-plate analysis, in which out-of-plane displacements of each strip satisfy eqn. (3) while in-plane displacements u and v satisfy the simultaneous equations:

$$\frac{E_x}{1-\mu_x\mu_y}\left[\frac{\partial^2 u}{\partial x^2} + \mu_x\frac{\partial^2 v}{\partial x\,\partial y}\right] + G_{xy}\left[\frac{\partial^2 v}{\partial x\,\partial y} + \frac{\partial^2 u}{\partial y^2}\right] = 0 \tag{9}$$

$$\frac{E_y}{1-\mu_x\mu_y}\left[\frac{\partial^2 v}{\partial y^2} + \mu_y\frac{\partial^2 u}{\partial x\,\partial y}\right] + G_{xy}\left[\frac{\partial^2 v}{\partial x^2} + \frac{\partial^2 u}{\partial x\,\partial y}\right] - \sigma'_x\frac{\partial^2 v}{\partial x^2} = 0 \tag{10}$$

If conditions of simple support are assumed at the ends of each strip ($v = \partial u/\partial x = 0$ and $w = \partial^2 w/\partial x^2 = 0$ at $x = 0, a$), solutions of eqns (4), (9) and (10) are periodic in x and may be obtained in a way which accounts rigorously for interactions between strips of laminate forming a stiffened panel. Details of this analysis method, together with generalized computer programs developed independently by several investigators, are contained in Refs 4–8.

Evaluation of non-linear and post-buckling behaviour, including the effects of imperfections, may be carried out using finite element (FE) analysis. Results described in this chapter were obtained using the general purpose non-linear FE program ASAS-NL, developed by Atkins R&D in collaboration with ARE, Dunfermline.[9] Eight-node isoparametric quadrilateral shell elements were employed with constant anisotropic

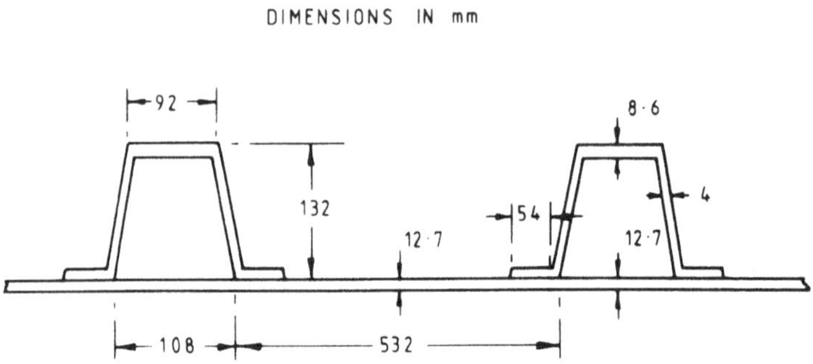

Mean material properties	Shell laminate	Stiffener tables	Stiffener webs and flanges
E_x (GPa)	15·0	19·5	15·0
E_y (GPa)	13·5	11·9	13·5
G_{xy} (GPa)	3·45	3·45	3·45
μ_x	0·135	0·11	0·135
μ_y	0·15	0·18	0·15
Flexural strength (MPa)	273	415	—

FIG. 2. Panel geometry and material properties.

material properties as defined in Fig. 2: large displacements were represented using an updated Lagrangean (moving coordinate) procedure with application of a modified Newton–Raphson iterative equilibrium correction. The adequacy of FE models was tested by obtaining estimates of initial buckling stresses and modes by linear eigenvalue analysis for comparison with exact folded-plate results.

SCOPE OF ANALYSIS

As a basis for evaluating interactions between local buckling and overall column-like buckling, analysis has been carried out for a series of longitudinally stiffened panels having the same cross-sectional geometry and material properties as the experimental panel L2 described in Ref. 2. By varying the span a between transverse frames a range of structures was obtained in which initial column buckling stresses varied from less than to substantially more than local laminate buckling stresses. Panel dimensions and material properties are indicated in Fig. 2. In addition to the experimental length ($a = 3060$ mm) spans of 1000, 1500, 2000, 2350, 2670 and 3500 mm were considered.

Initial buckling stresses were estimated approximately using eqns (4)–(8), with the breadth b taken equal to the unsupported span between stiffener webs, and were computed accurately using folded-plate and finite element analysis. Stiffener flanges, which have been shown to influence local buckling significantly,[2] were included in folded-plate and finite element models. Finite element analysis was also used to examine the non-linear, post-buckling behaviour of panels, including the influence of assumed imperfections. As shown in Fig. 3, the finite element model referred to a single longitudinal stiffener with attached shell laminate extending over two half-spans $a/2$. Conditions of simple support were assumed at the position of the transverse frame EF, with a transverse plane of symmetry at mid-span position CD and a 'moving' plane of symmetry at mid-span position AB: in this way account was taken of 'continuous beam' interactions between adjoining spans. Longitudinal planes of symmetry AC and BD were assumed to occur mid-way between longitudinal stiffeners.

For the purpose of non-linear analysis, initial deformation of each panel was assumed to have the form of the preferred local buckling mode with amplitude w_{op}, combined with an overall distortion of amplitude w_{os} having the form of the interframe column buckling mode. In each case two levels of imperfection were considered:

(i) $w_{op}/a = 0{\cdot}0005$, $w_{os}/a = 0{\cdot}0013$ (corresponding approximately to maximum distortions measured in experimental panel L2);

(ii) $w_{op}/a = 0{\cdot}000\,05$, $w_{os}/a = 0{\cdot}000\,13$.

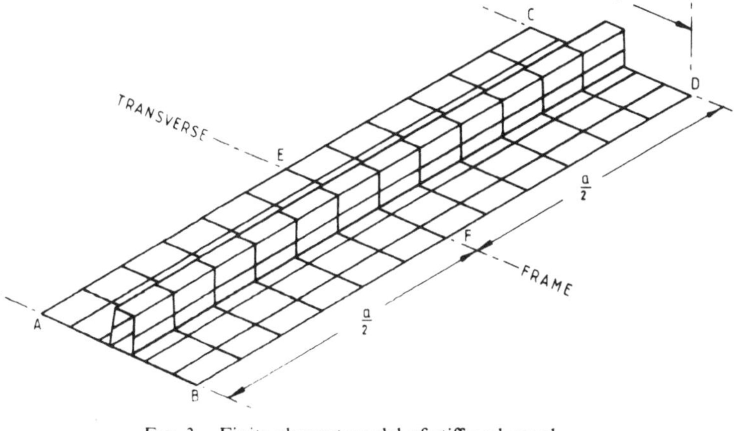

FIG. 3. Finite element model of stiffened panel.

Initial deformation was taken to be antisymmetric about EF, positive (towards the stiffener) at AB and negative at CD. Analysis was carried out by incremental application of uniform end-shortening displacement on the plane AB of each panel.

DISCUSSION OF RESULTS

Initial Buckling Behaviour

Initial buckling stresses obtained from the approximate formulae and by folded-plate and FE analysis are listed in Table 1. A selection of corresponding buckling modes, obtained from the folded-plate solutions, is shown in Fig. 4. Initial buckling stresses determined by FE eigenvalue analysis are generally within 5% of exact folded-plate values, lending some credibility to the FE model used in the non-linear analysis.

As observed previously, critical stresses corresponding to local buckling

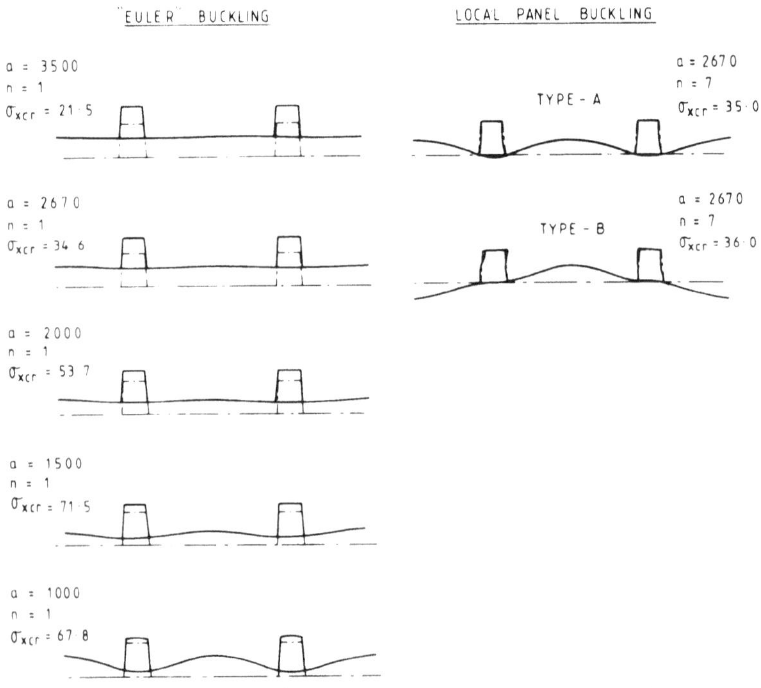

FIG. 4. Selected initial buckling modes computed by folded-plate analysis.

TABLE 1
Initial buckling modes and stresses (MPa)

Span a (mm)	Mode		Computed results		Approximate formulae		
	n	Type	Folded plate	Finite element	Eqn. (4)	Eqn. (6)	Eqn. (8)
3 500	1	Euler	21·5	22·2			22·9
	8	A	35·0	34·3	26·6 ($n = 7$)		
	8	B	35·4	35·2			
	9	A	34·9	35·8			
	9	B	35·5	36·0			
	10	A	35·9			50·1 ($n = 10$)	
	10	B	37·0				
3 060	1	Euler	27·4	28·2			29·4
	6	A	36·1	34·1	26·6 ($n = 6$)		
	6	B	37·0	36·2			
	7	A	35·0	33·9			
	7	B	35·4	34·7			
	8	A	35·0	35·3			
	8	B	35·9	35·4		50·1 ($n = 9$)	
2 670	1	Euler	34·6	35·5			37·6
	6	A	35·1	35·0	26·6 ($n = 5$)		
	6	B	35·4	34·5			
	7	A	35·0	35·5			
	7	B	36·0	35·7			
	8	A	36·7	38·0		50·1 ($n = 8$)	
	8	B	37·8				
2 350	1	Euler	42·6	43·5			47·0
	5	B	35·6	34·5	26·6 ($n = 4$)		
	5	A	35·7	35·5			
	6	A	34·9	35·1			
	6	B	35·8	35·4			
	7	A	36·6	37·4		50·1 ($n = 7$)	
	7	B	37·7	38·1			
2 000	1	Euler	53·7				61·8
	4	B	35·9	35·3	26·6 ($n = 4$)		
	4	A	36·7	36·6			
	5	A	34·8	34·9			
	5	B	35·7	35·2			
	6	A	36·7	37·7		50·1 ($n = 6$)	
	6	B	37·9	38·0			
1 500	1	Euler	71·5				96·5
	3	B	35·9		26·6 ($n = 3$)		
	3	A	36·7				
	4	A	35·1			50·1 ($n = 4$)	
	4	B	36·1				
	5	A	39·3				
	5	B	40·5				
1 000	1	B	54·2	49·6			
	1	A/Euler	67·8	67·5			161·4
	2	B	35·9	36·4	26·6 ($n = 2$)	50·1 ($n = 2$)	
	2	A	36·7	36·8			
	3	A	36·7	36·9			
	3	B	37·9	37·7			

of the shell laminate between stiffeners occur in closely spaced pairs; asymmetric modes (Type B) are usually (but not always) preceded by symmetric (Type A) modes. The latter generally involve virtually (but not exactly) zero displacement of stiffeners out of the panel plane. Type B modes involve distortion of stiffener cross-sections, including some sideways bending of the stiffener tables. As might be expected, local buckling stresses and associated modal wavelengths are insensitive to the span a.

In the longest panel examined ($a = 3500$ mm) overall buckling occurs with virtually no local deformation at a critical stress 6% less than that given by the Euler formula (eqn. (8)). As panel length is reduced the 'column' mode becomes increasingly influenced by local deformation of the shell laminate and the associated critical stress falls progressively below the Euler value. In the shortest panel examined ($a = 1000$ mm) local shell deformation dominates the symmetric $n = 1$ mode and the critical stress is only 42% of the Euler value (lower than the column buckling stress for $a = 1500$ mm). In the case of a panel of length 2670 mm, chosen to give approximate equality of local ($n = 7$) and column ($n = 1$) buckling stresses, the column mode is only slightly influenced by local deformation.

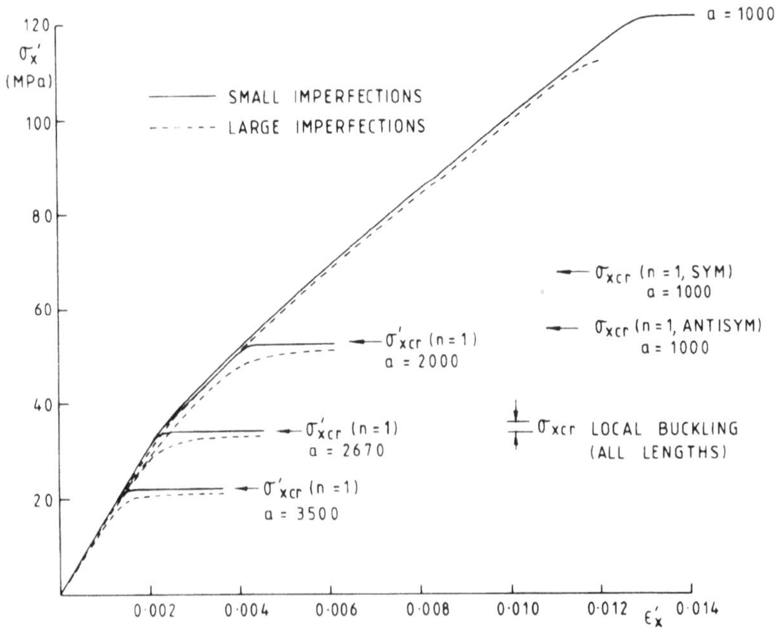

FIG. 5. Load-shortening curves for stiffened panels.

Non-linear and Post-buckling Behaviour

Non-linear load–deformation relationships have been computed for four panels with lengths $a = 3500, 2670, 2000$ and 1000 mm, supplementing results reported previously[2] for the experimental panel L2 ($a = 3060$ mm), which were found to correlate fairly well with test data. Computed load-shortening curves, i.e. average compressive stresses σ_x' plotted against average strains ε_x', are shown in Fig. 5. Mid-span lateral displacements of stiffeners are shown in Fig. 6: positive displacements (at section AB) are towards and negative displacements (at CD) away from the stiffener outstand.

For panels of length 3500, 2670 and 2000 mm, compressive loads reach their effective maxima at or just below initial overall buckling stress levels:

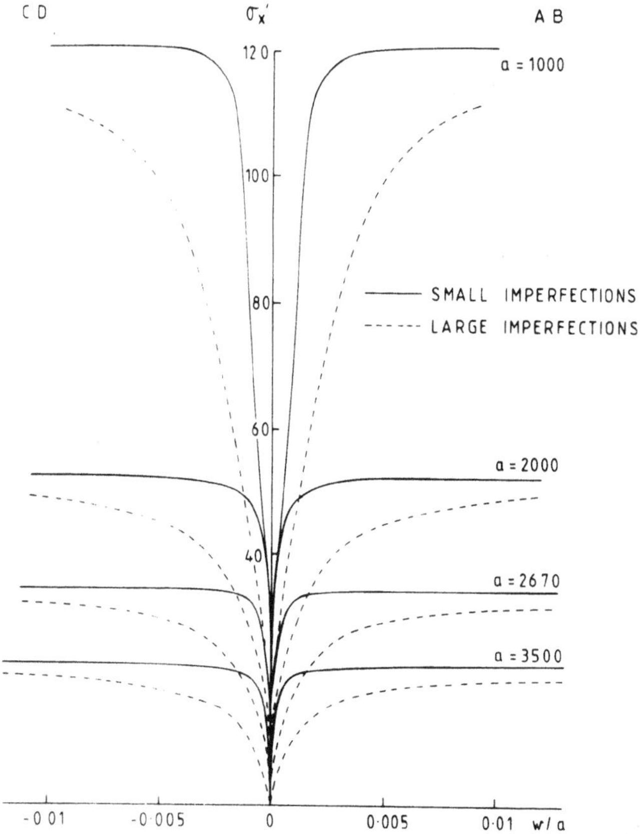

FIG. 6. Mid-span lateral displacement of stiffeners.

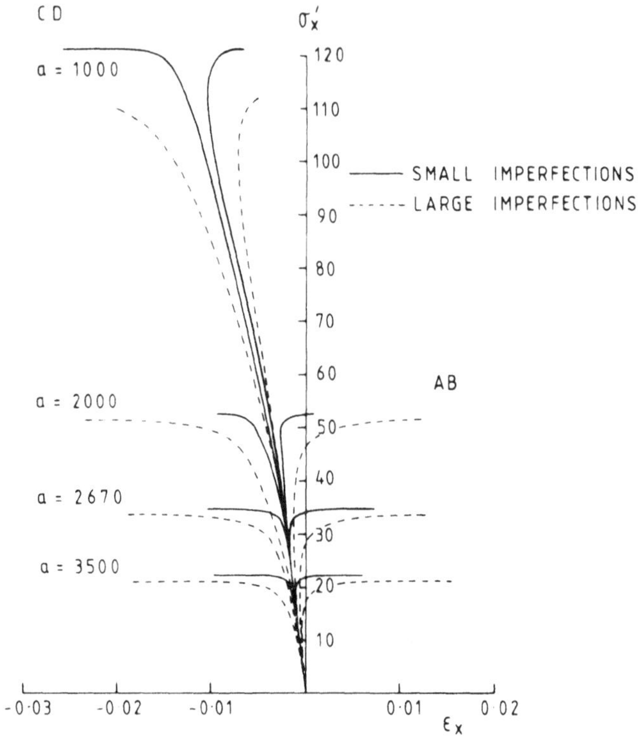

FIG. 7. Mid-span outer-fibre stiffener strains.

collapse might be expected to occur at these load levels as a result of outer-fibre material failure at a strain of about 0·02. Computed outer-fibre stiffener strains ε_x at mid-span positions (AB and CD) are shown for each panel in Fig. 7. Computed deformation of the longest panel at peak load is shown in Fig. 8.

In the shortest panel ($a = 1000$ mm), σ'_x reaches a plateau at about 120 MPa, substantially higher than the initial $n = 1$ buckling stresses but lower than the eqn. (8) Euler stress (161 MPa): it appears that this load corresponds to 'column' buckling in which overall flexural rigidity of the stiffened panel is reduced by loss of effective width caused by local buckling. Computed post-buckling deformations of the panel at average compressive stresses of 73 and 120 MPa are shown in Fig. 9. It is evident from this diagram and from Fig. 6 that significant 'column' displacements develop at the higher load level. Short-wavelength compressive buckling of the stiffener webs is also apparent.

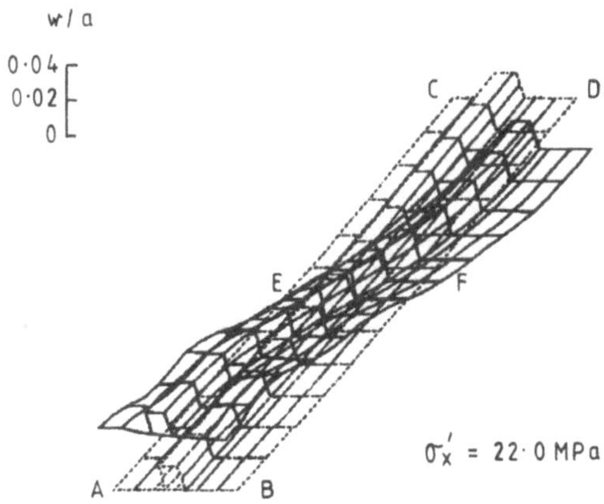

FIG. 8. Long panel ($a = 3500$ mm): computed deformation showing local buckling caused by secondary compression.

It may be noted from Fig. 5 that in the two shortest panels ($a = 2000$ and 1000 mm) occurrence of local buckling at an average stress of about 35 MPa results in a reduction of about 40% in the compressive stiffness ($d\sigma'_x/d\varepsilon'_x$) of each panel. A further reduction of about 10% in axial stiffness occurs at an average stress of about 65 MPa, corresponding approximately to the symmetric $n = 1$ buckling stress. Associated distributions of compressive stress over the cross-section and loss of effective width are indicated in Fig. 10. If the flexural rigidity EI of the cross-section is modified to reflect this loss of effective width, eqn. (8) gives a Euler buckling stress of 132 MPa, about 10% higher than the load plateau indicated by non-linear analysis.

Interactive Effects

Interactive buckling effects in a longitudinally stiffened FRP panel under longitudinal compression may be summarized as follows.

(i) Buckling and post-buckling behaviour will usually be strongly influenced by coupling between strips of laminate forming the shell and stiffeners: this effect invalidates the use of simple formulae for evaluation of local buckling stresses (except to provide crude upper and lower bounds) and can only be accounted for by use of folded-plate, finite-element or finite-strip analysis. In long panels for which the overall ($n = 1$) initial buckling stress is of the same order as or less than the local buckling stress and in which the preferred local buckling wavelength is a small fraction of the span a, interaction between overall and local initial buckling is slight

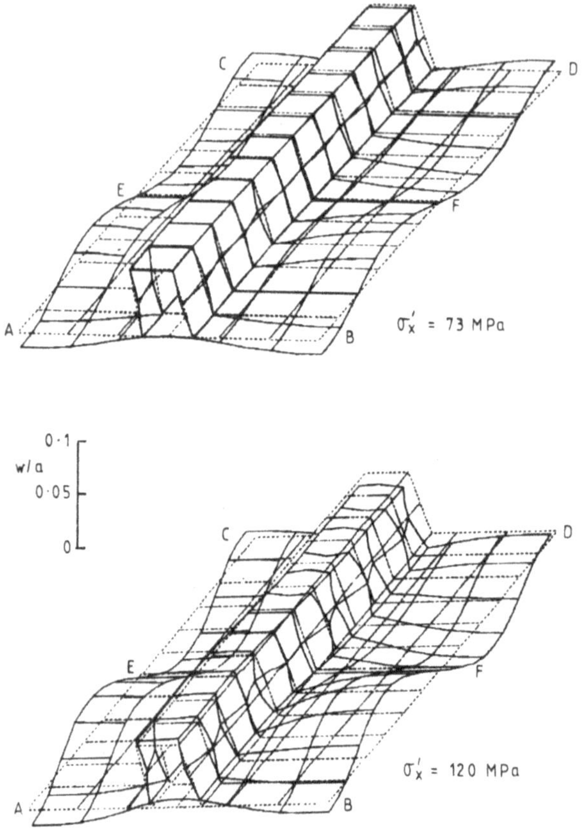

FIG. 9. Short panel ($a = 1000$ mm): computed post-buckling deformations.

and the former may be examined with reasonable accuracy using the Euler formula. In short panels, where the local initial buckling stress is substantially less than the overall buckling stress and the preferred number of buckling half-waves (n) is small, overall instability may be strongly influenced by local deformation and may occur at a stress substantially less than the Euler value.

(ii) Local buckling of the shell laminate between stiffeners is associated with loss of effective width, as illustrated in Fig. 10, causing an increase in stiffener stress which will tend to accelerate outer-fibre material failure, together with a reduction in the incremental (tangent) flexural rigidity of the cross-section affecting overall beam-column behaviour. This interactive

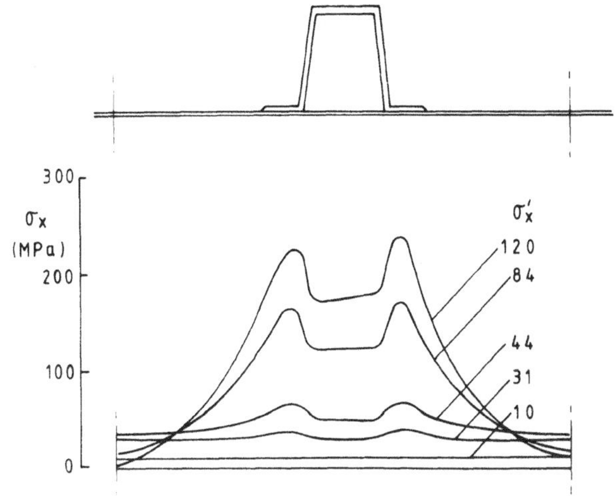

FIG. 10. Compressive stress distributions across shell laminate showing reduced effective width.

effect is likely to be most marked in short stiffened panels where local buckling precedes overall instability. It is part of the reason why computed load–displacement curves (Figs 5 and 6) for panels with $a = 2000$ mm and $a = 2670$ mm reach a plateau, where collapse is likely to occur as a result of outer-fibre material failure, somewhat below the initial overall buckling stress. In the case of the shortest panel examined ($a = 1000$ mm) the overall ($n = 1$) buckling mode is dominated by local deformation; despite loss of effective width associated with local buckling, computed load–displacement curves climb to a level well above the initial $n = 1$ buckling stress. It should be noted, however (as demonstrated in Ref. 2), that compressive failure of a short panel may be caused prematurely by debonding of stiffeners induced by local buckling. Failure may also be precipitated by outer-fibre material failure of the locally buckled shell laminate.

(iii) Bending of stiffeners associated with overall beam-column deformation may induce significant secondary compressive or tensile stress in the shell laminate between stiffeners. Where bending is towards the stiffener outstand such secondary stress is compressive, accelerating the occurrence of local buckling, loss of effective width and hence loss of overall flexural rigidity; where bending of stiffeners occurs away from the outstand, secondary stress in the shell is tensile, tending to inhibit local buckling and loss of effective width. This effect is illustrated in Fig. 8. A consequence is that lateral deformation of a multi-span stiffened panel tends to grow in an unsymmetrical manner with displacements towards the stiffener outstand exceeding displacements in the opposite direction. A two-span 'continuous beam' model, as shown in Fig. 3, is necessary to account for this form of behaviour in a numerical analysis. The effect is, however, found to be small for cases examined in this chapter: over the range of deformation shown in Fig. 6 the maximum difference between 'upward' and 'downward' displacements is 10%.

(iv) Approximate equality of critical stresses referring to local ($n = 7$) and column ($n = 1$) buckling modes (for the panel of length 2670 mm) does not appear to cause any special enhancement of the interaction effects discussed above.

DESIGN IMPLICATIONS

Current design recommendations for GRP ships[2] specify that factors of safety of 2·0 should be maintained against local and overall buckling of longitudinally stiffened panels under compressive load, reduced to 1·5 in the case of local buckling where positive measures, e.g. use of bolts or resilient adhesive with high peel strength, are employed to enhance stiffener attachment. As discussed in Ref. 2, these factors are intended to account for typical initial imperfections but not for creep or for variability and long-term degradation of material properties, which should be allowed for separately by application of supplementary partial safety factors.

Various forms of interaction between local and overall buckling have been identified in the present numerical study. For panels with cross-sectional geometry similar to that of Fig. 2 it appears that interaction effects are accounted for adequately by the margins indicated above and that no additional partial safety factor is necessary, even for the case where critical stresses referring to local and overall buckling modes are approximately equal. It seems likely that these conclusions will apply equally to most practical stiffened panels: scope clearly exists, however, for further exploration of interactive buckling effects in panels of differing geometry and for different patterns of initial deformation.

ACKNOWLEDGEMENT

Thanks are due to the authors' colleague, Mr W. C. Kirkwood, for assistance in carrying out computation.

REFERENCES

1. CHALMERS, D. W., OSBORN, R. J. and BUNNY, A., Hull construction of MCMVs in the United Kingdom, International Symposium on Mine Warfare Vessels and Systems, RINA, London, June 1984.
2. SMITH, C. S. and DOW, R. S., Compressive strength of longitudinally stiffened GRP panels, *Composite Structures—3* (Marshall, I. H. ed.). Elsevier Applied Science Publishers, London, 1985.
3. WITTRICK, W. H., Correlation between some stability problems for orthotropic and isotropic plates under biaxial and uniaxial direct stress, *Aeronaut. Quarterly*, **4**, No. 1 (1952), 83.
4. SMITH, C. S., Elastic analysis of stiffened plating under lateral loading, *Trans. RINA*, **108** (1966).
5. SMITH, C. S., Bending, buckling and vibration of orthotropic plate-beam structures, *J. Ship Res.*, **12**, No. 4 (1968), 249–268.
6. WITTRICK, W. H., A unified approach to the initial buckling of stiffened panels in compression, *Aeronaut. Quarterly*, **19** (1968), 265–283.
7. WITTRICK, W. H. and WILLIAMS, F. W., Buckling and vibration of anisotropic or isotropic plate assemblies under combined loadings, *Int. J. Mech. Sci.*, **16** (1974), 209–239.
8. VISWANATHAN, A. V., TAMEKUNI, M. and TRIPP, L. L., Elastic stability of biaxially loaded longitudinally stiffened composite structures, AIAA, ASME, SAE 14th Structures, Structural Dynamics and Materials Conference, Williamsburg, VA, 1973.
9. ATKINS R&D, ASAS-NL User Manual, Version 12, December 1984.

9

Post-buckled Stiffness of Rectangular Orthotropic Composite Laminates

Z. G. AZIZIAN and D. J. DAWE

Department of Civil Engineering, University of Birmingham, PO Box 363, Birmingham B15 2TT, UK

ABSTRACT

The finite strip method is used to study the immediate post-buckling behaviour of perfect, rectangular, orthotropic, symmetrically-laminated plates which are subjected to uniform end shortening. The analysis is based upon the use of first-order shear deformation plate theory. The loaded ends of the laminate are simply supported for out-of-plane behaviour and are free to expand in the plane of the plate. In numerical applications a particular type of finite strip is used in which all five reference displacement quantities are represented by cubic polynomial interpolation across the strip. Numerical studies reveal the good convergence properties of the strip model and show the result of including deformation effects in the analysis.

INTRODUCTION

Laminated composite rectangular plates are used frequently in situations where they are subjected to in-plane compressive loading and hence it is important to be able to predict their response to such loading. For perfect, flat laminates there is interest in knowing both the level of loading at which bifurcational instability, or buckling, occurs and the nature of the non-linear behaviour that follows buckling, particularly with regard to the membrane stiffness of the laminate immediately after buckling. The concern in the present paper is with demonstrating a method of determining buckling loads and immediate post-buckled stiffnesses of rectangular, orthotropic, symmetrically-laminated plates subjected to uniform end shortening. The loaded ends of such laminates are assumed to be simply supported for out-of-plane behaviour and free to expand in-plane.

In any analysis of laminates the choice of the theoretical model of plate behaviour is important. The great majority of past studies of the subject problem have been based on the use of the classical plate theory (CPT). Following the pioneering works of Levy[1] and Yamaki[2] for homogeneous isotropic plates, numerous CPT studies of the post-buckling behaviour of rectangular laminates have been made. These latter studies will not be itemised here but it is noted that many are detailed in the book by Chia[3] and that the paper by Harris[4] provides useful results which will be used for comparative purposes later in the present paper. The problem with the CPT is, of course, that due to the imposition of the Kirchhoff normalcy condition the laminate stiffness can be significantly overestimated. The first-order shear deformation plate theory (SDPT) relaxes the normalcy condition and hence accounts for the through-thickness shear deformation effects which can be of particular importance for laminates. In linear analyses it has been demonstrated clearly, for example in Refs 5–7, that it is often necessary to employ SDPT in preference to CPT so as to obtain a realistic theoretical model of plate behaviour. In non-linear analysis there is a paucity of information on this matter, though earlier work by the authors concerned with the large-deflection analysis of isotropic plates[8,9] and of laminates[10] again indicates the improvement associated with the use of SDPT.

In this paper description is given of the analysis of the immediate post-buckling behaviour of rectangular orthotropic laminates through use of the finite strip method (FSM) in the context of SDPT. This study follows in quite direct fashion the earlier large-deflection studies[8–10] though there are basic changes in the strip displacement field, of course, and furthermore attention is confined here to one particular strip model, with cubic interpolation of displacement quantities across a strip. Within the context of SDPT, and for laminates, the FSM does not appear to have been used heretofore for the analysis of post-buckling behaviour, though it is noted that CPT finite strip analyses of the post-buckling behaviour of isotropic plates have been made by Graves Smith and Sridharan[11] and by Hancock.[12] The present authors have also developed a CPT finite strip analysis of the post-buckling behaviour of rectangular orthotropic laminates and results based on the use of this are used for comparison purposes herein.

ANALYSIS

A flat rectangular plate, typically assumed to be composed of a number, *nl*, of layers of unidirectional fibre-reinforced composite material, is shown in Fig. 1. The plate edges running parallel to the y-axis (i.e. the plate ends) are assumed to be subjected to an in-plane compressive loading acting through rigid platens such that uniform end shortening occurs whilst free lateral expansion is allowed: these plate ends are simply supported. The plate edges running parallel to the x-axis (i.e. the longitudinal edges) are assumed to be free to move in the plane of the plate and to have arbitrary boundary conditions so far as out-of-plane behaviour is concerned.

When expanded to include extension/compression of the middle surface the basic assumptions for displacement behaviour in SDPT become[13-15]

$$\begin{aligned} \bar{u}(x,y,z) &= u(x,y) + z\psi_x(x,y) \\ \bar{v}(x,y,z) &= v(x,y) + z\psi_y(x,y) \\ \bar{w}(x,y,z) &= w(x,y) \end{aligned} \tag{1}$$

Here $\bar{u}$, $\bar{v}$ and $\bar{w}$ are the displacement components at a general point in the plate whilst u, v and w are these components for a point lying in the plate middle surface ($z = 0$). The quantities ψ_x and ψ_y are the rotations of initial plate normals along the x- and y-directions.

Substitution of eqns (1) into the standard Green's expressions for non-linear strains, and neglecting certain small-order terms,[8] gives the following non-linear strain-displacement relationships:

$$\begin{Bmatrix} \varepsilon_x \\ \varepsilon_y \\ \gamma_{yz} \\ \gamma_{zx} \\ \gamma_{xy} \end{Bmatrix} = \begin{Bmatrix} \dfrac{\partial u}{\partial x} + z\dfrac{\partial \psi_x}{\partial x} + \dfrac{1}{2}\left(\dfrac{\partial w}{\partial x}\right)^2 \\ \dfrac{\partial v}{\partial y} + z\dfrac{\partial \psi_y}{\partial y} + \dfrac{1}{2}\left(\dfrac{\partial w}{\partial y}\right)^2 \\ \psi_y + \dfrac{\partial w}{\partial y} \\ \psi_x + \dfrac{\partial w}{\partial x} \\ \dfrac{\partial u}{\partial y} + \dfrac{\partial v}{\partial x} + z\left(\dfrac{\partial \psi_x}{\partial y} + \dfrac{\partial \psi_y}{\partial x}\right) + \dfrac{\partial w}{\partial x}\dfrac{\partial w}{\partial y} \end{Bmatrix} \tag{2}$$

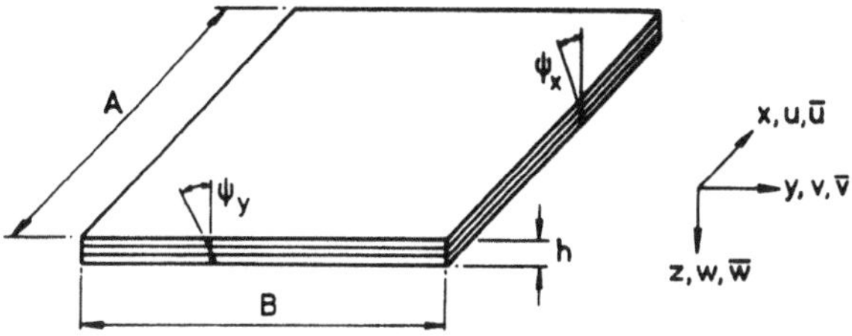

FIG. 1. A laminated rectangular plate.

Attention is now restricted to plates which are symmetrically laminated with respect to the middle surface and which have orthotropic properties. In this circumstance the strain energy, ΔU_p, per unit middle surface area of the laminate is

$$\Delta U_p = \frac{1}{2}\left[A_{11}\left(\frac{\partial u}{\partial x}\right)^2 + A_{22}\left(\frac{\partial v}{\partial y}\right)^2 + 2A_{12}\frac{\partial u}{\partial x}\frac{\partial v}{\partial y} + A_{66}\left(\frac{\partial u}{\partial y}+\frac{\partial v}{\partial x}\right)^2\right.$$
$$+ D_{11}\left(\frac{\partial \psi_x}{\partial x}\right)^2 + D_{22}\left(\frac{\partial \psi_y}{\partial y}\right)^2 + 2D_{12}\frac{\partial \psi_x}{\partial x}\frac{\partial \psi_y}{\partial y} + D_{66}\left(\frac{\partial \psi_x}{\partial y}+\frac{\partial \psi_y}{\partial x}\right)^2$$
$$\left. + A_{44}\left(\psi_y + \frac{\partial w}{\partial y}\right)^2 + A_{55}\left(\psi_x + \frac{\partial w}{\partial x}\right)^2\right]$$
$$+\frac{1}{2}\left[A_{11}\frac{\partial u}{\partial x}\left(\frac{\partial w}{\partial x}\right)^2 + A_{22}\frac{\partial v}{\partial y}\left(\frac{\partial w}{\partial y}\right)^2 + A_{12}\frac{\partial u}{\partial x}\left(\frac{\partial w}{\partial y}\right)^2\right.$$
$$\left. + A_{12}\frac{\partial v}{\partial y}\left(\frac{\partial w}{\partial x}\right)^2 + 2A_{66}\left(\frac{\partial u}{\partial y}+\frac{\partial v}{\partial x}\right)\frac{\partial w}{\partial x}\frac{\partial w}{\partial y}\right]$$
$$+\frac{1}{2}\left[\frac{1}{4}A_{11}\left(\frac{\partial w}{\partial x}\right)^4 + \frac{1}{4}A_{22}\left(\frac{\partial w}{\partial y}\right)^4 + \frac{1}{2}A_{12}\left(\frac{\partial w}{\partial x}\right)^2\left(\frac{\partial w}{\partial y}\right)^2\right.$$
$$\left. + A_{66}\left(\frac{\partial w}{\partial x}\right)^2\left(\frac{\partial w}{\partial y}\right)^2\right] \qquad (3)$$

In this equation the stiffness coefficients are defined as

$$A_{ij} = \sum_{l=1}^{nl}(Q_{ij})_l(h_l - h_{l-1}),\ D_{ij} = \frac{1}{3}\sum_{l=1}^{nl}(Q_{ij})_l(h_l^3 - h_{l-1}^3) \qquad i,j = 1,2,6 \quad (4)$$

$$A_{ij} = k_i k_j \sum_{l=1}^{nl}(Q_{ij})_l(h_l - h_{l-1}) \qquad i,j = 4,5 \quad (5)$$

where Q_{ij} for $i,j = 1, 2, 6$ are plane-stress reduced stiffnesses and Q_{ij} for $i,j = 4, 5$ are transverse shear stiffnesses.

The subscript l associated with Q_{ij} refers to the lth layer, and h_l is the distance from the plate middle surface to the lower surface of the lth layer. The parameters $k_i k_j$ in eqn. (5) are the prescribed shear correction factors, introduced to allow for the fact that the transverse shear strain distributions are not uniform through the plate thickness.

It is clear from eqn. (3) that the strain energy comprises three distinct contributions that depend upon quadratic, cubic and quartic functions of the five displacement-type quantities and their derivatives.

The concern here is with the response of the plate to a prescribed end shortening strain and the potential energy of the associated end load will thus be independent of the unprescribed degrees of freedom of the problem. Therefore the total potential energy is equal to the strain energy, i.e. is equal to the integral of ΔU_p over the middle surface.

To implement the Principle of Minimum Potential Energy requires the assumption of a displacement field consisting of separate expressions for u, v, w, ψ_x and ψ_y over the middle surface. In this study the finite strip method is used in which a localised displacement field is assumed for each strip. Detailed description of the FSM, in the context of SDPT, is available elsewhere[5-10] and so only brief description of points pertinent to the subject problem will be given here.

Under the action of the end shortening the type of plate considered here (of orthotropic material and with simply-supported ends) will buckle in a mode having q longitudinal half-waves with the half-wavelength $\lambda = A/q$: the corresponding nodal lines running across the plate will be straight and parallel to the plate ends and conditions at all such lines will repeat those at

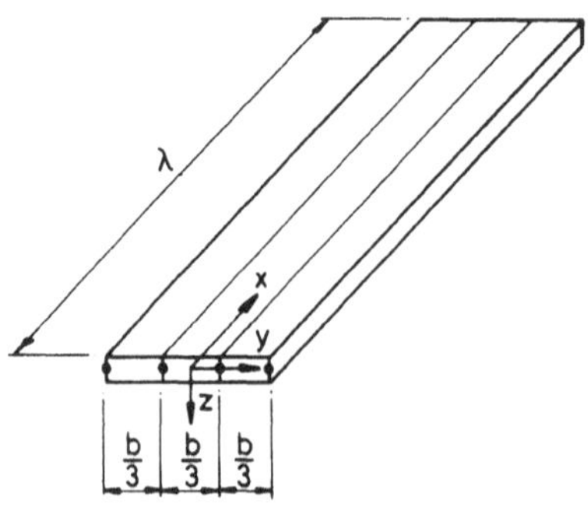

FIG. 2. A finite strip, based upon cubic crosswise interpolation.

the plate ends. Thus it is only necessary, in using the FSM, to consider a length of plate equal to λ. Figure 2 shows a typical finite strip which spans a length λ in the x-direction between lateral nodal lines and is of width b which is some portion of the plate width B.

For the particular post-buckling application considered in this study the strip displacement field is assumed to be of the form

$$
\begin{aligned}
u &= \varepsilon\left(\frac{\lambda}{2} - x\right) + \sin\frac{2\pi x}{\lambda}\sum u_m N_m(y) \\
v &= \alpha\varepsilon y + \sum v_{m0} N_m(y) + \cos\frac{2\pi x}{\lambda}\sum v_{m2} N_m(y) \\
w &= \sin\frac{\pi x}{\lambda}\sum w_m N_m(y) \\
\psi_y &= \sin\frac{\pi x}{\lambda}\sum \psi_{ym} N_m(y) \\
\psi_x &= \cos\frac{\pi x}{\lambda}\sum \psi_{xm} N_m(y)
\end{aligned}
\tag{6}
$$

Here ε is the prescribed end shortening strain and α is the appropriate value of Poisson's ratio, i.e. $\alpha = A_{12}/A_{22}$, to account for lateral plate expansion arising from the end shortening. All the indicated summations run from $m = 1$ to $m = n + 1$, where n is the assumed order of crosswise polynomial interpolation. The $N_m(y)$ are the shape functions for the m reference lines equi-spaced across a strip and the u_m, v_{m0}, v_{m2}, w_m, ψ_{ym} and ψ_{xm} are the corresponding displacement degrees of freedom of the strip. Although the post-buckling analysis has been programmed to allow a broad choice of the order of crosswise interpolation, in the applications described later this order is fixed at $n = 3$, i.e. cubic interpolation, and the strip thus has four reference lines as indicated in Fig. 2.

The assumed displacement field satisfies the desired conditions at the lateral nodal lines, namely

$$u = \pm\lambda\varepsilon/2,\ w = \psi_y = 0,\ v \text{ and } \psi_x \text{ free} \tag{7}$$

It is clear from eqns (6) that only a single longitudinal sine or cosine term is associated with each displacement quantity. This is sufficient for the present study of immediate post-buckled behaviour but for more advanced post-buckled behaviour extra terms would be desirable and the author's computer program in fact allows for this.

By substituting eqn. (6) into eqn. (5), integrating over the middle surface of the strip and recalling that the total potential energy equals the strain energy, it is ultimately possible to express the total potential energy of a strip, Π_p, as

$$\Pi_p = \tfrac{1}{2}\mathbf{d}^T(\mathbf{K} - \varepsilon\mathbf{K}^*)\mathbf{d} + \tfrac{1}{6}\mathbf{d}^T\mathbf{K}_1\mathbf{d} + \tfrac{1}{12}\mathbf{d}^T\mathbf{K}_2\mathbf{d} \tag{8}$$

Here $\mathbf{K}, \mathbf{K}_1, \mathbf{K}_2$ and $\mathbf{K}^*$ are square matrices which are each of order 24 when the displacement field is given by eqns (6) and cubic polynomial interpolation is used across a strip. The coefficients of $\mathbf{K}$ and $\mathbf{K}^*$ are constants whilst those of $\mathbf{K}_1$ and $\mathbf{K}_2$ are linear and quadratic functions, respectively, of the displacements. The column matrix $\mathbf{d}$ is the list of the strip degrees of freedom indicated in eqn. (6), i.e. it comprises displacements $u_1, v_{10}, v_{12}, w_1, \psi_{y1}, \psi_{x1}, u_2, v_{20}, \ldots, \psi_{x4}$.

In evaluating Π_p all integrations in the x-direction, of the trigonometric terms, are determined analytically whilst integrations in the y-direction are determined numerically using Gauss quadrature: however the quadrature scheme uses seven points across a strip and hence is effectively exact.

For the whole plate, comprising an assembly of finite strips, the total potential energy is obtained by summing the potential energies of the strips. After applying any appropriate zero-displacement boundary condition at longitudinal plate edges the plate potential energy is minimised with respect to each plate degree of freedom in turn and the result is the system of plate equilibrium equations

$$(\bar{\mathbf{K}} - \varepsilon\bar{\mathbf{K}}^* + \tfrac{1}{2}\bar{\mathbf{K}}_1 + \tfrac{1}{3}\bar{\mathbf{K}}_2)\bar{\mathbf{d}} = \mathbf{0} \tag{9}$$

The square matrices $\bar{\mathbf{K}}$, $\bar{\mathbf{K}}^*$, $\bar{\mathbf{K}}_1$ and $\bar{\mathbf{K}}_2$ and the column matrix $\bar{\mathbf{d}}$ are the whole plate equivalents of the individual-strip matrices $\mathbf{K}, \mathbf{K}^*, \mathbf{K}_1, \mathbf{K}_2$ and $\mathbf{d}$ respectively.

The system of equations given by eqn. (9) is, of course, non-linear in the post-buckled configuration, through the presence of the $\bar{\mathbf{K}}_1$ and $\bar{\mathbf{K}}_2$ matrices, and for a prescribed value of ε can only be solved for $\bar{\mathbf{d}}$ by an iterative procedure.

In the present study the concern is with the behaviour of ideal plates which are initially perfectly flat. In this circumstance the immediate post-buckled membrane stiffness is determined using the following procedure.

On applying increasing uniform end shortening from the initial unloaded state the plate remains perfectly flat until a critical end-shortening strain ε_c is reached at which bifurcational instability occurs. Up to this stage the u and v components of displacement are given simply by the terms dependent upon ε in eqns (6) and thus the u and v degrees of freedom are all zero as, of course, in the pre-buckled state are the w, ψ_x and ψ_y displacements. At buckling these latter displacements come into consideration and eqn. (9) reduces to

$$(\bar{\mathbf{K}}_0 - \varepsilon\bar{\mathbf{K}}_0^*)\bar{\mathbf{d}}_0 = \mathbf{0} \tag{10}$$

where the subscript 0 indicates that only out-of-plane displacements (w, ψ_x and ψ_y) are involved. This constitutes a standard linear eigenvalue problem which can be solved by a number of methods to yield the critical value ε_c and the associated mode shape of buckling, with freedoms $\bar{\mathbf{d}}_{0c}$. With ε_c known the actual physical end shortening over the half-wavelength is simply $e_c = \lambda\varepsilon_c$, the critical end load is $P_c = (A_{11} - A_{12}^2/A_{22})B\varepsilon_c$ and the pre-buckled membrane stiffness is $S = P_c/e_c$.

The end shortening strain is now prescribed at a value which is very slightly in excess of the critical value and, in fact, in generating the results presented in the next section the prescribed value is $\varepsilon_p = 1{\cdot}01\varepsilon_c$. This prescribed value is substituted into the non-linear system of equilibrium equations (eqn. (9)) together with an estimate of the post-buckled deformation pattern (i.e. of $\bar{\mathbf{d}}$) at this prescribed level of end-shortening strain, such estimate being based on the knowledge of the buckled mode. Solution of the non-linear system of equations, to yield the degrees of freedom $\bar{\mathbf{d}}$, is achieved, to within prescribed acceptable limits, using the Newton–Raphson iterative procedure. The tangent stiffness matrix $\bar{\mathbf{K}}_T$, which relates incremental forces to incremental displacements, is used in doing this. This matrix is given by

$$\bar{\mathbf{K}}_T = \bar{\mathbf{K}} - \varepsilon\bar{\mathbf{K}}^* + \bar{\mathbf{K}}_1 + \bar{\mathbf{K}}_2 \tag{11}$$

Once $\bar{\mathbf{d}}$ is known the end load P_p corresponding to the prescribed end-shortening strain ε_p is determined through integration of the derived longitudinal stress distribution over the plate cross-section. Then the immediate post-buckled membrane stiffness is defined simply as $S^* = (P_p - P_c)/(e_p - e_c)$.

NUMERICAL APPLICATIONS

All the results presented here relate to square plates of side length A. In using the finite strip method the conditions at the loaded plate ends ($x = 0$ and $x = A$) correspond to simply-supported ends so far as out-of-plane behaviour is concerned. At the unloaded edges four combinations of boundary condition are applied in turn. For out-of-plane behaviour both such edges can be either simply supported or clamped, when the plates are referred to as SSSS and SCSC plates respectively. For in-plane behaviour both edges are either completely free to move (i.e. are free to wave) or are free to move subject to the constraint that the edges remain straight.

The results presented are for values of a buckling factor K and for the ratio of immediate post-buckled membrane stiffness to pre-buckled stiffness, i.e. the ratio S^*/S. This ratio is denoted $(S^*/S)_1$ when the unloaded edges are free to wave in-plane and $(S^*/S)_2$ when the unloaded edges move in-plane but remain straight.

Consideration is given first to plates of homogeneous, isotropic material and then to symmetrically-laminated, cross-ply plates having orthotropic properties with respect to the plate axes. For all plates comparative results are presented which are based on the use of another type of finite strip analysis, this time within the context of the CPT. Details of this CPT FSM analysis will not be given here, except to say that the strip displacement field embraces cubic polynomial variations of u, v and w across the strip, and that quoted CPT FSM values are based on the use of sufficient strips (12 in the half-plate) to expect that full convergence has been achieved within very close limits. Where available, other CPT comparative results are also quoted.

Isotropic Plates

Two relative plate thicknesses are considered, corresponding to $A/h = 10$ and $A/h = 1000$. Of course, results presented for the thicker plate are rather hypothetical since elastic buckling of such a thick plate would not occur in practice, being preceded by material yielding.

The buckling factor K for the isotropic plates is defined as

$$K = 12(1 - \nu^2)A^2\sigma_c/\pi^2h^2E$$

where σ_c is the critical stress. The shear correction factors used in generating the SDPT results are $k_4^2 = k_5^2 = 5/6$.

In Table 1 results are given for the SSSS plates ($q = 1$) in the form of a study of the convergence of SDPT FSM values of K and S^*/S with respect to the number, n_s, of strips used in a symmetric half-width of the plate. It can be seen that convergence is very rapid, to such an extent that the two-strip solutions are almost exact. It is noted that no bound conditions are expected to apply to values of (S^*/S) whereas K values should, and do, converge from above with increasing n_s. The results given in Table 1 reveal that the calculated buckling factor is, of course, affected by the choice of plate theory, as evidenced by a 5·6% reduction in K for the thick plate, as compared to the CPT value. (All quoted CPT values in Tables 1–4 apply to

TABLE 1
Buckling factor and immediate post-buckled stiffness of SSSS square, isotropic plates: convergence of SDPT FSM results

A/h	n_s	K	$(S^*/S)_1$	$(S^*/S)_2$
10	1	3·789 7	0·414 0	0·502 2
	2	3·786 5	0·408 2	0·500 0
	3	3·786 5	0·408 2	0·500 0
	4	3·786 5	0·408 0	0·499 8
	5	3·786 5	0·408 0	0·499 8
100	1	4·008 6	0·416 8	0·503 5
	2	4·000 5	0·408 3	0·500 0
	3	4·000 1	0·408 1	0·499 9
	4	4·000 0	0·408 2	0·500 0
	5	4·000 0	0·408 2	0·500 0
CPT solution[2,4]		4·000 0	0·408	0·500 0
CPT FSM solution		4·000 0	0·408 2	0·500 0

any relative plate thickness.) However, the immediate post-buckled stiffness is not affected by the choice of (out-of-plane) plate theory for the SSSS plates, since this quantity is dependent only upon in-plane properties.[4] There is a significant difference in the values $(S^*/S)_1$ and $(S^*/S)_2$.

Results of a similar study for SCSC plates are given in Table 2 ($q = 2$). Again convergence of the SDPT FSM results is rapid, though less so than for the SSSS plates. The results given in Table 2 show that the predicted immediate post-buckled stiffness of SCSC plates (as well as the buckling factor) is affected by the choice of plate theory. For the thick SCSC plate the neglect of transverse shear effects (as in CPT) leads to predictions of immediate post-buckled stiffness which are almost 6% greater than are those based on the use of the SDPT. Correspondingly the overestimate of K is just over 20% when using the CPT. For the SCSC plates there is little difference between the values $(S^*/S)_1$ and $(S^*/S)_2$.

Orthotropic Laminates

The plates considered here are five-layer, orthotropic, cross-ply, square laminates. All plies have the same thickness and identical material properties. The latter correspond to a high-modulus fibre composite with

$$E_L/E_T = 30, \; G_{LT}/E_T = 0{\cdot}6, \; G_{TT}/E_T = 0{\cdot}5 \qquad \text{and} \qquad \nu_{LT} = 0{\cdot}25$$

TABLE 2
Buckling factor and immediate post-buckled stiffness of SCSC square, isotropic plates: convergence of SDPT FSM results

A/h	n_s	K	$(S^*/S)_1$	$(S^*/S)_2$
10	1	6·432 6	0·441 8	0·442 3
	2	6·387 0	0·421 9	0·423 2
	3	6·384 4	0·421 1	0·422 8
	4	6·383 9	0·421 3	0·422 6
	5	6·383 7	0·421 3	0·422 6
1 000	1	7·945 2	0·504 6	0·504 8
	2	7·716 3	0·449 4	0·450 4
	3	7·697 2	0·446 4	0·447 5
	4	7·693 2	0·445 9	0·447 1
	5	7·692 0	0·445 8	0·446 9
CPT FSM solution		7·691 3	0·445 8	0·446 9

where subscripts L and T refer to directions along and transverse to the fibres, respectively.

Relative to the plate axes, two types of ply arrangement are used in turn: these are a 0°/90°/0°/90°/0° laminate and a 90°/0°/90°/0°/90° laminate, where angles are measured from the x-axis of the plate. Shear correction factors used in the SDPT FSM analyses are calculated by the method of Chow[16] and are $k_4^2 = 0{\cdot}554\,11$, $k_5^2 = 0{\cdot}858\,57$ for the 0°/90°/0°/90°/0° laminate and vice versa for the 90°/0°/90°/0°/90° laminate. The buckling factor is defined as

$$K = \sigma_c A^2 h / \pi^2 D_{11}$$

Rather than conducting a further convergence study, the finite strip modelling is fixed and the relative thickness is changed in several stages between $A/h = 10$ and $A/h = 1000$. Guided by the results obtained in the convergence study for isotropic plates the strip modelling here is selected to correspond to $n_s = 4$ for SSSS plates and $n_s = 5$ for SCSC plates. Results calculated for K and S^*/S are given in Tables 3 and 4. It should be noted that the buckled (and immediately post-buckled) mode has two longitudinal halfwaves (i.e. $q = 2$) for the SCSC 90°/0°/90°/0°/90° laminates and, with a single exception, one for the other laminates. The exception is that for the SSSS 90°/0°/90°/0°/90° laminate with $A/h = 10$ the SDPT FSM prediction is that initial buckling corresponds to $q = 2$, and this should be

TABLE 3
Buckling factor and immediate post-buckled stiffness of 0°/90°/0°/90°/0° orthotropic square laminates

A/h	SSSS plate			SCSC plate		
	K	$(S^*/S)_1$	$(S^*/S)_2$	K	$(S^*/S)_1$	$(S^*/S)_2$
10	1·024 9	0·350 0	0·457 3	1·517 0	0·422 1	0·513 2
20	1·290 4	0·349 9	0·457 2	2·223 0	0·462 5	0·553 1
30	1·356 3	0·350 0	0·457 3	2·443 2	0·474 4	0·566 2
50	1·392 7	0·350 0	0·457 3	2·575 9	0·482 1	0·574 8
100	1·408 8	0·350 0	0·457 2	2·636 9	0·485 4	0·578 5
1 000	1·414 1	0·350 0	0·457 2	2·657 8	0·486 3	0·579 6
CPT solution[4]	1·414 2	—	0·457 2	—	—	—
CPT FSM solution	1·414 2	0·349 9	0·457 3	2·657 9	0·486 7	0·580 0

TABLE 4
Buckling factor and immediate post-buckled stiffness of 90°/0°/90°/0°/90° orthotropic, square laminates

A/h	SSSS plate			SCSC plate		
	K	$(S^*/S)_1$	$(S^*/S)_2$	K	$(S^*/S)_1$	$(S^*/S)_2$
10	3·016 7[a]	0·338 2[a]	0·352 7[a]	3·751 1	0·382 0	0·394 1
20	4·398 4	0·357 6	0·551 6	6·493 1	0·425 7	0·436 9
30	4·662 7	0·357 6	0·551 5	7·608 5	0·444 9	0·456 1
50	4·747 0	0·357 8	0·551 7	8·374 1	0·457 9	0·469 4
100	4·801 6	0·357 3	0·551 2	8·754 5	0·464 3	0·475 9
1 000	4·820 0	0·357 9	0·551 8	8·889 4	0·466 4	0·478 1
CPT solution[4]	4·820 1	—	0·551 6	—	—	—
CPT FSM solution	4·820 1	0·357 4	0·551 3	8·890 5	0·466 5	0·478 2

[a] These results for the SSSS plate with $A/h = 10$ correspond to two longitudinal half-waves: for one half-wave the equivalent values are $K = 3{\cdot}493\,4$, $(S/S^*)_1 = 0{\cdot}357\,9$ and $(S^*/S)_2 = 0{\cdot}551\,9$.

borne in mind when considering the quoted results for the SSSS plate in Table 4.

The results given in Tables 3 and 4 show, as expected, that the neglect of transverse shear effects in CPT leads to significant overestimation of the buckling factor for each of the four types of laminate when the laminate is other than very thin. For the SSSS laminates the immediate post-buckled stiffness is independent of the plate theory used, whilst for the SCSC laminates this quantity does depend quite significantly on the plate theory. For example, for the SCSC 90°/0°/90°/0°/90° laminate the CPT solution for $(S^*/S)_1$ overestimates the immediate post-buckled stiffness by 22·1, 9·6, 4·8 and 1·9% for values of $A/h = 10$, 20, 30 and 50 respectively.

CONCLUSIONS

The finite strip method, used in the context of shear deformation plate theory, has been shown to be an efficient procedure for determining the buckling load and the immediate post-buckled membrane stiffness of symmetric, orthotropic, rectangular laminates with simply-supported ends when subjected to uniform end shortening. The described procedures can be, and are being, extended to accommodate more general situations, including the analysis of prismatic plate assemblies.

The inclusions of through-thickness shear deformation effects in the mathematical model of plate behaviour does affect the calculated value of the buckling factor, often very significantly, whatever boundary conditions apply at the longitudinal edges. On the other hand it has been seen that if the longitudinal edges are simply supported there is no shear deformation effect on the immediate post-buckled membrane stiffness (unless this effect changes the form of the buckled mode). Such is not the case when the longitudinal edges are clamped, for then it is clear that there can be a significant reduction in the calculated immediate post-buckled stiffness when shear effects are taken into account.

ACKNOWLEDGEMENTS

The authors are pleased to acknowledge that the work described here has been supported financially both by the Science and Engineering Research Council and by Procurement Executive, Ministry of Defence.

REFERENCES

1. LEVY, S., Bending of rectangular plates with large deflection, NACA, TR No. 737, 1942.
2. YAMAKI, N., Post-buckling behaviour of rectangular plates with small initial curvature loaded in edge compression, *J. appl. Mech.*, **26** (1959), 407–414.
3. CHIA, C. Y., *Nonlinear Analysis of Plates*, New York, McGraw-Hill, 1980.
4. HARRIS, G. Z., The buckling and post-buckling behaviour of composite plates under biaxial loading, *Int. J. Mech. Sci.*, **17** (1975), 187–202.
5. CRAIG, T. J. and DAWE, D. J., Flexural vibration of symmetrically-laminated composite, rectangular plates, including transverse shear effects, *Int. J. Solids Struct.*, **22** (1986), 155–169.
6. DAWE, D. J. and CRAIG, T. J., The influence of shear deformation on the natural frequencies of laminated rectangular plates, in: *Composite Structures—3* (I. H. Marshall ed.), London, Elsevier Applied Science, 1985, pp. 660–676.
7. DAWE, D. J. AND CRAIG, T. J., The vibration and stability of symmetrically-laminated composite rectangular plates subjected to in-plane stresses, *Composite Structures*, **5** (1986), 281–307.
8. AZIZIAN, Z. G. and DAWE, D. J., Geometrically non-linear analysis of rectangular Mindlin plates using the finite strip method, *Computers and Structures*, **21** (1985), 423–436.
9. DAWE, D. J. and AZIZIAN, Z. G., The performance of Mindlin plate finite strips in geometrically nonlinear analysis, *Computers and Structures*, **23** (1986), 1–14.
10. AZIZIAN, Z. G. AND DAWE, D. J., Analysis of the large deflection behaviour of laminated composite plates using the finite strip method, in: *Composite Structures—3* (I. H. Marshall ed.), London, Elsevier Applied Science, 1985, pp. 677–691.
11. GRAVES SMITH, T. R. and SRIDHARAN, S., A finite strip method for the post-locally buckled analysis of plate structures, *Int. J. Mech. Sci.*, **20** (1978), 833–842.
12. HANCOCK, G. J., Nonlinear analysis of thin sections in compression, *J. Struct. Div., ASCE*, **107** (ST3), (1981), 455–471.
13. MINDLIN, R. D., Influence of rotary inertia and shear on flexural motions of isotropic elastic plates, *J. appl. Mech.*, **18** (1951), 31–38.
14. YANG, P. C., NORRIS, C. H. and STAVSKY, Y., Elastic wave propagation in heterogeneous plates, *Int. J. Solids Struct.*, **2** (1966), 665–684.
15. WHITNEY, J. M. and PAGANO, N. J., Shear deformation in heterogeneous anisotropic plates, *J. appl. Mech.*, **37** (1970), 1031–1036.
16. CHOW, T. S., On the propagation of flexural waves in an orthotropic laminated plate and its response to an impulsive load, *J. Comp. Mater.*, **5** (1971), 306–319.

10

Large Deflection of Anisotropic Bimodular Composite Material Plates

FARAMARZ GORDANINEJAD

Mechanical Engineering Department, University of Nevada—Reno, Reno, Nevada 89557, USA

ABSTRACT

This chapter presents the results for geometrically nonlinear bending of thin, anisotropic, single-layer, rectangular plates constructed of bimodular materials. The classical plate theory, which accounts for the bimodular action and large rotations, is employed. Finite element results for transverse deflection are presented for simply-supported, graphite/epoxy and polyester-rubber plates under uniform distributed load. Also, the effects of the aspect ratio and bimodularity ratio as well as the angle of fiber orientation on the maximum transverse deflection are discussed.

INTRODUCTION

Large deflection of composite structures constructed of bimodular materials has been of interest for more than a decade. Bimodular materials are those which exhibit different behaviors under tensile and compressive loads. The actual nonlinear stress–strain behavior of these materials are approximated with bilinear stress–strain curves.

The research on the nonlinear bending of bimodular material structures started with the work of Kamiya.[1,2] He analyzed the large deflection of thin circular and rectangular isotropic bimodular plates. The study of nonlinear flexure of orthotropic bimodular composite strips using the dynamic relaxation method is due to Turvey.[3] Reddy and Chao[4] addressed the large deflection of laminated orthotropic, bimodular material plates using a displacement finite element method. In the studies mentioned above, either isotropic or orthotropic plates have been analyzed. Recently, a small-deformation analysis of single-layer, anisotropic plates was presented by Gordaninejad.[5]

In this chapter the analysis of large deflection of thin anisotropic, single-layer, rectangular, bimodular composite material plates is presented. Bert's 'fiber-governed compliance' model[6] is used to determine the neutral surface position (NSP). Numerical results for transverse deflection are presented for a simply-supported plate under uniform distributed load, and the effects of plate aspect ratio, bimodularity ratio and angle of fiber orientation on the maximum deflection are discussed.

THEORY AND FORMULATION

The analysis of structures constructed of bimodular materials are more difficult than those made of unimodular (ordinary) materials. The calculations become even more complicated when the plate is assumed to be anisotropic. In this study, a displacement finite element model is implemented since obtaining a closed-form solution for nonlinear analysis is not possible. The finite element formulation of this work is discussed in detail in Ref. 5. Here the focus is on the determination of the NSP for the nonlinear analysis.

Consider a rectangular plate of dimensions a, b and h with a coordinate system taken such that xy-plane coincides with midplane of the plate. The thickness coordinate (z) is measured downward from the midplane. Considering the classical plate theory, the displacements are

$$U = u + zw_{,x} \qquad V = v + zw_{,y} \qquad W = w \tag{1}$$

where U, V and W are total displacements, and u, v and w are displacements of the midplane and $(\ )_{,x} = \partial(\ )/\partial x$. In the thin plate theory, the strain–displacement relations for large deflection are given by

$$\begin{aligned} \varepsilon_1 &= u_{,x} + zw_{,xx} + \tfrac{1}{2}w_{,x}^2 \\ \varepsilon_2 &= v_{,y} + zw_{,yy} + \tfrac{1}{2}w_{,y}^2 \\ \varepsilon_6 &= u_{,y} + v_{,x} + w_{,x}w_{,y} + 2zw_{,xy} \end{aligned} \tag{2}$$

The stretching (A), bending-stretching (B) and bending (D) stiffnesses for a single-layer, bimodular plate are

$$(A, B, D) = \left[\int_{-h/2}^{z_n} Q^t(1, z, z^2)\,dz + \int_{z_n}^{h/2} Q^c(1, z, z^2)\,dz\right] \tag{3}$$

where Q^{t} and Q^{c} are tensile and compressive reduced stiffnesses. The NSP for an anisotropic plate can be defined using the fiber-governed compliance model as follows:

$$Q' = \varepsilon_\theta = \varepsilon_1 \cos^2\theta + 2\varepsilon_6 \sin\theta\cos\theta + \varepsilon_2 \sin^2\theta \tag{4}$$

where θ is the angle of fiber orientation. Substitution of eqn. (2) into eqn. (4) yields

$$z_n = \frac{(u_{,x} + \frac{1}{2}w_{,x}^2)\cos^2\theta + (v_{,y} + \frac{1}{2}w_{,y}^2)\sin^2\theta + (u_{,y} + v_{,x} + w_{,x}w_{,y})\sin\theta\cos\theta}{w_{,xx}\cos^2\theta + 2w_{,xy}\sin\theta\cos\theta + w_{,yy}\sin^2\theta} \tag{5}$$

Here, z_n is the location of the neutral surface of the plate, and it is not known a priori due to its dependence on displacements. Therefore, an iterative technique must be used. This is done by first assuming initial values for z_n, then computing global stiffness coefficients and displacements. Next, eqn. (5) is used to update the location of the neutral surface of the plate. The above procedures are repeated until the convergence on the NSP is achieved.

The finite element formulation for a typical element has the following form:

$$[K]\{\Delta\} = \{F\} \tag{6}$$

where $\{\Delta\} = \{u, v, w\}$ is the displacement vector, $\{F\}$ is the generalized force vector, and $[K]$ is the element stiffness matrix. The entries of this matrix are given in the Appendix. Once the convergence on the NSP is achieved another iteration is required to obtain the convergence on the displacements $\{\Delta\}$.

NUMERICAL RESULTS AND DISCUSSION

Numerical results are presented below for thin, anisotropic, rectangular plates constructed of bimodular materials. The material properties of graphite/epoxy and polyester-rubber which are used in the computations are given in Table 1. Due to the symmetry, only a quarter of the plate with a 2×2 mesh of nine-node isoparametric elements was considered. This mesh was compared to a 5×5 four-node isoparametric element and the results were in agreement within 0·5%. However, the 2×2 mesh needed less computation time.

TABLE 1
Material properties of graphite/epoxy and polyester-rubber

Property	Graphite/epoxy		Polyester-rubber	
	Tensile	Compressive	Tensile	Compressive
E_{11} (psi)	24×10^6	22×10^6	$0{\cdot}089 \times 10^6$	$5{\cdot}351 \times 10^3$
E_{22} (psi)	$1{\cdot}2 \times 10^6$	$1{\cdot}1 \times 10^6$	$1{\cdot}160 \times 10^3$	$1{\cdot}537 \times 10^3$
ν_{12}	0·32	0·30	0·475	0·185

To validate the results, a comparison for the maximum deflection of a single-layer anisotropic plate was made between the results presented in Ref. 7 and those of this work. This is illustrated in Fig. 1. The results were also compared with those of Ref. 4 for the validity of the implementation of bimodular action in the analysis (Table 2).

The effects of the angle of fiber orientation and dimensionless load ($\bar{q} = qE^c_{22}a^4/h^4$) on the nondimensionalized nonlinear center deflection of thin, square plates are shown in Figs 2 and 3. The transverse deflection increases nonlinearly with load. By increasing the angle of fiber orientation, the plate stiffness decreases. As a consequence, the transverse deflection increases. However, for certain angles, the NSP falls outside of the plate thickness and the whole plate will be under tension. Since the tensile

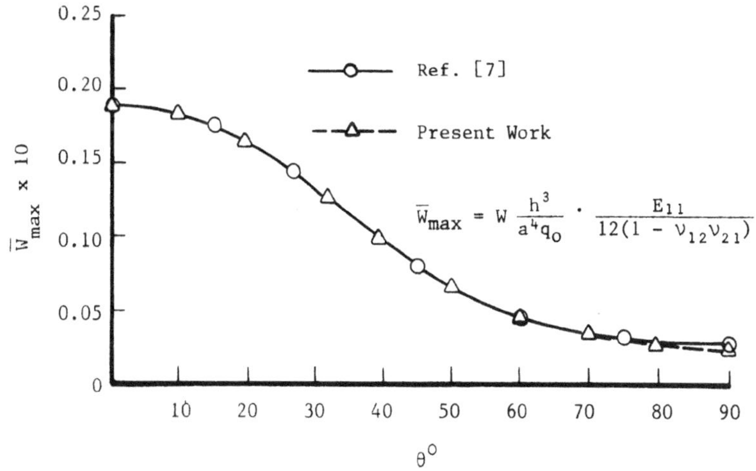

FIG. 1. Comparison between the results of Ref. 7 and the present work for nondimensionalized center deflection of a thin, single-layer, anisotropic, unimodular, clamped plate under uniform distributed load ($a/b = 2$, $E_{11}/E_{22} = 10$, $\nu_{12} = 0{\cdot}3$).

TABLE 2

Comparison between the results of Ref. 4 and present work for nondimensionalized, nonlinear center deflection of single-layer (0°), simply-supported, aramid-rubber and polyester-rubber plates under uniform distributed load (a/h = 100)

$\bar{q}^a$	$\bar{w}^b$			
	Aramid-rubber		Polyester-rubber	
	Ref. 4	This work	Ref. 4	This work
Linear	0·2959	0·2960	0·1683	0·1683
10	0·2720	0·2720	0·1654	0·1654
20	0·4963	0·4963	0·3233	0·3233
30	0·6821	0·6822	0·4718	0·4719
40	0·8417	0·8418	0·6106	0·6107
50	0·9798	0·9799	0·7399	0·7399

[a] $\bar{q} = qE_{22}^c a^4/h^4$.
[b] $\bar{w} = w/h$.

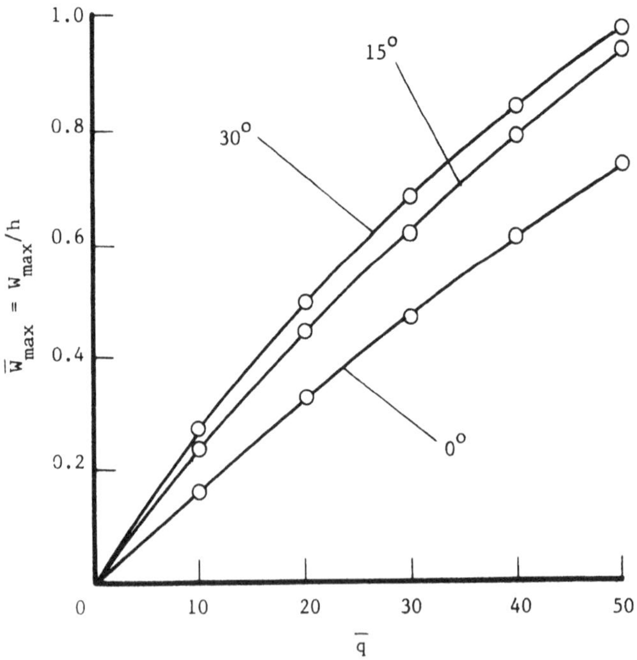

FIG. 2. Effect of the angle of fiber orientation on nondimensionalized nonlinear center deflection of a single-layer, simply-supported polyester-rubber plate under uniform distributed load ($a/h = 100$, $a/b = 1$).

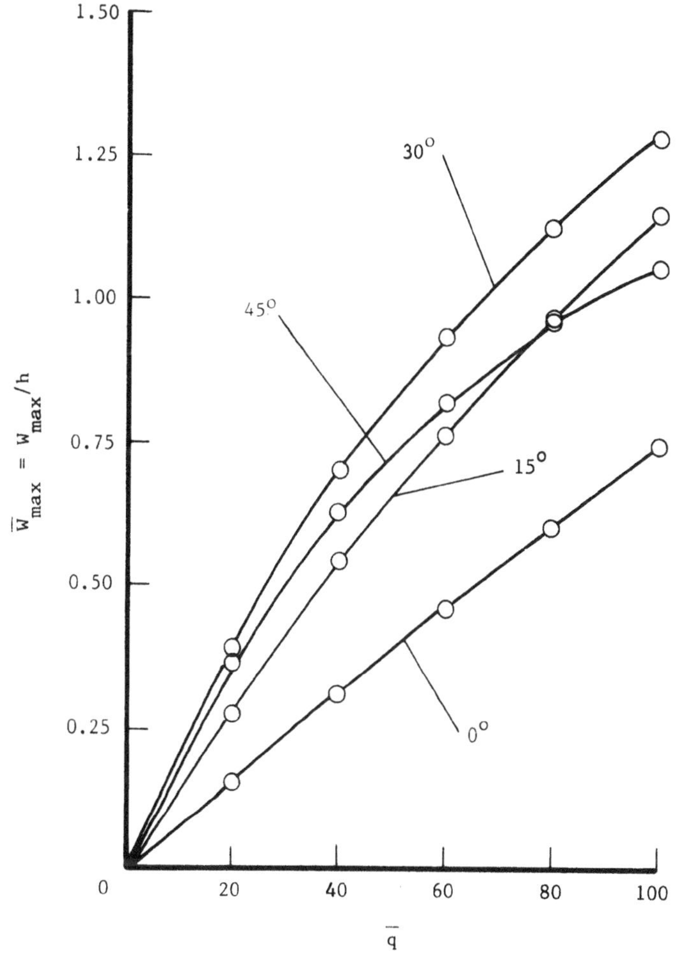

FIG. 3. Effect of the angle of fiber orientation on nondimensionalized nonlinear center deflection of single-layer, simply-supported, graphite/epoxy plate under uniform distributed load ($a/h = 100$, $a/b = 1$).

modulus is greater than the compressive modulus for graphite/epoxy, the plate stiffness becomes higher than the case where part of the plate is under compression. This increase in stiffness reduces the transverse deflection (see Fig. 2).

The effects of plate aspect ratio (b/a) and the angle of fiber orientation on the nondimensionalized center deflection for graphite/epoxy and polyester-rubber are shown in Fig. 4 and tabulated in Table 3, respectively. The

TABLE 3

Effect of plate aspect ratio on the nondimensionalized, nonlinear center deflection of polyester-rubber, single-layer, anisotropic, simply-supported plates under uniform distributed load ($\bar{q}=20$, $a/h=100$)

$\theta°$	$\bar{w}_{max}=w_{max}/h$			
b/a	0·5	1·0	2·0	3·0
0	0·1300	0·3232	0·3558	0·3510
15	0·1261	0·4414	0·5765	0·5834
30	0·1077	0·4957	0·8948	0·9950
45	0·0932	0·4674	1·0996	1·3984

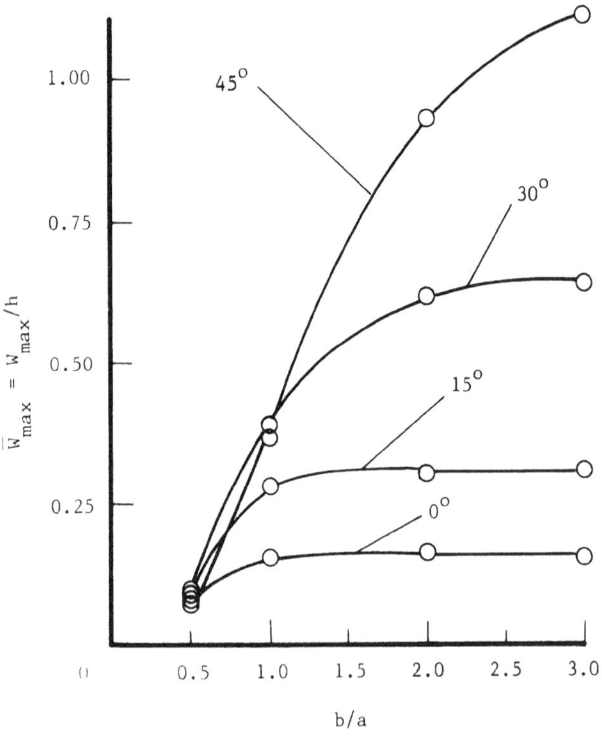

FIG. 4. Effect of plate aspect ratio on the nondimensionalized nonlinear center deflection of graphite/epoxy single-layer, anisotropic, simply-supported plates under uniform distributed load ($q=20$, $a/h=100$).

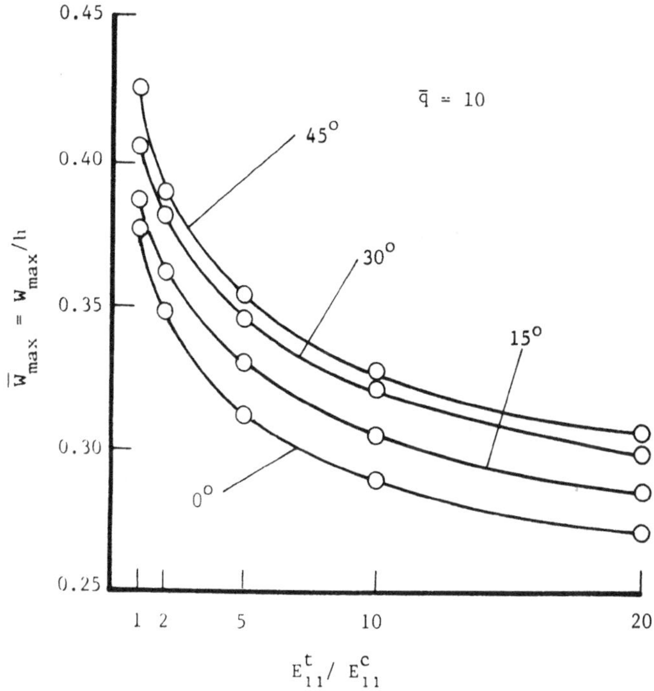

FIG. 5. Effect of bimodulatory ratio on nondimensionalized nonlinear center deflection of anisotropic, simply-supported square plates under uniform distributed load ($a/h = 100$).

nonlinear deflection increases with the aspect ratio and the angle of fiber orientation. It should be mentioned that since the load is fairly small, the NSP stays within the thickness of the plate which in turn results in more deflection for 45° angle compared to 30°.

Finally, the effect of bimodularity ratio (E^t_{11}/E^c_{11}) on the nonlinear deflection of a square plate is shown in Fig. 5. The transverse deflection decreases with the aspect ratio, and increases with the angle of fiber orientation.

CONCLUSIONS

An expression is developed for the determination of the location of the neutral surface to be used in the geometrically nonlinear analysis of anisotropic bimodular material plates. The neutral surface position depends on displacements, angle of fiber orientation as well as the applied load.

As long as this surface stays within the limit of the plate thickness, the deflection increases with an increase in the angle of fiber orientation. Otherwise, due to the increase in plate stiffness the deflection decreases with increasing the fiber angle. Also, the effects of bimodularity and plate aspect ratio on the transverse deflection are found to be significant.

REFERENCES

1. KAMIYA, N., Large deflection of a different modular circular plate, *ASME Journal of Engineering Materials and Technology*, **97H** (1976), 52–56.
2. KAMIYA, N., An energy method applied to large elastic deflection of a thin plate of bimodulus material, *Journal of Structural Mechanics*, **3** (1975), 317–329.
3. TURVEY, G. J., Study of the nonlinear elastic flexure of bimodular composite strips, *Fiber Science and Technology*, **16** (1982), 121–147.
4. REDDY, J. N. and CHAO, W. C., Nonlinear bending of bimodular-material plates, *International Journal of Solid and Structures*, **19** (1983), 229–237.
5. GORDANINEJAD, F., Finite-element analysis of thick, single-layer, rectangular anisotropic bimodular composite-material plates (submitted), also presented at the 10th US National Congress of Applied Mechanics, Austin, Texas, June 1986.
6. BERT, C. W., Models for fibrous composites with different properties in tension and compression, *ASME Journal of Engineering Materials and Technology*, **99H** (1977), 344–349.
7. ASHTON, J. E. and WHITNEY, J. M., *Theory of Laminated Plates*, Technomic Publishing Co., 1970.

APPENDIX: ELEMENTS OF STIFFNESS MATRIX

The matrix $[K]$ is a nonsymmetric matrix for geometrically nonlinear problems. For the present case the coefficients of this matrix are given by (for more details see Ref. 5)

$$\begin{aligned} &[K^{11}]_{NL} = [K^{11}]_L && [K^{22}]_{NL} = [K^{22}]_L \\ &[K^{12}]_{NL} = [K^{21}]_L^T && [K^{23}]_{NL} = \tfrac{1}{2}[K^{32}]_L^T \\ &[K^{13}]_{NL} = \tfrac{1}{2}[K^{31}]_L^T && [K^{33}]_{NL} = [K_1^{33}]_L + [K_2^{33}] \end{aligned} \tag{A.1}$$

$$L = \text{linear} \qquad NL = \text{nonlinear}$$

where

$$[K^{33}] = \tfrac{1}{2}\,\mathrm{Re}\,[N_1\phi_{i,x}\phi_{j,x} + N_6\phi_{i,x}\phi_{j,y} + N_2\phi_{i,y}\phi_{j,y}]\,\mathrm{d}x\,\mathrm{d}y \qquad \text{(A.2)}$$

and

$$\begin{aligned}
\bar{N}_1 &= A_{11}w_{,x}^2 + A_{12}w_{,y}^2 + 2A_{16}w_{,x}w_{,y} \\
\bar{N}_2 &= A_{12}w_{,x}^2 + A_{22}w_{,y}^2 + 2A_{26}w_{,x}w_{,y} \\
\bar{N}_6 &= A_{16}w_{,x}^2 + A_{26}w_{,y}^2 + 2A_{66}w_{,x}w_{,y}
\end{aligned} \qquad \text{(A.3)}$$

In eqn. (A.2) ϕ is the shape function.

11

Design of the ARALL F-27 Lower Wing Fatigue Panel

J. W. GUNNINK
Delft University of Technology, Faculty of Aerospace Engineering, Kluyverweg 1, 2629 HS Delft, The Netherlands

and

P. A. VAN DER SCHEE
Fokker Aircraft Company, Structural Design Office, P.O. Box 7600, 1117 ZJ Schipol, The Netherlands

ABSTRACT

To prove the excellent static and fatigue properties of ARALL (Aramid Aluminium Laminate) a realistic aircraft component was designed. This structure is being tested under flight simulation fatigue loading as well as under static loading. The high fatigue loaded area of the lower skin of the outer wing of the Fokker F-27 Friendship has been selected as a design study for this purpose. In this chapter the design philosophy and detail design, as well as results of the finite element analyses and CAD procedures will be presented. Also the preliminary fatigue test results will be discussed briefly.

INTRODUCTION

ARALL, developed by Delft University of Technology and commercialized by ALCOA, is an adhesive bonded laminate which combines the advantages of high strength isotropic aluminium sheet with the fatigue and fracture resistance of aramid fibres. It was developed principally to obtain a material with good fatigue strength, in which possible cracks grow very slowly.[1,2] After more than 5 years of extensive testing ARALL shows that it is highly insensitive to fatigue, has high strength, good residual strength and excellent durability properties.[1-4] These basic material properties of ARALL give a significant improvement over existing aluminium alloys, and are in almost all respects better even than advanced composites.[5] During the development of ARALL it became clear that, to prove its advantages in aircraft structures, it would be necessary to design and manufacture a realistic aircraft component, to test it under flight simulation fatigue loading as well as under static loading, and to compare the results with an equivalent aluminium structure. The lower skin of the outerwing of the Fokker F-27 Friendship was selected as a good example on which to carry out a design study in ARALL. The reason for choosing this particular structure was that all the essential information for the existing F-27 wing structure was readily available.[6] Furthermore, a major component of the existing structure of the Fokker F-27 lower wing had been extensively tested in fatigue.[7] The structure which Fokker tested was a double version of the most fatigue prone area of the wing structure, station 4155-5075 (Fig. 1). These tests were part of the continuing airworthiness programme of the F-27 and have contributed to the extended life of the aircraft (from 60 000 flights up to 90 000 flights). For a fair comparison it was necessary to redesign a significant part of the structure, which could replace directly the

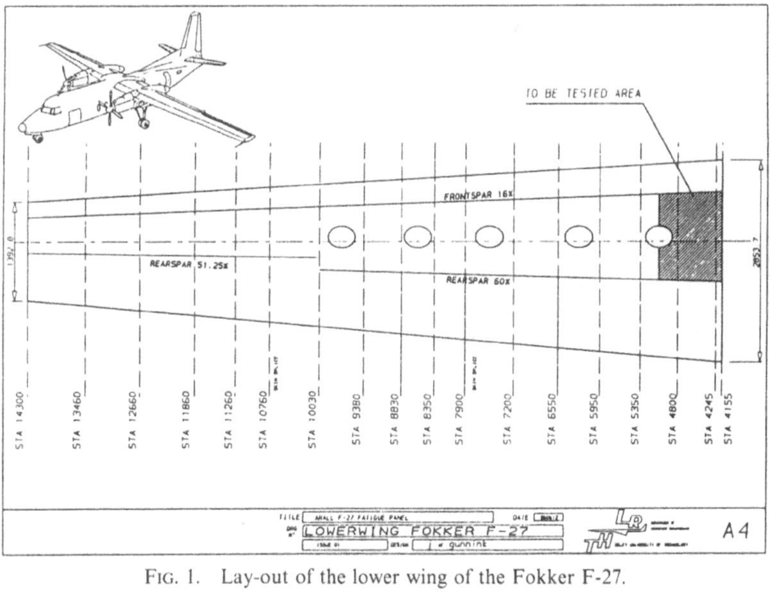

FIG. 1. Lay-out of the lower wing of the Fokker F-27.

aluminium structure. By this approach all the requirements of the original structure are taken into account. This means no hidden (dis)advantages can be attributed to one or other of the structures.

So from 1980 onwards, alongside development of the ARALL material itself, preliminary design studies on the lower wing structure were performed at Delft University of Technology with the then available material data.[1,2,5] Finally, these studies indicated that using ARALL in the lower wing of this particular aircraft, a weight saving of 30% is within reach.[8] This means that, by comparing the ARALL structure with the Fokker fatigue test panels, a good assessment can be made of the performance of ARALL in realistic structures. The results of the preliminary structural design studies gave the justification to go ahead with a full-scale detail design, and to carry out the testing of an ARALL lower wing fatigue panel. However, Delft University of Technology was not able to manufacture and to fatigue test a panel of the required size (1·8 by 1·2 m). Fokker, convinced of the possibilities of the material, decided to participate in the ARALL lower wing fatigue panel project of Delft University by taking care of the manufacturing and testing of the panel. Then Delft University, as well as having overall responsibility for the project and some structural detail testing, also undertook the design and analysis of the panel and the making of production drawings.[9] Thus in 1984 a joint F-27 wing panel program was started between Delft University of Technology and Fokker.

TARGETS

Originally in the test on the ARALL structure, the following targets were set:

1. 1-g stress level $\sigma_{1\text{-g}} = 100\ \text{N/mm}^2$.
2. Same fatigue loading as for the aluminium panel.
3. Average weightsaving of the ARALL panel of about 25%.
   While a fully successful panel should demonstrate:
4. Crackfree life of 45 000 flights.
5. Life without repairs of 75 000 flights.
6. Economic repair life of 90 000 flights.
7. Residual tensile strength 1·1 times the limit load.
8. Scatterfactor of 2 (later 3) must be applied.

Later on some changes were made as a result of further evaluation of the material. For instance, the targets 1 and 2 can lead to a conflicting situation, as in fact occurs. However, to obtain the most realistic comparison, it was decided to give a higher priority to target 2. This resulted in the decision to fatigue test the panel up to the economic repair life, with the fatigue loadings of the aluminium Fokker F-27 panels. If feasible, the structural fatigue tests will then continue at the increased 1-g stress level of $\sigma_{1\text{-}g} = 100\,\mathrm{N/mm^2}$. For several reasons the economic repair life has been raised to 270 000 flights (targets 6 and 8). The unknown behaviour of ARALL in a realistic full-scale structure was one of the most important reasons.

During the development stage of ARALL it was decided to standardize ARALL in two grades, i.e. ARALL 1, based on aluminium 7075-T6 and prestrained and ARALL 2, based on aluminium 2024-T3 and in the as-cured condition. Both grades are very good fatigue resistant materials with high static tensile strength. The application of ARALL 1 is intended for high fatigue loaded structures together with high compressive loadings, whereas ARALL 2, less expensive than ARALL 1, is primarily for high fatigue loaded structures with relative low compression loadings. An ARALL F-27 lowerwing fatigue panel will be manufactured and tested with both grades of ARALL. However this paper will be restricted to the panel made of ARALL 2. The results of this panel will be used to improve the second fatigue panel made of ARALL 1.

MECHANICAL PROPERTIES

In Table 1 the principal mechanical properties of both grades of ARALL are given. The properties of ARALL 1 are from the test results of ALCOA and some aircraft manufacturers[4] and results of Delft University,[1,2,5] whereas the properties of ARALL 2 are from the preliminary values of Delft University and ALCOA. ALCOA reported also preliminary figures for the bearing strength of ARALL 1.[4] It is found that the ultimate bearing strength of ARALL is equal to the summation of the bearing strengths of the aluminium layers. This corresponds with the approach of Gunnink and Rothwell about static strength of riveted joints.[10] They concluded that in general the existing rivet tables can be used for ARALL.

PRIMARY DESIGN ALLOWABLES

In Ref. 8 the primary structural design 'allowables' are considered. Four allowables are discussed, the design-limit, design-ultimate, durability and

TABLE 1
Mechanical properties

Average mechanical properties		ARALL 1 laminate	ARALL 2 laminate
Tensile ultimate strength (MPa)	L	772	703
	LT	379	317
Tensile yield strength (MPa)	L	655	351
	LT	324	234
Compressive yield strength (MPa)	L	365	240
	LT	379	270
Elongation (%)	L	0·6	1·3
	LT	6·3	12·4
Elastic tensile modulus (GPa)	L	69·6	65·5
	LT	48·3	44·8
Ultimate strain (%)	L	1·7	2·5
	LT	7·8	13·6
Blanking shear (MPa)		262	248
Density (g/cm^3)		2·29	2·29

damage tolerance. The ratio between the ultimate tensile strength and the tensile yield stress is for ARALL 1 less than 1·5 and for ARALL 2 more than 1·5. So for ARALL 1 the design-limit allowable has no influence on the static design. However the design limit is also not the decisive factor for structures of ARALL 2, due to the notch behaviour of ARALL (Fig. 2) by which the static ultimate allowable strength is reduced significantly in comparison to the unnotched tensile strength. It is found that the ratio between the static ultimate design allowable and the tensile yield strength of ARALL 2 is in practice always less than 1·5 (notch factor > 2·0). As design ultimate tensile strength for the re-design of the Fokker F-27 lower wing, a notch factor of 3·0 is taken. This results for ARALL 1 in an ultimate tensile design stress $\bar{\sigma}_t = 490\ \mathrm{N/mm^2}$ and for ARALL 2 $\bar{\sigma}_t = 440\ \mathrm{N/mm^2}$. The allowable compressive panel strength to be used in design is related to buckling and compressive yield stress, and is given by Fig. 3. Extensive fatigue testing of the material shows the good fatigue resistance of ARALL. It is obvious that by increasing the 1-g stress level crack initiation starts earlier and the final fatigue crack length (in the outer aluminium layer) is also increased. For practical application it was decided to take a 1-g stress level $\sigma = 100\ \mathrm{N/mm^2}$. Durability of ARALL has been investigated extensively.[3] Virtually no degradation in strength and stiffness, even in

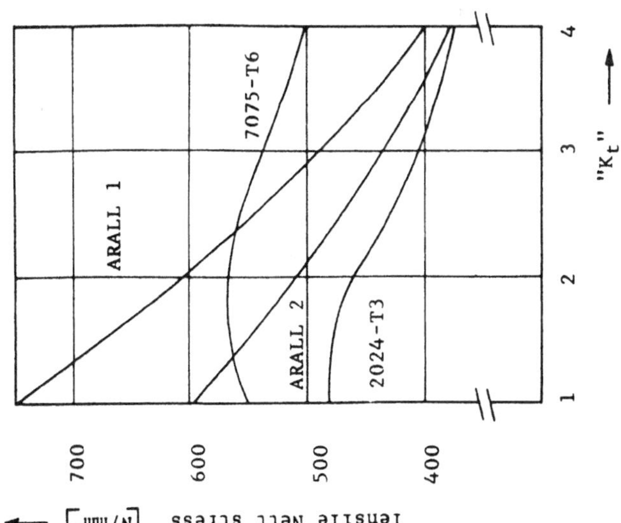

FIG. 2. Notch behaviour of materials.

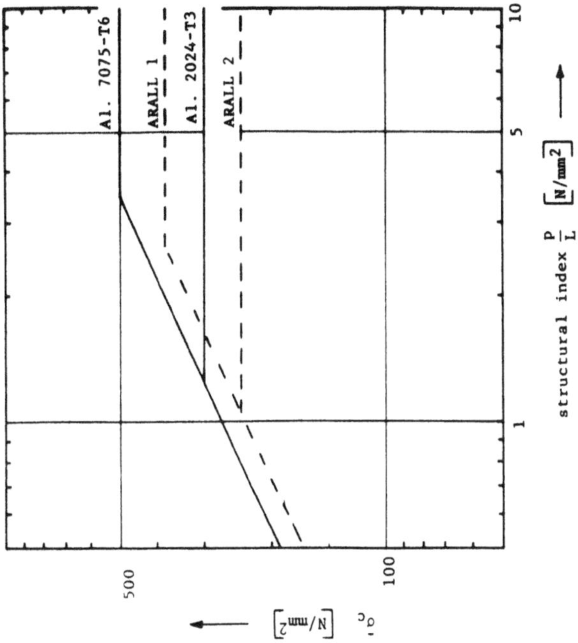

FIG. 3. Structural ultimate compressive strength.

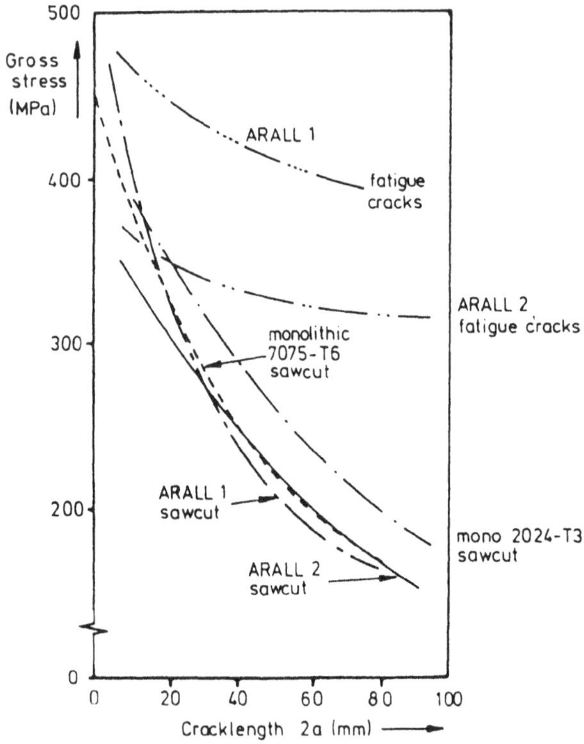

FIG. 4. Residual strength of unstiffened panels (panel width 160 mm).

aggressive environments, was observed. The static tensile strength of ARALL, even after fatigue loading to more than three aircraft lifetimes, reduces to not less than 90% of its initial strength (Fig. 4). Because the initial strength in almost all cases is (much) higher than the design ultimate strength, the reduction in strength still gives the structure the possibility to sustain the design ultimate load. In fact this reduction does not harm the structure. However this approach is based on the results for very small structural parts, and must still be proved for realistic structures. This is in fact the intention in the on-going ARALL F-27 lower wing fatigue panel project.

DESIGN

The lower wing of the Fokker F-27 can be subdivided into three basic parts: the access panel and two 'side' panels. The access panel contains the

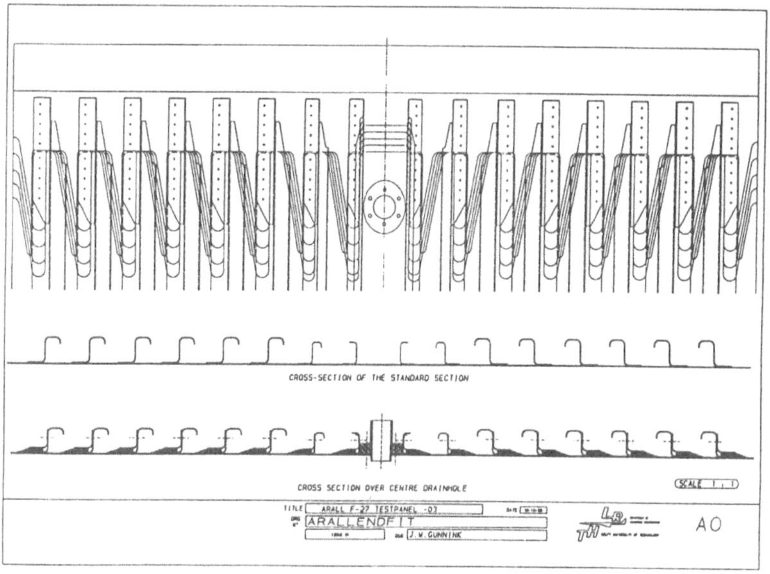

FIG. 5. End-fitting of the ARALL F-27 test panel.

manholes and the reinforcements around it, whereas the 'side' panels are basically undisturbed. According to the preliminary structural design studies it was decided to design a panel with a so-called 'semi-load carrying' access panel. This requires the cross-sectional area of the access panel to be as small as possible. According to the several loading conditions to which the panel has to be designed, it is found that one of the most critical factors for the access panel in the compressive load case, which determines the stiffener-skin configuration. In Fig. 5 cross-section of the panel is shown. The outer wing/inner wing connection at sta. 4155 (Fig. 1) is in the F-27 aircraft a shear type of connection. The panel represents the wing location at this section (sta. 4155 – 5075). In designing the end-fitting of the panel this type of connection has to be used. In accordance with the discussion in the previous section on bearing strength and strength of riveted joints, the end-fitting of the panel has been designed (Fig. 5). This detail of the panel has been extensively tested in fatigue, and finally statically tested to determine the residual strength (Fig. 6). It was found that after some small modifications the behaviour of these end-fitting test panels was excellent. In fatigue after the equivalent of 300 000 flights only very small cracks in the outer aluminium layer of the stringers (at the end of the fingertips) were

FIG. 6. Test of the first end-fitting panel.

observed. Residual strength tests showed that these panels did not fail at these cracks, but at the panel–rib connection. At this connection the panel has the smallest cross-sectional area and, due to the riveted connection of the rib-simulation to the panel, the panel has a notch factor which resulted in a reduction of the static strength of the panel below the residual strength of the fatigue cracked area. However, the residual strength was still about 5% higher than the required ultimate design strength. So the test results for these end-fitting test panels were satisfactory for the design as a whole, as well as validating the design approach mentioned in the previous section. It is obvious that the most elaborate part to design was the tank cover and the reinforcements around the manhole. Due to the several types of connection, and interference of structural elements with each other, this area is rather complicated. However, designing with the MEDUSA CAD-system in use at Delft University has introduced a high degree of flexibility

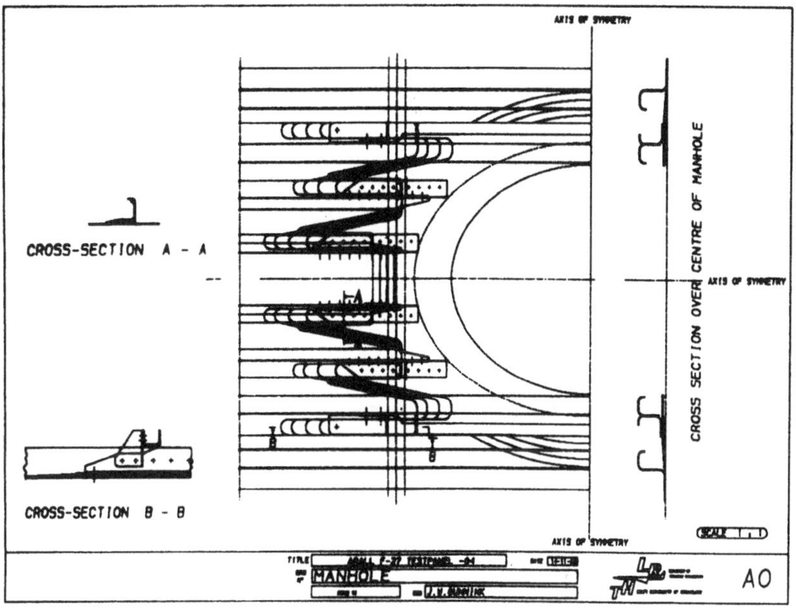

FIG. 7. Structural lay-out of the manhole area.

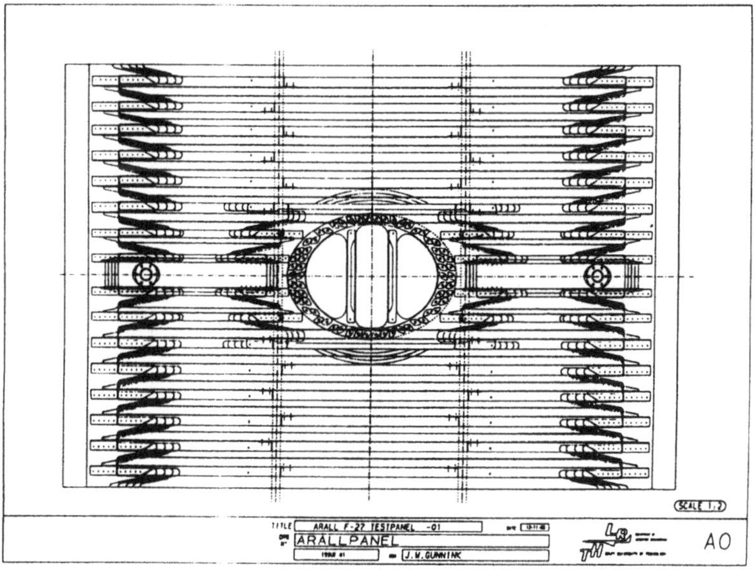

FIG. 8. Overall structural drawing of the ARALL F-27 test panel.

into the design procedure. Especially for the detail design of the manhole area the CAD system showed its advantages. To arrive at an optimized structure finite element (FE) calculations have been used extensively. In the next section these calculations will be discussed in more detail. Interactive use of FE calculations and the CAD system has resulted in the structure of the manhole area shown in Fig. 7. Finally this resulted in the overall structural drawing of the panel shown in Fig. 8.

FINITE ELEMENT ANALYSIS

The overall analysis of the panel is performed with the finite element program GIFTS (Graphics Interactive Finite-element Total System).[11] The test panel has two axes of symmetry, which means only one quarter of the panel has to be modelled. GIFTS also has the capability of substructuring. With this technique the end-fitting area (Fig. 5) is modelled separately from the manhole area (Fig. 7). As mentioned in the previous section, finite element analysis was used extensively to arrive at an optimized structure. Especially the structure around the manhole was optimized by this technique. In this section the analysis of this area will be discussed briefly. Mainly due to available computer storage, the analysis of the panel is performed on a two-dimensional model which does not include out-of-plane bending of the panel. The results of the detail design are incorporated in modelling the panel. For instance the end-fitting structure is primarily determined by the connection inner wing/outer wing and the fade-out of the stiffeners at this location. The geometry of the stringer-skin combinations for the access panel and the 'side' panels are determined by compressive buckling with the program OBSOZPAN.[12] The reinforcements and joints of the manhole are primarily determined by the assumptions that the load in the access panel has to be carried totally by the tank cover and its local reinforcements. In particular this last item was a complicating factor for the stress analysis. The bolted connection of the tank cover and the change in neutral axis in the tank cover area gives an uncertainty in the load transfer of this section. On the other hand from (past) practice it is known that a good design of the tank cover area allows a load transfer through the cover of 60–70% of the loading in the access panel. In the optimization procedure this behaviour is taken into account. For this purpose two configurations of the panel are analysed, a panel with reduced cover stiffness to simulate the reduced load transfer and a panel

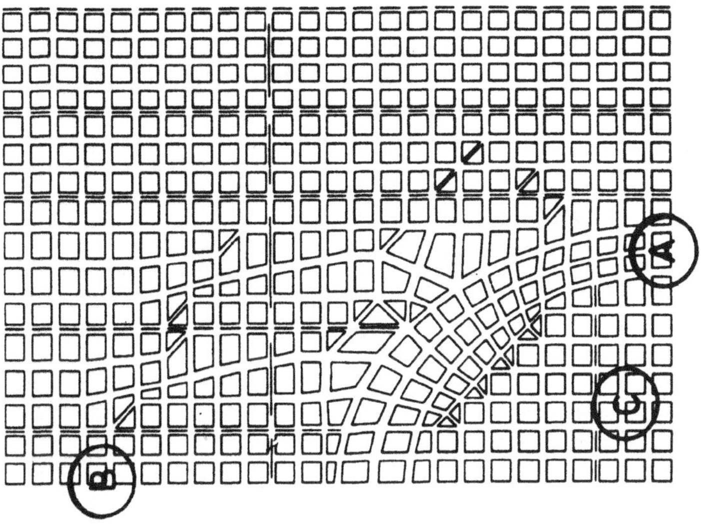

FIG. 9. FE-model of the manhole area.

with full cover stiffness to analyse the cover and its critical areas. These analyses show three critical areas (Fig. 9):

(A) the reinforcements of the manhole on the short axis of symmetry;
(B) the access panel between the manhole and the end-fitting;
(C) the cover.

It is obvious that the model with the reduced cover stiffness determines location A and the model with the full stiffness locations B and C. During the optimization of the panel several structural parameters (such as stringer orientation in the cover) were investigated. Finally, a panel was obtained with acceptable stresses in the three critical areas for both FE models.

RESULTS

The production drawings made on the MEDUSA CAD-system could be used directly in the manufacture, and Fokker was able in a relatively short time to produce the panel without serious difficulty (Fig. 10). The weight of the ARALL panel was 67% of the weight of the Fokker panels. This means the target, a weight saving of at least 25%, was met. Static testing of the panel at the Fokker Experimental Department shows good agreement with the FE calculations. As would be expected severe bending was observed in

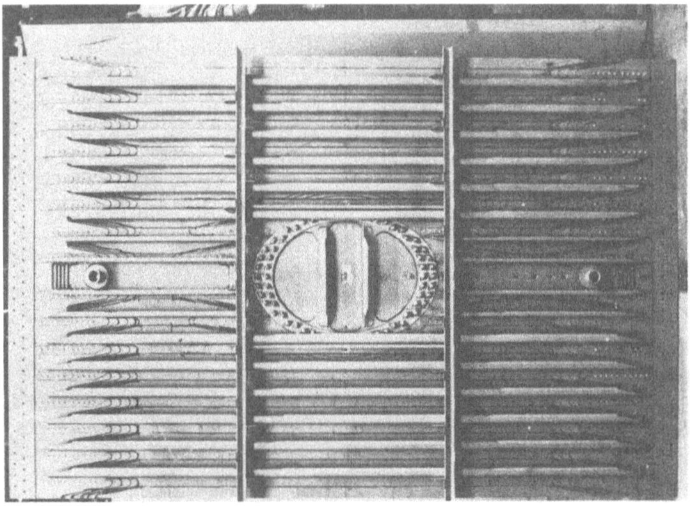

FIG. 10. ARALL F-27 lower wing fatigue panel.

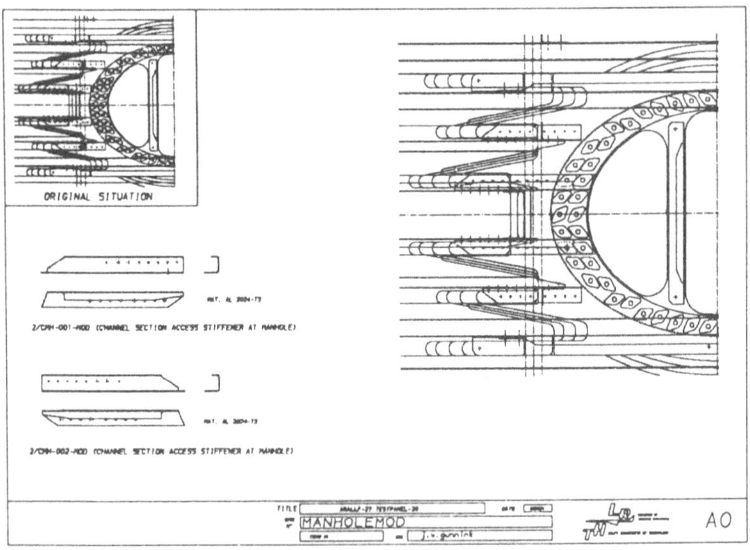

FIG. 11. Modification of the manhole area.

the tank cover area. This behaviour has caused fatigue cracking during the test. After 20 000 flights this critical area was reanalysed in more detail.[13] The GIFTS calculations showed significant bending stresses in the reinforcement at the longitudinal axis of the manhole. This could be improved significantly by extension of the stiffeners to the edge of the reinforcements (Fig. 11). This small modification reduced the bending stresses locally by about 50%. According to this result the panel was modified. Continuing fatigue testing showed that the problems had been solved. The panel is fatigue tested up to the economic fatigue life target of 270 000 flights.[14] Some small fatigue cracks were observed, especially at the bolt holes in the reinforcements of the manhole. However, after the cracks had started, they extended with decreasing crack growth rate and arrested after about 20 000 flights. In this case no modification was performed. It was remarkable that after the design modification hardly any growth of the cracks in the modified area was observed. This shows the excellent damage tolerance behaviour of ARALL.

CONCLUSIONS

With the design and testing of the ARALL F-27 test panel eight primary targets were set. The preliminary test results have shown that almost all these targets were met. However, the residual strength has still to be proved. This test will be performed in the first quarter of 1987. Only the target concerning the crackfree life of 45 000 flights needs some attention. As mentioned before, after 20 000 flights some cracks occurred at the reinforcement of the tank cover. This was due to severe local bending, and after some re-analysis and design the problem was solved. Without repairing the cracks fatigue testing of the structure was able to be continued up to 270 000 flights. This shows the excellent damage tolerance behaviour of ARALL, since hardly any crack growth was observed in the modified area. It is expected that small fatigue cracks in the structure can be tolerated without repair, and in this condition the structure can still sustain design ultimate load. This means the new design philosophy seems to be a good approach to the design of ARALL structures. However, this needs further development.

Finally it can be concluded through the design and by the preliminary test results of the ARALL F-27 test panel that this new material has the ability to reduce the weight and increase the life of aircraft structures and is a genuine competitor for advanced composite materials.

ACKNOWLEDGEMENTS

The contribution of all participants of the ARALL lower wing project of Delft University of Technology and Fokker is gratefully acknowledged. ALCOA is acknowledged for their efforts to deliver the ARALL material one year earlier than originally planned.

REFERENCES

1. Gunnink, J. W., Verbruggen, M. L. C. E. and Vogelesang, L. B., ARALL, a light weight structural material for impact and fatigue sensitive structures, Paper 89, *Tenth European Rotorcraft Forum*, August 28–31, 1984, The Hague, The Netherlands, also *Vertica*, **10**, No. 2 (1986), 241–254.
2. Gunnink, J. W., Vogelesang, L. B. and Schijve, J., Application of a new hybrid material (ARALL) in aircraft structures, ICAS-82-2.6.1, *13th ICAS-conference*, Seattle, August 1982, pp. 990–1000.
3. Verbruggen, M. L. C. E., Aramid reinforced aluminium laminates: ARALL adhesion problems and environmental effects, Ph.D. thesis, Faculty of Aerospace Engineering, Delft University of Technology, The Netherlands, 1987.
4. Bucci, R. J., Mueller, L. N., Schultz, R. W. and Prohaska, J. L., ARALL laminates—results from a cooperative test program, Alcoa Laboratories, presented at Advanced Materials Technology 87, *32nd International SAMPE Symposium and Exhibition*, Anaheim, CA, April 1987.
5. Vogelesang, L. B. and Gunnink, J. W., ARALL: a materials challenge for the next generation of aircraft, *Materials and Design*, 7, No. 6 (1986), November/December.
6. Anon., Type-record Fokker F-27 Friendship.
7. Schee, P. A. v. d., Outerwing lowerskin fatigue testing as a part of the continuing airworthiness programme of the Fokker F-27, paper at *10th ICAF Symposium*, Brussels, 1979.
8. Gunnink, J. W., Introductory re-design of the lowerwing of the Fokker F-27 (to be published), Faculty of Aerospace Engineering, Delft University of Technology, The Netherlands, 1987.
9. Gunnink, J. W. and Veggel, L. v., ARALL F-27 testpanel research project (Fokker involvement), Fokker Report R-3228, The Netherlands, 1984.
10. Gunnink, J. W. and Rothwell, A., An assessment of the static strength of flush-head aluminium alloy rivets in two grades of 'ARALL' sheet, Report LR-391, Faculty of Aerospace Engineering, Delft University of Technology, The Netherlands, 1983.
11. Verberne, C. J., The influence of the access door and the reinforcements around the access hole on the stress distribution in the ARALL testpanel, Thesis, Faculty of Aerospace Engineering, Delft University of Technology, The Netherlands, 1985.

12. GUNNINK, J. W., Overall buckling of specially orthotropic Z-stiffened panels, Part 1: Theory, Report LR-351, Faculty of Aerospace Engineering, Delft University of Technology, The Netherlands, 1982.
13. NOTENBOOM, R. P., GUNNINK, J. W. and VRIES, J. DE, Analysis of the modified manhole area of the ARALL F-27 lower wing panel (to be published), Faculty of Aerospace Engineering, Delft University of Technology, 1987.
14. VEGGEL, L. V., JONGEBREUR, A. and GUNNINK, J. W., Damage tolerance aspects of an experimental ARALL F-27 lower wing skin panel, paper at the ICAF Conference, Ottawa, 1987.

12

The Construction of a Postbuckled Carbon Fibre Wing Box

W. G. Brooks

College of Aeronautics, Cranfield Institute of Technology, Wharley End, Bedfordshire MK43 0AL, UK

ABSTRACT

A postbuckled carbon fibre structure requires special design considerations to avoid delamination failures. A practical design for a postbuckled wing box is presented. The design features incorporated have been proven in the author's testing programme. This is leading to the testing of a full-scale wing box for an aerobatic light aircraft. Various means of improving the strength of skin/stiffener junctions are described, including the use of Kevlar stitching at this critical area.

A co-curing technique is used to form all the joints made directly to the skin surfaces. Tooling suitable for low volume production has been developed to form these joints. Final assembly of the structure is by adhesive bonding. Different configurations of joint are evaluated. A type of slotted joint is developed which uses features incorporated into the co-cured mouldings. This type of joint allows the structure to be assembled without the use of expensive clamping jigs.

1. THE WING DESIGN

The wing has been designed as a box beam structure using ten ribs, as seen in Fig. 1. It has two full-depth spars. There are eight stiffeners across the chord at the root and three at the tip, including the spar caps. It is designed to withstand 13·5 g at the aircraft's aerobatic weight. It is intended to allow the wing skins to postbuckle in order to maximise the strength/weight ratio.

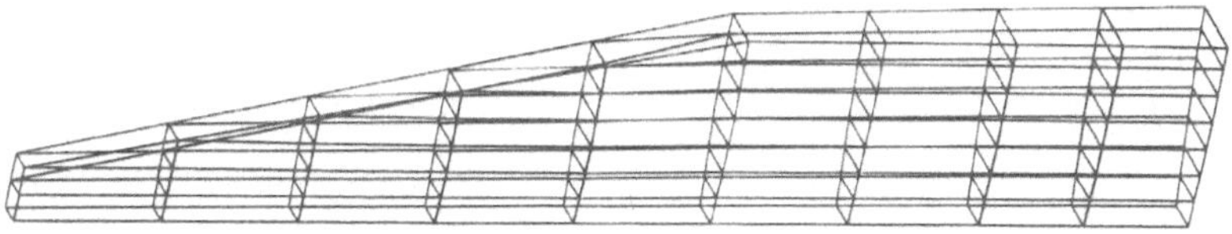

FIG. 1. Finite element wing box model.

The design methods used to size the structure and optimise the configuration have been described in Ref. 1.

The wing is constructed from a 120°C curing prepreg carbon fibre/epoxy system, Ciba-Geigy 913. High strength unidirectional fibres are used for the stiffener caps and skin mid-plies with a cured thickness of 0·127 mm per ply. A woven five-harness satin cloth reinforcement is used for the ±45° outer skin layers. The cured thickness of this material is 0·34 mm. A woven reinforcement was chosen for ease of handling.

2. PROBLEM AREAS OF POSTBUCKLED DESIGN IN COMPOSITES

2.1. Skin/Stiffener Joints

To avoid modal coupling between stiffener torsional buckling and plate local buckling a closed section stiffener has been used. The type of stiffener shown by the program PASCO (Ref. 2) to give the highest panel compression buckling load for a given weight is of a closed trapezium section (Fig. 2).

These stiffeners can be produced by means of a Rohacell foam core of the required cross-section. Some form of tooling is required to produce the external form of the stiffener. The foam then provides a reaction to the tooling, allowing the laminate to be consolidated. This foam core is designed to stay in place after curing of the panel.

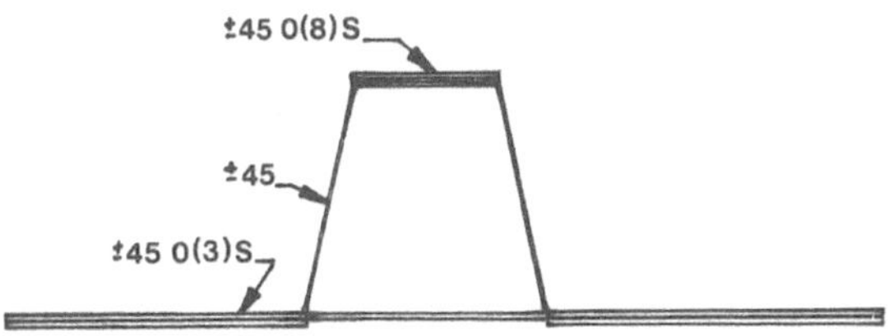

FIG. 2. Typical stiffener.

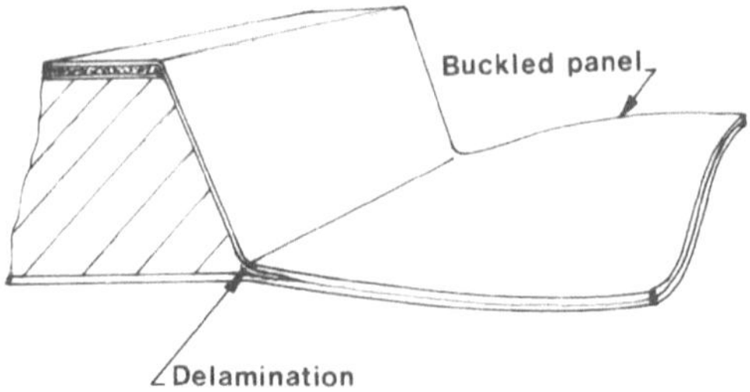

FIG. 3.

One cause of failure of the panels in Ref. 3 was separation of the skin from the stiffeners due to the effects of local buckling. As the skin panels buckle, both bending and peeling loads are generated at the skin/stiffener junction. These panels had the stiffeners co-cured to the skin as integral members. However, they were constructed in such a way as to produce a weak peel-loaded junction between the stiffener and the skin, as shown in Fig. 3. Three methods have been explored to prevent or delay this mode of failure.

2.1.1. The corrugated method

This method of panel lay-up is shown in Fig. 4. It was intended to prevent stiffener separation from the skin by putting most of the skin plies into the stiffener sidewalls. In this way continuous fibre paths would exist across the junction. This method was used in the construction of the first testbox. This box was representative of the outer wing and developed a postbuckling ratio of 5.

The main disadvantage of this scheme arises when the panel has more than two stiffeners. It then becomes extremely difficult to form the prepreg

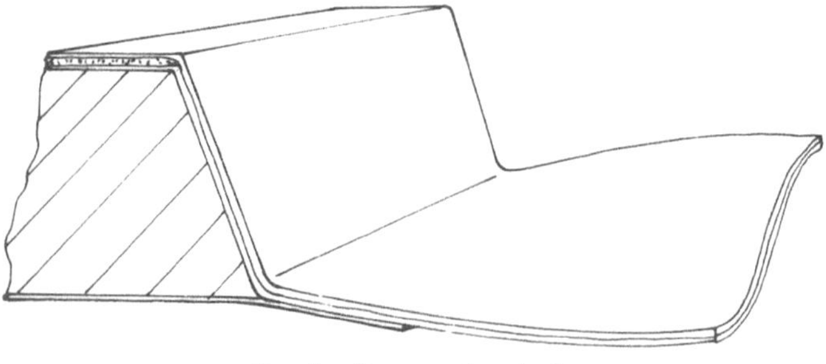

FIG. 4. Corrugated method.

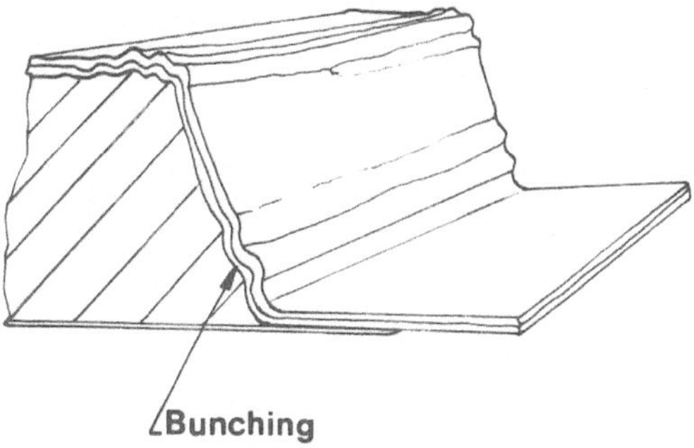

FIG. 5.

so that it conforms exactly to each stiffener profile. Excess material is forced to bunch together, usually at the stiffener corners (Fig. 5). This occurred despite the use of tooling that was completely successful when used on a single stiffener panel.

There are also a large proportion of $\pm 45°$ fibres placed into the stiffener walls by this method. These fibres are used in the skin panels to give a high buckling load. These are not required in the stiffeners, which are essentially beam-columns. Their function is to take up axial loading as the skin panels buckle, and to provide structural stability in the long column and torsional modes. For this purpose unidirectional material is required at the stiffener top and base. Thin shear webs are required between them to provide a torsionally stiff section.

2.1.2. The stiching method

Stitching of the panel at the skin/stiffener junction was seen as a means of achieving two aims. These were to make the prepreg conform tightly to the stiffener profile at the lay-up stage, as well as to improve the strength of the joint. A standard industrial sewing machine was used. A 70-tex spun Kevlar thread was used with an ultimate tensile load of 57 N. Due to the limited throat of the machine, a lay-up technique had to be developed which used subassemblies. These were then co-cured together. An interleaving technique was devised for this purpose, as shown in Fig. 6. This technique has the advantage of allowing adjustment of the stiffener spacing on the tool. Any excess material could move in the interleaves instead of bunching during the cure. The compression skin panel of testbox 2 was successfully produced and tested using this method. This testbox, as shown in Fig. 7, is representative of the wing root area and has a postbuckling ratio of 2·5.

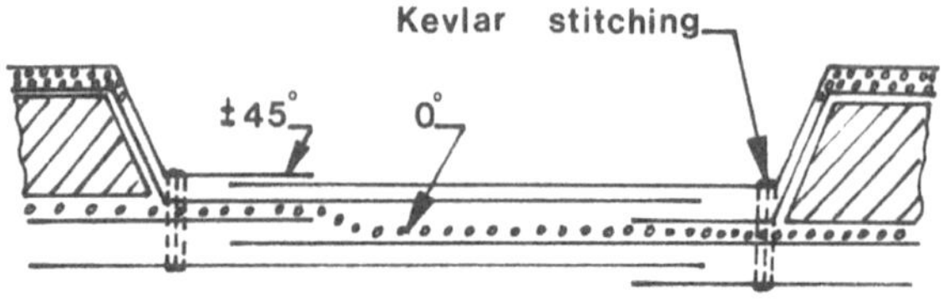

FIG. 6. Interleaving and stitching.

Although this method was successful, the manufacturing process is rather labour intensive. Slow stitching rates have to be used to allow the viscous resin to flow as the needle passes through, typically 150 stitches/min at 20°C. Faster speeds result in more damage as more fibres are broken rather than flowing around the needle. Increasing the temperature of the prepreg reduces the viscosity but increases the resin pick-up on the needle. A solvent pad can then be used to keep the needle clean but with the risk of the solvent affecting the prepreg.

2.1.3. Co-cured joint design

The function strength can be considerably improved by the incorporation of an anti-peel bridging strip, as shown in Fig. 8. A single ply of woven ±45° carbon has a great effect on the peeling strength compared with the simple joint. This bridging strip can be incorporated into the construction under the stiffener foam core, bending upwards into the stiffener sidewall. The void at the junction between the strip, the stiffener sidewall and the skin is filled with resin or a bundle of unidirectional fibres. Reference 4 shows that increasing the corner radii here increases the joint peeling strength.

FIG. 7. Testbox 2.

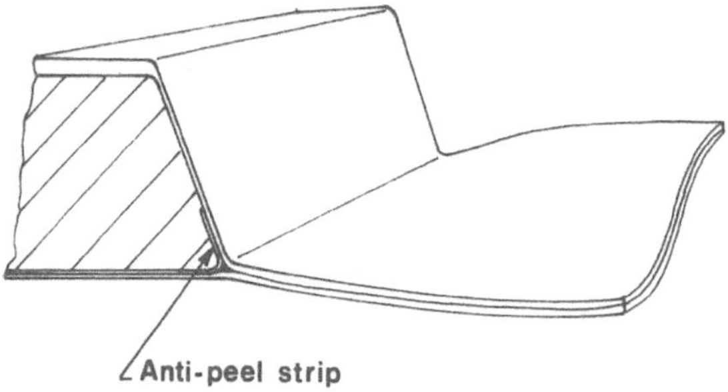

FIG. 8.

To solve the problems of stiffener spacing adjustment and float of excess material, the stiffener sidewall plies have been designed to overlap at the stiffener cap. The stiffener cap is a structurally stable area. A joint here is not liable to failure through delamination caused by local buckling effects.

2.2. Testing of the Skin/Stiffener Joints

A special test coupon has been devised to duplicate the peeling and bending loads at the junction, as seen in Fig. 9. Using this method, various configurations of co-cured joint were tested with results shown in Table 1. In the case of the stitched joints it was found that the stitching pitch had to be in the order of 1 mm or less to realise an advantage over the plain control specimen. The stitches had also to be made within 0·5 mm of the junction in order to provide an improvement.

The very great improvement in strength from the use of an anti-peel strip can be seen. It allows the interlaminar tensile loads to be distributed over a large area instead of the load acting on the stress concentration at the joint in the brittle matrix.

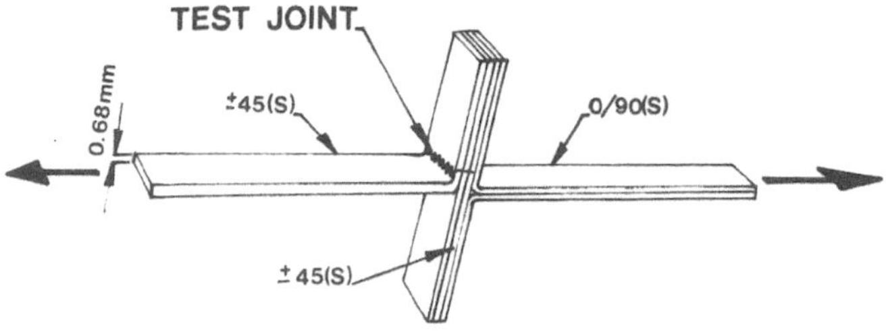

FIG. 9. Peel test coupon.

TABLE 1
Skin/stiffener junction results

Specimen	Load/width (N/mm)
Control-simple co-cured peeling joint	78
Zigzag stitched, 0·6 mm feed, 1·8 mm wide stitches using 70-tex spun Kevlar. Delaminated under the stitching	100
With anti-peel strip one ply woven CFRP at junction. Did not peel—delaminated under the joint	104
With anti-peel strip (two plies woven CFRP at junction). Delaminated under the junction	118

3. THE POSITIONING OF ASSEMBLY JOINTS

Correct location of bonded assembly joints in a postbuckled structure is of great importance. The prime consideration is to avoid placing these joints adjacent to buckled components. Good and bad methods of joining a spar to a postbuckled skin panel are shown in Fig. 10. Joint geometry should also be simple unless the two components to be joined have been produced

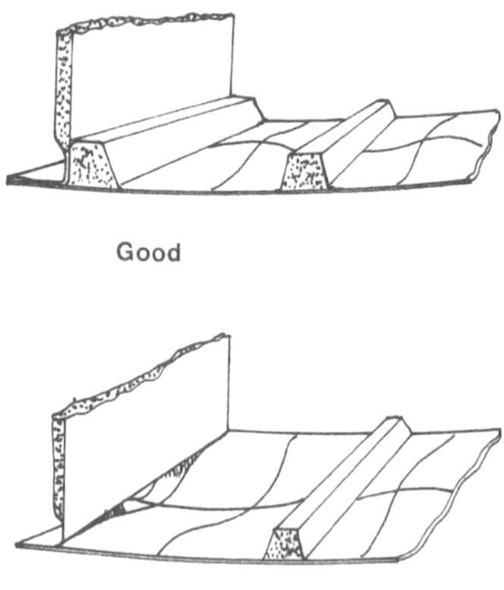

FIG. 10. Spar to skin joints.

on very precise matched tooling. This is one reason why it has been found better to co-cure the rib flanges with the skin. The box structure can then be completed by the use of rib shear webs.

3.1. Clamping of Simple Lap Joints

Clamping of lap joints for assembly, particularly if one side is blind, can be a major problem. This is particularly true for thin sections where the clamping force has to be distributed along the length of the component. The spar joints of testbox 2 were clamped by means of a partial vacuum drawn inside the box, as shown in Fig. 11. This technique is acceptable for

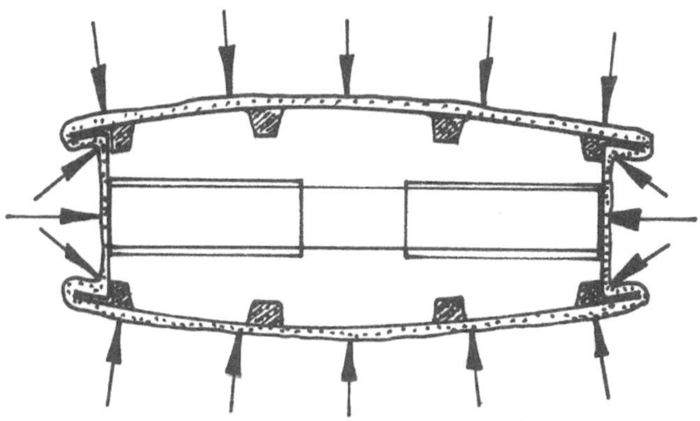

FIG. 11. Vacuum clamping.

boxes of near square section. Boxes of long chord require reinforcement to avoid crushing. An alternative solution is to design a large clamping jig, the cost of which is too high to justify for a one-off structure. This type of simple lap joint also incorporates no means of controlling the bondline thickness. Over-clamping can result. This results in the bondline being too thin with a consequent reduction in strength.

3.2. Slotted Joints

These disadvantages led the author to investigate slotted joints, as shown in Fig. 12. This type of joint can be co-cured as a feature into the structure. It can be designed to be self-clamping with bondline thickness control. It can also locate the components together in the desired position. These joints load the adhesive in pure shear without the peeling effects which can occur in single lap joints.

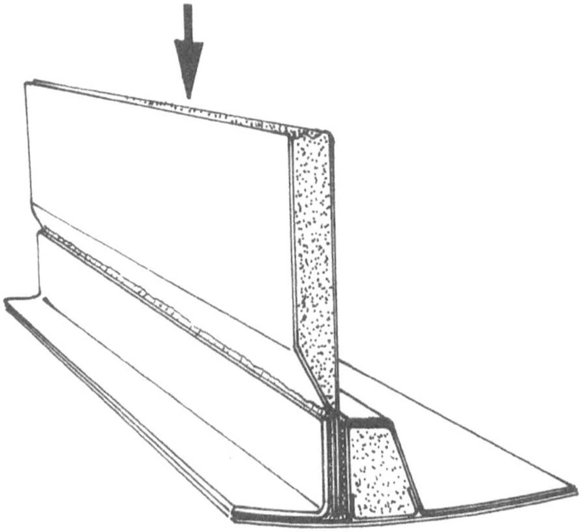

FIG. 12. Slotted joint.

3.3. Spar Joints

For these reasons continuous slotted joints are used to join each spar shear web to the stiffened skin panels. In this design a sandwich stabilised shear web is used. This is designed not to buckle. This 'slot together' joint concept could work particularly well with a sine wave stabilised shear web, requiring very little strength from the adhesive.

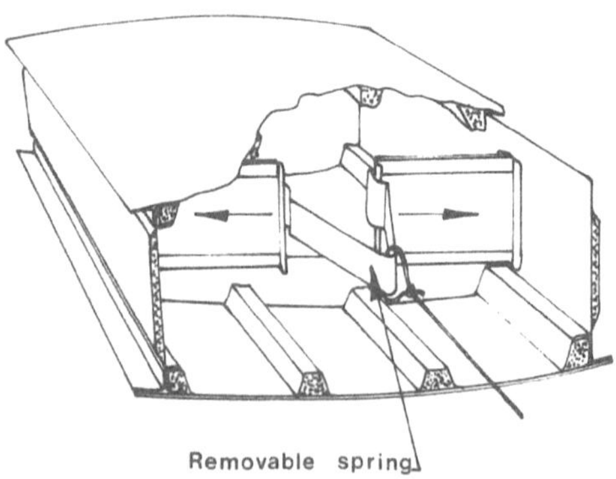

FIG. 13. Rib to spar bonding.

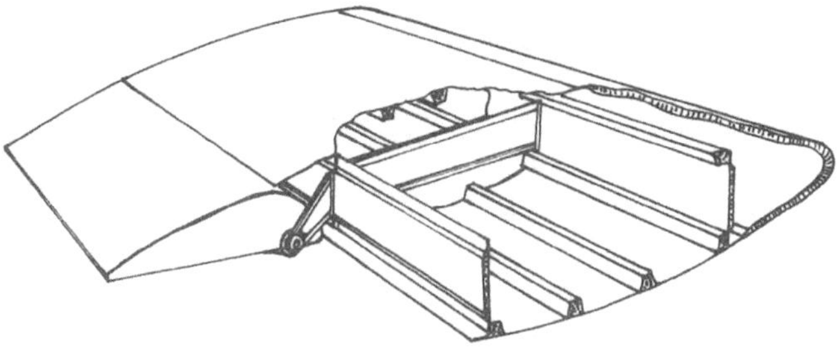

FIG. 14. Pickup rib detail.

3.4. Rib Joints

The ribs are joined together by means of a slotted joint similar to that of the spars. However, in the case of the ribs there can be a need to form a bond between the ribs and the spar shear web (Fig. 13). This is accomplished by cutting the rib into two sections. The bond clamping force is then provided by means of a removable hairpin spring. This spring slides the two rib sections apart along their slotted joints.

Where pickup points such as those for the aileron hinges are required, the solution is to cut the spar at this point. This allows the rib web to come through as a continuous member. The rib then diffuses the load from the ribs into skin shear through the slotted joints and co-cured rib flanges (Fig. 14). Since only the spar shear web has been cut, no problems exist with discontinuity of the spar caps.

4. TOOLING

There are many advantages of using composite materials in the tooling. Complex forms such as the rib tooling can be produced at low cost. The coefficient of thermal expansion can be the same as that of the component to be produced. Tooling can be produced which is of high stiffness with comparatively low weight. The tooling system for this test structure is shown in Fig. 15. For this one-off test structure all the composite tooling is produced by the wet laminating process. A high temperature resin system, SP690/590, is used which cures at room temperature. A slow postcuring cycle is then required to give the resin a heat deflection temperature of 130°C. By using a 120°C curing prepreg for the final structure, this wet laminated tool can be used directly to produce the component. This was a major reason for choosing Ciba-Geigy 913 prepreg material.

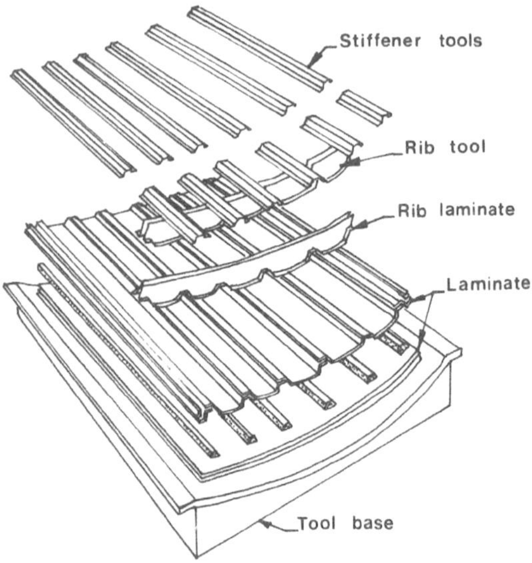

FIG. 15. Tooling system.

4.1. Skin Tooling

The tooling used to produce the co-cured skin panels is of carbon/epoxy. A specially orthotropic laminate 6 mm thick of woven material at 0/90° to the rear spar is used. An 'egg box' type of stiffening structure has been laminated to the undersurface of the tool. The depth of this should be at least $\frac{1}{30}$ of the tool length for adequate stiffness.

4.1.1. Ribs

The tooling for the rib flanges was produced from wet laminated mouldings taken from the tool surface. Using a dummy rib and stiffeners, a carbon/epoxy moulding was made, as shown in Fig. 16. These mouldings were cut at the apex of each stiffener. This allows each part of the moulding to distribute the autoclave pressure evenly on to the co-curing panel.

4.1.2. Stiffeners

The stiffeners were originally produced using silicone pressure intensifiers. This material can cause contamination of the CFRP surface, and so is no longer used. A system of folded steel sections is now used to consolidate the prepreg.

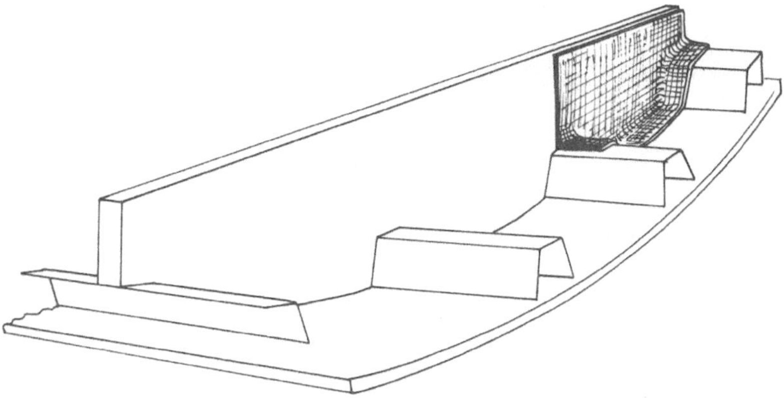

FIG. 16. Rib tool moulding technique.

4.1.3. Jointing features

The slotted joints are formed by inserting PTFE-coated light alloy strips into the moulding. These are removed after cure to reveal a slot. The slot is then prepared for bonding with an abrasive. A layer of textured peel-ply could be used here for the purpose of forming a good bond.

5. CONCLUSIONS

Critical problem areas of postbuckled design in composites have been identified. These are mainly in joint areas between stiffeners and buckled skin panels. Three methods of preventing failure by skin/stiffener separation have been described. Of these, the method of including an anti-peel strip into the junction is the simplest and most effective.

Stitching of the laminate in this area is effective but is a laborious process due to the small stitching pitch required. The technique was not used in the final design on these grounds. However, an upper bound to the process where failure occurs by fibre breakage at the junction has not yet been found.

The location and type of bonded assembly joints has been found to be important for success. It has been shown that joints should be made at structurally stable parts of the structure to avoid debonding.

Other features of the design have been developed to simplify production. These include the overlapping technique at the stiffener caps to avoid bunching; the slotted joints at the ribs and the slotted spar joints. These last two features allow assembly of a bonded structure without the need for expensive clamping systems.

ACKNOWLEDGEMENT

This work is supported by the Royal Aircraft Establishment, Farnborough.

REFERENCES

1. Brooks, W. G., The design and construction of a post buckled carbon fibre wing box structure, *ICAS Conference Proceedings*, College of Aeronautics, Cranfield Institute of Technology.
2. Belgrano, G., The design, manufacture and testing of stiffened aircraft compression panels using advanced materials, M.Sc. structural design thesis, Cranfield Institute of Technology, 1984.
3. Anderson, M. *et al.*, PASCO panel analysis and sizing code, capability and analytical foundations, NACA Technical Memorandum 30181, January 1980.
4. Copes, R. and Pipes, R., Design of the spar to wingskin joint, Centre for Composite Materials, University of Delaware; Lenoe *et al.*, *Fibrous Composites in Structural Design*, New York, Plenum Press, 1978.

13

Design of 32-m Diameter GRP/Composite Blade

CHEN YUN-CHENG, CHENG YU-YUE and WU PEI-CHAO

Shanghai GRP Research Institute of the State Administration of Building Materials Industry, 700 Yao Hua Road, Shanghai, People's Republic of China

ABSTRACT

This paper gives a general review of the development of GRP/composite blades in China, together with their application and advantages.

This paper mainly discusses structural design of 32-m diameter GRP/composite blade for 200-kW wind turbine. Description of the main formulae of aerodynamic loads and centrifugal forces is given. The strength and stiffness of the GRP/composite blade can be calculated and tailored according to the load characteristics. The main amount of fibreglass should be laid in the axial direction of the blade.

The rotating blade is subjected to periodic loads. Because fatigue is serious in large blades, we calculate the structural response by using Friedmann's aeroelastic modelling.

Blades are a key part of a rotor and their performances have a great influence on the performances of the rotor machine. GRP/composite will be very satisfactory for the aerodynamic and structural requirements of the blade.

INTRODUCTION

GRP/composite, as a new type of structural material, has been widely used in various blades. A GRP/composite blade has four remarkable advantages. (1) A reasonably constructed GRP/composite blade can sustain loads effectively. The blade strength and stiffness can be tailored in accordance with external forces exerted on the blade. GRP is a heterogeneous material consisting of glass fibre and resin and can be treated as a kind of structure; its properties depend on those of glass fibre and resin, and their ratio and interface. (2) A sophisticated airfoil shape can be easily fabricated, thus increasing the efficiency of the rotor. (3) A GRP blade has good fatigue resistance and effectively sustains the dynamic forces. (4) A GRP blade has excellent corrosion resistance and weatherability. Many experts regard GRP/composite as an ideal material for the manufacture of the blade.

Early in the 1960s, led by Professor Zhu Yi-Ling,[1] our institute began to study, design and manufacture GRP blades. They are axial flow fan blades, cooling tower blades, blower blades, hovercraft propeller blades (tip speed 300 m/s), wind tunnel blades and wind turbine blades. Seven types of wind turbine blades (rated powers 1–50 kW, diameters 3–19 m) have been developed successfully. A 32-m diameter blade for a 200-kW wind turbine is being built at present. This will be the largest blade in China.

AERODYNAMIC DESIGN

Aerodynamic performance of a blade is one of the important properties of a wind turbine. An optimum wind turbine not only requires a high power coefficient but also lower noise and operating safety. In order to design optimum wind turbine blades, several methods are available, such as Glauert's, Wilson's, Griffiths's, etc.[2–4] The experiences are very important. The configuration obtained from theoretical calculation should be modified to meet the requirements of both aerodynamic and structural performances. Due to the high strength and lower elastic moduli of GRP, we should choose larger thickness/chord ratios to enhance the stiffness of the blade.

STRUCTURE DESIGN

1. Shapes of the Blade Cross-section

The structural shape of the blade section has a great influence upon the property of the whole structure; a hollow thin-wall structure which can both increase stiffness and reduce weight is usually used. In order to increase the strength and prevent instability, spar, rib and foam which are filled in hollow are used according to various structural requirements. Typical blade sections of hollow thin-wall structure are shown in Fig. 1.

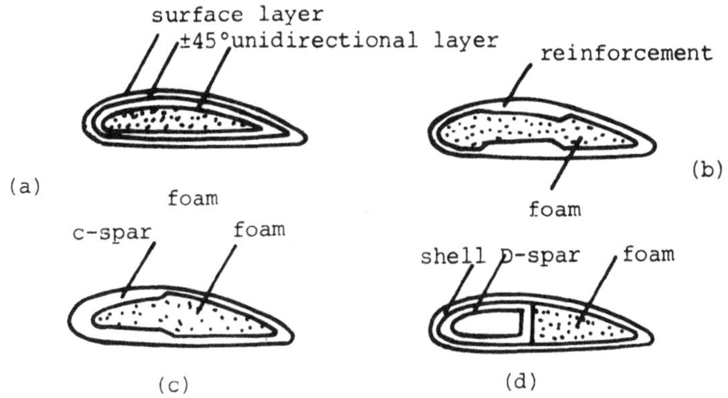

FIG. 1. Some typical sections of hollow thin-wall blades.

Unidirectional layers which are made of 4:1 or 7:1 or 11:1 glass cloth (the ratios of warp glass content to woof glass content) are subjected to axial forces (the bending moments and centrifugal force); $\pm 45°$ layers which are made of 1:1 glass cloth are subjected to torsional moments and shear forces. The main loads on the blade are longitudinal, so it is suitable to use unidirectional GRP. The surface layer is the cloth which has a rich resin or resin coating so as to increase aerodynamic efficiency and corrosion resistance.

2. Cloth Lay-out

The strength and stiffness of a GRP/composite blade can be tailored according to the load characteristics. The major amount of fibre glass should be laid in axial direction of the blade. As the bending moments, torsional moments and centrifugal force become greater from tip to root, so does the thickness of thin-wall structure. The strength of the blade must be considered as well as the deformation condition. As every other dimension of the blade section is much shorter than its length, the calculation of strength and stiffness can be simplified by treating as a cantilever thin-wall beam rather satisfactorily. Cloth used in the blade are orthotropic materials. Some mechanical properties of GRP are listed in Table 1. The properties of GRP are rather complex. They are related to fibre properties and its direction, resin properties and its content, surface treatment of fibre and manufacturing technology.

In practical design, we first assume thickness distribution and calculate the strength and deformation, then modify the thickness. Thus, if this is repeated several times, we can obtain an optimum thickness distribution.

TABLE 1
Some mechanical properties of GRP

Types	E_L (MPa)	E_T (MPa)	G_{LT} (MPa)	σ_L (MPa)	σ_T (MPa)	A^a (%)
1:1 cloth GRP	$1{\cdot}8 \times 10^4$	$1{\cdot}8 \times 10^4$	$3{\cdot}6 \times 10^3$	300	300	45
4:1 cloth GRP	$2{\cdot}6 \times 10^4$	$1{\cdot}2 \times 10^4$	$2{\cdot}9 \times 10^3$	500	140	45
7:1 cloth GRP	$4{\cdot}2 \times 10^4$	$1{\cdot}8 \times 10^4$	$5{\cdot}5 \times 10^3$	900	170	22

[a] σ_L is longitudinal strength; σ_T is transverse strength. A is resin content.

Increasing the $\pm 45°$ layers in laminate, the longitudinal elastic moduli of the laminate is reduced and the shear modulus increased. The longitudinal elastic modulus E and shear modulus G in laminate approximately are[1]

$$E = E_0(1 - K) + E_{45°}K \quad (1)$$

$$G = G_0(1 - K) + G_{45°}K \quad (2)$$

where E_0 and G_0 are longitudinal tensile elastic modulus and shear modulus in the unidirectional layers, respectively. $E_{45°}$ and $G_{45°}$ are the longitudinal tensile and shear modulus in the $\pm 45°$ layers, respectively. K is the ratio of 45° layers to the laminate. Thus, we can easily estimate the deformation of blade and average stresses in the laminate by means of classical beam theory. For further calculation, the composite laminate theory[5] is needed.

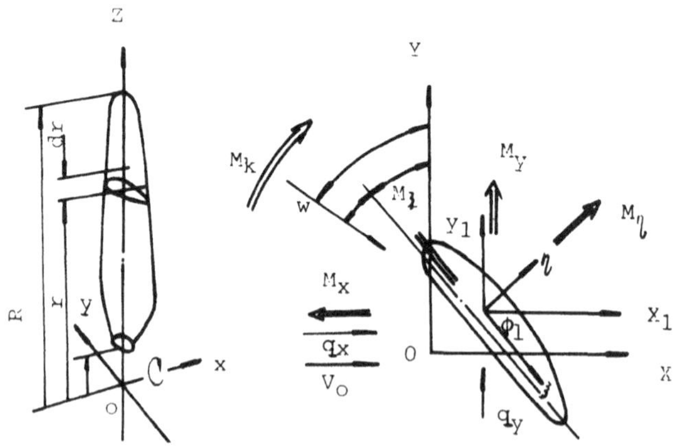

FIG. 2. Blade subjected to loads (ϕ_1 is the centre of gravity section).

3. Load Calculation

Under operating conditions, the loads acting on the blade include inertial force, aerodynamic force and gravity, and are a function of time. They are especially influenced by gust and wind direction, therefore the aerodynamic load is random.

In addition, the loads are not the same under different operating conditions. Thus it is difficult to define the operating conditions rationally and calculate the loads accurately. The static and rated operating conditions are considered. In the static condition, we assume that the wind speed is a maximum and the blade is stationary.

The loads acting on the blade are shown in Fig. 2.

Aerodynamic loads

(a) Aerodynamic loads (per unit length)

$$q_{xa} = 0{\cdot}5cw^2c_x \qquad q_{ya} = 0{\cdot}5cw^2c_y \qquad m_c = 0{\cdot}5c^2w^2c_m \tag{3}$$

(b) Shears

$$Q_{xa} = \int_r^R q_{xa}\,dr \qquad Q_{ya} = \int_r^R q_{ya}\,dr \tag{4}$$

(c) Bending moments

$$M_{xa} = \int_r^R q_{ya}(r_1 - r)\,dr_1 \qquad M_{ya} = \int_r^R q_{xa}(r_1 - r)\,dr_1 \tag{5}$$

(d) Torques

$$M_{ka} = \int_r^R L(X_c - c/4)\,dr + \int_r^R m_c\,dr \tag{6}$$

where c is airfoil chord; $C_x = C_L \cos\phi + \sin\phi$, $C_y = C_L \sin\phi - C_D \cos\phi$ and C_L, C_D, C_m are profile, life and drag coefficient, respectively; L is lift; and X_c is elastic centre.

Gravity loads

(a) Shear

$$Q_{yw} = \int_r^R \rho_s g F \sin\psi\,dr \tag{7}$$

(b) Tension

$$P_w = \int_r^R \rho_s g F \cos\psi\,dr \tag{8}$$

(c) Bending moment

$$M_{xw} = \int_r^R \rho_s g F(r_1 - r) \sin\psi \, dr_1 \tag{9}$$

(d) Torque

$$M_{kw} = \int_r^R \rho_s g F(X_0 - X_c) \sin\theta \sin\psi \, dr \tag{10}$$

where ρ_s is equivalent GRP mass density, F is equivalent blade cross-sectional area, X_0 and Y_0 are equivalent gravitational centres, ψ is azimuthal angle and θ is blade twist angle.

Centrifugal loads

(a) Tension

$$P = \Omega^2 \int_r^R \rho_s F r \, dr \tag{11}$$

(b) Shear

$$Q_{yp} = \Omega^2 \int_r^R \rho_s F Y_0 \, dr \tag{12}$$

(c) Bending moments

$$M_{xp} = \Omega^2 \left[r \int_r^R \rho_s F Y_0(r_1) \, dr_1 - Y_0(r) \int_r^R \rho_s F r_1 \, dr_1 \right]$$

$$M_{yp} = \Omega^2 \int_r^R [X_0(r) - X_0(r_1)] \rho_s F r_1 \, dr_1 \tag{13}$$

(d) Torque

$$M_{kp} = -\Omega^2 \left\{ \int_r^R \rho_s J_{x_1 y_1} \, dr + \int_r^R \rho_s F Y_0(r_1)[X_0(r_1) - X_0(r)] \, dr_1 \right\} \tag{14}$$

where $J_{x_1 y_1}$ is the product of inertia about the x_1, y_1 axes. According to the loads, we can conveniently obtain the stresses and deformations in each cross-section of the blade.

4. Calculating Stresses and Deformations

Supposing ξ, η are the principal inertia axes of section and J_ξ, J_η are the principal inertia moments, then the relations are:

$$M_\xi = M_x \sin\theta + M_y \cos\theta \qquad M_\eta = -M_x \cos\theta + M_y \sin\theta \tag{15}$$

The tensile stress and bending stress are:

$$t = \frac{\Sigma P}{F} \qquad \text{(tensile stress)} \tag{16}$$

$$b = \frac{M_\xi}{J_\xi}\eta + \frac{M_\eta}{J_\eta}\xi \qquad \text{(bending stress)} \tag{17}$$

Because J_η is sufficiently large, we can neglect deformation in direction ξ. We can calculate deformations in directions y and x:

$$\Delta y = \int_{r_0}^{R} \frac{M_\xi}{EJ_\xi}(R-r)\sin\theta\,\mathrm{d}r \qquad \Delta x = \int_{r_0}^{R} \frac{M_\xi}{EJ_\xi}(R-r)\cos\theta\,\mathrm{d}r$$

Total deformation is

$$\Delta = \sqrt{\Delta x^2 + \Delta y^2} \tag{18}$$

This is a first-order approximation. In further calculations, the interaction between the centrifugal force and deformation must be considered.

5. Joint Design of GRP Blade

Loads of a blade are maximum at the root. Transferring the loads which are exerted on the blade to the hub is an important problem for designers. Joint design must solve this problem. Since GRP interlaminar shear strength is lower, we must study suitable forms of joint which can avoid this weakness of GRP. We always pay great attention to this topic.

6. Analysis of Vibration

The rotating blade is subjected to periodic loads. The significant periodic loads are gravity, gyroscopic and aerodynamic forces. These periodic loads may excite the structural response. If the exciting frequency is approaching one of the natural frequencies of the blade, resonance may happen at that frequency. In order to avoid resonance, designers must predict the frequencies of the blade. It is also important to predict the aeroelastic stability boundaries to avoid flutter of the blade.

Friedmann[6] has presented a detailed discussion of the problem about aeroelastic responses and the stability of large wind turbines. In the modelling, effects of the wind shear, tower shadow and gust, etc., were considered. We use this aeroelastic modelling and derive fundamental equations. Due to the complexity of the equations, Galerkin's energy method is used and ordinary differential equations are derived:

$$M\ddot{\boldsymbol{\delta}} + (C + C_g + C_a)\dot{\boldsymbol{\delta}} + (K + K_g + K_a)\boldsymbol{\delta} = \mathbf{F} \tag{19}$$

where M, C and K are mass, structural damping and stiffness matrices, respectively; C_g is Coriolis matrix; K_g is centrifugal force matrix; C_a and K_a are aerodynamic damping and stiffness matrices, respectively; $\boldsymbol{\delta}$ is the generalized displacement vector; and $\mathbf{F}$ is the generalized exciting force vector. Equation (16) is based on the assumption that the airfoil lift coefficient is proportional to the angle of attack.

The steady-state responses are important for fatigue life estimation. In order to obtain the steady-state response, an integration must be carried out. Stability boundaries can be obtained from homogeneous equations:

$$M\ddot{\boldsymbol{\delta}} + (C + C_g + C_a)\dot{\boldsymbol{\delta}} + (K + K_g + K_a)\boldsymbol{\delta} = 0 \tag{20}$$

DESIGN OF 32-m DIAMETER GRP/COMPOSITE BLADE

1. Main Data for the Blade

Rotor data

Rated power	200 kW
Blade number	3
Rotating speed	41 rpm
Working wind speed	4·5 ~ 20 m/s
Length of the blade	15·1 m
Chord (minimum)	0·6 m
Mass (per D-spar)	700 kg
First-order frequency	2·4 Hz
Diameter	32 m
Airfoil	NACA 4410·5 ~ 29·5
Rated wind speed	11·5 m/s
Survival wind speed	60 m/s
Chord (maximum)	1·77 m
Twist	10·6°
Mass (per blade)	1451 kg

Blade configuration

For blade configuration, see Table 2 and Fig. 3.

TABLE 2
Blade configuration

Radius r (m)	Chord c (m)	Twist θ (°)	Thickness t (m)
3	1·77	10·6	0·5220
6	1·5	5·9	0·3290
8	1·32	4·2	0·2575
10	1·14	2·9	0·1993
12	0·96	1·8	0·1490
14	0·78	0·87	0·1040
16	0·60	0	0·0630

2. Aerodynamic Characteristics of the Rotor (see Table 3)

TABLE 3
Aerodynamic characteristics[a]

V_0 (m/s)	P (kW)	T (kN)	C_P	C_T
5	11·2	12·6	0·18	1·03
7	67·4	21·4	0·40	0·89
9	157·5	31·0	0·44	0·78
11	277·9	40·5	0·42	0·68
12	344·4	44·8	0·40	0·63
14	480·4	52·4	0·36	0·54
16	598·1	58·2	0·30	0·46
20	734·8	65·6	0·19	0·33

[a] $n = 41$ rpm; $\theta_{tip} = 2°$.

3. Structural Design of 32-m Diameter Blade

Design conditions

The determination of design conditions and calculating loads is very important, but it is also very difficult. In the designing, we mainly use the quasi-static method. This method considers inertial force, aerodynamic force and alternative gravity. We can calculate quasi-static loads by using eqns (3)–(14). The three main design conditions must be considered. They are shown in Table 4.

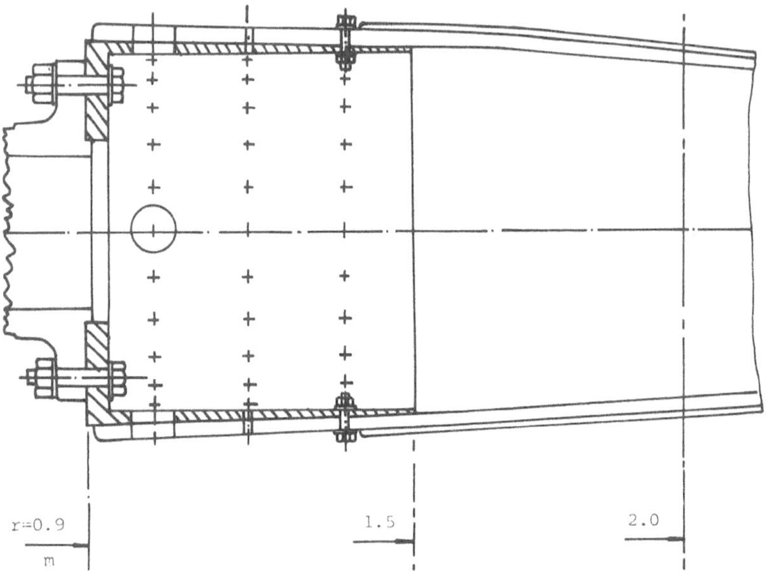

FIG. 3. The root steel structure is glued into the blade.

TABLE 4
Main design condition

Design condition	Definition	Blade place and wind speed	Loads: Bending moment $M_{y,o}$ $(N-m)$	$M_{x,o}$ $(N-m)$	Thrust $F_{x,o}$ (N)
1	Normal operating condition, $\theta_{tip}=2°$, $n_R=41$ rpm; infinite life	Upper, $V_o=20$ m/s	$1{\cdot}9\times10^5$	$7{\cdot}3\times10^4$	$2{\cdot}1\times10^4$
		Lower, $V_o=10$ m/s	$1{\cdot}2\times10^5$	$1{\cdot}0\times10^4$	$1{\cdot}3\times10^4$
2	Transient dangerous condition, overspeed $n=120\%$, $n_R=50$ rpm, wind deviation $=45°$, $\theta_{tip}=2°$, $C_L=1{\cdot}5$, $N=10^4$ cycles	Upper, $V_o=27$ m/s	$M_{f,o}=5{\cdot}5\times10^5$		$F_{f,o}=6{\cdot}3\times10^4$
		Lower, $V_o=14$ m/s	$M_{f,o}=1{\cdot}5\times10^5$		$F_{f,o}=1{\cdot}6\times10^4$
3	Strength check condition		$M_{f,o}=7{\cdot}8\times10^5$		$F_{f,o}=1{\cdot}1\times10^5$

Note: f, due to flapping; o, at root.

Structural design specifications

(*a*) *Main targets of the blade design*

— The mass per blade ≤ 1500 kg.

— The gravity centre of the section from leading edge ≤ 30% C.

— Main structures are not unstable and buckling in any operating condition.

— The deflection in the normal condition ≤ 0·64 m.

— The deflection in any condition ≤ 1·5 m.

(*b*) *Mechanical properties of GRP used in the spar* (*see Table 5*)

TABLE 5
Mechanical properties of GRP

Items	MPa	MPa
Tension	$E_L = 2 \times 10^4$	$E_T = 1 \times 10^4$
	$\sigma_L = 350$	$\sigma_T = 100$
Compression	$\sigma_L = 260$	
Shear	$G = 3 \times 10^3$	$\tau = 55$
Bending	$E_L = 2{\cdot}4 \times 10^4$	$E_T = 1 \times 10^4$
	$\sigma_L = 400$	$\sigma_T = 110$
Fatigue ($N = 10^7$)	$\sigma_{-1} = 100$	$\sigma_0 = 140$

Main results

(*a*) *Main properties of blade sections* (*see Table 6*)

TABLE 6
Main properties of blade sections

r(m)	1·5	3	4	5	6	7	8	9
F(m^2 × 10^{-4})	744·5	680·1	588·9	517·3	460·8	404	351·5	304·2
J(m^4 × 10^{-4})	42·8	24·4	14·7	9·7	6·7	4·6	3·2	2·2
J(m^4 × 10^{-4})	42·8	67·8	52·9	42·1	33·9	25·5	20·2	15·9
J_K(m^4 × 10^{-4})	85·5	61·1	47·3	33·2	23·8	16·7	11·2	7·6

r(m)	10	11	12	13	14	15	16
F(m^2 × 10^{-4})	260·9	219·5	193·5	132·9	96·6	72·1	56·2
J(m^4 × 10^{-4})	1·44	0·92	0·54	0·32	0·15	0·08	0·03
J(m^4 × 10^{-4})	12·3	9·17	5·95	4·02	2·58	1·61	0·97
J(m^4 × 10^{-4})	4·93	3·17	1·95	1·14	0·58	0·28	0·14

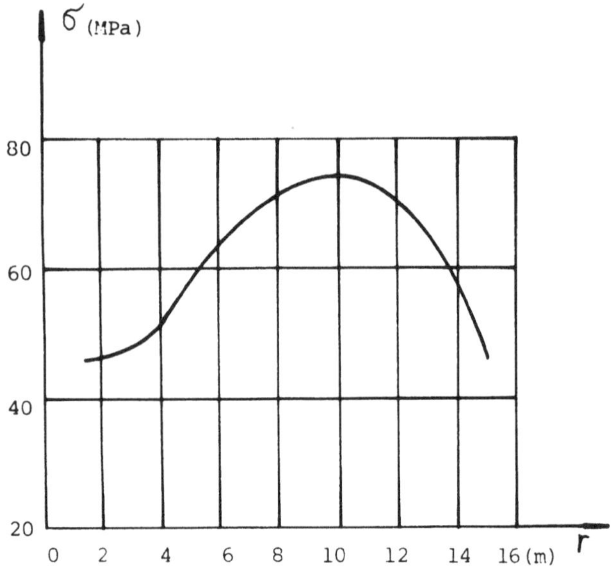

FIG. 4. Bending stresses for design condition 2.

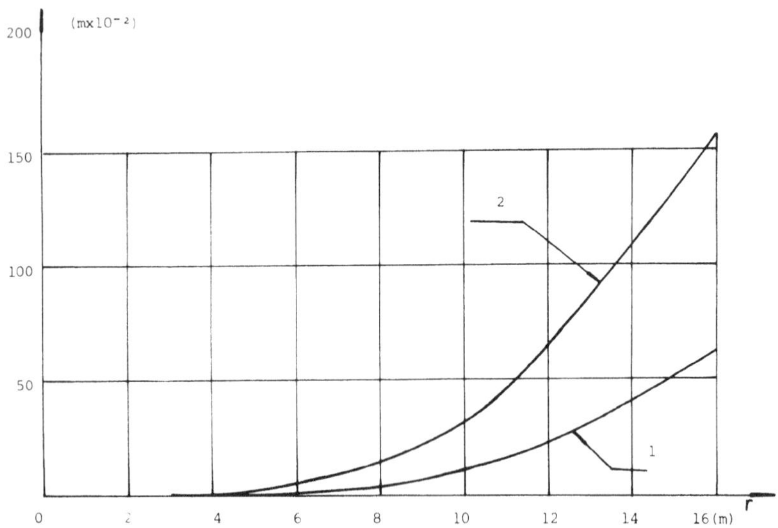

FIG. 5. Deflection for design conditions 1 and 2.

(b) Mass distribution

D-spar	700 kg	Steel	228 kg
Foam	110 kg	Root structure	388 kg
Others	25 kg		
Altogether	1451 kg		

(c) Strength and stiffness

— The bending stresses for design condition 2 is shown in Fig. 4.
— The deflections for design conditions 1 and 2 are shown in Fig. 5.

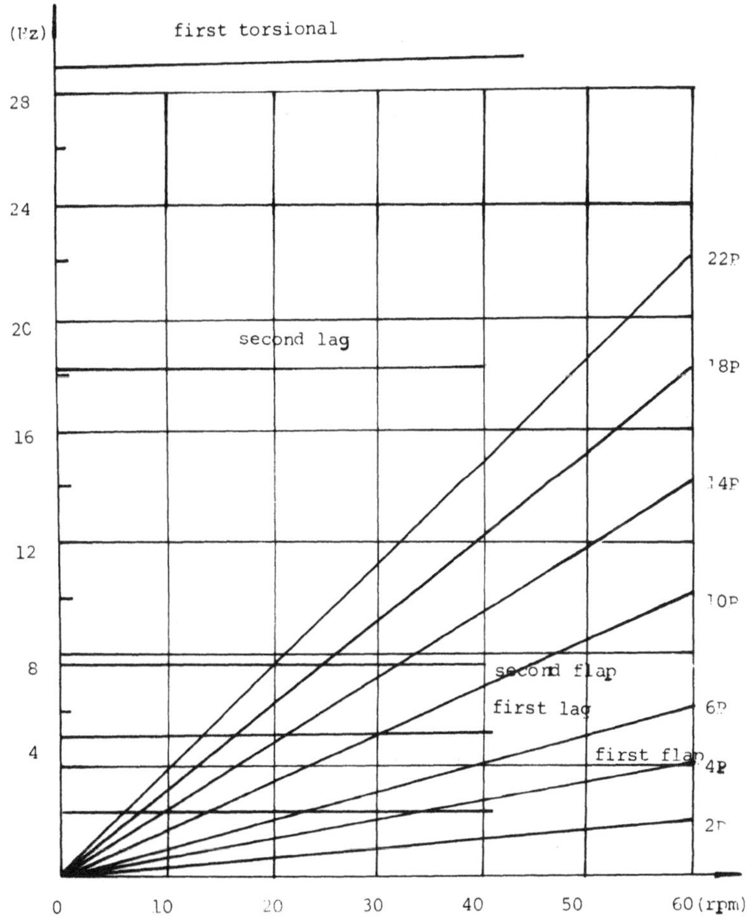

FIG. 6. Campbell diagram of the blade.

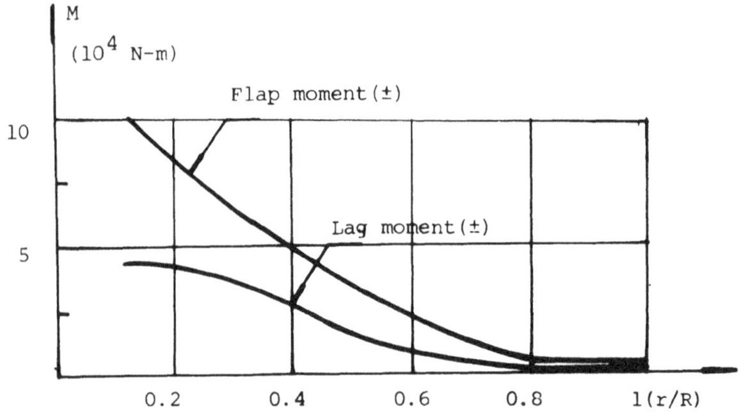

FIG. 7. Cyclic loading under design condition 1.

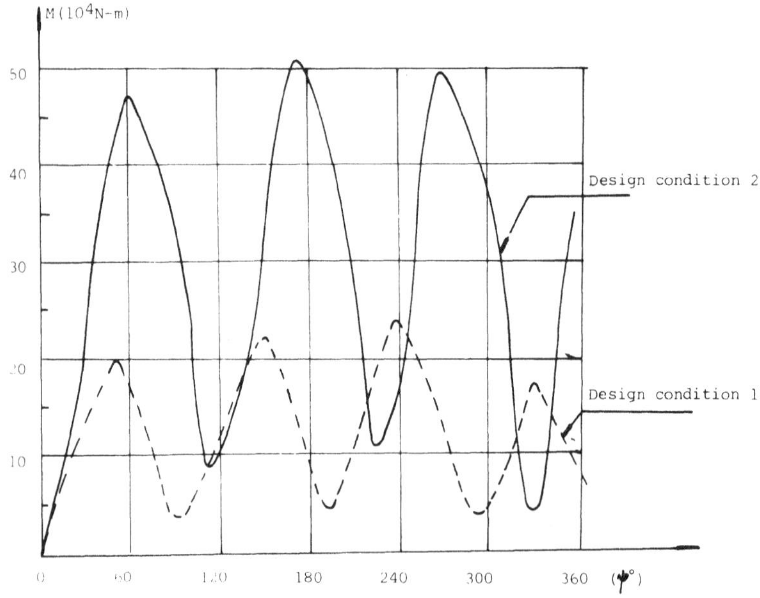

FIG. 8. Cyclic loading at root under design condition 2.

4. Dynamic Characteristics of the Blade

Figure 6 shows a Campbell diagram of the blade. The first flap natural frequencies of the blade are 2·4 Hz in the stationary condition and 2·6 Hz in the rated condition. The first lag natural frequency is 5·2 Hz. Figure 7 shows cyclic loading under design condition 1. Figure 8 shows cyclic loading at root under design condition 2.

ACKNOWLEDGEMENTS

I am very grateful to Professor Zhu Yi-Ling, Helge Petersen (Denmark, Risø Library, design adviser of 200-kW wind turbine) and Deng Zhao-Huai (senior engineer), who gave me a lot of help. I also give many thanks to my colleagues.

REFERENCES

1. ZHU YI-LING, *GRP Structural Design*, China, Building Engineering Publishing House, 1980.
2. GLAUERT, H., *Airplane Propeller*, Berlin, Julius Springer, 1935.
3. WILSON, R. E. and WALKER, Performance-optimized horizontal axis wind turbines, International Symposium on Wind Energy Systems, 1976.
4. GRIFFITHS, R. T., *Applied Energy*, **13**(3) (1983), 183–194.
5. JONES, R. M., *Mechanics of Composite Materials*, New York, McGraw-Hill, 1975.
6. FRIEDMANN, P. P., Aeroelastic stability and responses analysis of large horizontal-axis wind turbines, *J. Industrial Aerodynamics*, **5** (1980), 373–401.

14

Design, Manufacturing and Tests of Large Wind Turbine Rotor Blades

J. A. GÜEMES

ETSI Aeronáuticos, Universidad Politécnica de Madrid, 28040-Madrid, Spain

and

F. AVIA

Instituto de Energías Renovables, Avda. Complutense 22, 28040-Madrid, Spain

ABSTRACT

The main requirements influencing the design of wind turbine rotor blades are discussed. The key element in achieving an economically successful WECS is the ability to manufacture the rotor blades at low cost in a configuration which will meet the severe structural requirements, produce a high level of performance and have a long life with low maintenance. A new manufacturing method has been conceived and successfully tested.

Details on the process and the ongoing qualification program are given.

1. INTRODUCTION

Starting with the oil embargo, a large number of activities has been undertaken looking for new energy sources. Wind energy conversion systems (WECS) are thought as the most effective to produce electrical energy. Every range of power has been investigated. Small machines, up to 50 kW, mainly intended for individual or isolated applications. Medium size (typically 300 kW) or large machines (more than 1000 kW), connected to utility networks, could afford approximately 3% of the overall electric energy consumption at a cost competitive with fossil or nuclear fueled sources.

Large machines are potentially the most effective in cost terms, due to the scale effect on the cost of the components. At the present time, technology for small or medium size machines has attained its maturity, running for long periods without need for special attention. The costs of these systems are still high, needing institutional help to develop its market.

Large systems are at the demonstration stage. Contributing to the international efforts, European Communities are sponsoring several projects in this area, one of which, called AWEC-60 (Advanced Wind Energy Converter, 60 m diameter), is being developed by German and Spanish firms. This chapter deals with some of the design requirements, materials and processes considered, and the concept selected for this AWEC-60 rotor blade.

2. DESIGN REQUIREMENTS

The external blade shape is dictated by aerodynamics requirements, to achieve the maximum annual energy output, the wind variability being taken into account. Minor adjustments in external geometry are allowed, for structural or manufacturing reasons, but their effects on performances have to be carefully monitored.

Internal blade geometry has to be defined to satisfy three structural criteria:

(A) High strength to withstand the aerodynamics, gravity and inertial loads resulting from operational and extreme loading conditions. Fatigue and environmental effects have to be considered.

(B) Adequate stiffness and structural damping to avoid strong resonances. Natural frequencies must be higher than rotor speed, and non-coincident with its harmonics, in order to avoid dynamic load amplification.

(C) Aeroelastic stability requirements. Rotary wings are prone to flap-lap and torsional modes of instability. Gravity centers in each section must be located properly, and high torsional stiffness is required. For this reason the concept of a structural D spar is usually preferred over the two halves of a monocoque structure.

For quality and cost requirements, this chapter mainly concerns materials and process selection. Early designs had tried metallic structures, dural or steel welded constructions, but it was soon recognized that composite materials offered better characteristics for the production of long, slender, tapered beams as wind turbine blades are.[1] The high directionality of stresses, the good environmental resistance needed and the complicated shape to reproduce makes this application ideally suited for composites. The better properties/cost ratio is found with E-glass/polyester, so this material is the most commonly used.

To be economically competitive with other energy sources, the recurrent cost for a system such as AWEC-60 should be around 200 Mptas (1·5 MECUS approximately) (Ref. 2). Most of the elements of the machine are already conventional, well-developed products, so cost reductions due to engineering advances are rather difficult for them. To satisfy cost objectives, the three rotor blades should cost no more than 15% of the aforementioned quantity.

The weight of each blade, satisfying structural requirements, is found to be near 4000 kg (in E-glass/polyester, assuming properties at a fiber volume content of 50%). That means that the cost of the finished blade, in serial production, should be around 2600 Ptas/kg (200 ECU/kg), a rather difficult target for the quality of laminate required. Reproducibility, homogeneity, etc., are other factors.

3. DESIGN AND ANALYSIS

The design preferred was a continuous load carrying D spar, adapted to the external shape of the airfoils from the leading edge to the maximum thickness (30% of the chord, approximately), and then close in circular form, for manufacturing reasons which will later be explained. At the root end a D spar has a cylindrical shape, 1·8 m diameter and 40 mm wall thickness, where it is bolted to a metallic flange, joining the hub and blade pitch movement system. With the exception of the blade root attachment, where the configuration is 50% 0°, 25% 90°, 25% ±45°, the laminate of the D spar is mainly longitudinal fibers; only 20% of fibers are at the ±45° direction to achieve the minimal transversal properties needed. Specimen tests, conditioned at water saturation, have given a minimum tensile strength of 400 MPa in the longitudinal direction with an elastic moduli of 28 GPa.

The secondary structures are sandwich panels, with skins of quasi-isotropic GRP and nuclei of PVC foam, regularly supported by ribs and joined to the spar by adhesive and screws. Its strains being imposed by the spar deformations, the design concern for these elements is structural stability.

Dynamic stresses at each point of the structure, for each load case, were obtained by a computer program called EOLO (Ref. 3).

Inputs for this program are external geometry, stiffness, pretwist and mass distribution, position of centers in each section, operational and wind speed conditions. It calculates the aerodynamic loads produced by the fluctuating wind speed or consequent to blade vibrations, and inertial loads. As a result, maximum stress envelopes are obtained.

Cumulative fatigue produced by normal operations should not exceed contours of 10^8 cycles, equivalent to 20 years of service. Extreme conditions, such as hurricanes or emergency feathering, should not exceed the maximum strength. The safety factor is 2·5 in all cases.

4. BLADE MANUFACTURING PROCESS

This is the most debatable matter in the development of large blades. Up to now, two methods have proven to be technically successful:

(a) Filament winding. A sequential procedure is needed to accommodate wall thickness variation; fiber orientation will be near $\pm 25°$.
(b) Transverse filament tape. The tape is circumferentially wound on to the mandrel.

In both cases, the mandrel must have adequate stiffness and need to be extracted. Investment is extremely high, making the price fourfold of the target.

We have conceived and developed a novel procedure, making use of a rigid external mould and an internal conflatable mandrel able to meet the aforementioned requirements. At the time of submission of this chapter, it has been successfully tested for producing inner and outer sections of the blade, in lengths up to 4 m.

Raw materials are longitudinal E-glass tapes of 300 mm width, thixotropic polyester resin of medium viscosity, and long pot life at room temperature.

The process consists of the following steps:

(1) Tapes are longitudinally layed on to a female mould, reproducing the leading edges, by an automated mechanism running along the length of the blade. Previously, the tape went through a bath of resin, where consumption of resin is controlled for the desired fiber content; local

inhomogeneities in resin distribution will be corrected at a later stage. Thickness variations can be easily produced by programming the number and distribution of tapes at each section.

(2) After completion of the laminate for the leading edge, a conflatable mandrel is extended on to the cavity; a slight overpressure is given, adapting the mandrel to the walls of the lower mould and leaving a cylindrical free surface.

(3) Lamination proceeds on to the free surface of the mandrel, closing the spar, by a process similar to that described in (1). After finishing, an upper metallic mould is put on to and secured to the lower mould.

(4) Internal pressure increased to 0·3 MPa. A closed loop for heating affords a smooth increase of temperature; a combination of pressure and the lower viscosity associated with higher temperatures ensure a good compaction of the laminate. Gel transition takes place at 90°C; this temperature is maintained for 1 h to achieve a fully cured product.

(5) Removing the upper mould, ribs and secondary structure are positioned and joined. The procedure is the same for the root flange.

5. QUALITY ASSURANCE AND QUALIFICATION TEST

Blade sections manufactured by this process show a very uniform resin distribution (fiber volume $= 55 \pm 5\%$) and low void content ($<2\%$), even bad resin distributions were intentionally obtained; pressure and temperature greatly helps in achieving this uniformity. Advance of curing and wall thickness is monitored at selected places by dynamic dielectric analysis.

Tests of 1:10 scale specimens simulating the attachment are fatigue tested by rotary bending, to evaluate the fatigue strength and obtain information about reduction of the initial stresses in the bolts.

A part of the outer blade has been loaded to ultimate strength by a bending test, to measure the buckling limit of the shell construction and debonding problems.

A full blade ultimate load test is planned for mid 1987.

Testing of subsequent blades will consist of:

Visual inspection and dimensional verification.

Measurement of mass, gravity center, etc.

Loading at limit load and measurement of deflections; evaluation of damage, if any, and calibration of extensometry.

Measurement of natural frequencies. Acceptance limits.

REFERENCES

1. BOWEN, D. H., MAXKELL, C. W. A. *et al.*, Materials aspects of large aerogenerator blades, *European Wind Energy Conf.*, Oct. 1984, pp. 281–286.
2. Plan de Energías Renovables, publ. Ministerio de Industria y Energía, 1985.
3. GÜEMES, J. A., EOLO, Manual de Usario, publ. ETSI Aeronáuticos, 1983.

15

Wind Turbines—The Potential for Composites

GEORGE ELLIOT and ALAN J. WOOTTON

Department of Trade and Industry, National Wind Turbine Centre, National Engineering Laboratory, East Kilbride, Glasgow G75 0QU, Scotland, UK

ABSTRACT

The paper provides an overview of the emerging wind turbine industry particularly in the UK. It explores the potential for introducing composite materials into a whole range of components which are obtained in wind turbines and why traditional materials solutions can be the source of problems.

Wind energy has been harnessed by many through the ages both for transportation in the form of sailing ships and for power in the form of wind-driven machines. Wind-driven machines have been used for many purposes, mainly for grinding grain and pumping water; indeed parts of the world, like Holland, relied on them for thier very existence in draining the large areas of land which lie below sea level. In Britain the story was much the same. Windmills were widely used for grinding grain throughout the country and evidence of small wind pumps are still plentiful for agricultural land drainage which was a typical use for wind machines throughout the civilised world.

Wind-driven machines did not feature to any great extent in the industrial evolution of Britain or the western world. Energy was supplied from thermal sources such as wood, coal and oil. More recently nuclear power has been used. Little had been done to harness the energy of the wind for electrical generation although a few machines were operated in the 1940s and 1950s by the North of Scotland Hydro Electric Board in some of their remoter areas. It was not until the 1973 oil price increase, forced on the western world by the OPEC countries with the threats of an energy-scarce world that interest in wind energy and other alternative sources became rekindled. That a vital commodity such as energy could be under the control of another nation or group was a difficult pill to swallow. The need to be self-sufficient in energy became very dominant in the thinking of the time and indeed if a source of energy could be everlasting or renewable then that seemed a sensible way forward. In the UK research was initiated around that time into the complete range of renewable energies ranging from solar, biomass, wind, wave, tidal to geothermal as well as a renewed look at energy efficiency and conservation.

On the various energy research programmes sponsored by the UK Department of Energy it is believed that the most promising of these sources of energy is wind energy. Most promising in the context that it can be harnessed and utilised on a national level at a price that is within the range of existing forms of electrical power generation. The research story over the last decade or so has been similar in a few other countries notably the United States who launched a major government programme into wind generating devices. This was also true of some countries in Europe such as West Germany, The Netherlands and Denmark who identified the need for a renewable source of energy.

During this time Britain was capitalising on its newly found oil and gas reserves in the North Sea. The oil, together with a plentiful supply of coal, and the industrial recession sparked off by the 1973 oil crisis, caused Britain to be relatively complacent in developing renewable sources of energy. Nevertheless, in Britain there are now good prospects of a new industry emerging to manufacture and sell wind turbines. At present it involves some 25 companies either manufacturing or developing machines. As well as this there are almost 100 other UK organisations involved in supplying components or some other service associated with wind turbines.

In order that an industry can become established many factors need to be right. The principal ones are as follows. Firstly, there has to be a market, that is people must have a need to buy the product; and secondly, the product offered for sale has to be able to compete with other manufacturers, with foreign competition in the same business, or from whatever other product, can supply the market with the commodity, energy in this case. Let us now consider these two particular factors in slightly more detail.

THE MARKET

The world market for energy continues to grow. In the developed countries as they become more dependent on manufactured commodities, there is a greater and greater need for energy. In the third world countries, as they develop, their hunger for a supply of energy also continues to grow. Traditionally energy has been tapped from indigenous fuel sources; primarily wood, coal, then oil and in some of the developed countries from nuclear energy. This pattern, of course, has not been open to those countries who do not have their own indigenous resources. They have had to depend on importing fuels with the inherent high costs involved which can prove to be crippling to the economy of a developing nation.

Nuclear energy, however, presents other problems. Those countries already committed to nuclear energy have put a halt to their operations in order that they can be re-examined in the light of the 1986 'Chernobyl disaster', following on the 'Three Mile Island' accident. Of those countries who have not yet entered into the nuclear energy arena, many have chosen not to and it is they, together with the developing countries, who primarily provide the potential market for wind generated energy. The UK, however, which has all the energy it needs available to it, remains committed to nuclear energy, so the potential for wind energy for main grid generation is somewhat limited and is only considered on hard economic terms.

Wind turbines can be used to supply energy in various ways. By far their greatest potential is probably in the form of electricity connected to a main power generation network, where wind energy simply becomes a further input in the mix of energy sources that stable generating grids have come to rely on. In the UK, however, this sector of the market is liable to be limited mainly due to the mix of energies already available and to the commercial interests already invested in them, although it is technically feasible that as much as 25% of Britain's energy needs could be supplied from alternative sources such as wind energy.

The other method where wind energy can be utilised is in the 'self-help' market. Here, private individually-owned machines would supply individual farms or community centres or even groups of machines would supply whole communities. This is likely to be the major market in terms of numbers of machines that would be realised in the United Kingdom in the short term. It has been made possible by the removal of the monopoly of electricity supply from the main utilities which, as a result of the 1983 Energy Act, favours this sort of activity.

Autonomous systems are another area of great interest; these would be a combination of an electrical generating wind turbine with some other source of electrical energy. It would usually be the combination of wind turbines and diesel generating sets, although in suitable climates it could be wind and solar energy in the form of photovoltaics. Wind diesel systems have the advantage of being totally independent. The diesel generating sets operating when there is insufficient wind energy to meet the demand, the wind energy component thereby saving valuable and high cost diesel fuel. This type of system is likely to be the first input into a totally new undeveloped area to provide the first sources of electricity for that community.

Wind turbines, however, need not produce electricity and some machines have been built which produce heat. This is generally achieved by some means of converting mechanical energy to heat. The benefits of these machines is that they can be relatively low cost and be used whatever the wind climate. Such machines could be used for heating greenhouses in the horticultural industry or swimming pools or providing space heating in community applications. The major benefit of heat is that the energy can to some extent be stored, usually in the form of hot water to make up the shortfall for those occasions when the wind is not available.

Wind turbines have also been used for thousands of years for pumping water. Although these have become largely unfashionable and, in the western world, been displaced by other means of pumping water for drainage or irrigation, they are now becoming fashionable once again because of the high cost of fuel. Currently in the UK there is no single company manufacturing water-pumping machines, although the potential for export seems to be very large in particular to the Arab countries and to northern and eastern Africa.

FINANCIAL ASPECTS

The second criterion is that the energy must be able to compete with whatever else is available. Indeed in many parts of the world, especially in remote and island communities, wind energy is readily able to undercut imported fuel oil which can cause the price of electricity to be as high as 60p per kWh, a factor of ten greater than the selling price in the United Kingdom.

In the larger grid connected systems, experience over the past four or five years from California has brought to light some interesting statistics. Wind-turbine manufacturers hoping to sell into that market can now install machines for £800 per kW and can guarantee performance and reliability of 95% with a life expectancy of 30 years. This means that, on the best sites in Britain, electricity could be generated from wind turbines for 2p per kWh, which is cheaper than any other form of power generation possible today. Part of the reason comes from the short-term interest payments that are necessary for wind turbines as compared with the major conventional power station installations.[1] Even in the self-help market, machines which have been in operation connected to individual farms have shown a pay-back time of around six years. The prospect of having free electricity for some 20–25 years could seem a very attractive proposition to a farmer. With potential prospects as rosy as this, then it is surprising that widespread application of wind-turbine generators has not already taken place. The basic reason is that the machines at present are largely still under development with a few technical hurdles still to be overcome.

THE TECHNICAL SCENE

Modern wind generators have evolved in various forms largely led by work carried out in the United States and Denmark. The engineering of wind turbines is a difficult area where the vagaries of the wind and its effect on the

FIG. 1. A horizontal axis wind turbine.

wind turbines are largely unknown and a considerable variety of wind harnessing devices have emerged. However, the most favoured type tends to be the conventional or horizontal axis type which falls into two main categories: pitch controlled and stall regulated (Fig. 1).

Pitch-controlled machines whereby the blade pitch is varied to suit the wind speed are the ideal engineering solution for a variable climate, as one can maintain close control of the machine. A considerable number of machines of this type already exist but are expensive. However, having the pitch-control mechanism, which is usually hydraulically operated, can in some instances :: use problems in maintenance. In order to simplify the machine, it is desirable to have fixed blades, thereby reducing the cost of the blade arrangement. However, this depends on being able to design the blade profile very accurately in order that the machine will start up in light winds and shed load at the appropriate high wind speed or stop aerodynamically. Composite materials have been used for the manufacture of blades of this type, commonly in GRP, however several manufacturers are now also using blades of wood epoxy laminates with a GRP skin.

FIG. 2. A vertical axis wind turbine.

Vertical axis machines have enjoyed some success. Being able to place the power-generating machinery at ground level is a major advantage mainly for maintenance purposes. But, the development of vertical axis machines has trailed behind that of horizontal axis machines and a great deal of further research and development is still required. Some successful machines of this type are currently in operation in the United States and several innovators in the United Kingdom favour this concept. This design concept favours blades manufactured by pultrusion since uniform aerofoil cross-sections are used along the blade length but at present are mainly manufactured from aluminium alloys (Fig. 2).

OPPORTUNITIES FOR COMPOSITES

Components for wind generators must be reliable as a great number will exist in remote locations and be expected to operate for long periods between maintenance visits in order to be economical. Target lives of 25 or 30 years in harsh environments make composites good candidate materials.

Clearly whatever material is used to manufacture the blades must enable them to withstand the most severe loading expected without failure or excessive deflection. However low weight composites simplify erection problems and have 'knock-on' benefits for the design of the tower, shaft, bearing and hub. Other advantages of composites that can be exploited by the industry are that they cause less electromagnetic interference and therefore cause less interaction with radio transmitted signals than metal which is a major problem area in the siting of wind turbines. Composites have the added benefit that the failure modes are benign resulting in reduced risk of catastrophic failure and blade throw, another main area of concern; foreign object damage is usually easily repairable and smooth, aerodynamically efficient and aesthetically pleasing surfaces are possible.

The ideal blade should be light but with high strength and no fatigue limitations. Carbon fibre reinforced plastics satisfy these requirements exceptionally well but these are ruled out on cost grounds and glass fibre reinforcement is used instead as well as wood polymer composites.

The potential for composites is not limited to just the blades of wind generators or other obvious applications such as the nacelle housing and nose cones, but a whole range of components which go to make up a wind turbine. The wind-turbine industry is in its infancy and is ripe for the application of innovative ideas and materials providing they can do the job cost effectively. Take blades for example and consider providing variable blade geometry by using rigid but aeroelastic composites, this would optimize the machine performance and minimize structural wind loading. These techniques are being applied to advanced aircraft design where wings undergo aeroelastic tailoring and with time could find their way into wind turbine blades.

Towers are another main wind-turbine component which could benefit from composites, whether they be of a lattice or mono (tubular) type, particularly for ease of handling and erection especially in remote communities where neither cranes nor access is readily available. Similar GRP structures are already used for other applications.

Vertical axis machines in general favour the generating machinery to be at ground level and in some designs necessitate the use of drive shafts which in some cases could be tens of metres in length. The appropriate tubular composite structure could enormously reduce shaft weight and bearing loads. As well as the main nacelle housing, the possibility of gearboxes, switch-gear and other protective casings being made from composites also exists, with the benefits of zero maintenance and weight saving.

Autonomous wind diesel systems are already capturing significant attention world-wide and some developers favour an energy storage component in the system to smooth the energy supply at switchover between wind and diesel generation. The Kinetic Energy Storage System (KESS) developed by BP, which is a composite flywheel, is already the subject of considerable interest and experimental evaluation.

TECHNICAL SUPPORT

In keeping with other countries who have established a wind turbine manufacturing industry, the UK Government created within the Department of Trade and Industry a unit based and operated by the National Engineering Laboratory at East Kilbride in Scotland called the National Wind Turbine Centre (NWTC), whose function is to provide technical support to the emerging manufacturing industry. The NWTC is able to bring the vast range of engineering resources within the National Engineering Laboratory to the aid of UK manufacturers in order to tackle their development problems. In addition, special test facilities have been set up by the NWTC, with the help of the Scottish Development Agency, in the form of a unique test site located on a Scottish moorland some 340 m (1100 ft) above sea level on which manufacturers can have their machines tested in a relatively severe wind environment (Fig. 3). The NWTC is also

FIG. 3. The national wind turbine centre—test site.

able to provide for manufacturers independent accreditation of their machines which can be used for marketing or purposes of raising finance or gaining credibility.

The NWTC is also responsible for the creation of standards pertaining to wind turbines which will assist their acceptability into the community. This involves dealing with the safety of wind turbines, with the noise levels that they emit and with the effects that they could have on radio and TV transmissions. In addition to this work, research is also being undertaken into the major cost areas of current wind-turbine design in order to make future machines more competitive.

BUSINESS POTENTIAL

The business potential for wind turbines around the world is substantial. Not only does it involve the manufacturer but it also involves the installation and the servicing aspects as well as transporting machines perhaps half way across the world. The potential becomes firmer as more and more countries adopt national policies to install wind turbines into their electricity network. At a recent international conference held in

Rome, 34 countries presented technical papers on the subject of wind energy. Five countries have now announced substantial targets to install major contributions to their grid network totalling 7300 MW of additional wind turbine capacity up to the year 2000. This does not include a US Government target of 800 MW by 1988 which was exceeded because of market led activities in California with more than 1600 MW being operational at the beginning of 1986. Based on the current size range available targets represent somewhere between 7000 and 140 000 individual wind turbines depending on the size of machines that are selected. This is a considerable number of products by any standards and a major part of the wind turbines could be constructed from composite materials if the industry were able to provide competitive products to meet the market demand.

In the UK for example if wind energy were to be taken up for the purposes of central generation it would be possible to supply around 25% of the total national electricity consumption from wind turbines. This represents some £10 000–£20 000 million worth of wind turbines. In addition private generators in the self-help market could also create business of roughly the same order. A survey carried out within the European Economic Community estimates that 500 000 machines could be installed by the year 2000, representing between £50 000 and £100 000 million of business. Estimates for the United States put the potential business there at around £50 000 million and conservatively for the rest of the world at around £70 000 million. Substantial financial rewards are thus available if a share of the market could be for composite components. Indeed some successes have already been achieved by two of the larger British companies who have already exported machines valued at around £40 million to the United States, with the potential of expanding this business in the future, approximately 25% of this was from composite blades.

Denmark, who started manufacturing wind turbines seriously in 1981, have now achieved a domestic operation of 1800 machines. By 1985 they had expanded their business to being able to export £200 million worth annually, and specialist blade manufacturers share in this business. The Danish example provides a great deal of encouragement to others who might wish to enter this rapidly developing industry.

THE FUTURE

This chapter has tried to paint a broad picture of the wind-turbine industry, particularly in the UK as it is at pressent. With the courage to develop new products, the composite industry has considerable potential for obtaining a substantial share of this new and exciting market.

However, the situation has to develop in several ways at the same time. Not only does the market have to be stimulated to being aware that energy could be supplied from wind turbines, but manufacturers have to make available well designed reliable products in order to meet the market need. The difficulty is that these two activities have to occur in parallel and be carefully nurtured if they are to succeed. If wind turbines or components are developed ahead of the market potential then that could lead to financial ruin. This delicately balanced situation has to be carefully maintained and the careful nurturing, perhaps by increased government incentives, that was done in the leading countries, such as Denmark and the United States, could help to tip the balance favourably. Nevertheless wind energy does provide an exciting possibility for a range of new products in composite materials that manufacturing industry could turn to and it would appear that all the ingredients for a successful future industry already exist.

ACKNOWLEDGEMENT

This chapter is published by permission of the Director, National Engineering Laboratory. It is Crown copyright.

REFERENCE

1. Wind power for the UK, A British Wind Energy Association Position Paper, January 1987.

16

Structural Performance of Centrifugally Cast Glass Reinforced Plastic Pipes

M. A. Kanona and R. D. Currie

Johnston Pipes Limited, Doseley, Telford, Shropshire TF4 3BX, UK

ABSTRACT

The centrifugally cast glass reinforced plastic (GRP) pipes have been used in the sewage and drinking water supply pipelines for a number of years, the rate of use has increased progressively as confidence in the material and the process has been gained. In this chapter the mechanical and structural performance is discussed based on short-term and long-term testing of the actual pipe. The results of the short-term testing include burst pressure, specific stiffness and longitudinal strength. Long-term test results include pressure regression, strain corrosion, stress corrosion, flexural creep testing and other specialised testing. A brief outline of the company, the quality control system and usage of the pipes is also presented.

1. INTRODUCTION

The process of centrifugally cast glass reinforced plastic (GRP) pipes was originally developed some 30 years ago by Hobas in Switzerland. Fifteen years ago it was introduced into the UK by Johnston Pipes under the trade name of ARMAFLO pipes. Over the years, as user confidence has increased in the product, the use of the process has been spread worldwide. Currently these pipes are manufactured in Switzerland, UK, Austria, Sweden, Italy, USA, Australia and Japan, and current negotiations are taking place for a plant in the Eastern bloc. In the UK the demand has also increased due to confidence in the pipes and due to their competitiveness in the pipe market. Their use in the UK water industry has increased over the last decade to currently over 70 km per annum, primarily for water mains and also for sewerage applications.

2. PIPE WALL BUILD-UP

The pipes are manufactured by depositing a pre-set weight of unsaturated polyester resin, chopped glass fibre rovings and aggregate into a rotating steel mould. Due to the centrifugal forces (which are in the region of 35 g) the glass fibres and aggregates are fully impregnated by the resin and air voids are removed due to the compaction. The final inside surface of the pipe is finished with a relatively thick gel coat (1 mm) of chemically resistant flexibilised resin which is capable of sustaining high levels of surface strain when exposed for long durations in septic conditions. The outside surface is made using a high percentage of aggregate to resist handling, loading and laying.

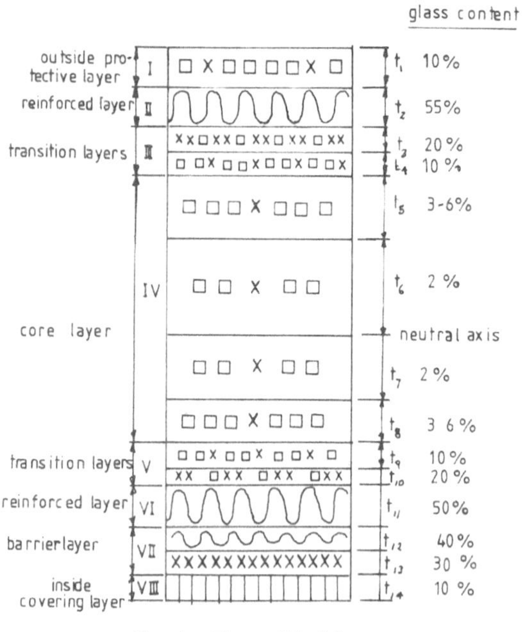

FIG. 1. Pipe wall build-up.

The structural parts are the barrier layer (40% glass) just beneath the gel coat to enhance the gel coat followed by the inside glass layer and outer glass layer and an aggregate/resin/glass layer sandwiched between them. Pipe thickness varying between 5 and 65 mm can be normally manufactured with a typical pipe wall build-up, as shown in Fig. 1.

The standard pipe diameters range from 300 to 2500 mm nominal diameter and are usually classified by their specific stiffness and their pressure rating.

3. RAW MATERIALS

For the pipes the essential ingredients are E-type glass fibre to BS 3691, main barrel resin (B type) to BS 3431 and silica aggregate to subangular/round shape graded to AF 60.[1] The joints are made using continuous glass fibres/chopped roving and an ultra-violet (UV) cured resin by a filament wound technique.

4. MANUFACTURING PROCESS

The consistency and the high quality of ARMAFLO pipes stems from the automated nature of the process, where the operator has little to do in the actual manufacture, hence there is minimum human error. The main advantages of the process in comparison to other GRP pipe processes are: (i) the ability to utilise relatively thick gel coats; (ii) the ability to introduce relatively high levels of aggregate, hence thicker pipes of high specific stiffness at an economic cost; (iii) the process results in a high quality finish on both the inside and the outside of the pipe and a uniform thickness is obtained; and (iv) the ability to tailor axial and hoop strength where required.

The computer pipe wall build-up is converted into the required quantities of raw materials delivered into the mould in steps. The delivery of raw materials is carried out by the feeder (see Fig. 2) which travels in and out of the mould. All raw materials are fed into the mould by weight automatically, but quite separately the glass and resin quantities are continuously verified by a digital read-out from a weight table for the glass and flow meter for the resin. These coordinated activities are schematically shown in Fig. 3. The feeder itself can be programmed to deposit up to 20 kg/min of raw material. After the pipe is pulled out of the mould secondary

FIG. 2. Pipe making.

procedures are carried out. These include cutting and sealing the ends, marking, weighing and jointing (for details see Refs 1 and 2).

5. TESTING AND STRUCTURAL PERFORMANCE OF PIPES

ARMAFLO GRP pipes are manufactured and tested in accordance with the procedures laid down by BS 5480, Parts 1 and 2 (Ref. 3). It is agreed by both users and manufacturers that the criteria laid down by BS 5480 give confidence in 50 years' service requirements. It is for this reason that at least 90% of the testings carried out by Johnston Pipes is in accordance with BS 5480 requirements. Other additional in-house testing, such as fatigue, abrasion and product development, are also carried out. Testing is carried out for long- and short-term properties. The first are accelerated types of tests to ensure the suitability of the pipe for long term whilst the second are mainly for daily quality requirements. These are as follows:

(a) All pressure pipes with their joints are tested to at least $1{\cdot}5 \times$ working pressure (WP) in the factory for at least 5 min against

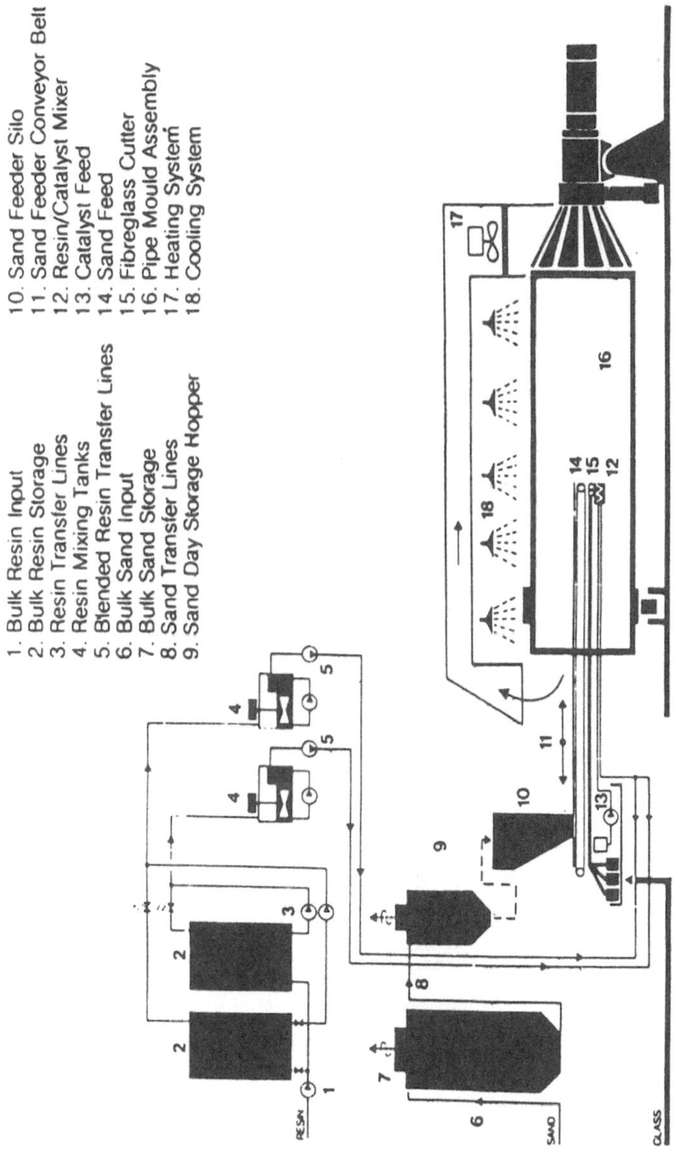

FIG. 3. Centrifugal casting process.

leak tightness. As a design verification circumferential strain measurements are taken on each and every pipe at 1·0 WP and every pipe is individually identified. The records also include the corresponding numbers and date of make of both the pipe and collar joint.

Longer durations of tests (over two years) were carried out at 2 × WP with the pipe showing an increase of circumferential strain from 0·38 to 0·58%. Naturally pipes in service which are operating at 1 × WP will have far lower rates and levels of creep.

(b) For pipes which are classified as gravity pipes (non-pressure application), three samples are tested in each batch to satisfy 10% Acceptable Quality Limited (AQL) as required by BS 5480.

(c) Daily visual inspection of all pipes manufactured in their finished form is carried out by experienced personnel, where certain undesirable effects are highlighted, such as impact damage due to improper handling, faulty marking, etc. Pipes are also visually inspected prior to loading and delivery to site.

(d) Specific stiffness of pipes, which is a destructive test, is carried out as required in BS 5480. A ring sample is deflected initially to a low level of deflection and the pipe stiffness is calculated and checked against that specified. Then the actual structural integrity such as bore cracking, delamination, etc., are checked for initial deflections above three times the predicted long-term deflection. Every test is given a number and filed. The photograph in Fig. 4 shows a 450-mm diameter pipe deflected to 15% under a parallel plate test with no sign of structural damage.

The corresponding long-term testing for specific stiffness is carried out by line loading a pipe ring totally immersed in water and the deflection is monitored for at least 10 000 h (1·14 years). The creep factor (aging plus semi-permanent set as defined by BS 5480) is later assessed, as shown in Fig. 5.

The current test procedure is pessimistic since the test is very much affected by geometrical status of the ring, i.e. an elliptical ring loaded at its minor axis will undoubtedly be less stiff than a similar circular ring loaded in the same way. Nevertheless, until a more realistic testing method is proposed, this creep factor is currently utilised in design.

(e) Tensile longitudinal testing. This is carried out on samples obtained from the stiffness ring and the average of three must be above the minimum stated in BS 5480 for normal handling.

FIG. 4. 15% ovalisation of 450-mm pipe ring.

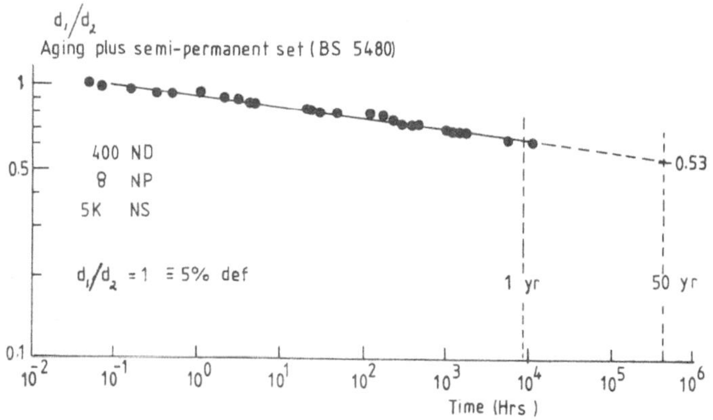

FIG. 5. Long-term flexural testing.

(f) Burst testing. This is a short-term destructive test carried out on a batch basis to ensure that the burst pressure is above the minimum required. Figure 6 shows a typical fracture of a GRP pipe. It has been often proven that although the liner resin does not contribute to the structural integrity of the pipe under pressure loading, it is vital in preventing what is termed as weepage, which is a probable

FIG. 6. Pipe burst.

type of failure in pipes manufactured by other techniques.[4,5] The mere fact that centrifugally cast pipes do include an excellent gel coat, barrier layer and are made with chopped fibres means the pipes do not weep prior to burst. The burst pressure and in turn the design working pressure is predicted based on long-term pressure regression first using the method in BS 5480 and the regression line is predicted using least squares analysis. All the pipes, whose results

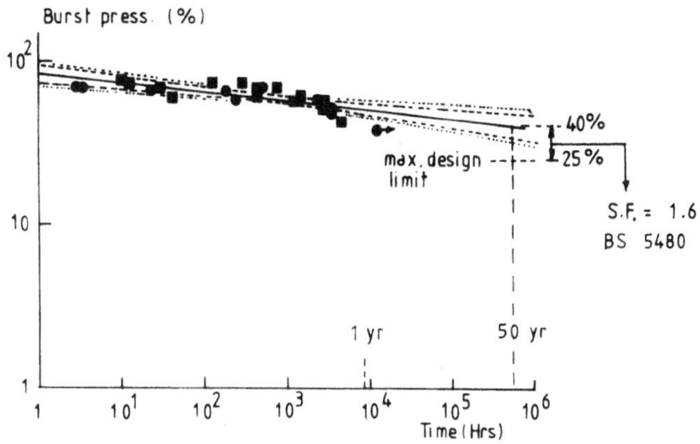

FIG. 7. Long-term pressure testing.

are shown in Fig. 7, have failed where a longitudinal crack has developed throughout the pipe thickness. Current pipe design employs a minimum safety factor of 1·6 of the extrapolated burst pressure over 50 years.

(g) Strain corrosion. For pipes intended to be used in septic conditions, the quality control method suggested by BS 5480 is employed, where three rings are set to a pre-set strain in 1·0 normality H2 SO4. These are tested as shown in Table 1. This is equivalent to 0·53% strain imposed for 100 000 h. The suitability of ARMAFLO pipes is confirmed by years of testing, as indicated for example by the ring shown in Fig. 8. which includes a pipe joint after nine years on test. Intensive academic work carried out elsewhere confirmed the criteria developed by Johnston Pipes for liner resins.[6,7]

TABLE 1

QC strain limited (%)	Minimum hours of duration
0·76	10
0·7	100
0·64	1 000

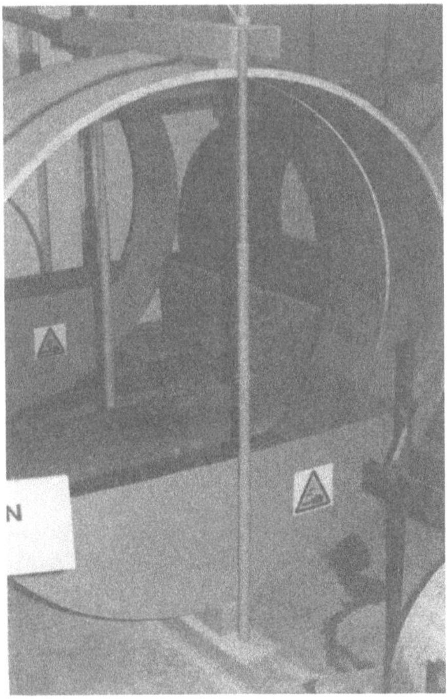

FIG. 8. Strain corrosion.

6. OTHER LONG-TERM TESTS

6.1. Fatique Test

In co-operation with Boulton Fabrications,[8] Johnston Pipes is developing new equipment for full-scale pressure fatigue testing of ARMAFLO GRP pipes. The pressure fatigue test rig is shown in Fig. 9. A pipe of 12 bar WP has been cycled from 0 to 12 bar over 2 400 000 cycles with no change in its elastic circumferential integrity as shown from circumferential strain measurements carried out before and after cycling, as shown in Fig. 10. An internal check of inside and outside surfaces revealed no cracks or micro-cracks. This test is ongoing at 18 700 cycles/day and already exceeds the normal likely fatigue loading in service conditions. Fatigue performance is not currently included in BS 5480, and the method developed by Johnston Pipes is to enhance the credibility of ARMAFLO pipes in service.

FIG. 9. Pipe pressure cyclic test rig.

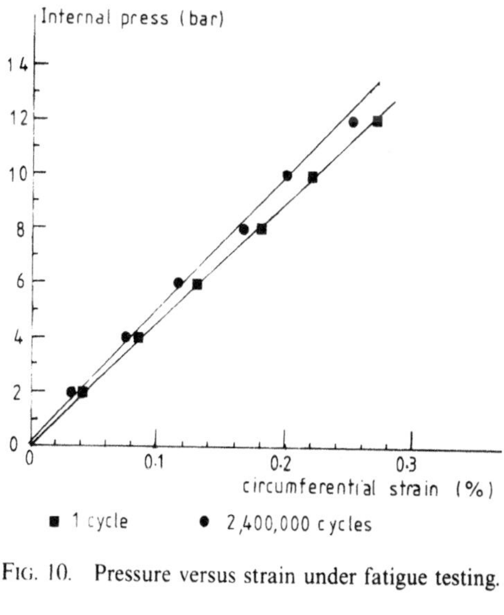

FIG. 10. Pressure versus strain under fatigue testing.

6.2. Abrasion Tests

These are carried out on full-scale pipe samples abraded using $\frac{3}{4}$-in gravel for over 500 000 cycles both on the inside and outside with no change in their burst strength in comparison with bursts of virgin pipes.

6.3. Impact Testing

Several tests carried out on 5000 and 10 000 N/m^2 specific stiffness have successfully met BS 5480 test requirements. Severer impact tests were also carried out on pipes using a 2-kg steel ball dropped from 2 m height with no leakage at 3 × WP after 24 h. A pressure pipe of 5000 N/m^2 specific stiffness and 1000 mm diameter was found to have a star crack after being installed; the repair procedure used has been proved in the laboratory by impacting a similar pipe at two places and using a 20-lb hammer. When the pipe was repaired it was pressurised to 65% of its burst pressure and left for 24 h with no leakage (95% lower confidence limit), then when taken to failure its burst pressure was that of an intact sample and not at the point of repair.

7. SUMMARY AND CONCLUSIONS

After various tests carried out on short-term and long-term bases and, more important, based on excellent performances of pipes already buried for more than 10 years under the ground, ARMAFLO GRP pipes made by the centrifugally casting process have proved themselves as outstanding candidates in the water industry.

REFERENCES

1. ARMAFLO centrifugally cast GRP pipe technical publication.
2. Currie, R. D. and Kanona, M., Johnston centrifugally cast GRP pipes (the Hobas process), 1st Int. Conf. Automated Composites, Sept. 1986, University of Nottingham, pp. 14/1–14/15.
3. BS 5480, Parts 1 and 2, British Standards Institution.
4. Kanona, M., Flexibility and stress intensification factors of GRP pipe bends, Ph.D. thesis, Aston University, 1985, 92 pp.
5. Hull, D., Legg, M. J. and Spencer, B., Failure of glass/polyester filament wound pipe, *Composites*, Jan. 1978, pp. 17–24.
6. Hogg, P. J. and Hull, D., Role of matrix properties on the stress corrosion of GRP, 13th Reinforced Plastics Congress, BPF Conf., Nov. 1982.
7. CIRIA Report 112, GRP sanitary sewer pipes—integral corrosion-resistance linings.
8. Pipe fatigue test rig—Boulton Fabrications, Stoke-on-Trent, UK—Mr J. Reeves.

17

The Support of Horizontal Cylindrical GRP Vessels—Saddles or Longitudinal Beams?

ANDREW J. WARRENDER, JAMES WOOD

Dept. of Mechanical and Production Engineering, Paisley College of Technology, Paisley PA1 2BE, Scotland, UK

and

ALWYN S. TOOTH and WILLIAM M. BANKS

Dept. of Mechanical and Offshore Engineering, University of Strathclyde, Glasgow G1 1XJ, Scotland, UK

ABSTRACT

Horizontal cylindrical GRP vessels are normally supported on either twin saddles or longitudinal beams which extend the full length of the vessel. To examine the way in which these supports influence the strain distribution in the vessel, results are presented from a series of experimental investigations. These have been carried out on a 1 m diameter, 4 m long, CSM reinforced polyester vessel subject to hydraulic loading. In the first series of tests the vessel was supported on twin saddles of various geometries and vessel/saddle interface conditions. In a later series, longitudinal beams, extending the length of the vessel were used to provide support.

Particular attention was given to providing theoretical analyses relevant to the longitudinal beam case, which are compared with the appropriate experimental results. The analysis for the saddle case is given in Ref. 1. On the basis of these results the two types of support are compared and their strengths and weaknesses enumerated. It is concluded that the beam support provides a good viable alternative to the twin saddle support for a large range of storage applications

1. INTRODUCTION

The correct choice of tankage required to store hazardous liquids in the process plant environment requires assiduous consideration. A corrosion resistant barrier is required to protect the vessel to ensure its long term structural integrity. In the case of carbon steel vessels this is often provided by using a rubber or glass lining. However, in more recent years GRP vessels have found general acceptance since they are more economical to manufacture than the steel lined vessels. Furthermore, the continuous development of new resins and glass reinforcements should enable the designer to tailor the material properties to the process conditions to further improve the integrity of the plant.

A great deal of such storage is provided in vertical GRP vessels, however, horizontally mounted vessels are often used when there is a restriction on the available height, when the plant is subject to internal pressure or when storage is required at elevated locations. These vessels are usually supported by either rings or saddles (Fig. 1) located a short distance from the vessel ends. The approach used is invariably the same as that employed for the steel vessels despite the fact that:

(a) typical steel vessels are much longer, ($4 \leq$ length/diameter ≤ 20) than GRP vessels, which are in the range ($1 \leq$ length/diameter ≤ 4), and

(b) the use of saddle supports induces high tensile stresses on the inside surface of the vessel at the uppermost point of the support, which it is considered, may lead to local cracking of the corrosion resistant barrier of a GRP vessel, allowing hostile liquid ingress and possible failure due to stress corrosion cracking.

It could well be that the GRP vessel is at greater risk than the corresponding steel vessel from an inadequate method of support. The question of the most suitable support for a particular installation is therefore addressed in this chapter. Consideration is given to saddle supports of various widths, angles and vessel/saddle interface materials, and to the use of longitudinal beam supports which extend the full length of the vessel (Fig. 2).

The use of longitudinal beams as an alternative method of support is illustrated in BS 4994 (Ref. 2). These supports consist of a pair of beams which are either constructed entirely in reinforced plastic or are formed over stiffening sections such as carbon steel angles.

This chapter presents the results of a series of hydraulic tests conducted

FIG. 1. Twin saddle supported 1 m diameter vessel under hydraulic test.

on a 1 m diameter, 4 m cylindrical length and 14 mm average thickness, CSM reinforced polyester vessel. In the first instance the vessel was supported on saddle supports of various geometries (widths, angles) and interface materials and secondly on longitudinal beams. Particular attention is given to the results obtained when the vessel is supported by the longitudinal beams. These are compared with results obtained for the saddle supported case which are presented in detail in Ref. 3.

FIG. 2. Longitudinal beam supports on the 1 m diameter vessel.

Theoretical analyses of the longitudinal supported vessel are presented using both a simple ring approach and the Finite Element Method in which the vessel is more accurately modelled. These predictions are compared with the experimentally obtained stress distributions.

On the basis of these results the relative merits of the saddle and longitudinal beam type supports are compared and their strengths and weaknesses enumerated.

2. EXPERIMENTAL ARRANGEMENT FOR SADDLE SUPPORT TESTS

The test vessel was constructed by the hand lay-up method using glass-fibre in the form of CSM and polyester resin. The vessel was 1 m inside diameter and 4 m long (tan to tan) with semi-ellipsoidal ends. On completion of the tests, extensive destructive testing of the vessel was undertaken to determine the wall thickness and glass content and to evaluate the elastic properties. The average values determined by these tests are: modulus of elasticity (tension) = 7258 N/mm^2; Poisson's ratio = 0·34; wall thickness (of cylinder) = 14 mm and glass content = 5·06 kg/m^2.

Two basic steel saddles of rigid construction were fabricated with a saddle angle of 180° and width of 140 mm. The inside radius of the saddles was 6 mm greater than the outside radius of the vessel in order to allow various 6 mm thick insert plates to be bolted to the saddles. In this way, a number of different saddle widths, angles and interface materials could be tested using the same basic saddle. Figure 1 shows the overall view of the test arrangement.

The vessel was extensively strain-gauged around profiles at distances of 150, 450, 750 and 1050 mm from one tangent line and also close to the vessel mid-span position on both the inner and outer surfaces. Particular attention was given to the saddle horn regions in which strips of 10 axial and 10 circumferential gauges of 1 mm gauge length were fitted straddling the saddle horn. Details of the strain gauge locations are given in Ref. 3.

Preliminary tests (Ref. 4) indicated that the quality of the saddle/vessel interface fit significantly influences the resulting strain distribution in the saddle region. In order to provide a uniform seating for the saddle, a smooth layer of unreinforced polyester resin was therefore formed around the saddle profile encapsulating the strain gauges on the outer surface. For each test, the saddles were individually grouted to this layer using a filled polyester resin. Adhesion of the saddle and vessel was prevented by polishing the existing resin moulding and the saddle interface and then coating these surfaces with a mould release agent prior to grouting. This procedure enabled the saddle to be removed from the vessel after each test was completed.

For each of the tests, the vessel was progressively filled with water and the strain levels recorded at intervals corresponding to one tenth by weight of the full vessel. Strains were recorded during both filling and emptying. For each of the tests the vessel was filled and emptied twice. No significant difference in the strains was observed for the two tests.

3. EXPERIMENTAL RESULTS FOR THE SADDLE SUPPORTED CASE

In all of the tests it was found that the maximum strain occurred in the immediate saddle horn region when the vessel was full. The maximum value was compressive and occurred in the circumferential direction on the outer surface. A high tensile value occurred on the inner surface at this point indicating that, in this region, the vessel was predominantly in a bending mode. A typical distribution of circumferential strain around the saddle centre profile on both the inner and outer surfaces is shown in Fig. 3. An extensive programme of tests was undertaken in order to investigate the influence of the following on the maximum strain levels: the saddle width; the saddle angle; the distance of the saddles from the ends and the use of a

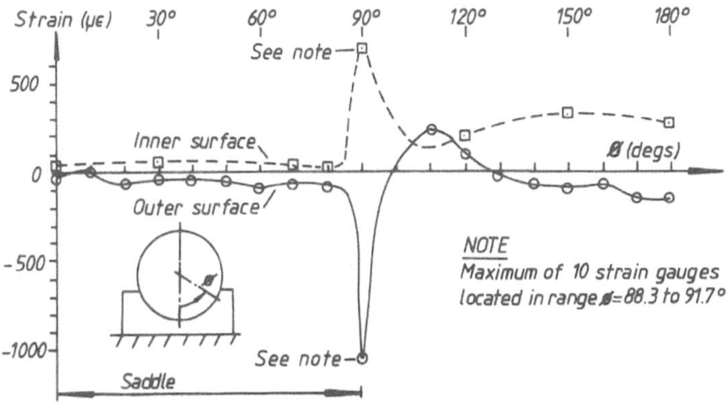

FIG. 3. Typical experimental variation of circumferential strain on the inner and outer surfaces around the saddle centre profile.

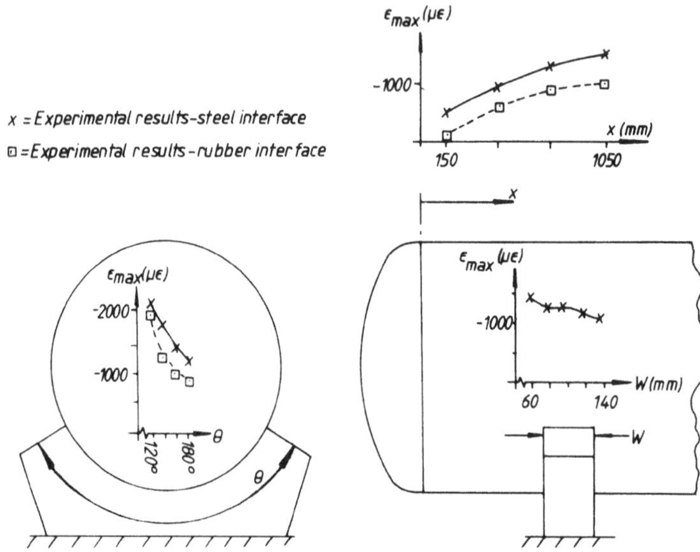

FIG. 4. Maximum experimental circumferential strains in the saddle supported vessel for different saddle angles, widths and locations.

layer of rubber between the saddle and vessel. Details of the tests undertaken and the experimental results obtained are given in Ref. 3. The test results are summarised in Fig. 4.

The significant observations made from these are as follows:

(i) The reduction in strain with saddle width is modest possibly because the distribution of interface pressure between the saddle and vessel is such that the edge regions of the saddle carry the major part of the load. The maximum strain recorded for 180°, 60 mm wide steel saddles located 750 mm from the tangent lines was $-1440\ \mu\varepsilon$. A reduction in this value of 29% occurred when the saddle width was increased to 140 mm.

(ii) The magnitude of the peak strain is highly dependent on the included angle of the supports. The peak strain for 120°, 100 mm wide steel saddles located 750 mm from the tangent lines was $-2270\ \mu\varepsilon$. This reduced by 43% with an increase in angle from 120° to 180°.

(iii) Locating the saddles close to the ends can result in greatly reduced saddle horn strains. A reduction of the peak strain of 66% was observed when the saddles were moved from a distance of 1050 mm to 150 mm from the tangent lines. Since typical horizontal GRP vessels are relatively short, locating the saddles near the ends is unlikely to result in high strains at the vessel mid-span.

(iv) The use of a rubber interface introduces an element of radial flexibility into the saddle design which resulted, on average, in a reduction of 25% in the maximum strains recorded for the corresponding steel interface.

The experimental results for the saddle supported vessel are compared with a theoretical approach, previously used to analyse metallic vessels, in Ref. 1. In general good agreement was observed between the experimental and analytical results, provided that the saddles were located more than approximately 200 mm (2/5 of the vessel radius) from the ends.

4. EXPERIMENTAL ARRANGEMENT FOR LONGITUDINAL BEAM TESTS

On completion of the tests in which the vessel was supported by saddles, the unreinforced polyester resin layer, which formed the basis of the seating for the saddles, was removed and full length channels were designed in accordance with BS 4994 (Ref. 2).

Carbon steel angles 100 mm × 100 mm × 11 mm thick were incorporated into the supports. These steel angles were located on the vessel using a simple jig such that the corner of the angles corresponded to the $\phi = 45°$ position. The vessel surface was abraided and all dust removed prior to forming the additional reinforcing laminate in order to ensure an adequate bond. Since the existing vessel surface was uneven, there were a number of gaps between the steel angles and the vessel into which a filled polyester resin was intruded. An overlay laminate of CSM reinforced plastic approximately 3 mm thick was formed over the steel angles extending 130 mm to either side in the circumferential direction. The region around the bottom of the vessel between the supports was then reinforced with a layer approximately 3 mm thick. Finally a further reinforcement approximately 4 mm thick was formed over the steel angle extending 120 mm to either side in the circumferential direction. The various overlay laminates are shown diagrammatically in Fig. 5. The resulting distance between the outer vertical surfaces of the angles was 845 mm which is slightly greater than the minimum distance of 750 mm (corresponding to 0·75 times the inside diameter) as specified in BS 4994. Once the reinforcing

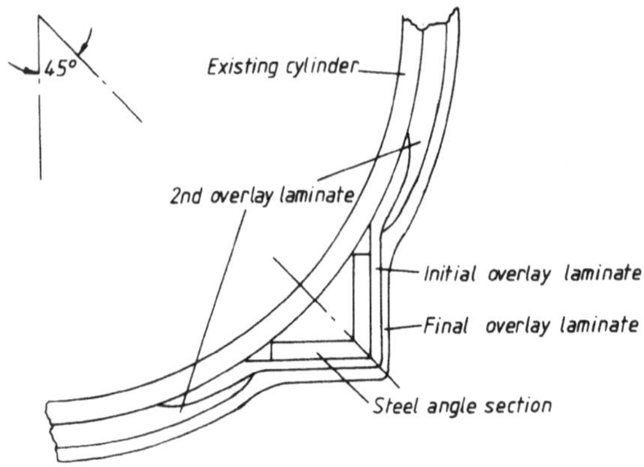

FIG. 5. Reinforcing laminates for the longitudinal beam supported case.

laminate of the support had cured, the vessel was located on a pair of steel I beams as shown in Fig. 2.

The vessel was strain gauged on one side, from the zenith ($\phi = 180°$) to the nadir ($\phi = 0°$). Attention was given to the two profiles 750 mm from the tangent line and 245 mm from the mid-span. These profiles correspond to the saddle centre profile previously used and also the profile near the vessel mid-span. The latter profile was chosen since it was anticipated that the maximum strain values would occur at the vessel mid-span.

Strain gauges of 3 mm gauge length were used throughout. At each of the two profiles described above, gauge pairs (axial and circumferential) were located every 15° from the nadir to the zenith on both the inner and outer surfaces. The exception to this was at the location of the support ($\phi = 45°$) where additional gauge pairs were located on the outer surface at the junctions of the angle support with the vessel and at the joins of the final overlay laminate with the cylindrical shell. The corresponding locations on the inner surface were also gauged.

5. EXPERIMENTAL RESULTS FOR BEAM SUPPORTED VESSEL

The method of laminating over the steel beam supports described in the previous section resulted in a significant local increase in thickness of the overlay laminate at the base of the beam at either end. Consequently, gaps of up to 10 mm existed between the supports and the I-beams on which they rested. As an initial attempt to compensate for these gaps, a layer of neoprene rubber identical to that used as an interface between the saddle and vessel in earlier tests was placed between the beam supports and the I-beams. The vessel was progressively filled with water and the strain levels recorded in the same manner as for the saddle support tests.

It was found that the experimental results obtained from two identical filling and emptying cycles were significantly different. It was considered that this was caused by the lack of fit between the supports and the associated I-beams. In order to improve the fit, a similar approach to that adopted for the saddle/vessel interface fit previously described was used. A layer of filled polyester resin was placed along the top surface of the I-beams and the vessel lowered horizontally thus causing extrusion of the excess resin. The vessel was then half-filled with water to ensure that it was uniformly seated and left until the resin had cured.

The strains were recorded during two filling and emptying cycles as before. In this case no significant difference was observed between the results obtained for the two loading cycles. The circumferential strain distributions obtained around the outer and inner surfaces at the profile 245 mm from the vessel centre line are presented in Fig. 9. This figure shows that the maximum strain of $-289\,\mu\varepsilon$ was recorded on the outer surface at $\phi = 49°$. The strain distribution obtained on the profile 750 mm from the end was of the same form as that shown in Fig. 9.

6. ANALYSES OF BEAM SUPPORTED CASE

6.1. Ring Analysis

A simple ring approach assuming uniform wall thickness throughout, was considered as a first approximation to the longitudinal beam supported vessel. This method assumes that each profile of the vessel behaves in a similar manner, i.e. the effect of the semi-ellipsoidal ends is neglected together with the axial stress resultants M_x and N_x. A unit length of the vessel is considered as a ring and subjected to radial loading corresponding to that caused by hydrostatic pressure. The ring is supported vertically at the mid-point of the base of the angle support. The ring material was assumed to be homogeneous, isotropic and linear elastic. Since the model is symmetrical about the vertical axis, only one half of the cross-section was considered. The internal moment and force at A are given by M_A and T_A. The notation used in the analysis is defined in Fig. 6.

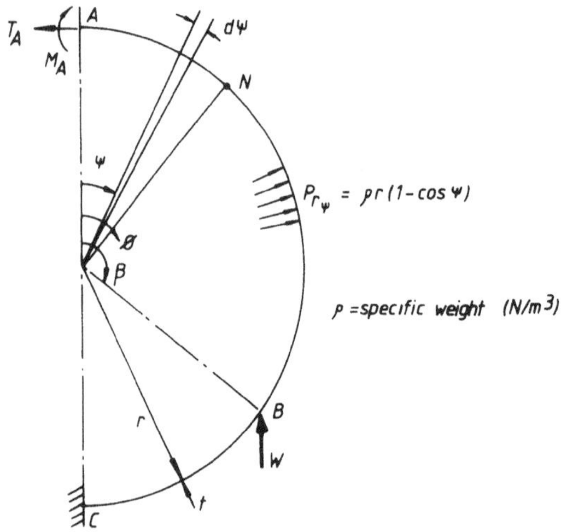

FIG. 6. Ring analysis for the beam supported vessel.

The bending moment at any point N (located at angle ϕ) in the ring, due to the pressure, P_{r_ψ}, is given by:

$$M_N = \int_0^\phi P_{r_\psi} r \, d\psi \, r \sin(\phi - \psi) \tag{1}$$

The applied radial loading due to hydrostatic pressure is:

$$P_{r_\psi} = \rho r(1 - \cos\psi) \tag{2}$$

where ρ = specific weight of fluid in N/m^3.

Substituting (2) into (1) and simplifying gives:

$$M_N = \rho r^3\left(1 - \cos\phi - \frac{\phi}{2}\sin\phi\right) \tag{3}$$

The forces at A also contribute to the circumferential bending moment. The total value of this at any point, N, is given by:

$$M_\phi = \rho r^3\left(1 - \cos\phi - \frac{\phi}{2}\sin\phi\right) + M_A - T_A r(1 - \cos\phi) \tag{4}$$

between A and B.

$$M_\phi = \rho r^3\left(1 - \cos\phi - \frac{\phi}{2}\sin\phi\right) + M_A - T_A r(1 - \cos\phi) - Wr(\sin\beta - \sin\phi) \tag{5}$$

between B and C.

The unit load method may be employed to solve for M_A and T_A as follows. Consider a unit moment, m, applied at point A. The rotation due to m is given by:

$$\theta = \int \frac{M_\phi m}{EI} r\,\mathrm{d}\phi \tag{6}$$

From symmetry, $\theta = 0$ hence

$$\int M_\phi r\,\mathrm{d}\phi = 0 \tag{7}$$

The sections A to B and B to C are considered separately by using eqns (4) and (5) as appropriate in the integration process of eqn. (7). After simplification the first equation relating M_A and T_A is obtained:

$$M_A = \frac{-\rho r^3}{2} + T_A r + \frac{Wr}{\pi}[(\pi - \beta)\sin\beta - 1 - \cos\beta] \tag{8}$$

A similar procedure can be adopted to derive a second equation involving M_A and T_A, since the horizontal displacement of A relative to C is zero. In this a unit force p is applied at A. The resulting equation is given by:

$$M_A = \frac{-7\rho r^3}{8} + \frac{3T_A r}{2} + \frac{Wr}{\pi}\left[(\pi - \beta)\sin\beta - \frac{5}{4} - \cos\beta + \sin^2\beta + \frac{1}{4}\cos 2\beta\right] \tag{9}$$

Equating (8) and (9) and solving for T_A and M_A gives:

$$T_A = \frac{W}{\pi}\left[\frac{1}{2} - 2\sin^2\beta - \frac{1}{2}\cos 2\beta\right] + \frac{3\rho r^2}{4} \tag{10}$$

$$M_A = \frac{\rho r^3}{4} + \frac{Wr}{\pi}\left[-\frac{1}{2} - 2\sin^2\beta - \frac{1}{2}\cos 2\beta + (\pi - \beta)\sin\beta - \cos\beta\right] \tag{11}$$

From the expressions for M_A and T_A given in eqns (10) and (11) the circumferential moment M_ϕ at any angle ϕ, can be derived, using eqns (4) and (5).

In addition to bending moments, a direct circumferential force, T, exists in the ring at any angle ϕ. This is given by:

$$T_\phi = T_A\cos\phi \qquad \text{between A and B}$$

$$T_\phi = T_A\cos\phi + W\sin\phi \qquad \text{between B and C}$$

The appropriate values of T_ϕ can be obtained by substituting for T_A from eqn. (10).

The resulting strains are given by

$$\varepsilon_\phi = \frac{1}{E}\left[\frac{T_\phi}{t} \pm \frac{6M_\phi}{t^2}\right] \tag{12}$$

These are presented on Fig. 9 along with the corresponding experimental values.

6.2. Two-dimensional Plane Strain Variable Thickness Finite Element Model

In an initial attempt to analyse the beam supported case by the Finite Element Method, the effects of the vessel ends and the longitudinal stresses due to hydrostatic loading were neglected. It was further assumed that each profile was similar and any longitudinal thickness variations were negligible, and also that the vessel was symmetrical about its vertical axis. In this case, the analysis was considerably simplified as it was only

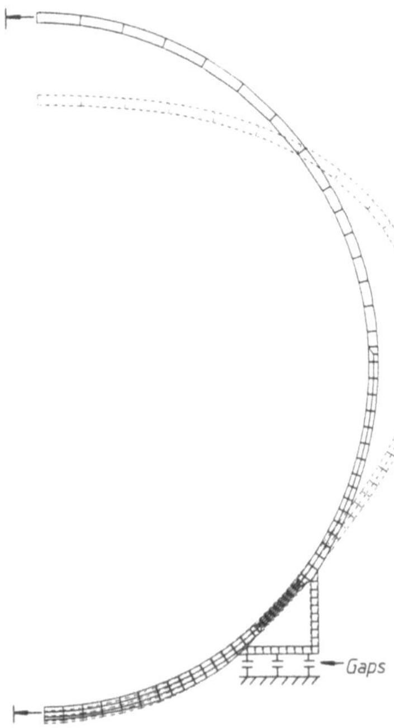

FIG. 7. Two-dimensional, variable wall thickness FE model, showing the displaced shape dotted, under hydraulic loading.

necessary to model the cross-section of one half of the cylinder and support. The thickness variation (as determined by destructive testing of the vessel) around one half of the strain gauged profile, 750 mm from the vessel end, was accurately modelled. For this, 8-noded, plane strain, isoparametric quadrilateral elements (Ref. 5) were mainly employed as shown in Fig. 7. The material properties given in Section 2 were used assuming material isotropy.

Symmetrical boundary conditions were imposed on the vertical diameter of the model and gap elements were used at the base of the support to allow the appropriate vertical reactions to develop. No horizontal restraint was imposed in the transverse direction at this position. Hydrostatic pressure, corresponding to the vessel full of water, was imposed on the inner surface of the cylinder. Various checks were made to assess the integrity and validity of the analysis and these were found to be satisfactory.

A comparison of the experimental results with the strain distribution predicted by this analysis is presented in Fig. 9.

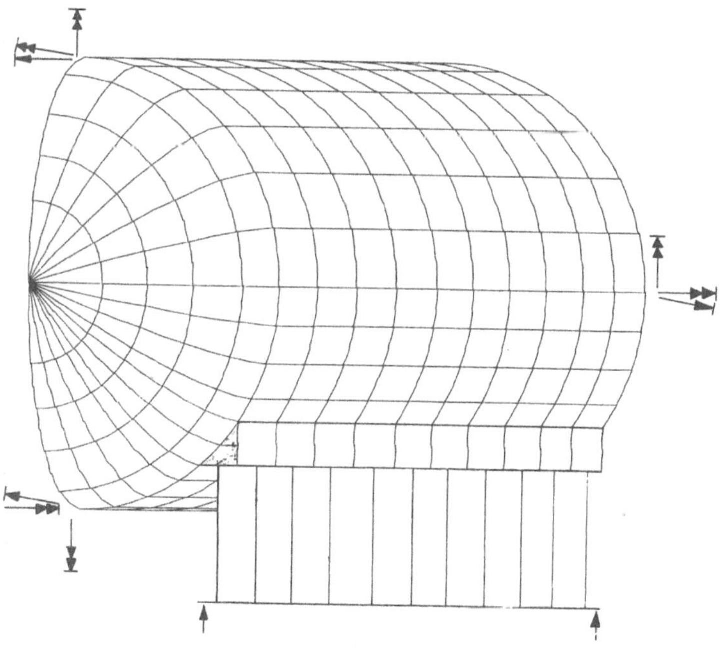

FIG. 8. Three-dimensional, variable wall thickness FE model of one quarter of the vessel, with I-beam supports.

6.3. Three-dimensional Thin Shell Variable Thickness Finite Element Model

The strains and displacements induced in a horizontal cylindrical vessel by the support reactions are no doubt influenced near to the vessel ends by the stiffening effect. This fact was clearly demonstrated for the twin saddle support case in Section 3. It was therefore considered that neglecting this effect in the previous analysis for the beam supported case was a significant source of inaccuracy. A further Finite Element Model of one quarter of the vessel, including the end, was therefore created as shown in Fig. 8. With this

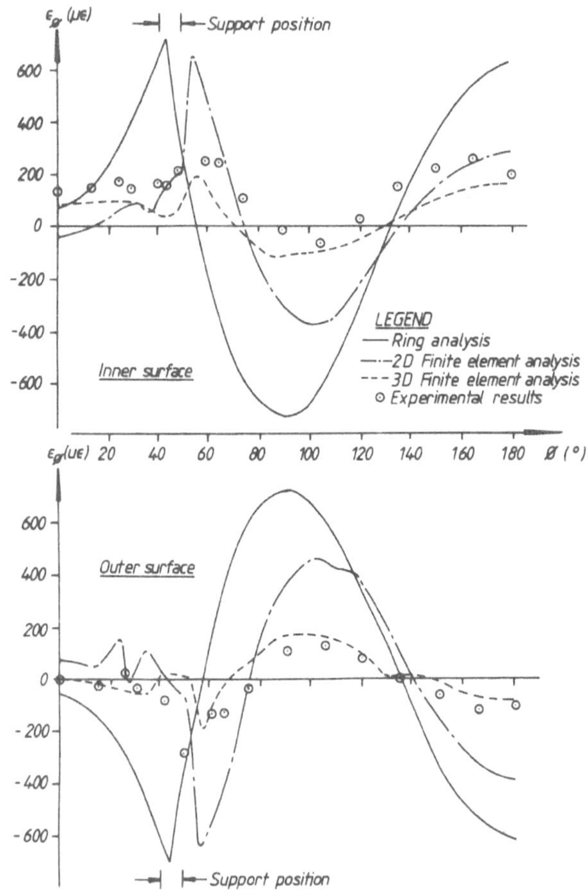

FIG. 9. Comparison of the inner and outer surface strains obtained experimentally and analytically.

model, longitudinal effects due to hydrostatic pressure on the vessel ends were introduced. The thickness variation in the circumferential direction, previously used for the two-dimensional analysis was modelled and, as before, any longitudinal wall thickness variation was neglected. The thickness of the vessel end was assumed to be a uniform 14 mm throughout. In this case, the I-beam on which the vessel rested was also modelled. Symmetrical boundary conditions were imposed where appropriate and the nodes along the centre of the I-beam lower flange were restrained in the vertical direction.

The elements used in this model were primarily 8-noded quadrilateral semi-loof variable thickness thin shell elements as implemented in the PAFEC finite element suite.[5] With this type of thin shell element, the mid-surface of all shell/plate structures are modelled. The nodal thickness parameters are only used in the derivation of the element matrices. This representation can lead to several problems as highlighted in Section 7.

As with the previous analysis, checks were made to assess the integrity and validity of the analysis and these were found to be satisfactory. The predicted strain distribution on the profile near the vessel centre is compared with the corresponding experimental results in Fig. 9.

7. COMPARISON OF THE VARIOUS ANALYTICAL APPROACHES WITH THE EXPERIMENTALLY MEASURED STRAINS

7.1. The Ring Analysis

This analytical model neglects the beneficial influence of the ends and the presence of both the axial membrane and bending effects. Furthermore the beam support is represented as a single line. Not surprisingly therefore, the predicted strains are much higher than the measured values. The approach, however, does provide the designer with a simple method incorporating a generous factor of safety.

7.2. The Two-dimensional Finite Element Analysis

Like the ring approach, this analysis neglects the axial effects but does introduce the local variation in thickness of the vessel and more realistic modelling in the support region. Although the predicted strain distribution is somewhat nearer to the measured values than the ring approach, the differences in the peak values of the two models are small. From a design point of view, the simplicity of the ring approach outweighs the slight improvement in accuracy of the results obtained using this finite element model.

7.3. The Three-dimensional Finite Element Model

The thin shell model incorporating the semi-ellipsoidal ends and the I-beam supports does provide much better agreement with the measured values. The model reproduces the significant trends and features of the strain distribution fairly well. It could be used for further investigations to examine the influence of beam stiffness, beam location and the case where the beam is only supported at certain discrete points. In addition, the effect of thickness variation and out of circularity could also be investigated. The model does, however, contain a number of limitations (see Ref. 6). For example, it is difficult to accurately model the additional reinforcing in the nadir region between the supports with what is essentially a centre line theory. Furthermore, the modelling of the support region is likely to underestimate the peak strain—see the value at $\phi = 49°$ on the outer surface (Fig. 9). Nevertheless this FE model can provide useful comparative results whereby the optimum support location and stiffness can be obtained.

8. SADDLES OR LONGITUDINAL BEAMS?

GRP storage vessels are in general considerably shorter (i.e. $1 \leq$ length/diameter ≤ 4) than their metallic counterparts. The longitudinal beam support, extending the full length of the vessel, is therefore a viable economic proposition which could perhaps be used instead of the more common saddle type support.

The results obtained for the 1 m diameter vessel supported on twin saddles show that the maximum strain is highly dependent on the saddle angle, saddle location and the interface material used. For the range of tests considered these factors are reflected in the magnitude of the maximum strain which varied from $-150\,\mu\varepsilon$ to $-2100\,\mu\varepsilon$. In comparison, the maximum strain recorded in the longitudinal beam arrangement was $-289\,\mu\varepsilon$. It could well be argued that with care the 'best' saddle case would match the 'best' longitudinal beam case. However, this parity may be difficult to achieve in practice in each and every installation. Ideally, a method of support is required that is not critically dependent upon certain installation procedures, and which will remain trouble-free over the life of the plant. In Table 1 the pros and cons are tabulated to compare the strengths and weaknesses of the two methods of support.

TABLE 1

Design consideration \ Support	Saddle support	Longitudinal beam
Fitting support to vessel	Rather critical. Rubber interface useful. Peak strain over small region which in time may deteriorate locally	No problem
Fitting support to foundation	No problem	Care needed, otherwise bolting to foundation could induce high strains in vessel. Should be set down on resin base prior to bolting
Access to vessel for pipework in base region	Very easy	Rather restrictive
Manufacture of support	Requires specialised equipment, may be necessary to sub-contract to either steel fabricator or Civil Engineering Contractor (if concrete)	Easily done by GRP Laminator
Corrosion of support	Regular maintenance required for steel saddle	No maintenance as no metallic surfaces
Thermal effects due to high or low temperature of contents	Not much of a problem with rubber interface	If steel stiffeners used, differential temperature expansion may cause high axial strain
Long-term effects	Some deflection between supports can occur	No problem likely as continuously supported

On the basis of the work presented in this chapter the longitudinal beam type of support offers a very good alternative to the twin saddle support in almost every case. It is only deficient in the case of thermal load where special care would be required to avoid the use of steel stiffening in the beam support and also to provide freedom in the axial direction.

9. CONCLUDING COMMENTS

This chapter presents the results of experimental work carried out on a 1 m diameter vessel subject to hydraulic loading. Two different types of support are compared. It is found that the saddle support which is extensively used for metallic vessels, imposes high forces on the vessel at the uppermost point of the saddle, known as the horn. Although these forces are more 'kindly' distributed when rubber is used as an interface, they do cause high values of strain in a localised region of the vessel. The longitudinal beam support, on the other hand, largely avoids these regions of high localised strain and provides a continuous support along the length of the vessel. Although care should be exercised with this latter support in those cases where significant temperature differences may occur between the stored liquid and ambient conditions, it is considered that the longitudinal beam provides a good alternative to the saddle support.

ACKNOWLEDGEMENTS

The authors are indebted to the Science and Engineering Research Council (UK), Polymer Engineering Group for sponsorship, encouragement and support. Also to the associated industrial sponsors who willingly contributed to the programme.

REFERENCES

1. Tooth, A. S., Warrender, A. J. and Banks, W. M., A design procedure for horizontal cylindrical GRP vessels supported on twin saddles. To be presented at the 6th International Conference for Composite Materials, Imperial College, London, July 20–25, 1987.
2. British Standards Institution, Design and construction of vessels and tanks in reinforced plastics, BS 4994: 1986.
3. Warrender, A. J., Tooth, A. S. and Banks, W. M., The effect of saddle geometry on the behaviour of horizontal cylindrical GRP storage vessels under various fill conditions. The British Plastics Federation, Reinforced Plastics Congress, Nottingham, 17–19 September 1986, paper 34, pp. 167–170.
4. Tooth, A. S., Warrender, A. J. and Banks, W. M., The support of horizontal RP storage vessels—some preliminary tests, *Proc. of the 40th Annual Conference, Reinforced Plastics/Composites Institute*, The Society of the Plastics Industry Inc., Jan 28–Feb. 1, 1985, Session 15-D.
5. PAFEC Ltd. PAFEC Level 6.1 Data Preparation Manual, PAFEC Ltd, Strelley Hall, Strelley, Nottingham 1986.
6. Wood, J., Observations on shell intersections, 4th World Congress on Finite Element Methods, Interlaken, Switzerland, Sept. 1984.

18

How to Predict the Burst Pressure of a FRP Pipe

XUE YUAN-DE

Department of Engineering Mechanics, Tongji University, Shanghai, People's Republic of China

and

CHEN CHIN-KUNG

Shanghai GRP Research Institute, Shanghai, People's Republic of China

ABSTRACT

The purpose of this paper is to illustrate how to predict the burst pressure of a FRP pipe or a cylindrical vessel, using the split-D tension test of the circular ring cut from them. Two modification factors must be regarded. The first one results from the additional bending stress in the region near the mid-plane. Two new mechanical models are suggested for calculating the bending stress mentioned above. The second factor is the influence of the material's variability, which also results in the differences of the experimental data between the ring and the pipe. After considering these two factors, we give the predicted value from the split-D test of the ring, which satisfactorily agrees with the hydrostatic pressure of the pipe.

INTRODUCTION

The thin-walled pipes and the cylindrical vessels made up of FRP composites are widely used in engineering. In order to measure their hoop tensile strength and modulus of elasticity, the split-D tension test of the circular ring cut from them is often used. The ring is mounted on two aligned D fixtures which are shaped semicircularly and in contact with most

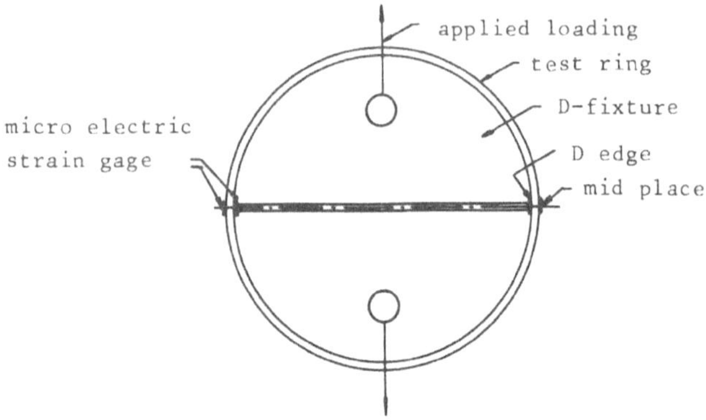

FIG. 1. Schematic of the split-D tension test.

of the inner surface of the ring. When these two D fixtures are pulled apart by the tensile loading machine, the load is applied to the ring, producing an equivalent average internal pressure loading (Fig. 1).

The results obtained from micro-electric strain gauges (0·2 × 0·2 mm) prove that the total strain in the mid-plane of the test ring consists of two parts, i.e. the tensile strain which is proportional to the applied load and the bending strain which decreases when the deformation of the ring is increased (Fig. 2). So the measured hoop tensile strength of the test ring from the split-D test is less than that of either the pipe or the cylindrical

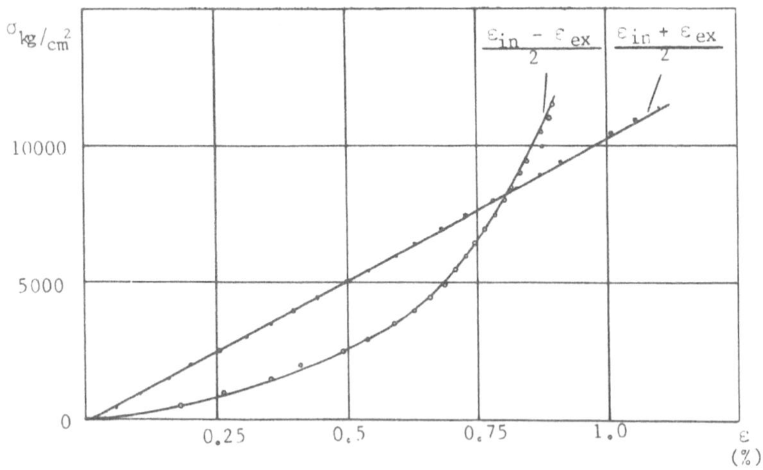

FIG. 2. The stress–strain curve of the ring by the split-D tension test.

vessel from the hydrostatic burst test, if we ignore the influence of the additional bending stress.

MECHANICAL MODEL FOR ANALYSIS OF STRESSES IN THE TEST RING

Although it is difficult to determine the internal pressure loading between the test ring and the D fixtures, two new methods of calculation are suggested in this paper which are different from the one in Ref. 1.

The first one assumes that most of the points which are situated on the inner surface of the test ring and in contact with the D fixtures before deformation, are still in contact with them after deformation except for a few points which are located at the D edge; the geometrical condition, shown in Fig. 3, can be written as follows:

$$\frac{U_{\theta_i}}{W_1 - W_{\theta_i}} = \operatorname{ctg}\theta_i$$

This geometrical condition can also be expressed in the following form:

$$(B)\{\delta\} = \{C\}$$

where (B) is a matrix related to boundary constrained condition, $\{\delta\}$ are the displacements of the contacted nodes and $\{C\}$ are the equivalent external

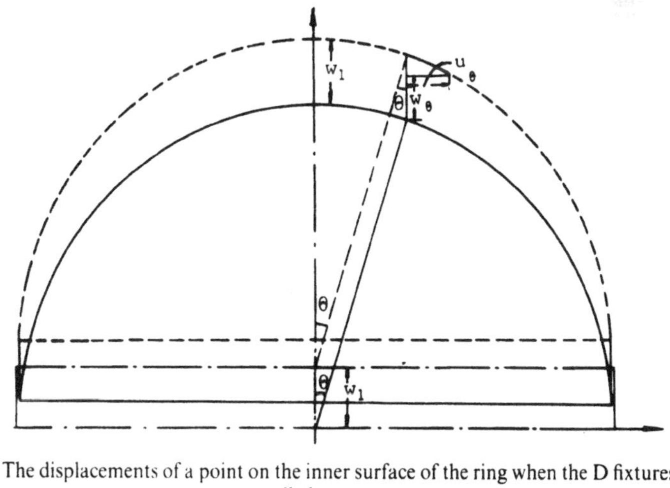

FIG. 3. The displacements of a point on the inner surface of the ring when the D fixtures are pulled apart.

forces of nodes, here equal to zero. For this problem, for example, when $i = 1$, it can be given by

$$(1, -\mathrm{ctg}\,\theta_1, -1)\begin{bmatrix} W_1 \\ U_{\theta_1} \\ W_{\theta_1} \end{bmatrix} = 0$$

and when $i = 7$,

$$\underset{7\times 15}{(B)} = \left[\begin{array}{ccccccc} 1 & -\mathrm{ctg}\,\theta_1 & -1 & 0 & 0 & 0 & 0 \\ 1 & 0 & 0 & -\mathrm{ctg}\,\theta_2 & -1 & 0 & 0 \\ 1 & 0 & 0 & 0 & 0 & -\mathrm{ctg}\,\theta_3 & -1 \\ 1 & 0 & 0 & 0 & 0 & 0 & 0 \\ 1 & 0 & 0 & 0 & 0 & 0 & 0 \\ 1 & 0 & 0 & 0 & 0 & 0 & 0 \\ 1 & 0 & 0 & 0 & 0 & 0 & 0 \end{array}\right.$$

$$\left.\begin{array}{cccccccc} 0 & 0 & 0 & 0 & 0 & 0 & 0 & 0 \\ 0 & 0 & 0 & 0 & 0 & 0 & 0 & 0 \\ 0 & 0 & 0 & 0 & 0 & 0 & 0 & 0 \\ -\mathrm{ctg}\,\theta_4 & -1 & 0 & 0 & 0 & 0 & 0 & 0 \\ 0 & 0 & -\mathrm{ctg}\,\theta_5 & -1 & 0 & 0 & 0 & 0 \\ 0 & 0 & 0 & 0 & -\mathrm{ctg}\,\theta_6 & -1 & 0 & 0 \\ 0 & 0 & 0 & 0 & 0 & 0 & -\mathrm{ctg}\,\theta_7 & -1 \end{array}\right]$$

$$\underset{15\times 1}{(\delta)} = (W_1 \quad U_{\theta_1} \quad W_{\theta_1} \quad U_{\theta_2} \quad W_{\theta_2} \quad \cdots \quad U_{\theta_7} \quad W_{\theta_7})^{-1}$$

If we can find a stiffness matrix element $(K)^e$ which satisfies the following equation:

$$(K)^e = (B)^T(D)(B)$$

where

$$(D) = d_{\text{large}}(I)$$

superimposing $(K)^e$ on the total stiffness element (K) of this structure, then the geometrical condition will also be satisfied. For this problem, for example, when $i = 7$, $(K)^e$ can be given by the matrix shown opposite.

$$
\underset{15\times 15}{(\boldsymbol{K})^e} =
\begin{bmatrix}
7 & -\operatorname{tg}\theta_1 & -1 & -\operatorname{tg}\theta_2 & -1 & -\operatorname{tg}\theta_3 & -1 & -\operatorname{tg}\theta_4 & -1 & -\operatorname{tg}\theta_5 & -1 & -\operatorname{tg}\theta_6 & -1 & -\operatorname{tg}\theta_7 & -1 \\
 & \operatorname{tg}^2\theta_1 & \operatorname{tg}\theta_1 & 0 & 0 & 0 & 0 & 0 & 0 & 0 & 0 & 0 & 0 & 0 & 0 \\
 & & 1 & 0 & 0 & 0 & 0 & 0 & 0 & 0 & 0 & 0 & 0 & 0 & 0 \\
 & & & \operatorname{tg}^2\theta_2 & \operatorname{tg}\theta_2 & 0 & 0 & 0 & 0 & 0 & 0 & 0 & 0 & 0 & 0 \\
 & & & & 1 & 0 & 0 & 0 & 0 & 0 & 0 & 0 & 0 & 0 & 0 \\
 & & & & & \operatorname{tg}^2\theta_3 & \operatorname{tg}\theta_3 & 0 & 0 & 0 & 0 & 0 & 0 & 0 & 0 \\
 & & & & & & 1 & 0 & 0 & 0 & 0 & 0 & 0 & 0 & 0 \\
 & & & & & & & \operatorname{tg}^2\theta_4 & \operatorname{tg}\theta_4 & 0 & 0 & 0 & 0 & 0 & 0 \\
 & & & & & & & & 1 & 0 & 0 & 0 & 0 & 0 & 0 \\
 & & & & & & & & & \operatorname{tg}^2\theta_5 & \operatorname{tg}\theta_5 & 0 & 0 & 0 & 0 \\
 & & & & & & & & & & 1 & 0 & 0 & 0 & 0 \\
 & & & & & & & & & & & \operatorname{tg}^2\theta_6 & \operatorname{tg}\theta_6 & 0 & 0 \\
 & & & & & & & & & & & & 1 & 0 & 0 \\
 & & & & & & & & & & & & & \operatorname{tg}^2\theta_7 & \operatorname{tg}\theta_7 \\
 & & & & & & & & & & & & & & 1
\end{bmatrix}
$$

The second one, we call it a combined structure model, can be explained as follows. The D fixture is simplified as if it consists of a series of rods which can rotate about the centre of the circle and will be compressed when they are deformed. At the same time, those points which are situated on the inner surface of the test ring and in contact with the D fixture turn round along a curve formed by the ends of these rods. When loaded, the amount of rods decreases gradually, by one piece at a time, to simulate the process of separating a few contact points on the inner surface of the ring and on the D fixture at the D edge. Then the finite element method can be used on this combined structure with the rod elements and the 2-D plane strain elements, as shown in Fig. 4.

Based on the calculations mentioned above, the deformation curve and the stress distribution of the circular ring are obtained, as shown in Figs 5 and 6. The results indicate that the stress consists of two parts, i.e. the tensile stress which is uniform along the ring and equal to $P/2bt$ and the bending stress which appears in a part of the region near the mid-plane and reaches its maximum value at the mid-plane.

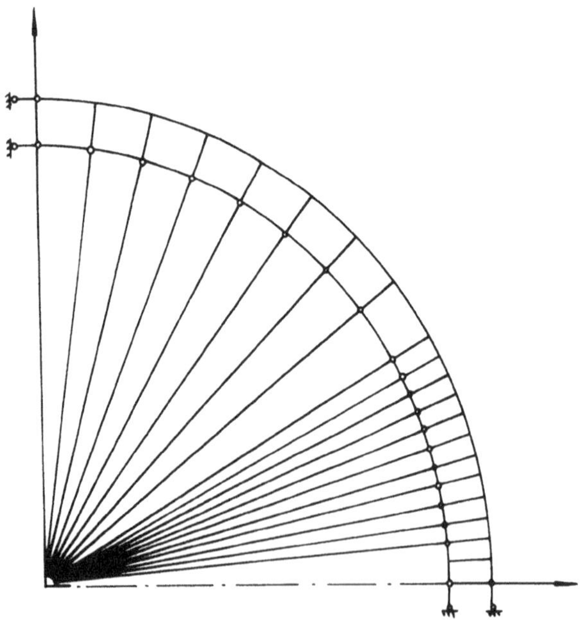

FIG. 4. Schematic of the combined structure model.

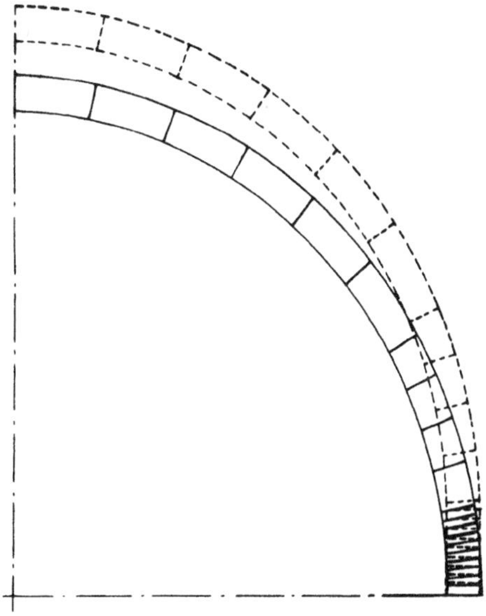

FIG. 5. The deformation curve of the ring.

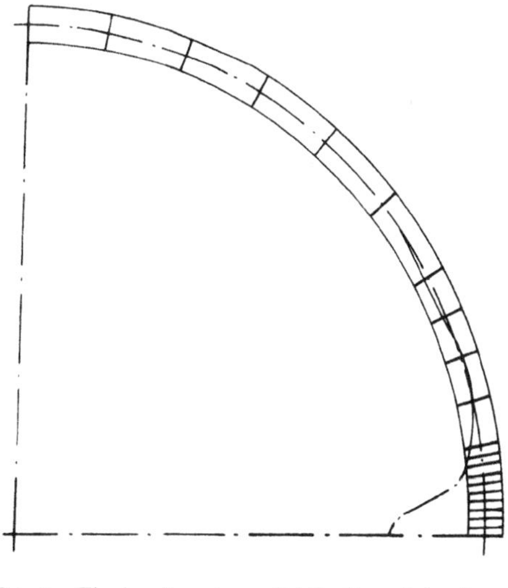

FIG. 6. The bending stress distribution of the ring.

TABLE 1
Experimental tensile strengths for hydrostatic burst and split-D tension test

Test methods	N	σ_b (kg/cm^2)	C_v (%)
Hydrostatic burst	31	15 840	11·6
Split-D tension	390	11 960	23·8

N = number of specimens.

PREDICTION OF THE BURST PRESSURE OF A FRP PIPE

The hydrostatic tensile strength of a kind of FRP pipe and the split-D tensile strength of the circular ring cut from the pipe is given in Table 1.

If we want to predict the burst pressure of the FRP pipe from the experimental results of the ring, two modification factors must be regarded.

The first is the bending stress mentioned above and for this kind of ring the maximum calculated bending stress equals about 70% of the average tensile stress. The results obtained from micro-electric strain gauges indicate that the bending stress is greatly influenced according to whether the D fixture precisely matches the ring or not. If the diameter of the D

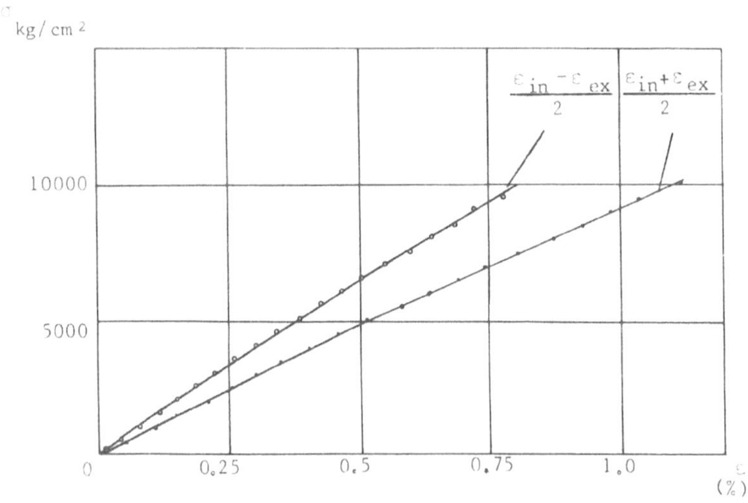

FIG. 7. The stress–strain curve of the ring by the split-D tension test when the D fixture precisely matches the ring.

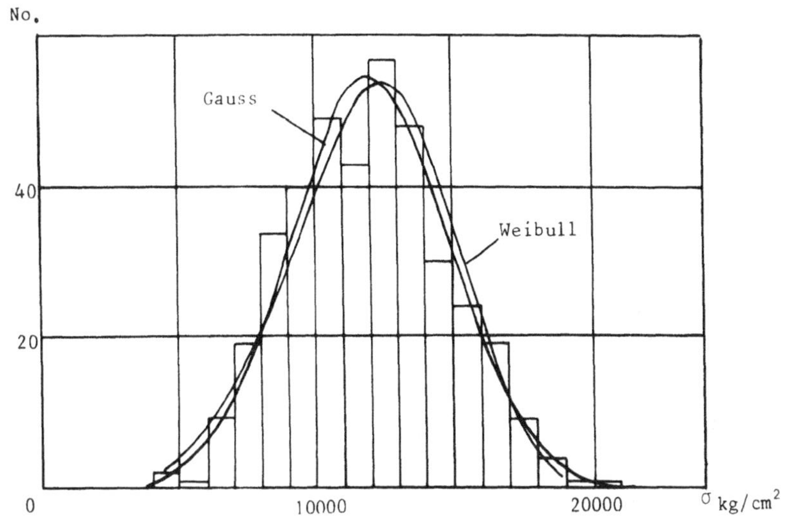

FIG. 8. The statistical distribution of the tensile strength of the ring for split-D test.

fixture is less than the inner diameter of the ring, there is a larger initial bending stress in the ring. When both are approximately equal, the initial bending stress in the ring is small and nearly equals the calculated one, as shown in Fig. 7.

The second one is the influence of the different dimensions between the pipe and the test ring, for the burst failure of the pipe may take place at any point in the whole circumference L, while fracture of the test ring always takes place in the uncontacted length l. Thus, the modification coefficient k is given by

$$k = \left(\frac{l}{L}\right)^{1/\alpha_0}$$

where α_0 is the Weibull statistical parameter which expresses variability of tested composites, as shown in Fig. 8.

After considering previous factors, we obtain the predicted value of burst pressure of this kind of pipe, which agrees with the hydrostatic pressure of the pipe.

REFERENCE

1. KNIGHT, C. E. Jr, Failure analysis of the split-D test method, Composite Materials: Testing and Design (Fourth Conference), ASTM STP 617, American Society for Testing and Materials, 1977, pp. 201–204.

19

A Refined Theory of Laminated Doubly Curved Shells

CAI SI-WEI and CAI MIN
Hefei Polytechnic University, Hefei,
People's Republic of China

ABSTRACT

The doubly curved laminated shell is considered in this paper. The shell may be composed of an arbitrary number of bounded layers, each of which may possess different thicknesses, orientations and anisotropic elastic properties.

In the theory, the displacement component between layers are considered as fundamental unknown functions. The transverse shear deformation as well as expansional strain in each lamina is included. Equations of equilibrium for laminae are first derived. Continuity of displacement and stress have been prescribed at the interfaces between laminae before the final governing equations are derived. The theory is applied to the problem of a four-layered laminated cylindrical shell and the validity of it is assessed by comparing with solutions obtained from the exact theory of elasticity.

We assume the displacement components of the jth lamina in axis ξ_1, ξ_2 directions respectively

$$u = \frac{1}{\alpha_{1j}} L_{1j} u^j + \zeta \phi_{1j} \qquad v = \frac{1}{\alpha_{2j}} L_{2j} v^j + \zeta \phi_{2j} \tag{1}$$

where

$$L_{1j} = \alpha_{1j}\left(1 + \frac{\zeta}{R_{1j}}\right) \qquad L_{2j} = \alpha_{2j}\left(1 + \frac{\zeta}{R_{2j}}\right) \tag{2}$$

ξ_1–ξ_2–ζ denote the orthogonal curvilinear coordinates, such that the ξ_1–ξ_2 curves are lines of curvature on the middle surface of the jth lamina.

$\zeta = 0$, ζ-curve are straight lines perpendicular to this middle surface and is measured positive outward. The lines of principal curvature coincide with the coordinate lines. The values of the principal radii of curvature of this middle surface are denoted by R_{1j}, R_{2j} or $1/K_{1j}$, $1/K_{2j}$, L_{1j}, L_{2j} are Lamé coefficients. α_{1j}, α_{2j} are those coefficients of this layer middle surface. u^j, v^j are displacement components of the jth middle surface in ξ_1, ξ_2 directions respectively. ϕ_{1j}, ϕ_{2j} are the rotation angles of normal sections about ξ_2, ξ_1 axis respectively. h is the thickness of jth layer.

From (1) we get

$$u^j = \tfrac{1}{2}(u_{j+1} + u_j) \qquad \phi_{1j} = \left(\frac{1}{h_j} - \frac{K_{1j}}{2}\right) u_{j+1} - \left(\frac{1}{h_j} + \frac{K_{1j}}{2}\right) u_j$$

$$v^j = \tfrac{1}{2}(v_{j+1} + v_j) \qquad \phi_{2j} = \left(\frac{1}{h_j} - \frac{K_{2j}}{2}\right) v_{j+1} - \left(\frac{1}{h_j} + \frac{K_{2j}}{2}\right) v_j \tag{3}$$

where u_j, v_j, u_{j+1}, v_{j+1} are displacement components on the upper and lower surfaces of the jth lamina. Substituting (3) into (1), u, v will be expressed in terms of u_{j+1}, u_j, v_{j+1}, v_j.

Again, we assume the transverse displacement

$$w^j = w(\xi_1, \xi_2) \tag{4}$$

Since the strains of the jth lamina are known as

$$\varepsilon_1^j = \varepsilon_1^0)^j + \zeta K_1^j \qquad \varepsilon_2^j = \varepsilon_2^0)^j + \zeta K_2^j$$

$$\varepsilon_4^j = \varepsilon_4^0)^j \qquad \varepsilon_5^j = \varepsilon_5^0)^j \qquad \varepsilon_6^j = \varepsilon_6^0)^j + \zeta K_6^j \tag{5}$$

where

$$\varepsilon_1^0)^j = \frac{1}{\alpha_{1j}}\frac{\partial u^j}{\partial \xi_1} + \frac{w}{R_{1j}} \qquad \varepsilon_2^0)^j = \frac{1}{\alpha_{2j}}\frac{\partial v^j}{\partial \xi_2} + \frac{w}{R_{2j}}$$

$$\varepsilon_4^0)^j = \frac{1}{\alpha_{2j}}\frac{\partial w}{\partial \xi_2} + \phi_2 - \frac{v^j}{R_{2j}} \qquad \varepsilon_5^0)^j = \frac{1}{\alpha_{1j}}\frac{\partial w}{\partial \xi_1} + \phi_1 - \frac{u^j}{R_{1j}} \tag{6}$$

$$\varepsilon_6^0)^j = \frac{1}{\alpha_{1j}}\frac{\partial v^j}{\partial \xi_1} + \frac{1}{\alpha_{2j}}\frac{\partial u^j}{\partial \xi_2}$$

$$K_1 = \frac{1}{\alpha_{1j}}\frac{\partial \phi_{1j}}{\partial \xi_1} \qquad K_2 = \frac{1}{\alpha_{2j}}\frac{\partial \phi_{2j}}{\partial \xi_2}$$

$$K_6 = \frac{1}{\alpha_{1j}}\frac{\partial \phi_{2j}}{\partial \xi_1} + \frac{1}{\alpha_{2j}}\frac{\partial \phi_{1j}}{\partial \xi_2} + \frac{1}{2}\left(\frac{1}{R_{2j}} - \frac{1}{R_{1j}}\right)\left(\frac{1}{\alpha_1}\frac{\partial v^j}{\partial \xi_1} - \frac{1}{\alpha_2}\frac{\partial u^j}{\partial \xi_2}\right) \tag{7}$$

From virtual work theory we get the equilibrium equations for the jth lamina

$$
\begin{gathered}
-\frac{\partial N_1^j}{\alpha_{1j}\partial\xi_1}-\frac{\partial N_6^j}{\alpha_{2j}\partial\xi_2}+\frac{1}{2}\left(\frac{1}{R_{2j}}-\frac{1}{R_{1j}}\right)\frac{\partial M_6^j}{\alpha_{2j}\partial\xi_1}-\frac{Q_1^j}{R_{1j}}-q_{\xi_1}^j=0\\
-\frac{\partial N_2^j}{\alpha_{2j}\partial\xi_2}-\frac{\partial N_6^j}{\alpha_{1j}\partial\xi_1}-\frac{1}{2}\left(\frac{1}{R_{2j}}-\frac{1}{R_{1j}}\right)\frac{\partial M_6^j}{\alpha_{1j}\partial\xi_1}-\frac{Q_2^j}{R_{2j}}-q_{\xi_2}^j=0\\
-\frac{N_1^j}{R_{1j}}+\frac{N_2^j}{R_{2j}}-\frac{\partial Q_1^j}{\alpha_{1j}\partial\xi_1}-\frac{\partial Q_2^j}{\alpha_{2j}\partial\xi_2}-q^j=0\\
-\frac{\partial M_1^j}{\alpha_{1j}\partial\xi_1}-\frac{\partial M_6^j}{\alpha_{2j}\partial\xi_2}+Q_1^j-m_1^j=0\\
-\frac{\partial M_2^j}{\alpha_{2j}\partial\xi_2}-\frac{\partial M_6^j}{\alpha_{1j}\partial\xi_1}+Q_2^j-m_2^j=0
\end{gathered}
\tag{8}
$$

where

$$
\begin{bmatrix} N_1\\ N_2\\ N_{12}\\ N_{21}\\ Q_1\\ Q_2\\ M_1\\ M_2\\ M_{12}\\ M_{21}\end{bmatrix}^j
=\int_{-h_j/2}^{h_j/2}
\begin{bmatrix}
\sigma_1(1+\zeta/R_2)\\
\sigma_2(1+\zeta/R_1)\\
\sigma_6(1+\zeta/R_2)\\
\sigma_6(1+\zeta/R_1)\\
\sigma_5(1+\zeta/R_2)\\
\sigma_4(1+\zeta/R_1)\\
\zeta\sigma_1(1+\zeta/R_2)\\
\zeta\sigma_2(1+\zeta/R_1)\\
\zeta\sigma_6(1+\zeta/R_2)\\
\zeta\sigma_6(1+\zeta/R_1)
\end{bmatrix}^j \mathrm{d}\zeta
\tag{9}
$$

$$
q^j=\left(1+\frac{h_j}{2R_{1j}}\right)\left(1+\frac{h_j}{2R_{2j}}\right)\sigma_\zeta\left(\frac{h_j}{2}\right)
-\left(1-\frac{h_j}{2R_{1j}}\right)\left(1-\frac{h_j}{2R_{2j}}\right)\sigma_\zeta\left(-\frac{h_j}{2}\right)
$$

$$
q_\xi^j=\left(1+\frac{h_j}{2R_{1j}}\right)\left(1+\frac{h_j}{2R_{2j}}\right)\tau_{\xi_1\zeta}\left(\frac{h_j}{2}\right)
-\left(1-\frac{h_j}{2R_{1j}}\right)\left(1-\frac{h_j}{2R_{2j}}\right)\tau_{\xi_1\zeta}\left(-\frac{h_j}{2}\right)
$$

$$q^j_{\xi_2} = \left(1+\frac{h_j}{2R_{1j}}\right)\left(1+\frac{h_j}{2R_{2j}}\right)\tau_{\xi_2\zeta}\left(\frac{h_j}{2}\right) - \left(1-\frac{h_j}{2R_{1j}}\right)\left(1-\frac{h_j}{2R_{2j}}\right)\tau_{\xi_2\zeta}\left(-\frac{h_j}{2}\right) \tag{10}$$

$$m^j_1 = \frac{h_j}{2}\left[\left(1+\frac{h_j}{2R_{1j}}\right)\left(1+\frac{h_j}{2R_{2j}}\right)\tau_{\xi_1\zeta}\left(\frac{h_j}{2}\right) - \left(1-\frac{h_j}{2R_{1j}}\right)\left(1-\frac{h_j}{2R_{2j}}\right)\tau_{\xi_1\zeta}\left(-\frac{h_j}{2}\right)\right.$$

$$m^j_2 = \frac{h_j}{2}\left[\left(1+\frac{h_j}{2R_{1j}}\right)\left(1+\frac{h_j}{2R_{2j}}\right)\tau_{\xi\zeta}\left(\frac{h_j}{2}\right) - \left(1-\frac{h_j}{2R_{1j}}\right)\left(1-\frac{h_j}{2R_{2j}}\right)\tau_{\xi_2\zeta}\left(-\frac{h_j}{2}\right)\right.$$

σ_1, σ_2 are member stresses in ξ_1, ξ_2 directions respectively. σ_6 are shearing stress in ξ_1–ξ_2 plane. σ_5, σ_4 are transverse shearing stresses.

For shallow shells, however, we can neglect ζ/R in comparison with unity. One has $N_{12} = N_{21} = N_6$, $M_{12} = M_{21} M_6$ in (8).

From (3), (4), (5), (9) and Hooke's law by substituting successively we can express (8) in terms of unknown displacement functions $u_{j+1}, u_j, v_{j+1}, v_j, \ldots$

In our further discussion we would rather transform (8) by substituting the equilibrium conditions between layers. This procedure will be illustrated in the following.

From (8) we can solve

$$\tau^j_{\xi_1\zeta}\left(\frac{h_j}{2}\right) = \left(\frac{h}{2}\left[-\frac{\partial N_1}{\alpha_1\partial\xi_1} - \frac{\partial N_6}{\alpha_2\partial\xi_2} + \frac{1}{2}\left(\frac{1}{R_2}-\frac{1}{R_1}\right)\frac{\partial M_6}{\alpha_2\partial\xi_2} - \frac{Q_1}{R_1}\right] - \frac{\partial M_1}{\alpha_1\partial\xi_1} - \frac{\partial M_6}{\alpha_2\partial\xi_2} + Q_1\right\}\frac{1}{h(1+h/2R_1)(1+h/2R_2)}\Bigg)_j \tag{11}$$

$$\tau^j_{\xi_1\zeta}\left(-\frac{h_j}{2}\right) = \left(\frac{h}{2}\left[-\frac{\partial N_1}{\alpha_1\partial\xi_1} - \frac{\partial N_6}{\alpha_2\partial\xi_2} + \frac{1}{2}\left(\frac{1}{R_2}-\frac{1}{R_1}\right)\frac{\partial M_6}{\alpha_2\partial\xi_2} - \frac{Q_1}{R_1}\right] + \frac{\partial M_1}{\alpha_1\partial\xi_1} + \frac{\partial M_6}{\alpha_2\partial\xi_2} - Q_1\right\}\frac{-1}{h(1-h/2R_1)(1-h/2R_2)}\Bigg)_j \tag{12}$$

In the same manner from the $j+1$th layer we get

$$\tau^{j+1}_{\xi_1\zeta}\left(-\frac{h_{j+1}}{2}\right) = \left(\frac{h}{2}\left[-\frac{\partial N_1}{\alpha_1\partial\xi_1} - \frac{\partial N_6}{\alpha_2\partial\xi_2} + \frac{1}{2}\left(\frac{1}{R_2}-\frac{1}{R_1}\right)\frac{\partial M_6}{\alpha_2\partial\xi_2} - \frac{Q_1}{R_1}\right] + \frac{\partial M_1}{\alpha_1\partial\xi_1} + \frac{\partial M_6}{\alpha_2\partial\xi_2} - Q_1\right\}\frac{-1}{h(1-h/2R_1)(1-h/2R_2)}\Bigg)_{j+1} \tag{13}$$

From equilibrium conditions between layers

$$\tau^{j}_{\xi_1\zeta}\left(\frac{h_j}{2}\right)=\tau^{j+1}_{\xi_1\zeta}\left(-\frac{h_{j+1}}{2}\right) \tag{14}$$

we get

$$\begin{aligned}&\left(\left\{\frac{h}{2}\left[-\frac{\partial N_1}{\alpha_1\,\partial\xi_1}-\frac{\partial N_6}{\alpha_2\,\partial\xi_2}+\frac{1}{2}\left(\frac{1}{R_2}-\frac{1}{R_1}\right)\frac{\partial M_6}{\alpha_2\,\partial\xi_2}-\frac{Q_1}{R_1}\right]\right.\right.\\&\qquad\left.\left.-\frac{\partial M_1}{\alpha_1\,\partial\xi_1}-\frac{\partial M_6}{\alpha_2\,\partial\xi_2}+Q_1\right\}\frac{1}{h(1-h/2R_1)(1-h/2R_2)}\right)_j\\&=\left(\left\{\frac{h}{2}\left[-\frac{\partial N_1}{\alpha_1\,\partial\xi_1}-\frac{\partial N_6}{\alpha_2\,\partial\xi_2}+\frac{1}{2}\left(\frac{1}{R_2}-\frac{1}{R_1}\right)\frac{\partial M_6}{\alpha_2\,\partial\xi_2}-\frac{Q_1}{R_1}\right]\right.\right.\\&\qquad\left.\left.+\frac{\partial M_1}{\alpha_1\,\partial\xi_1}+\frac{\partial M_6}{\alpha_2\,\partial\xi_2}-Q\right\}\frac{-1}{h(1-h/2R_1)(1-h/2R_2)}\right)_{j+1}\\&\qquad\qquad j=1,2,\ldots,n-1\end{aligned} \tag{15}$$

n is the total number of layers of the shell. For the cases where there are no shear loads of the upper and lower surfaces of the whole shell, we have

$$\tau^{1}_{\xi_1\zeta}\left(-\frac{h_1}{2}\right)=0 \qquad \tau^{n}_{\xi_1\zeta}\left(\frac{h_n}{2}\right)=0$$

and we get

$$\begin{aligned}&\frac{h}{2}\left[-\frac{\partial N_1}{\alpha_1\,\partial\xi_1}-\frac{\partial N_6}{\alpha_2\,\partial\xi_2}+\frac{1}{2}\left(\frac{1}{R_2}-\frac{1}{R_1}\right)\frac{\partial M_6}{\alpha_2\,\partial\xi_2}-\frac{Q_1}{R_1}\right]\\&\qquad\left.+\frac{\partial M_1}{\alpha_1\,\partial\xi_1}+\frac{\partial M_6}{\alpha_2\,\partial\xi_2}-Q_1\right)_1=0\end{aligned} \tag{16}$$

$$\begin{aligned}&\frac{h}{2}\left[-\frac{\partial N_1}{\alpha_1\,\partial\xi_1}-\frac{\partial N_6}{\alpha_2\,\partial\xi_2}+\frac{1}{2}\left(\frac{1}{R_2}-\frac{1}{R_1}\right)\frac{\partial M_6}{\alpha_2\,\partial\xi_2}-\frac{Q_1}{R_1}\right]\\&\qquad\left.-\frac{\partial M_1}{\alpha_1\,\partial\xi_1}-\frac{\partial M_6}{\alpha_2\,\partial\xi_2}+Q_1\right)_n=0\end{aligned} \tag{17}$$

In the same manner, from (8), (8) and

$$\tau^{j}_{\xi_2\zeta}\left(\frac{h_j}{2}\right)=\tau^{j+1}_{\xi_2\zeta}\left(-\frac{h_{j+1}}{2}\right) \qquad \tau^{1}_{\xi_2\zeta}\left(-\frac{h_1}{2}\right)=0 \qquad \tau^{n}_{\xi_2\zeta}\left(\frac{h_n}{2}\right)=0$$

we get

$$\left\{\frac{h}{2}\left[-\frac{\partial N_2}{\alpha_2\,\partial\xi_2}-\frac{\partial N_6}{\alpha_1\,\partial\xi_1}-\frac{1}{2}\left(\frac{1}{R_2}-\frac{1}{R_1}\right)\frac{\partial M_6}{\alpha_2\,\partial\xi_2}-\frac{Q_2}{R_2}\right]\right.$$
$$\left.\left.-\frac{\partial M_2}{\alpha_2\,\partial\xi_2}-\frac{\partial M_6}{\alpha_1\,\partial\xi_1}+Q_2\right\}\Big/\left[h\left(1+\frac{h}{2R_1}\right)\left(1+\frac{h}{2R_2}\right)\right]\right)_j$$
$$\left\{\frac{h}{2}\left[-\frac{\partial N_2}{\alpha_2\,\partial\xi_2}-\frac{\partial N_6}{\alpha_1\,\partial\xi_1}-\frac{1}{2}\left(\frac{1}{R_2}-\frac{1}{R_1}\right)\frac{\partial M_6}{\alpha_2\,\partial\xi_2}-\frac{Q_2}{R_2}\right]\right.$$
$$\left.\left.+\frac{\partial M_2}{\alpha_2\,\partial\xi_2}+\frac{\partial M_6}{\alpha_1\,\partial\xi_1}-Q_2\right\}\Big/\left[-h\left(1-\frac{h}{2R_1}\right)\left(1-\frac{h}{2R_2}\right)\right]\right)_{j+1}$$
$$j=1,2,\ldots,n-1 \tag{18}$$

$$\frac{h}{2}\left[-\frac{\partial N_2}{\alpha_2\,\partial\xi_2}-\frac{\partial N_6}{\alpha_1\,\partial\xi_1}-\frac{1}{2}\left(\frac{1}{R_2}-\frac{1}{R_1}\right)\frac{\partial M_6}{\alpha_2\,\partial\xi_2}-\frac{Q_2}{R_2}\right]$$
$$\left.+\frac{\partial M_2}{\alpha_2\,\partial\xi_2}+\frac{\partial M_6}{\alpha_1\,\partial\xi_1}-Q_2\right)_1=0 \tag{19}$$

$$\frac{h}{2}\left[-\frac{\partial N_2}{\alpha_2\,\partial\xi_2}-\frac{\partial N_6}{\alpha_1\,\partial\xi_1}-\frac{1}{2}\left(\frac{1}{R_2}-\frac{1}{R_1}\right)\frac{\partial M_6}{\alpha_2\,\partial\xi_2}-\frac{Q_2}{R_2}\right]$$
$$\left.-\frac{\partial M_2}{\alpha_2\,\partial\xi_2}-\frac{\partial M_6}{\alpha_1\,\partial\xi_1}+Q_2\right)_n=0 \tag{20}$$

Again, (8) must be satisfied in each layer. Let $j=1,2,\ldots,n$. From (8) we have n questions. Summing up all those equations we can eliminate the peeling stresses between layers and we get one equation

$$\left(\frac{N_1}{R_1}+\frac{N_2}{R_2}-\frac{\partial Q_1}{\alpha_1\,\partial\xi_1}-\frac{\partial Q_2}{\alpha_2\,\partial\xi_2}\right)_n\frac{1}{b_n}\cdot\frac{a_{n-1}}{b_{n-1}}\cdots\frac{a_2}{b_2}$$
$$+\left(\frac{N_1}{R_1}+\frac{N_2}{R_2}-\frac{\partial Q_1}{\alpha_1\,\partial\xi_1}-\frac{\partial Q_2}{\alpha_2\,\partial\xi_2}\right)_{n-1}\frac{1}{b_{n-1}}\cdot\frac{a_{n-2}}{b_{n-2}}\cdots\frac{a_2}{b_2}+\cdots$$
$$+\left(\frac{N_1}{R_1}+\frac{N_2}{R_2}-\frac{\partial Q_1}{\alpha_1\,\partial\xi_1}-\frac{\partial Q_2}{\alpha_2\,\partial\xi_2}\right)_2\frac{1}{b_2} \tag{21}$$
$$+\left(\frac{N_1}{R_1}+\frac{N_2}{R_2}-\frac{\partial Q_1}{\alpha_1\,\partial\xi_1}-\frac{\partial Q_2}{\alpha_2\,\partial\xi_2}\right)_1\frac{1}{a_1}$$
$$+\sigma_\zeta\left(\frac{h_n}{2}\right)\frac{a_n}{b_n}\cdots\frac{a_2}{b_2}-\sigma_\zeta\left(-\frac{h_1}{2}\right)\frac{b_1}{a_1}=0$$

where

$$a_j=\left(1+\frac{h_j}{2R_{1j}}\right)\left(1+\frac{h_j}{2R_{2j}}\right)\qquad b_j=\left(1-\frac{h_j}{2R_{1j}}\right)\left(1-\frac{h_j}{2R_{2j}}\right)\tag{22}$$

(15)–(21) are equilibrium equations of the shell used instead of (8) in this theory.

Let all the resultant forces and moments or twisting moments in (8) be expressed in unknown displacement components $u_{j+1}, u_j, v_{j+1}, v_j, \ldots, w$, we have (assume each layer of the shell is orthotropic)

$$\frac{1}{2(1+h/2R_1)(1+h/2R_2)}\left(\left[\frac{1}{\alpha_1}\frac{\partial}{\partial\xi_1}\quad 0\quad \frac{1}{\alpha_2}\frac{\partial}{\partial\xi_2}\quad \frac{1}{R_1}-\frac{2}{h}\quad 0\quad \frac{2}{\alpha_1 h}\frac{\partial}{\partial\xi_1}\quad 0\quad -\frac{1}{\alpha_2}\times\left(\frac{1}{2R_2}-\frac{1}{2R_1}-\frac{2}{h}\right)\frac{\partial}{\partial\xi_2}\right][A][u]\right)_j$$

$$=\frac{1}{2(1-h/2R_1)(1-h/2R_2)}\left(\left[\frac{1}{\alpha_1}\frac{\partial}{\partial\xi_1}\quad 0\quad \frac{1}{\alpha_2}\frac{\partial}{\partial\xi_2}\quad \frac{1}{R_1}+\frac{2}{h}\quad 0\quad \frac{2}{\alpha_1 h}\frac{\partial}{\partial\xi_1}\quad 0\quad -\frac{1}{\alpha_2}\times\left(\frac{1}{2R_2}-\frac{1}{2R_1}+\frac{2}{h}\right)\frac{\partial}{\partial\xi_2}\right][A][u]\right)_{j+1}\qquad j=1,2,\ldots,n-1\tag{23}$$

$$\left(\left[-\frac{h}{2\alpha_1}\frac{\partial}{\partial\xi_1}\quad 0\quad -\frac{h}{2\alpha_2}\frac{\partial}{\partial\xi_2}\quad -\frac{h}{2}\left(\frac{1}{R_1}-\frac{2}{h}\right)\quad 0\quad -\frac{1}{\alpha_1}\frac{\partial}{\partial\xi_1}\quad 0\quad \frac{1}{\alpha_2}\times\left(\frac{h}{4R_2}-\frac{h}{4R_1}-1\right)\frac{\partial}{\partial\xi_2}\right][A][u]\right)_n=0\tag{24}$$

$$\left(\left[-\frac{h}{2\alpha_1}\frac{\partial}{\partial\xi_1}\quad 0\quad -\frac{h}{2\alpha_2}\frac{\partial}{\partial\xi_2}\quad -\frac{h}{2}\left(\frac{1}{R_1}+\frac{2}{h}\right)\quad 0\quad -\frac{1}{\alpha_1}\frac{\partial}{\partial\xi_1}\quad 0\quad \frac{1}{\alpha_2}\times\left(\frac{h}{4R_2}-\frac{h}{4R_1}+1\right)\frac{\partial}{\partial\xi_2}\right][A][u]\right)_1=0\tag{25}$$

$$\frac{1}{2(1+h/2R_1)(1+h/2R_2)}\Bigg[0 \quad \frac{1}{\alpha_2}\frac{\partial}{\partial\xi_2} \quad \frac{1}{\alpha_1}\frac{\partial}{\partial\xi_1} \quad 0 \quad \frac{1}{R_2}-\frac{2}{h} \quad 0 \quad \frac{2}{\alpha_2 h}\frac{\partial}{\partial\xi_2} \quad \frac{1}{\alpha_1}$$

$$\times\left(\frac{1}{2R_2}-\frac{1}{2R_1}+\frac{2}{h}\right)\frac{\partial}{\partial\xi_1}\Bigg][A][u]\Bigg)_j$$

$$=\frac{1}{2(1-h/2R_1)(1-h/2R_2)}\Bigg[0 \quad \frac{1}{\alpha_2}\frac{\partial}{\partial\xi_2} \quad \frac{1}{\alpha_1}\frac{\partial}{\partial\xi_1} \quad 0 \quad \frac{1}{R_2}+\frac{2}{h} \quad 0 \quad \frac{2}{\alpha_2 h}\frac{\partial}{\partial\xi_2} \quad \frac{1}{\alpha_1}$$

$$\times\left(\frac{1}{2R_2}-\frac{1}{2R_1}-\frac{2}{h}\right)\frac{\partial}{\partial\xi_1}\Bigg][A][u]\Bigg)_{j+1} \qquad j=1,2,\ldots,n-1 \quad (26)$$

$$\Bigg[0 \quad \frac{h}{2\alpha_2}\frac{\partial}{\partial\xi_2} \quad \frac{h}{2\alpha_1}\frac{\partial}{\partial\xi_1} \quad 0 \quad \frac{h}{2R_2}-1 \quad 0 \quad \frac{1}{\alpha_2}\frac{\partial}{\partial\xi_2} \quad \frac{1}{\alpha_1}$$

$$\times\left(\frac{h}{4R_2}-\frac{h}{4R_1}+1\right)\frac{\partial}{\partial\xi_1}\Bigg][A][u]\Bigg)_n=0 \quad (27)$$

$$\Bigg[0 \quad \frac{h}{2\alpha_2}\frac{\partial}{\partial\xi_2} \quad \frac{h}{2\alpha_1}\frac{\partial}{\partial\xi_1} \quad 0 \quad \frac{h}{2R_2}+1 \quad 0 \quad \frac{1}{\alpha_2}\frac{\partial}{\partial\xi_2} \quad \frac{1}{\alpha_1}$$

$$\times\left(\frac{h}{4R_2}-\frac{h}{4R_1}-1\right)\frac{\partial}{\partial\xi_2}\Bigg][A][u]\Bigg)_1=0 \quad (28)$$

$$\frac{a_2a_3\ldots a_{n-1}}{b_2b_3\ldots b_n}[T]^n[A]^n[u]^n+\frac{a_2a_3\ldots a_{n-2}}{b_2b_3\ldots b_{n-1}}[T]^{n-1}[A]^{n-1}[u]^{n-1}+\cdots$$

$$+\frac{a_2}{b_2b_3}[T]^3[A]^3[u]^3+\frac{1}{b_2}[T]^2[A]^2[u]^2+\frac{1}{a_1}[T]^1[A]^1[u]^1+$$

$$+\sigma_\zeta\left(\frac{h_n}{2}\right)\frac{a_2a_3\ldots a_n}{b_2b_3\ldots b_n}-\sigma_\zeta\left(-\frac{h_1}{2}\right)\frac{b_1}{a_1}=0 \quad (29)$$

in which

$$[T]=\left[\frac{1}{R_1}\frac{1}{R_2} \quad 0 \quad -\frac{1}{\alpha_1}\frac{\partial}{\partial\xi_1} \quad -\frac{1}{\alpha_2}\frac{\partial}{\partial\xi_2} \quad 0 \quad 0 \quad 0\right]$$

$$[u]=[u_{j+1} \quad u_j \quad v_{j+1} \quad v_j \quad w]^T \qquad j=1,2,\ldots,n \quad (30)$$

The elements of the matrix $[A]$

$$A(1,1)=\left(\frac{h}{2}+\frac{h^2}{12R_2}-\frac{h^3}{24R_1R_2}\right)\frac{C_{11}\partial}{\alpha_1\partial\xi_1}$$

$$A(2,1)=\left(\frac{h}{2}+\frac{h^2}{12R_1}-\frac{h^3}{24R_1^2}\right)\frac{C_{12}\partial}{\alpha_1\partial\xi_1}$$

$$A(3,1)=\frac{h\,C_{66}\partial}{2\alpha_2\partial\xi_2} \qquad A(4,1)=\left(1-\frac{h}{R_1}\right)C_{55}$$

$$A(5,1)=0 \qquad A(6,1)=\left(\frac{h^2}{12}-\frac{h^3}{24R_1}+\frac{h^3}{24R_2}\right)\frac{C_{11}\partial}{\alpha_1\partial\xi_1}$$

$$A(7,1)=\frac{h^2\,C_{12}\partial}{12\alpha_1\partial\xi_1} \qquad A(8,1)=\left(\frac{h^2}{12}-\frac{h^3}{48R_1}-\frac{h^3}{48R_2}\right)\frac{C_{66}\partial}{\alpha_1\partial\xi_2}$$

$$A(1,2)=\left(\frac{h}{2}-\frac{h^2}{12R_2}-\frac{h^3}{24R_1R_2}\right)\frac{C_{11}\partial}{\alpha_1\partial\xi_1}$$

$$A(2,2)=\left(\frac{h}{2}-\frac{h^2}{12R_1}-\frac{h^3}{24R_1^2}\right)\frac{C_{12}\partial}{\alpha_1\partial\xi_1}$$

$$A(3,2)=\frac{h\,C_{66}\partial}{2\alpha_2\partial\xi_2} \qquad A(4,2)=\left(-1-\frac{h}{R_1}\right)C_{55}$$

$$A(5,2)=0 \qquad A(6,2)=\left(-\frac{h^2}{12}-\frac{h^3}{24R_1}+\frac{h^3}{24R_2}\right)\frac{C_{11}\partial}{\alpha_1\partial\xi_1}$$

$$A(7,2)=-\frac{h^2\,C_{12}\partial}{12\alpha_1\partial\xi_1} \qquad A(8,2)=\left(-\frac{h^2}{12}-\frac{h^3}{48R_1}-\frac{1}{48R_2}\right)\frac{C_{66}\partial}{\alpha_2\partial\xi_2}$$

$$A(1,3)=\left(\frac{h}{2}+\frac{h^2}{12R_2}-\frac{h^3}{24R_2^2}\right)\frac{C_{12}\partial}{\alpha_2\partial\xi_2}$$

$$A(2,3)=\left(\frac{h}{2}+\frac{h^2}{12R_1}-\frac{h^3}{24R_1R_2}\right)\frac{C_{22}\partial}{\alpha_2\partial\xi_2}$$

$$A(3,3)=\frac{h\,C_{66}\partial}{2\alpha_1\partial\xi_1} \qquad A(4,3)=0$$

$$A(5,3)=\left(1-\frac{h}{R_2}\right)C_{44} \qquad A(6,3)=\frac{h^2\,C_{12}\partial}{12\alpha_2\partial\xi_2}$$

$$A(7,3)=\left(\frac{h^2}{12}+\frac{h^3}{24R_1}-\frac{h^3}{24R_2}\right)\frac{C_{22}\,\partial}{\alpha_2\,\partial\xi_2}$$

$$A(8,3)=\left(\frac{h^2}{12}-\frac{h^3}{48R_1}-\frac{h^3}{48R_2}\right)\frac{C_{66}\,\partial}{\alpha_1\,\partial\xi_1}$$

$$A(1,4)=\left(\frac{h}{2}-\frac{h^2}{12R_2}-\frac{h^3}{24R_2^2}\right)\frac{C_{12}\,\partial}{\alpha_2\,\partial\xi_2}$$

$$A(2,4)=\left(\frac{h^2}{12}-\frac{h^2}{12R_1}-\frac{h^3}{24R_1R_2}\right)\frac{C_{22}\,\partial}{\alpha_2\,\partial\xi_2}$$

$$A(3,4)=\frac{h}{2}\frac{C_{66}\,\partial}{\alpha_1\,\partial\xi_1} \qquad A(4,4)=0$$

$$A(5,4)=\left(-1-\frac{h}{R_2}\right)C_{44} \qquad A(6,4)=-\frac{h^2}{12}\frac{C_{12}\,\partial}{\alpha_2\,\partial\xi_2}$$

$$A(7,4)=\left(-\frac{h^2}{12}+\frac{h^3}{24R_1}-\frac{h^3}{24R_2}\right)\frac{C_{22}\,\partial}{\alpha_2\,\partial\xi_2}$$

$$A(8,4)=\left(-\frac{h^2}{12}-\frac{h^3}{48R_1}-\frac{h^3}{48R_2}\right)\frac{C_{66}\,\partial}{\alpha_1\,\partial\xi_1}$$

$$A(1,5)=\frac{h}{R_1}C_{11}+\frac{h}{R_2}C_{12} \qquad A(2,5)=\frac{h}{R_1}C_{12}+\frac{h}{R_2}C_{22}$$

$$A(3,5)=0 \qquad A(4,5)=hC_{35}\frac{\partial}{\alpha_1\,\partial\xi_1}$$

$$A(5,5)=hC_{44}\frac{\partial}{\alpha_2\,\partial\xi_2} \qquad A(6,5)=\frac{h^3C_{11}}{12R_1R_2}+\frac{h^3C_{12}}{12R_2^2}$$

$$A(7,5)=\frac{h^3C_{12}}{12R_1^2}+\frac{h^3C_{22}}{12R_1R_2} \qquad A(8,5)=0$$

(23)–(29) are the governing equations in our refined theory. Here both the equilibrium conditions and displacement continuity between layers are prescribed. There are $2(n+1)+1$ equations to solve $2(n+1)+1$ unknowns $u_1,\ldots,u_{n+1},v_1,\ldots,v_{n+1},w$.

The boundary conditions are that one member of each of the following products must be prescribed on each side of the shell. Side along ξ_2 direction: $N_1^j, u_1^j, N_6^j, v_2^j; Q_1^j, w; M_1^j, \phi_{1j}; M_6^j, \phi_{2j}$. Side along ξ_1 direction: N_2^j, u_2^j; N_6^j, v_1^j; Q_2^j, w; M_2^j, ϕ_{2j}; M_6^j, ϕ_{1j}.

As an example, we consider the special case—a cylindrical shell with each layer orthotropic. We have the governing equations:

$$\frac{1}{h_j(1+h_j/2R_j)}\left[0 \quad \frac{-\partial}{\partial\chi} \quad 0 \quad \frac{-h_j}{2}\frac{\partial}{\partial x} \quad 1 \quad 0 \quad 0 \quad \frac{-h_j}{2R_j}\frac{\partial}{\partial\theta} \quad 0 \quad \frac{-1}{R_j}\frac{\partial}{\partial\theta}\right][A]^j[u]^j$$

$$= \frac{1}{h_{j+1}(1-h_{j+1}/2R_{j+1})}\left[0 \quad \frac{-\partial}{\partial\chi} \quad 0 \quad \frac{h_{j+1}}{2}\frac{\partial}{\partial\chi} \quad 1 \quad 0 \quad 0 \quad \frac{h_{j+1}}{2R_{j+1}}\frac{\partial}{\partial\theta} \quad 0 \quad \frac{-1}{R_{j+1}}\frac{\partial}{\partial\theta}\right]$$

$$\times[A]^{j+1}[u]^j \qquad j=1,2,\ldots,n-1$$

$$\left[0 \quad \frac{-\partial}{\partial\chi} \quad 0 \quad \frac{-h_n}{2}\frac{\partial}{\partial\chi} \quad 1 \quad 0 \quad 0 \quad \frac{-h_n}{2}\frac{\partial}{R_r\partial\theta} \quad 0 \quad \frac{-\partial}{R_n\partial\theta}\right][A]^n[u]^n=0$$

$$\left[0 \quad \frac{-\partial}{\partial\chi} \quad 0 \quad \frac{h_1}{2}\frac{\partial}{\partial\chi} \quad 1 \quad 0 \quad 0 \quad \frac{h_1}{2}\frac{\partial}{R_1\partial\theta} \quad 0 \quad \frac{-\partial}{R_1\partial\theta}\right][A]^1[u]^1=0$$

$$\frac{-1}{h_j(1+h_j/2R_j)}\left[\frac{1}{R_j}\frac{\partial}{\partial\theta} \quad 0 \quad \frac{1}{2R_j}\frac{\partial}{\partial\theta} \quad 0 \quad 0 \quad \frac{h_j}{2R_j}-1 \quad \frac{h_j}{2}\frac{\partial}{\partial\chi} \quad 0 \quad \frac{\partial}{\partial\chi} \quad 0\right]$$

$$\times[A]^j[u]^j = \frac{1}{h_{j+1}(1-h_{j+1}/2R_{j+1})}$$

$$\times\left[\frac{-\partial}{R_{j+1}\partial\theta} \quad 0 \quad \frac{h_{j+1}\partial}{2R_{j+1}\partial\theta} \quad 0 \quad 0 \quad \frac{h_{j+1}}{2R_{j+1}}+1 \quad \frac{h_{j+1}}{2}\frac{\partial}{\partial\chi} \quad 0 \quad \frac{-\partial}{\partial\chi} \quad 0\right]$$

$$\times[A]^{j+1}[u]^{j+1} \qquad j=1,2,\ldots,n-1$$

$$\left[\frac{1}{R_n}\frac{\partial}{\partial\theta} \quad 0 \quad \frac{h_n}{2R_n}\frac{\partial}{\partial\theta} \quad 0 \quad 0 \quad \frac{h_n}{2R_n}-1 \quad \frac{h_n}{2}\frac{\partial}{\partial\chi} \quad 0 \quad \frac{\partial}{\partial\chi} \quad 0\right][A]^n[u]^n=0$$

$$\left[\frac{-1}{R_1}\frac{\partial}{\partial\theta} \quad 0 \quad \frac{h_1}{2R_1}\frac{\partial}{\partial\theta} \quad 0 \quad 0 \quad \frac{h_1}{2R_1}+1 \quad \frac{h_1}{2}\frac{\partial}{\partial\chi} \quad 0 \quad \frac{-\partial}{\partial\chi} \quad 0\right][A]^1[u]^1=0$$

$$\frac{a_2.a_3\ldots a_{n-1}}{b_2.b_3\ldots b_n}-[T]^n[A]^n[u]^n+\frac{a_2.a_3\ldots a_{n-2}}{b_2.b_3\ldots b_{n-1}}[T]^{n-1}[A]^{n-1}[u]^{n-1}$$

$$+\frac{a_2}{b_2b_3}[T]^3[A]^3[u]^3+\frac{1}{b_2}[T]^2[A]^2[u]^2+\frac{1}{a_1}[T]^1[A]^1[u]^1$$

$$+\sigma_\zeta\left(\frac{h_n}{2}\right)\frac{a_2a_3\ldots a_n}{b_2b_3\ldots b_n}-\sigma_\zeta\left(-\frac{h_1}{2}\right)\frac{b_1}{a_1}=0$$

In which

$$[T]^i=\left[0 \quad 0 \quad -\frac{1}{R_i} \quad 0 \quad \frac{\partial}{\partial\chi} \quad \frac{1}{R_i}\frac{\partial}{\partial\theta} \quad 0 \quad 0 \quad 0 \quad 0\right]$$

$$[U]^i=[u_{i+1} \quad u_i \quad v_{i+1} \quad v_i \quad w]^T$$

$$i=1,2,\ldots,n$$

$$[A] = \begin{vmatrix}
\frac{C_{12}h^2}{12}\frac{\partial}{\partial\chi} & -\frac{C_{12}h^2}{12}\frac{\partial}{\partial\chi} & C_{22}\left(\frac{h^2}{12}-\frac{h^3}{24R}\right)\frac{\partial}{R\,\partial\theta} & -C_{22}\left(\frac{h^2}{12}+\frac{h^3}{24R}\right)\frac{\partial}{R\,\partial\theta} & -\frac{C_{22}h^3}{12R} \\
\frac{C_{11}h^2}{12}\left(1+\frac{h}{2R}\right)\frac{\partial}{\partial\chi} & \frac{C_{11}h^2}{12}\left(\frac{h}{2R}-1\right)\frac{\partial}{\partial\chi} & \frac{C_{12}}{12}\frac{h^2}{R}\frac{\partial}{\partial\theta} & -\frac{C_{12}h^2}{12}\frac{\partial}{R\,\partial\theta} & 0 \\
\frac{C_{12}h}{2}\frac{\partial}{\partial\chi} & \frac{C_{12}h}{2}\frac{\partial}{\partial\chi} & C_{22}\left(\frac{h}{2}+\frac{h^3}{24R^2}-\frac{h^2}{12R}\right)\frac{\partial}{R\,\partial\theta} & C_{22}\left(\frac{h}{2}+\frac{h^3}{24R^2}-\frac{h^2}{12R}\right)\frac{\partial}{R\,\partial\theta} & C_{22}\left(\frac{h}{R}+\frac{h^3}{12R^3}\right) \\
C_{11}\left(\frac{h}{2}+\frac{h^3}{12R}\right)\frac{\partial}{\partial\chi} & C_{11}\left(\frac{h}{2}-\frac{h^2}{12R}\right)\frac{\partial}{\partial\chi} & \frac{C_{12}h}{2}\frac{\partial}{R\,\partial\theta} & \frac{C_{12}h}{2}\frac{\partial}{R\,\partial\theta} & C_{12}\frac{h}{R} \\
C_{55} & -C_{55} & 0 & 0 & C_{55}h\frac{\partial}{\partial\chi} \\
0 & 0 & C_{44}\left(h+\frac{h^3}{12R^2}\right)\left(\frac{1}{h}-\frac{1}{2R}\right) & -C_{44}\left(h+\frac{h^3}{12R^3}\right)\left(\frac{1}{h}-\frac{1}{2R}\right) & C_{44}\left(h+\frac{h^3}{12R^2}\right)\frac{\partial}{R\,\partial\theta} \\
C_{66}\frac{h}{2}\frac{\partial}{R\,\partial\theta} & C_{66}\frac{h}{2}\frac{\partial}{R\,\partial\theta} & \left(C_{66}\frac{h}{2}+C_{66}\frac{h^2}{12R}\right)\frac{\partial}{\partial\chi} & C_{66}\left(\frac{h}{2}-\frac{h^2}{12R}\right)\frac{\partial}{\partial\chi} & 0 \\
C_{66}\left(\frac{h}{2}+\frac{h^3}{24R^2}-\frac{h^2}{24R}\right)\frac{\partial}{R\,\partial\theta} & C_{66}\left(\frac{h}{2}+\frac{h^3}{24R^2}-\frac{h^2}{24R}\right)\frac{\partial}{R\,\partial\theta} & C_{66}\frac{h}{2}\frac{\partial}{\partial\chi} & C_{66}\frac{h}{2}\frac{\partial}{\partial\chi} & 0 \\
C_{66}\frac{h^2}{12}\frac{\partial}{R\,\partial\theta} & -C_{66}\frac{h^2}{12}\frac{\partial}{R\,\partial\theta} & C_{66}\frac{h^3}{12}\left(\frac{1}{h}+\frac{1}{2R}\right)\frac{\partial}{\partial\chi} & -C_{66}\frac{h^3}{12}\left(\frac{1}{h}+\frac{1}{2R}\right)\frac{\partial}{\partial\chi} & 0 \\
C_{66}\left(\frac{h^2}{12}-\frac{h^3}{24R}\right)\frac{\partial}{R\,\partial\theta} & -C_{66}\left(\frac{h^2}{12}+\frac{h^3}{24R}\right)\frac{\partial}{R\,\partial\theta} & C_{66}\frac{h^2}{12}\frac{\partial}{\partial\chi} & -C_{66}\frac{h^2}{12}\frac{\partial}{\partial\chi} & 0
\end{vmatrix}$$

Boundary conditions are that one member of each of the following products must be prescribed on each lamina end: N_χ, u_0: $N_{\chi\theta}, v_0$; Q_χ, w; M_χ, ψ_χ; $M_{\chi\theta}$; ψ_0; χ; θ measured along the longitudinal and circumferential directions respectively, where

$$u_0 = \frac{u_{j+1} + u_j}{2} \qquad \psi_\chi = \frac{u_{j+1} - u_j}{2}$$

$$v_0 = \frac{v_{j+1} + v_j}{2} \qquad \psi_\theta = \frac{v_{j+1} - v_j}{2}$$

take care that $M_{\chi\theta} \neq M_{\theta\chi}$, $N_{\chi\theta} \neq N_{\theta\chi}$ in this case.

Consider a [90°, 0°, 0°, 90°] circular cylinder of infinite extent subject to the following internal pressure loading $\sigma_z(-h_1/2) = \cos\theta$ with all other surface tractions vanishing. The properties of each play relative to the fibre directions are $E_L = 20 \times 10^6$ psi, $E_T = 10^6$ psi, $G_L = 0{\cdot}6 \times 10^6$ psi, $G_{TT} = 0{\cdot}5 \times 10^6$ psi, $\nu_{LT} = \nu_{TT} = 0{\cdot}25$, where L signifies the direction parallel to the fibres and T the transverse direction. As $\nu_{LT} = \nu_{TL} \times E_T/E_L = 0{\cdot}0125$, for the two 0°-layers $c_{11} = 20{\cdot}0627 \times 10^6$ psi, $c_{22} = 0{\cdot}25078 \times 10^6$ psi, $c_{12} = 0{\cdot}6 \times 10^6$ psi, $c_{66} = 0{\cdot}6 \times 10^6$ psi, $c_{44} = c_{55} = 0{\cdot}5 \times 10^6$ psi, for the outer two 90°-layers $c_{11} = 1{\cdot}00314 \times 10$ psi, $c_{22} = 20{\cdot}0627 \times 10$ psi. $c_{12}, c_{66}, c_{44}, c_{55}$ are the same for all four layers respectively.

The inner radius of the cylinder is known to be 8 in. The thickness of each layer is 1 in. We have 11 governing equations with 11 unknown displacement components $u_1, u_2, \ldots, u_5, v_1, v_2, \ldots, v_5, w$. A solution of the problem is of the form $u_i = 0$, $v_i = c_i \sin 2$, $w = E\cos 2\theta$, $i = 1, 2, \ldots, 5$, c_i, E are unknown constants. The unique solution is $c_1 = -9{\cdot}969431 \times 10^{-6}$ in, $c_2 = -8{\cdot}313900 \times 10^{-6}$ in, $c_3 = -8{\cdot}24644 \times 10^{-6}$ in, $c_4 = -7{\cdot}92781 \times 10^{-6}$ in, $c_5 = -5{\cdot}9206208 \times 10^{-6}$ in, $E = 1{\cdot}667366 \times 10^{-5}$ in. Reasonably good agreement with the exact elasticity solution is observed.[2]

REFERENCES

1. Reddy, J. N. Exact solutions of moderately thick laminated shells, *J. Engng Mech.*, **110**, No. 5, May 1984.
2. Whitney, J. M. and Sun, C.-T., A refined theory for laminated anisotropic cylindrical shells, *J. appl. Mech.*, June 1974, 471–480.
3. Cai Si-Wei, A refined theory for laminated anisotropic cylindrical shells, *Proceedings of International Symposium on Composite Materials and Structures*, Technomic Publishing Co. Inc., U.S.A., 1986, pp. 154–162.

20

On the Behavior of Conical Shells Composed of Quasi-isotropic Composite Shells

JACK R. VINSON

Department of Mechanical Engineering, University of Delaware, Newark, Delaware 19716, USA

and

HOWARD S. KLIGER

H. S. Kliger and Associates, Inc., 12 Lench Avenue, Edison, New Jersey 08820, USA

ABSTRACT

Methods of analysis are presented for the design and analysis of a conical shell subjected to axially symmetric lateral and inplane loads, for any axially symmetric boundary conditions, composed of a quasi-isotropic composite material. Included are the thermoelastic effects, and the effects of thermal thickening and transverse shear deformation.

NOTATION

A_{ij}	Laminate inplane stiffnesses
C	$Eh/(1-\nu^2)$
D	$Eh^3/12(1-\nu^2)$
D_{ij}	Laminate bending stiffnesses
E	Elastic modulus in isotropic plane
G_c	Shear modulus in transverse plane
H	Horizontal stress resultant (perpendicular to z)
h	Shell thickness defined by (4)
i	$(-1)^{1/2}$

L	Slant length of cone
m_k	$h/2[(1 + h/2R^*)\tau^+ + (1 - h/2R^*)\tau^-]$
m_y	Surface shear $[(1 + \zeta/R^*)\zeta\sigma_{yz}]_{-h/2}^{+h/2}$
M_θ, M_y	Stress couples
$M_{T\theta}, M_{Ty}$	Thermal stress couples
N_θ, N_y	Stress resultants
$N_{T\theta}, N_{Ty}$	Thermal stress resultant
p	Lateral pressure
$\bar{p}_V$	$q_y \cos\beta - p \sin\beta$
$\bar{p}_H$	$q_y \sin\beta + p \cos\beta$
Q_y	Shear stress resultant
q_y	Surface shear $[(1 + \zeta/R^*)\sigma_{y\zeta}]_{-h/2}^{+h/2}$
R	Conical radius normal to axis of revolution
R^*	Conical radius normal to meridional coordinate
T	Temperature measured above stress-free temperature of shell
ΔT	$T(+h/2) - T(-h/2)$
U	Middle surface displacement parallel to z coordinate
u_0	Middle surface inplane displacement
V	Vertical stress resultant (parallel to z)
W	Middle surface displacement perpendicular to z coordinate
w_0	Lateral deflection of middle surface
y	Meridional coordinate
z	Coordinate axis of revolution
α	Coefficient of thermal expansion in plane of isotropy
α_ζ	Coefficient of thermal expansion in ζ-direction
β	Half cone angle
β_y	Rotation of the normal to the undeformed middle surface in the meridional direction
ε_{ij}^0	Middle surface strains
ζ	Coordinate outward normal from the middle surface
θ	Circumferential coordinate
ν	Poisson's ratio defined by (5)
σ_{ij}	Stress
$\tau \pm (\pm h/2)$	$\sigma_{y\zeta}(\pm h/2)$

INTRODUCTION

Numerous studies have been performed on cylindrical shells composed of composite materials; much less has been done with conical shells of composite materials. It is desirable to obtain analytical solutions for conical shells of composite materials which include in detail the behavior of the bending boundary layers at each end, wherein lie the largest stresses, due to the superposition of bending stresses on the membrane stresses.

In conical shells of composite materials, due to the geometry and the biaxiality of many loadings, quite often stacking sequences are employed which result in a quasi-isotropic composite laminate. Therefore, the methods of analysis presented are specifically directed toward conical shells of quasi-isotropic composite laminates.

Since almost all composite materials involve a cure cycle at an elevated temperature during the fabrication process, it is necessary to include in any analysis the effects of temperature. Thus, the methods of analysis developed herein include thermoelastic effects, with their attendant problems associated with satisfying nonhomogeneous boundary conditions.

Additionally, quite often the thermal properties in the shell thickness direction differ markedly from those parallel to the plane of the shell midsurface, because the latter quite often are fiber dominated properties, while in the thickness direction they are matrix dominated. Thus, in the methods presented, thermal strains in the thickness direction, as well as those in the other shell principal directions, are included.

Because in many composites the properties parallel to the shell midplane are fiber dominated, the ratio of inplane elastic stiffness to the transverse shear modules can be as high as 50 or 60, compared to two or three in an isotropic material; the structure appears to be weak or flexible in transverse shear. Again, the methods presented here include the effects of transverse shear deformation.

Because the methods of analysis are analytically general, they are applicable to polymer, metal and ceramic matrix composites, if their elastic and thermal properties are known, or can be estimated.

TRANSFORMATION OF QUASI-ISOTROPIC LAMINATE TO A TRANSVERSELY ISOTROPIC MATERIAL

It is convenient to transform the quasi-isotropic laminate of properties A_{ij} and D_{ij} and thickness $\bar{h}$ to an equivalent transversely isotropic material of thickness h, for briefness of presentation. Thus

$$A_{11}(=A_{22}) = \frac{Eh}{1-\nu^2} = C \tag{1a}$$

$$D_{11}(=D_{22}) = \frac{Eh^3}{12(1-\nu^2)} = D \tag{1b}$$

From the above the equivalent (fictitious) E and h are seen to be

$$E = \frac{A_{11}^{3/2}(1 - \nu^2)}{D_{11}^{1/2} 2\sqrt{3}} \tag{1c}$$

$$h = \left(\frac{D_{11}}{A_{11}}\right)^{1/2} 2\sqrt{3} \tag{1d}$$

It can also be shown that for the equivalent isotropic material

$$\nu = A_{12}/A_{11} \tag{1e}$$

GEOMETRY AND NOTATION

The geometry and notation of a truncated conical shell, as well as positive directions of displacements, stress resultants, stress couples and shear resultants are presented in Fig. 1. It is noted that, except for minor differences, the notation of Reissner[1] is used throughout this chapter. From Fig. 1, the geometric relations are noted:

$$R = y \sin\beta \qquad R^* = R/\cos\beta \qquad z = y\cos\beta$$

The plane of quasi-isotropy is the $y\theta$ plane.

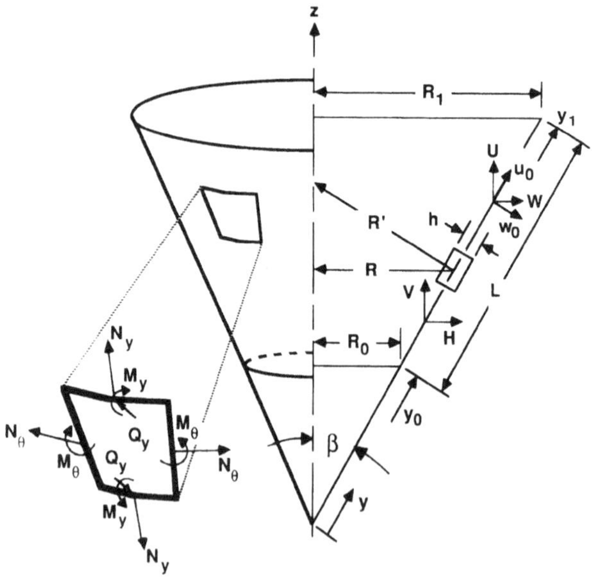

FIG. 1.

GOVERNING EQUATIONS

The governing equations for a conical shell of a quasi-isotropic composite material under an axially-symmetric load are given as follows:

$$N_\theta = C[\varepsilon^0_{\theta\theta} + \varepsilon^0_{yy}] - \frac{N_{T\theta}}{(1-\nu)} + \frac{E}{R^*(1-\nu^2)} \int_{-h/2}^{+h/2} \bar{w}\,\mathrm{d}\zeta \tag{2}$$

$$N_y = C[\varepsilon^0_{yy} + \nu\varepsilon^0_{\theta\theta}] - \frac{N_{Ty}}{(1-\nu)} + \frac{E\nu}{R^*(1-\nu^2)} \int_{-h/2}^{+h/2} \bar{w}\,\mathrm{d}\zeta \tag{3}$$

$$M_\theta = D\left[\frac{\beta_y}{R}\frac{\mathrm{d}R}{\mathrm{d}y} + \nu\frac{\mathrm{d}\beta_y}{\mathrm{d}y}\right] - \frac{M_{T\theta}}{(1-\nu)} + \frac{E}{R^*(1-\nu^2)} \int_{-h/2}^{+h/2} \bar{w}\zeta\,\mathrm{d}\zeta \tag{4}$$

$$M_y = D\left[\frac{\mathrm{d}\beta_y}{\mathrm{d}y} + \frac{\nu\beta_y}{R}\frac{\mathrm{d}R}{\mathrm{d}y}\right] - \frac{M_{Ty}}{(1-\nu)} + \frac{E\nu}{R^*(1-\nu^2)} \int_{-h/2}^{+h/2} \bar{w}\zeta\,\mathrm{d}\zeta \tag{5}$$

$$Q_y = \tfrac{5}{3}G_c h\varepsilon^0_{y\zeta} + \frac{m_k}{6} + \frac{5G_c}{4}\int_{-h/2}^{+h/2}\left[1-\left(\frac{\zeta}{h/2}\right)^2\right]\frac{\mathrm{d}\bar{w}}{\mathrm{d}y}\,\mathrm{d}\zeta$$
$$+ \frac{5}{4R}\int_{-h/2}^{h/2}\left(\frac{\zeta}{h/2}\right)^2\int_{-h/2}^{\zeta}\Theta\,\mathrm{d}\zeta\,\mathrm{d}\zeta \tag{6}$$

$$\varepsilon^0_{yy} = \frac{\mathrm{d}u_0}{\mathrm{d}y} \tag{7a}$$

$$\varepsilon^0_{\theta\theta} = \frac{w_0}{R^*} + \frac{u_0}{R}\frac{\mathrm{d}R}{\mathrm{d}y} \tag{7b}$$

$$\varepsilon^0_{y\zeta} = \frac{1}{2}\left[\beta_y + \frac{\mathrm{d}w_0}{\mathrm{d}y}\right] \tag{7c}$$

$$R\frac{\mathrm{d}N_y}{\mathrm{d}y} + N_y\frac{\mathrm{d}R}{\mathrm{d}y} - N_\theta\frac{\mathrm{d}R}{\mathrm{d}y} = -Rq_y \tag{8a}$$

$$R\frac{\mathrm{d}Q_y}{\mathrm{d}y} + Q_y\frac{\mathrm{d}R}{\mathrm{d}y} - N_\theta\frac{R}{R^*} = -Rp \tag{8b}$$

$$R\frac{\mathrm{d}M_y}{\mathrm{d}y} + (M_y - M_\theta)\frac{\mathrm{d}R}{\mathrm{d}y} - RQ_y = -Rm_y \tag{8c}$$

Note that these equations are similar to those of the classical thin shell theory except for the factors $\bar{w}$, $\varepsilon_{y\zeta}^0$ and θ. Here, $\bar{w}$ accounts for thermal expansions across the shell thickness and is defined as

$$\bar{w} = \int_0^\zeta \alpha_\zeta T \, d\zeta \tag{9}$$

$\varepsilon_{y\zeta}^0$ represents the inclusion of transverse shear deformation. It is noted that unlike the theory of Reissner[2] transverse normal strain is neglected, herein the assumption being that even though the shell is extremely weak in shear it is, nevertheless, a thin shell. θ is a thermal term defined as

$$\begin{aligned}\Theta = \frac{dR}{dy}\Bigg\{\Bigg[\frac{N_{T\theta}}{h(1-\nu)} + \frac{12\zeta}{h^3}\frac{M_{T\theta}}{(1-\nu)} - \frac{E\alpha T}{(1-\nu)}\Bigg] \\ - \frac{E}{R^*(1-\nu)}\Bigg[\frac{1}{h}\int_{-h/2}^{h/2} \bar{w}\, d\zeta + \frac{12\zeta}{h^3}\int_{-h/2}^{h/2} \bar{w}\zeta\, d\zeta - \bar{w}\Bigg]\Bigg\} \\ - \frac{d}{dy}\Bigg[\Bigg[R\Bigg\{\Bigg[\frac{N_{Ty}}{h(1-\nu)} + \frac{12\zeta}{h^3}\frac{M_{Ty}}{(1-\nu)} - \left(1+\frac{\zeta}{R^*}\right)\frac{E\alpha T}{(1-\nu)}\Bigg] \\ - \frac{E\nu}{R^*(1-\nu^2)}\Bigg[\frac{1}{h}\int_{-h/2}^{h/2} \bar{w}\, d\zeta + \frac{12\zeta}{h^3}\int_{-h/2}^{h/2} \bar{w}\zeta\, d\zeta - \bar{w}\Bigg]\Bigg\}\Bigg]\Bigg]\end{aligned} \tag{10}$$

Using the method of De Silva and Naghdi[3] to transform the governing equations into one second-order complex differential equation, the radial and axial stress resultants (H and V) and displacements (W and U) are first defined in Fig. 1. The following transformations are evident:

$$N_y = V\cos\beta + H\sin\beta \qquad Q_y = -V\sin\beta + H\cos\beta \tag{11}$$

$$u = U\cos\beta + W\sin\beta \qquad w = -U\sin\beta + W\cos\beta \tag{12}$$

Substituting (11) into (8) and integrating yields

$$RV = C_9 - \sin\beta \int_{y_0}^{y} y\bar{p}v(y)\, dy \tag{13}$$

Proper substitution among the strain displacement equations transformed to the new coordinate system will result in the compatibility equation:

$$y\frac{d}{dy}\varepsilon_{\theta\theta}^0 + \varepsilon_{\theta\theta}^0 = (2\varepsilon_{y\zeta}^0 - \beta_y)\cot\beta + \varepsilon_{yy}^0 \tag{14}$$

If the surface shear q_y is specified, $V(y)$ is determined from (13) by evaluating C_9 at the edge y_0. W and U are determined once β_y, N_θ and N_y are known. N_θ and N_y are derived from H. By proper substitution among all the previous equations, two coupled second-order differential equations result:

$$\beta_y'' + \left(\frac{R'}{R}\right)\beta_y' + \left(\frac{R'}{R}\right)^2 \beta_y - \frac{\cot\beta}{Dy}(RH) = \bar{F} \quad (15)$$

$$\frac{1}{Eh}\left[(RH)'' + \frac{1}{y}(RH)' - \frac{1}{y^2}(RH) - \frac{6E\cot\beta}{5G_c y} H\cos\beta\right] + \frac{\beta_y \cot\beta}{y} = \bar{G} \quad (16)$$

where

$$\bar{F} = -\frac{RV}{D_y} - \frac{m_y}{D} + \frac{1}{D_y(1-\nu)}[yM'_{Ty} + M_{Ty} - M_{T\theta}] + \frac{E\nu}{R^* D_y(1-\nu^2)}$$
$$\times\left[-\int_{-h/2}^{h/2} \bar{w}\zeta\,d\zeta + \frac{1}{\nu}\int_{-h/2}^{h/2} \bar{w}\zeta\,d\zeta - R^* y\frac{d}{dy}\left(\frac{1}{R^*}\int_{-h/2}^{h/2} \bar{w}\zeta\,d\zeta\right)\right] \quad (17)$$

$$\bar{G} = \frac{1}{Eh}\left[\frac{\cot\beta}{y^2}\left(1 - \frac{6E}{5G_c}\right)\right](RV) - \frac{1}{Eh}\left[\nu\bar{p}_v\cos\beta + \frac{(R^2\bar{p}_H)'}{R} + \nu R'\bar{p}_H\right]$$
$$-\frac{m_k\cot\beta}{5G_c hy} - \frac{1}{Eh(1-\nu)}[N'_{T\theta} - \nu N'_{Ty}] - \frac{1+\nu}{Ehy(1-\nu)} \quad (18)$$
$$\times[N_{T\theta} - N_{Ty}] + \frac{1}{hR^*}\int_{-h/2}^{h/2} \frac{d\bar{w}}{dy}\,d\zeta - \frac{3\cos\beta}{2Rh}$$
$$\times\int_{-h/2}^{h/2}\left[1 - \left(\frac{\zeta}{h/2}\right)^2\right]\frac{d\bar{w}}{dy}\,d\zeta - \frac{3\cos\beta}{2G_c hR^2}\int_{-h/2}^{h/2}\left(\frac{\zeta}{h/2}\right)^2\int_{-h/2}^{\zeta}\Theta\,d\zeta\,d\zeta$$

and prime denotes differentiation with respect to y. Introducing the notation

$$m = [12(1-\nu^2)]^{1/2} \qquad \frac{\hat{m}\cot\beta}{hy} = 2\mu^2 f(y) \qquad \psi = \frac{\hat{m}RH}{Eh^2}$$
$$f(y) = 1/y \qquad \mu^2 = \text{constant} \quad (19)$$

and defining the operator

$$\mathscr{L}(\,) = \frac{1}{f(y)}\left[(\,)'' + \frac{1}{y}(\,)' - \frac{1}{y^2}(\,)\right] \quad (20$$

eqns (15) and (16) reduce to

$$\mathscr{L}(\beta_y) - 2\mu^2\psi = \frac{\bar{F}}{f(y)} \quad (21)$$

$$\mathscr{L}(\psi) + 2\mu^2\beta_y - 2\mu^2\psi\frac{6Eh\cos\beta}{5G_c\hat{m}R} = \frac{\bar{G}}{f(y)} \quad (22)$$

Defining the complex variable

$$\Psi(y) = -\beta_y + ik\psi \tag{23}$$

and multiplying (22) by ik and subtracting (21) from it results in

$$\mathscr{L}(\Psi) - 2\mu^2 ik\left[-\beta_y + \left(-\frac{1}{ik} + \frac{6Eh\cos\beta}{5G_c R\hat{m}}\right)\psi\right] = -\frac{\bar{F}}{f(y)} + ik\frac{\bar{G}}{f(y)} \tag{24}$$

if k is considered to be constant.

To reduce (24) to the classical form

$$\mathscr{L}(\Psi) - 2\mu^2 ik\Psi = -\frac{\bar{F} + ik\bar{G}}{f(y)} \tag{25}$$

the Meissner condition must be of the form

$$k = -\frac{\gamma i}{2} + \left(1 - \frac{\gamma^2}{4}\right)^{1/2} \tag{26}$$

where

$$\gamma = \frac{6E\cos\beta}{5G_c\hat{m}}\left(\frac{h}{R}\right) \tag{27}$$

For the case of $(E/G_c)(h/R) \ll 1$, this will reduce to the classical solution of $k = \pm 1$. It is seen in (26) that k is neither real nor constant, but in fact $k = k(y)$. However, the asymptotic integration of (25) requires that k be constant. It therefore is necessary to investigate the conditions under which k can be considered constant. To this end, let

$$\gamma(y) = \frac{6Eh\cos\beta}{5G_c\hat{m}} \frac{1}{R_0(1 + \varepsilon/R_0)} \tag{28}$$

where

$$\varepsilon = (y - y_0)\sin\beta \tag{29}$$

A binomial expansion of (28), assuming $\Delta R/R \ll 1$, will yield

$$k = -\frac{\gamma_0 i}{2} + \left(1 - \frac{\gamma_0^2}{4}\right)^{1/2} \qquad \gamma \approx \gamma_0 = \frac{6E\cos\beta}{5G_c\hat{m}}\left(\frac{h}{R_0}\right) \tag{30}$$

Thus, for a very nearly constant k, it can be shown from (30) that for a maximum percentage change in γ, a maximum critical length L_{max} may be specified for a given cone geometry. Examining the complex value of k, it is found that if k is restricted to $\mathrm{Re}(k) \geq 10\,\mathrm{Im}(k)$, then $(k)^{1/2}$ can be considered real only and that the theory is valid for large R. Transforming (25) to its normal form using

$$\Phi = (y\sin\beta)^{1/2}\Psi \tag{31}$$

one arrives at

$$\Phi'' + \Phi\left[2\mu^2 i^3 \frac{k}{y} - \frac{3}{4y^2}\right] = (y \sin\beta)^{1/2}(\bar{G}ik - \bar{F})f(y) \tag{32}$$

HOMOGENEOUS SOLUTION: $\bar{F} = \bar{G} = p = V = 0$

Utilizing the classical method of asymptotic integration as outlined by Hildebrand[4] and performed explicitly for isotropic conical shells by Vinson,[5] the solution to (32) may be written:

$$\begin{aligned}\Phi_0 = {} & C_0 \cosh\delta\cos\delta + C_1 \sinh\delta\cos\delta + C_2\cosh\delta\sin\delta + C_3\sinh\delta\sin\delta \\ & - i[C_3\cosh\delta\cos\delta + C_2\sinh\delta\cos\delta - C_1\cosh\delta\sin\delta - C_0\sinh\delta\sin\delta]\end{aligned} \tag{33}$$

where

$$\delta = 2\mu y^{1/2} \tag{34}$$

With the aid of eqns (33), (31) and (23), Ψ may be reduced to its real and imaginary parts:

$$\mathrm{Re}(\Psi) = (y\sin\beta)^{-1/2}\chi_y = -\beta_y + \frac{\gamma}{2}\left(\frac{\hat{m}RH}{Eh^2}\right) \tag{35}$$

$$\mathrm{Im}(\Psi) = (y\sin\beta)^{-1/2}\chi_1 = \left(1 - \frac{\gamma^2}{4}\right)^{1/2}\left(\frac{\hat{m}RH}{Eh^2}\right) \tag{36}$$

Note that the complex value of k is accounted for in calculating (35) and (36). Having H, β_y can be found from (35). Then, using (4) and (5), M_y and M_θ are easily found. Assuming no surface shears, proper substitution among eqns (37), (11) and (8) will give the shear and stress resultants. Finally, having N_y and N_θ, the radial displacement is found from eqns (2), (3), (7) and (12). These quantities are tabulated as follows:

$$H = \frac{Eh^2}{\hat{m}}\frac{(y\sin\beta)^{-3/2}}{(1-\gamma^2/4)^{1/2}}\chi_1 \tag{37}$$

$$M_y = D[-\chi_2\Gamma_y + \chi_4\Delta_y - \chi_1\Lambda_y + \chi_3\Sigma_y] \tag{38}$$

$$M_\theta = D[\chi_4 \mathrm{I}_y - \chi_1 \mathrm{II}_y - \nu\chi_2\Gamma_y + (\nu/S)\chi_3\Sigma_y] \tag{39}$$

$$N_y = y^{-3/2}(\sin\beta)^{-1/2}\chi_1\rho_y \tag{40}$$

$$N_\theta = \rho_y(\sin\beta)^{-1/2}[\chi_1 \mathrm{III}_y + (\mu/y)\chi_3] \tag{41}$$

$$Q_y = \rho_y(y\sin\beta)^{-3/2}\chi_1\cos\beta \tag{42}$$

$$\mathrm{W} = \frac{\rho_y(\sin\beta)^{1/2}}{Eh}[\chi_1\eta_y + \mu\chi_3] \tag{43}$$

where the following definitions apply:

$$\chi_1 = C_1 \cosh\delta \sin\delta + C_0 \sinh\delta \sin\delta - C_3 \cosh\delta \cos\delta - C_2 \sinh\delta \cos\delta$$

$$\chi_2 = (C_3 - C_0)\cosh\delta \sin\delta + (C_3 + C_0)\sinh\delta \cos\delta + (C_2 - C_1)\sinh\delta \sin\delta + (C_2 + C_1)\cosh\delta \cos\delta$$

$$\chi_3 = (C_0 + C_3)\cosh\delta \sin\delta + (C_0 - C_3)\sinh\delta \cos\delta + (C_1 + C_2)\sinh\delta \sin\delta + (C_1 - C_2)\cosh\delta \cos\delta$$

$$\chi_4 = C_0 \cosh\delta \cos\delta + C_1 \sinh\delta \cos\delta + C_2 \cosh\delta \sin\delta + C_3 \sinh\delta \sin\delta$$

$$S = \frac{6Eh\cos\beta}{10\hat{m}G_c \sin\beta} \qquad \rho_y = \frac{Eh^2}{\hat{m}(1 - S^2/y^2)^{1/2}} \qquad \Gamma_y = \frac{\mu}{y(\sin\beta)^{1/2}}$$

$$\Delta_y = -y^{3/2}\left[\frac{1}{2(\sin\beta)^{3/2}} - \frac{\nu}{(\sin\beta)^{1/2}}\right]$$

$$\Lambda_y = \left[\frac{S(y\sin\beta)^{-1/2}}{(1 - S^2/y^2)^{1/2}}\right]\left[\frac{3-\nu}{y^2} + \frac{S^2 y^{-4}}{(1 - S^2/y^2)}\right]$$

$$\Sigma_y = \frac{S\mu}{y^2(1 - S^2/y^2)^{1/2}(\sin\beta)^{1/2}}$$

$$\mathrm{I}_y = y^{-3/2}\left[\frac{\nu}{2(\sin\beta)^{3/2}} - (\sin\beta)^{-1/2}\right]$$

$$\mathrm{II}_y = \left[\frac{S(y\sin\beta)^{-1/2}}{(1 - S^2/y^2)^{1/2}}\right]\left[\frac{1-4\nu}{y^2} - \frac{\nu S^2 y^{-4}}{(1 - S^2/y^2)}\right]$$

$$\mathrm{III}_y = \left[-(y^{3/2}/2) - \frac{S^2 y^{-7/2}}{(1 - S^2/y^2)}\right]$$

$$\eta_y = \left[-(\tfrac{1}{2} + \nu)y^{-1/2} - \frac{S^2 y^{-5/2}}{(1 - S^2/y^2)}\right]$$

For the classical case, since $\varepsilon^0_{y\zeta} = 0$, the lateral deflection of the middle surface could be found directly from (7c). In the present case, it is convenient to integrate (7a) to find u_0 first. Then, with the aid of (12) and (43), the lateral deflection w_0 can be found. From (7a), (2), (3) and (11), one has

$$u_0 = \int_{y_0}^{y} \frac{1}{Eh}(N_y - \nu N_\theta)\,\mathrm{d}y \tag{44}$$

and upon proper substitution of (40) and (41) into this and taking a two-term binomial expansion of the term

$$\left[1-\frac{S^2}{(\delta/2\mu)^4}\right]^{-1/2}$$

about δ_0, the following expression for the inplane middle surface deflection is found:

$$\begin{aligned}
u_0 = {} & E_1[\sin\delta(C_7 e^{\delta} + C_{10} e^{-\delta}) + \cos\delta(-C_5 e^{\delta} + C_4 e^{-\delta})] \\
& - E_2[\sin\delta(C_7\delta e^{\delta} + C_{10}\delta e^{-\delta} + C_{13} e^{\delta} + C_{14} e^{-\delta}) \\
& \qquad + \cos\delta(-C_5\delta e^{\delta} + C_6\delta e^{-\delta} + C_{11} e^{\delta} - C_{12} e^{-\delta})] \\
& - E_3[C_5\{\delta e^{\delta}(\sin\delta - \cos\delta) + e^{\delta}\cos\delta\} \\
& \qquad - C_{10}\{\delta e^{-\delta}(\sin\delta - \cos\delta) + e^{-\delta}\sin\delta\} \\
& \qquad - C_6\{\delta e^{-\delta}(\sin\delta + \cos\delta) + e^{-\delta}\cos\delta\} \\
& \qquad + C_7\{\delta e^{\delta}(\sin\delta + \cos\delta) - e^{\delta}\sin\delta\}] \\
& + E_4[C_5\{\delta^2 e^{\delta}(\sin\delta - \cos\delta) + 2\delta e^{\delta}\cos\delta - e^{\delta}(\sin\delta + \cos\delta)\} \\
& \qquad + C_6\{-\delta^2 e^{-\delta}(\sin\delta + \cos\delta) - 2\delta e^{-\delta}\cos\delta + e^{-\delta}(\sin\delta - \cos\delta\} \\
& \qquad + C_7\{\delta^2 e^{\delta}(\sin\delta + \cos\delta) - 2\delta e^{\delta}\sin\delta + e^{\delta}(\sin\delta - \cos\delta)\} \\
& \qquad - C_{10}\{\delta^2 e^{-\delta}(\sin\delta - \cos\delta) + 2\delta e^{-\delta}\sin\delta \\
& \qquad\qquad + e^{-\delta}(\sin\delta + \cos\delta)\}] + C_8
\end{aligned} \tag{45}$$

where

$$\begin{aligned}
4C_5 &= C_0 + C_1 + C_2 + C_3 & \quad 4C_{11} &= C_0 + C_1 \\
4C_6 &= C_0 - C_1 - C_2 + C_3 & \quad 4C_{12} &= C_1 - C_0 \\
4C_7 &= C_0 + C_1 - C_2 - C_3 & \quad 4C_{13} &= C_2 + C_3 \\
4C_{10} &= C_0 - C_1 + C_2 - C_3 & \quad 4C_{14} &= C_2 - C_3
\end{aligned} \tag{46}$$

$$\begin{aligned}
E_1 &= D_1\lambda^{-1/2}(1 + 2\delta_0^4\lambda^{-1}) + D_2\lambda^{-3/2}(1 + 6\delta_0^4\lambda^{-1}) \\
E_2 &= D_1(2\delta_0^3\lambda^{-3/2}) + D_2(6\delta_0^3\lambda^{-5/2}) \\
E_3 &= D_3\lambda^{-1/2}(1 + 2\delta_0^4\lambda^{-1}) \\
E_4 &= D_3(2\delta_0^3\lambda^{-3/2})
\end{aligned} \tag{47}$$

$$\lambda = \delta_0^4 - S^2(2\mu)^4 \qquad \delta_0 = 2\mu y_0^{1/2} \qquad E^* = Eh^2/\hat{m}(\sin\beta)^{1/2}$$

$$D_1 = \left(\frac{4\mu E^*}{Eh}\right)\left(1 + \frac{\nu}{2}\right) \qquad D_2 = \frac{64\mu^5\nu S^2 E^*}{Eh} \qquad D_3 = \frac{2\mu\nu E^*}{Eh} \tag{48}$$

The lateral deflection of the middle surface becomes

$$\begin{aligned}
w_0 = {} & \frac{2\rho_y(\sin\beta)^{1/2}}{Eh\cos\beta}\{\eta_y[\sin\delta(C_{11} e^{\delta} + C_{12} e^{-\delta}) + \cos\delta(-C_{13} e^{\delta} + C_{14} e^{-\delta})] \\
& + \mu[\sin\delta(C_5 e^{\delta} - C_6 e^{-\delta}) + \cos\delta(C_7 e^{\delta} - C_{10} e^{-\delta}]\} - u_0\tan\beta
\end{aligned} \tag{49}$$

The homogeneous solution provides five constants of integration: C_0, C_1, C_2, C_3 and C_8. The first four are associated with the deflection, slope, moment and shear boundary conditions at each end; C_8 is related to a reference point for measuring inplane deflection. C_9, which comes from (13), is related to the axial load to which the cone is subjected. These six constants must be evaluated through the satisfaction of six boundary conditions, three at each end.

It is noted that if $G_c = \infty$ (i.e. no transverse shear deformation), these reduce to the classical equations for an isotropic material and are identical to those of Vinson.[5] Also, C_8 and C_9 correspond to similar values of that work. Note also that if $E/G_c = 2(1+\nu)$, the equations derived are the homogeneous solution for a conical shell (including transverse shear deformation) composed of an isotropic material.

Edge Load Solutions

Case I. Consider the conical shell bounded by $y_0 \leq y \leq \infty$ with the boundary conditions at y_0 being

$$M_y(y_0) = M_0 \qquad Q_y(y_0) = 0 \tag{50}$$

From (48), for finite displacements

$$C_5 = C_{11} = C_7 = C_{13} = 0 \tag{51}$$

and

$$C_1 = -C_0 \qquad C_2 = -C_3 \tag{52}$$

Using (51) and (52), M_y and Q_y may be written as

$$Q_y = 2\rho_y \cos\beta(y\sin\beta)^{-3/2}(C_{12}\,e^{-\delta}\sin\delta + C_{14}\,e^{-\delta}\cos\delta) \tag{53}$$

$$\begin{aligned} M_y = 2D[&\Gamma_y(C_{10}\,e^{-\delta}\sin\delta + C_6\,e^{-\delta}\cos\delta) \\ &+\Delta_y(C_{14}\,e^{-\delta}\sin\delta - C_{12}\,e^{-\delta}\cos\delta) \\ &-\Lambda_y(C_{12}\,e^{-\delta}\sin\delta + C_{14}\,e^{-\delta}\cos\delta) \\ &+\Sigma_y(C_6\,e^{-\delta}\sin\delta - C_{10}\,e^{-\delta}\cos\delta)] \end{aligned} \tag{54}$$

From the boundary conditions (50), one can solve for C_{2M_0}, the subscript on C_2 referring to that value of C_2 for the particular edge load being discussed:

$$C_{2M_0} = \frac{M_0\,e^{\delta_0}}{D(\sin\delta_0 + \cot\delta_0\cos\delta_0)(\Gamma_{y0} + \Delta_{y0} - \Sigma_{y0})} \tag{55}$$

where the subscript y_0 refers to that value of Γ (or Δ, or Σ, etc.) taken at $y = y_0$.

The displacement equations become

$$\begin{aligned} w_{0M_0} &= \rho_y(\sin\beta)^{1/2}\,e^{-\delta}[\eta_y(C_{12M_0}\sin\delta + C_{14M_0}\cos\delta) \\ &\quad + \mu(C_{6M_0}\sin\delta - C_{10M_0}\cos\delta] - e^{-\delta}(u_{0M_0} - C_8)\tan\beta - C_8\tan\beta \end{aligned} \tag{56}$$

$$\begin{aligned} u_{0M_0} &= (-C_{10M_0}\sin\beta - C_{6M_0}\cos\beta) \\ &\qquad \times[-E_1 + E_2\delta - E_3(\delta+1) + E_4(\delta^2+2\delta+1)] \\ &\quad + (C_{6M_0}\sin\delta - C_{10M_0}\cos\delta)[E_3\delta - E_4(\delta^2-1)] \\ &\quad - (C_{14M_0}\sin\delta - C_{12M_0}\cos\delta)E_2 + C_8 \end{aligned} \tag{57}$$

where

$$C_{1M_0} = -C_{2M_0}\cot\delta_0 \qquad C_{6M_0} = C_{2M_0}\left(\frac{\cot\delta_0 - 1}{2}\right) \qquad C_{12M_0} = C_{1M_0}/2$$

$$C_{14M_0} = C_{2M_0}/2 \qquad C_{10M_0} = C_{2M_0}\left(\frac{\cot\delta_0 + 1}{2}\right) \tag{58}$$

Case II. Consider the conical shell bounded by $y_0 \le y \le \infty$ with the boundary conditions at y_0 being

$$M_y(y_0) = 0 \qquad Q_y(y_0) = Q_0 \tag{59}$$

In a similar manner to Case I,

$$C_5 = C_7 = C_{11} = C_{13} = 0 \tag{60}$$

$$C_1 = -C_0 \qquad C_2 = -C_3 \tag{61}$$

$$C_{2Q_0} = \frac{Q_0\,e^{\delta_0}(y_0\sin\beta)^{3/2}}{\rho_{y_0}\cos\delta_0\cos\beta(1 + J\tan\delta_0)} \tag{62}$$

where

$$J = \frac{(\Gamma_{y0} + \Delta_{y0} - \Sigma_{y0})\sin\delta_0 - (\Gamma_{y0} + \Lambda_{y0} + \Sigma_{y0})\cos\delta_0}{(\Gamma_{y0} + \Lambda_{y0} + \Sigma_{y0})\sin\delta_0 + (\Gamma_{y0} + \Delta_{y0} - \Sigma_{y0})\cos\delta_0} \tag{63}$$

The displacement equations become

$$\begin{aligned} w_{0Q_0} &= \frac{2\rho_y(\sin\beta)^{1/2}\,e^{-\delta}}{Eh\cos\beta}[\eta_y(C_{12Q_0}\sin\delta + C_{14Q_0}\cos\delta) \\ &\quad + \mu(C_{6Q_0}\sin\delta - C_{10Q_0}\cos\delta)] - e^{-\delta}(u_{0Q_0} - C_8)\tan\beta - C_8\tan\beta \end{aligned} \tag{64}$$

$$\begin{aligned} u_{0Q_0} &= (-C_{10Q_0}\sin\delta - C_{6Q_0}\cos\delta) \\ &\qquad \times[-E_1 + E_2\delta - E_3(\delta+1) + E_4(\delta^2+2\delta+1)] \\ &\quad + (-C_{10Q_0}\cos\delta + C_{6Q_0}\sin\delta)[E_3\delta - E_4(\delta^2-1)] \\ &\quad - (C_{14Q_0}\sin\delta - C_{12Q_0}\cos\delta)E_2 + C_8 \end{aligned} \tag{65}$$

where

$$C_{1Q_0} = JC_{2Q_0} \qquad C_{12Q_0} = \tfrac{1}{2}C_{1Q_0} \qquad C_{14Q_0} = \tfrac{1}{2}C_{2Q_0}$$
$$C_{6Q_0} = -\tfrac{1}{2}(C_{2Q_0} + C_{1Q_0}) \qquad C_{10Q_0} = -\tfrac{1}{2}(C_{1Q_0} - C_{2Q_0}) \tag{66}$$

Case III. Consider the conical shell bounded by $0 \le y \le y_1$ with the boundary conditions at y_1 being

$$M_y(y_1) = M_1 \qquad Q_y(y_1) = 0 \tag{67}$$

In a similar fashion to Case I,

$$C_6 = C_{10} = C_{14} = C_{12} = 0 \qquad C_1 = C_0 \qquad C_2 = C_3 \tag{68}$$

Using (39), (42) and (68), Q_y and M_y may be written as

$$Q_y = \frac{2\rho_y \cos\beta\, e^{\delta}}{(y\sin\beta)^{3/2}}(C_{11}\sin\delta - C_{13}\cos\delta) \tag{69}$$

$$M_y = 2D[-\Gamma_y e^{\delta}(-C_7\sin\delta + C_5\cos\delta) + \Delta_y e^{\delta}(C_{13}\sin\delta + C_{11}\cos\delta)$$
$$- \Lambda_y e^{\delta}(C_{11}\sin\delta - C_{13}\cos\delta) + \Sigma_y e^{\delta}(C_5\sin\delta + C_7\cos\delta)] \tag{70}$$

Applying the boundary conditions (67),

$$C_{2M_1} = \frac{M_1 e^{-\delta_1}}{DK_2\left(1 - \dfrac{K_1}{K_2}\cot\delta_1\right)} \tag{71}$$

where

$$K_1 = (\Gamma_{y_1} - \Lambda_{y_1} + \Sigma_{y_1})\sin\delta_1 + (-\Gamma_{y_1} + \Delta_{y_1} + \Sigma_{y_1})\cos\delta_1 \tag{72}$$

$$K_2 = (-\Gamma_{y_1} + \Delta_{y_1} - \Sigma_{y_1})\sin\delta_1 + (-\Gamma_{y_1} - \Lambda_{y_1} + \Sigma_{y_1})\cos\delta_1 \tag{73}$$

The displacements become

$$w_{OM_1} = 2\rho_y(\sin\beta)^{1/2} e^{\delta}[\eta_y(C_{11M_1}\sin\delta - C_{13M_1}\cos\delta)$$
$$+ \mu(C_{5M_1}\sin\delta + C_{7M_1}\cos\delta)] - e^{-\delta}(u_{OM_1} - C_8)\tan\beta - C_8\tan\beta \tag{74}$$

$$u_{OM_1} = (C_{7M_1}\sin\delta - C_{5M_1}\cos\delta)[E_1 - E_2\delta - E_3(\delta - 1) + E_4(\delta^2 - 2\delta + 1)]$$
$$+ (C_{7M_1}\cos\delta + C_{5M_1}\sin\delta)[-E_3\delta + E_4(\delta^2 - 1)]$$
$$- (C_{13M_1}\sin\delta + C_{11M_1}\cos\delta)E_2 + C_8 \tag{75}$$

where

$$C_{5M_1} = \tfrac{1}{2}C_{2M_1}(1 - \cot\delta_1) \qquad C_{11M_1} = -\tfrac{1}{2}C_{2M_1}\cot\delta_1$$
$$C_{7M_1} = -\tfrac{1}{2}C_{2M_1}(1 + \cot\delta_1) \qquad C_{13M_1} = \tfrac{1}{2}C_{2M_1} \tag{76}$$

Case IV. Consider the conical shell bounded by $0 \le y \le y_1$ with the boundary conditions at y_1 being

$$M_y(y_1) = 0 \qquad Q_y(y_1) = Q_1 \tag{77}$$

As in Case III, $C_6 = C_{10} = C_{12} = C_{14} = 0$ and $C_1 = C_0$, $C_2 = C_3$. Similarly,

$$C_{2Q_1} = -\frac{Q_1 \, \mathrm{e}^{-\delta_1}(y_1 \sin\beta)^{3/2}}{\rho_{y_1} \cos\beta\left(\dfrac{K_2}{K_1}\sin\delta_1 + \cos\delta_1\right)} \tag{78}$$

$$\begin{aligned} w_{0Q_1} = {} & \frac{2\rho_y(\sin\beta)^{1/2}\,\mathrm{e}^{\delta}}{Eh\cos\beta}\,[\eta_y(C_{11Q_1}\sin\delta - C_{13Q_1}\cos\delta) \\ & + \mu(C_{5Q_1}\sin\delta + C_{7Q_1}\cos\delta)] - \mathrm{e}^{\delta}(u_{0Q_1} - C_8)\tan\beta - C_8\tan\beta \end{aligned} \tag{79}$$

$$\begin{aligned} u_{0Q_1} = {} & (C_{7Q_1}\sin\delta - C_{5Q_1}\cos\delta) \\ & \times [E_1 - E_2\delta - E_3(\delta - 1) + E_4(\delta^2 - 2\delta + 1)] \\ & + (C_{7Q_1}\cos\delta + C_{5Q_1}\sin\delta)[-E_3\delta + E_4(\delta^2 - 1)] \\ & - (C_{13Q_1}\sin\delta + C_{11Q_1}\cos\delta)E_2 + C_8 \end{aligned} \tag{80}$$

where

$$\begin{aligned} & C_{1Q_1} = -C_{2Q_1}(K_1/K_2) \qquad C_{11Q_1} = +\tfrac{1}{2}C_{1Q_1} \\ & C_{5Q_1} = \tfrac{1}{2}(C_{1Q_1} + C_{2Q_1}) \qquad C_{13Q_1} = \tfrac{1}{2}C_{2Q_1} \\ & C_{7Q_1} = \tfrac{1}{2}(C_{1Q_1} - C_{2Q_1}) \end{aligned} \tag{81}$$

The length of the bending boundary layer is arbitrarily chosen such that when only 1% of an effect produced at one end is felt at the other end, the edge effects are considered uncoupled. Thus

$$\mathrm{e}^{(-\delta + \delta_0)} < 0{\cdot}01$$

And the critical slant length may be calculated[5] as

$$L_{13} = 3{\cdot}5(R^*_{\max}h)^{1/2} \tag{82}$$

PARTICULAR SOLUTIONS

Particular solution for lateral pressure. The case of an arbitrary pressure distribution $p(y)$ with zero surface shears is considered here. Setting $T = 0$, and using (17), (18) and (32), one arrives at

$$\Phi = 2\mu^2 y^{3/2}\frac{\mathrm{i}a_{\mathrm{p}}}{k} + y^{3/2}b_{\mathrm{p}} + \frac{\mathrm{i}}{2\mu^2 k}\left[\frac{3}{4y}\Phi - y\Phi''\right] \tag{83}$$

where

$$a_{\mathrm{p}} = \left(\frac{RV}{y}\right)\frac{\tan^2\beta(\sin\beta)^{1/2}}{Eh} \tag{84}$$

$$b_{\mathrm{p}} = -\frac{(\sin\beta)^{5/2}}{Eh}\left[\frac{(RV)(1-6E/5G_{\mathrm{c}})}{(y\sin\beta)^2} - 2p - y\frac{\mathrm{d}p}{\mathrm{d}y}\right] \tag{85}$$

The solution to (83) is assumed to be an asymptotic expansion in powers of the large constant $2\mu^2$, and is of the form

$$\Phi = 2\mu^2\left[\Omega_0 + \frac{\Omega_1}{2\mu^2} + \frac{\Omega_2}{(2\mu^2)^2} + \cdots\right] \tag{86}$$

Substituting (86) into (83) and equating like powers of $2\mu^2$, one arrives at

$$\Phi = 2\mu^2\left[\frac{\mathrm{i}y^{3/2}a_{\mathrm{p}}}{k} + \left(\frac{1}{2\mu^2}\right)\left(y^{3/2}b_{\mathrm{p}} - \frac{3y^{3/2}a'_{\mathrm{p}}}{k^2} - \frac{y^{5/2}a''_{\mathrm{p}}}{k^2}\right)\right] \tag{87}$$

As in the homogeneous solution, Φ may be divided into its complex parts and the following quantities become evident:

$$H_{\mathrm{p}} = E^*A_1/\sin\beta \tag{88}$$

$$V_{\mathrm{p}} = \frac{\sin\beta}{y}\int_{y_1}^{y} yp(y)\,\mathrm{d}y \tag{89}$$

$$Q_{y_{\mathrm{p}}} = H_{\mathrm{p}} - V_{\mathrm{p}}\sin\beta \tag{90}$$

$$N_{y_{\mathrm{p}}} = H_{\mathrm{p}}\sin\beta + V_{\mathrm{p}}\cos\beta \tag{91}$$

$$N_{\theta\mathrm{p}} = yp(y)\sin\beta\cos\beta + \frac{E^*A_2}{\gamma^2-2} + \frac{EhA_3\cot\beta}{(\sin\beta)^{1/2}} \tag{92}$$

$$M_{y_{\mathrm{p}}} = \frac{D}{(\sin\beta)^{1/2}}\left[-b_{\mathrm{p}} - yb'_{\mathrm{p}} + 2\mu^2\gamma A_3 - \nu b_{\mathrm{p}} + 2\mu^2\nu\gamma a_{\mathrm{p}} + \left(\frac{\gamma}{\gamma^2-2}+1\right)\left(\frac{A_2}{2} + \frac{\nu(3a'_{\mathrm{p}} + ya''_{\mathrm{p}})}{2}\right)\right] \tag{93}$$

$$M_{\theta\mathrm{p}} = \frac{D}{(\sin\beta)^{1/2}}\left[-b_{\mathrm{p}} + 2\mu^2\gamma a_{\mathrm{p}} + 2\mu^2\nu\gamma A_3 - \nu(b_{\mathrm{p}} + yb'_{\mathrm{p}}) + \left(\frac{\gamma}{\gamma^2-2}+1\right)\left(\frac{\nu A_2}{2} + \frac{3a'_{\mathrm{p}} + ya''_{\mathrm{p}}}{2}\right)\right] \tag{94}$$

$$u_{0p} = \frac{\sin\beta\cos\beta}{Eh}\int_{y_0}^{y}\frac{1}{y}\int_{y_0}^{y} yp(y)\,\mathrm{d}y\,\mathrm{d}y + \frac{h}{\hat{m}(\sin\beta)^{1/2}}$$

$$\times\int_{y_0}^{y} A_1\,\mathrm{d}y - \frac{v\cot\beta}{(\sin\beta)^{1/2}}\int_{y_0}^{y} A_3\,\mathrm{d}y - \frac{v\sin\beta\cos\beta}{Eh}$$

$$\times\int_{y_0}^{y} yp(y)\,\mathrm{d}y - \frac{vh}{\hat{m}(\gamma^2-2)(\sin\beta)^{1/2}}\int_{y_0}^{y} A_2\,\mathrm{d}y \quad (95)$$

$$w_{0p} = \frac{y\sin\beta}{Eh}\left[yp(y)\sin\beta\cos\beta + \frac{EhA_3\cot\beta}{(\sin\beta)^{1/2}} + \frac{E^*A_2}{\gamma^2-2}\right.$$

$$\left. - \frac{v\sin\beta\cos\beta}{y}\int_{y_0}^{y} yp(y)\,\mathrm{d}y - vE^*A_1\right] - u_{0p}\sin\beta \quad (96)$$

where

$$A_1 = 2\mu^2 a_p + \left(\frac{3a_p' + ya_p''}{\gamma^2-2}\right)$$

$$A_2 = 3a_p' + 5ya_p'' + y^2 a_p''' \quad (97)$$

$$A_3 = a_p + ya_p'$$

If a constant lateral pressure p_0 is specified, (84) and (85) reduce to

$$a_{po} = \frac{yp_0\tan^2\beta(\sin\beta)^{5/2}}{2Eh} \qquad b_{po} = \frac{p_0(\sin\beta)^{5/2}}{Eh}\left(\frac{3}{2} + \frac{6E}{10G_c}\right) \quad (98)$$

Particular solution for temperature. In a procedure analogous to that of the pressure solution, one finds that when $p(y) = 0$, the following solution results:

$$a_T = -\frac{\tan^2\beta(\sin\beta)^{1/2}}{Ehy}$$

$$\times\left\{\frac{1}{1-v}(yM_{Ty}' + M_{Ty} - M_{T\theta}) + \frac{vE}{R^*(1-v^2)}\right.$$

$$\left.\times\left[\left(\frac{1}{v}-1\right)\int_{-h/2}^{h/2}\bar{w}\zeta\,\mathrm{d}\zeta - R^*y\frac{\mathrm{d}}{\mathrm{d}y}\left(\frac{1}{R^*}\int_{-h/2}^{h/2}\bar{w}\zeta\,\mathrm{d}\zeta\right)\right]\right\} \quad (99)$$

$$b_{\mathrm{T}} = \frac{\tan\beta(\sin\beta)^{1/2}}{Eh} \times \left\{\frac{1}{1-\nu}(N'_{\mathrm{T}\theta} - \nu N'_{\mathrm{T}y}) + \frac{1+\nu}{1-\nu}(N_{\mathrm{T}\theta} - N_{\mathrm{T}y}) - \frac{E}{R^*} \times \int_{-h/2}^{h/2} \frac{\mathrm{d}\bar{w}}{\mathrm{d}y}\,\mathrm{d}\zeta + \frac{3E\cot\beta}{2y}\int_{-h/2}^{h/2}\left[1-\left(\frac{\zeta}{h/2}\right)^2\right]\frac{\mathrm{d}\bar{w}}{\mathrm{d}y}\,\mathrm{d}\zeta + \frac{3E\cos\beta}{2G_{\mathrm{c}}y^2\sin^2\beta}\int_{-h/2}^{h/2}\left(\frac{\zeta}{h/2}\right)^2\int_{-h/2}^{\zeta}\Theta\,\mathrm{d}\zeta\,\mathrm{d}\zeta\right\} \tag{100}$$

$$N_{y\mathrm{T}} = \frac{C_9\cot\beta}{y} + E^*B_1 \tag{101}$$

$$N_{\theta\mathrm{T}} = \frac{\hat{m}E^*B_3\cot\beta}{h} + \frac{E^*B_2}{\gamma^2-2} \tag{102}$$

$$V_{\mathrm{T}} = C_9/y\sin\beta \tag{103}$$

$$M_{y\mathrm{T}} = \frac{D}{(\sin\beta)^{1/2}}\left[-b_{\mathrm{T}} - yb'_{\mathrm{T}} + 2\mu^2\gamma B_3 - \nu b_{\mathrm{T}} + 2\mu^2\gamma\nu a_{\mathrm{T}} + \left(\frac{\gamma}{\gamma^2-2}+1\right)\left(\frac{B_2}{2} + \frac{\nu(3a'_{\mathrm{T}} + ya''_{\mathrm{T}})}{2}\right)\right] \tag{104}$$

$$M_{\theta\mathrm{T}} = \frac{D}{(\sin\beta)^{1/2}}\left[-b_{\mathrm{T}} + 2\mu^2\gamma a_{\mathrm{T}} - \nu(b_{\mathrm{T}} + yb'_{\mathrm{T}}) + 2\mu^2\nu\gamma B_3 + \left(\frac{\gamma}{\gamma^2-2}+1\right)\left(\frac{\nu B_2}{2} + \frac{3a'_{\mathrm{T}} + ya''_{\mathrm{T}}}{2}\right)\right] \tag{105}$$

$$Q_{y\mathrm{T}} = -C_9/y + E^*B_1\cos\beta \tag{106}$$

$$u_{0\mathrm{T}} = \frac{C_9\cot\beta}{Eh}\ln(y/y_0) + \frac{E^*}{Eh}\int_{y_0}^{y} B_1\,\mathrm{d}y - \frac{\nu\cot\beta}{(\sin\beta)^{1/2}}\int_{y_0}^{y} B_3\,\mathrm{d}y - \frac{\nu E^*}{Eh(\gamma^2-2)}\int_{y_0}^{y} B_2\,\mathrm{d}y \tag{107}$$

$$w_{0\mathrm{T}} = \frac{y\sin\beta}{Eh}\left[\frac{\hat{m}E^*B_3}{h\tan\beta} + \frac{E^*B_2}{\gamma^2-2} - \frac{\nu C_9\cot\beta}{y} - \nu E^*B_1\right] - u_{0\mathrm{T}}\sin\beta \tag{108}$$

where B_1, B_2 and B_3 correspond in definition to A_1, A_2 and A_3 except that a_{p} is now replaced by a_{T}.

For a linear temperature distribution

$$T = T_0 + \frac{\Delta T}{h}\zeta \tag{109}$$

and the simplifications on a_T and b_T become

$$a_{TL} = -\frac{\alpha_\zeta T_0 h^2 \tan\beta(\sin\beta)^{1/2}}{12y^2(1-\nu^2)} \qquad b_{TL} = -\frac{\alpha\Delta Th\tan\beta}{12y^2(1-\nu)(\sin\beta)^{1/2}} \tag{110}$$

TOTAL SOLUTION

The entire solution for a conical shell subjected to a thermal and pressure loading is simply the addition of the individual solutions. Thus, one may write

$$(\ ) = (\ )_{M_0} + (\ )_{Q_0} + (\ )_{M_1} + (\ )_{Q_1} + (\ )_p + (\ )_T \tag{111}$$

where ( ) can represent w, u, W, N_y, M_y, etc.

From (111) all displacements may be found, and the maximum displacement determined. Likewise, all stress resultants and stress couples can be found. Then, using standard laminate theory[6] for composite materials, the stresses can be found in each lamina and compared with a failure theory for analysis or design.

REFERENCES

1. REISSNER, E., On the theory of thin elastic shells, H. Reissner Anniversary Volume, 1944, pp. 231–247.
2. REISSNER, E., On a variational theorem in elasticity, *Journal of Mathematics and Physics*, **29** (1950), 90–95.
3. DE SILVA, C. and NAGHDI, P., Asymptotic solutions of a class of elastic shells of revolution with variable thickness, *Quarterly Journal of Applied Mathematics*, **15**(2) (1956), 169–182.
4. HILDEBRAND, F., On asymptotic integration in shell theory, *Proceedings of the Third Symposium in Applied Mathematics*, Vol. 3, 1950, pp. 53–66.
5. VINSON, J., Edge load solutions for conical shells, *Journal of Engineering Mechanics Division*, ASCE, February (1966).
6. VINSON, J. R. and SIERAKOWSKI, R. L., *The behavior of structures composed of composite materials*, Dordrecht, The Netherlands, Martinus-Nijhoff Publishers, 1986.

21

Delamination Buckling of Pressure-loaded Thin Cylinders and Panels

G. J. SIMITSES and Z. Q. CHEN

Georgia Institute of Technology, Atlanta, Georgia 30332, USA

ABSTRACT

Delamination buckling of long complete cylindrical shells and curved panels, when acted upon by uniform lateral loading, is being investigated. The geometry is such that it covers a wide range of length to radius ratios, as well as panels of different widths. Results, though, are presented only for very long configurations and a limited number of widths. The boundaries of the panels are either clamped or simply supported, along the straight edges. Finally, the geometry is free of initial geometric imperfections, and all parts of the shell are assumed to be virtually isotropic. The emphasis of the investigation concerns the effect of several factors and parameters on the critical load. These include, load behavior during buckling, width of delamination, position of delamination and panel width.

1. INTRODUCTION

Delamination is one of the most common failure modes in composite materials. Delamination develops as a result of imperfections in the production technology or due to the effect of certain factors during the operational life of the laminate, such as impact by foreign objects. The presence of delamination in a laminated composite material may cause early buckling and consequently reduction in the overall stiffness of the structure, which may lead to early failure.

Delamination of composite structures has received considerable attention in recent years. As far as the buckling response of delaminated plates is concerned, an extensive review is given in Ref. 1.

The problem of buckling of delaminated cylindrical shells has not received the deserved attention. Very few investigations have been carried out in this area. Kulkarni and Frederick[2] use a 'branched integration' technique to solve the problem of buckling of a two-layered cylindrical shell, partially debonded, and subjected to axial compression. Results are reported[2] for different lengths of debonding and inner to outer layer thickness ratios. A significant decrease in the critical load is observed. The buckling of stiffened circular cylindrical shells, with two unbonded orthotropic layers, is reported by Jones.[3] Jones[3] assumes that the layers do not separate during buckling, i.e. the deformation of both layers is assumed to be the same. Troshin[4] examines the effect of longitudinal delamination in a laminar cylindrical shell on the critical external pressure. The delamination is assumed to extend over the entire length of the shell. He[4] reports on the effect of the cylinder length and position of the delamination on the critical external pressure. Sallam and Simitses[5] report results on delamination buckling of thin cylindrical shells of perfect geometry, when subjected to uniform axial compression. The delamination region is of constant width covering the entire circumference.

The present paper deals with delaminated long, thin cylindrical shells and panels, when subject to lateral uniform pressure. The delaminated region is of constant width and it extends along the entire length of the thin shell. The geometry is assumed to be quasi-isotropic and several parametric studies are performed. First, the effect of load behavior during the buckling process (see Ref. 6) is presented. Moreover, the effect of position and size of delamination on the critical conditions is fully assessed. These studies use two sets of boundary conditions, simply supported and clamped.

2. MATHEMATICAL FORMULATION

The Koiter–Budiansky[7] buckling equations have been deduced from those given in the Appendix of Ref. 7. The version given below corresponds to a set obtained from Sanders[8]-type of kinematic relations. This version, in terms of stress, N_{ij}, and moment, M_{ij}, resultants, is

$$\begin{aligned} N_{xx,x} + N_{xy,y} + q^x &= 0 \\ N_{xy,x} + N_{yy,y} + (M_{yy,y} + M_{xy,x})^*/R - q\phi_y^* + q^y &= 0 \qquad (1) \\ M_{xx,xx} + 2M_{xy,xy} + M_{yy,yy} - N_{yy}/R - qR\phi_{y,y} + q^z &= 0 \end{aligned}$$

where ϕ_y is the rotation about the y-axis; q^x, q^y and q^z are corrections to surface loading due to load behavior (modelling) during buckling; q is the applied pressure; and R is the radius of the reference cylindrical surface. Note that if terms marked by an asterisk are dropped one obtains the well known Donnell-type of equation.

Since there exist three possibilities of load behavior during the buckling process,[6] the corrections to surface loading assume three distinctly different expressions.

Linear kinematic relations are employed for the additional strains, rotations, and changes in curvature and torsion in terms of the small additional displacement components u, v and w.[8] Moreover, the usual lamination theory[9] is employed, and it is assumed that all shell (laminates) parts are symmetric with respect to their midsurface (no coupling between extension and bending).

Use of the above in eqns (1) yields

$$A_{11}u_{,xx} + 2A_{13}u_{,xy} + A_{33}u_{,yy} + (A_{12} + A_{33})v_{,xy} + A_{13}v_{,xx} + A_{23}v_{,yy} + A_{12}\frac{w_{,x}}{R} + A_{23}\frac{w_{,y}}{R} + q^x = 0$$

$$A_{13}u_{,xx} + (A_{33} + A_{12})u_{,xy} + A_{23}u_{,yy} + \left(A_{33} + \frac{D_{33}}{R^2}\right)v_{,xx} + \left(2A_{23} + \frac{2D_{23}}{R^2}\right)v_{,xy} + \left(A_{22} + \frac{D_{22}}{R^2}\right)v_{,yy} - \frac{qv}{R} + A_{23}\frac{w_{,x}}{R} + \left(\frac{A_{22}}{R} + q\right)w_{,y} - \frac{1}{R}[D_{13}w_{,xxx} + (D_{12} + 2D_{33})w_{,xxy} + 3D_{23}w_{,xyy} + D_{22}w_{,yyy}] + q^y = 0 \quad (2)$$

$$-\frac{A_{12}}{R}u_{,x} - \frac{A_{23}}{R}u_{,y} - \frac{A_{23}}{R}v_{,x} - \frac{1}{R}(A_{22} + qR)v_{,y} + \left[\frac{D_{13}}{R}v_{,xxx} + \left(\frac{D_{12}}{R} + \frac{2D_{33}}{R}\right)v_{,xxy} + \frac{3D_{23}}{R}v_{,xyy} + \frac{D_{22}}{R}v_{,yyy}\right] - A_{22}\frac{w}{R^2} - [D_{11}w_{,xxxx} + 4D_{13}w_{,xxxy} + (2D_{12} + 4D_{33})w_{,xxyy} + 4D_{23}w_{,xyyy} + D_{22}w_{,yyyy}] + qRw_{,yy} + q^z = 0$$

where

$$A_{ij} = \sum_{k+1}^{N} \bar{Q}_{ij}^{(k)}(h_k - h_{k-1}); \qquad D_{ij} = \frac{1}{3}\sum_{k=1}^{N} \bar{Q}_{ik}^{(k)}(h_k^3 - h_{k-1}^3) \quad (3)$$

and h_k and h_{k-1} denote the z-coordinates of the upper and lower surfaces of the kth lamina in each laminate, respectively.

2.1. Description of the Delamination Problem

Thin circular cylindrical shells and panels with longitudinal delamination over the entire length are next considered. The geometry, loading and coordinate systems are shown in Fig. 1. The straight edges of the panel are either clamped or simply supported. The location and size of the delamination is arbitrary. Angle α denotes the region of the delamination,

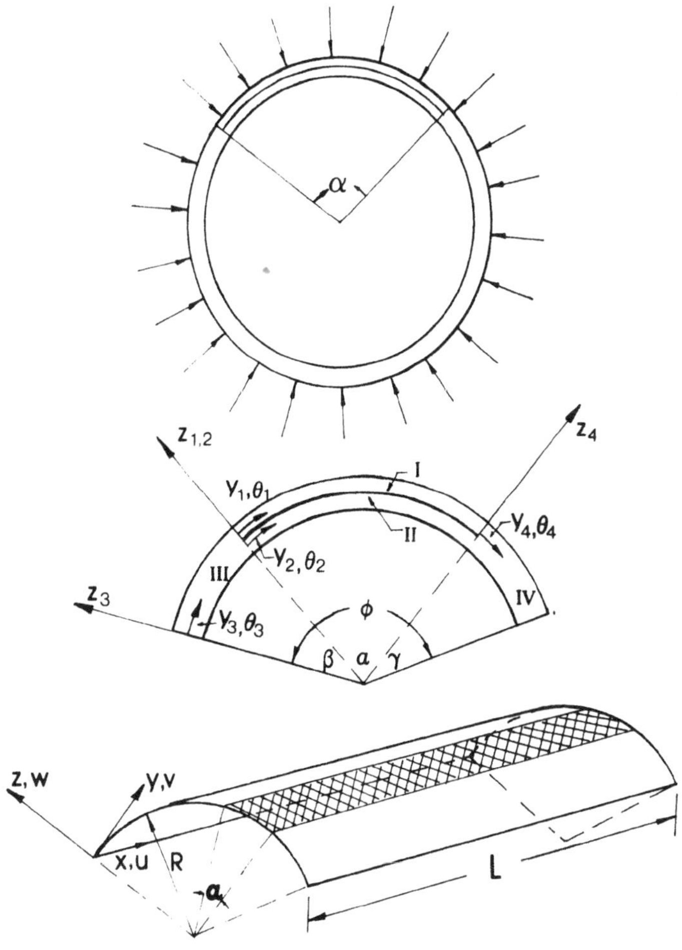

FIG. 1 Geometry, loading and sign convention.

while β and γ denote the location of it from left end and right end, respectively.

It is assumed that under subcritical loads the delamination does not expand. The panel is separated into four parts (four panels) by the delamination. Each part has a set of coordinates attached to it (Fig. 1) and the natural plane of the panel lies on the xy-plane. The panel is subjected to uniform external pressure, q, over the entire outer surface. Let h^i (i = I, II, III, IV) denote the thickness of the ith panel (see Fig. 1). The nondimensional parameter $\bar{h} = h^{\text{I}}/h^{\text{III}}$ is used to describe the thickness of the delamination. Let u^i, v^i, w^i (i = I, II, III, IV) be the displacement components of material points on the midplane of each part in the x, y and z directions, respectively.

The geometry is such that a membrane primary state exists ($B^i_{ij} = 0$) for all participating parts. Therefore, the buckling equations, eqns (1), apply to each part. They are subject to boundary conditions at $\theta_3 = 0$ and $\theta_4 = \gamma$ (see Fig. 1), and certain kinematic continuity conditions as well as force and moment local equilibrium conditions at the common boundaries for the various parts ($\theta_3 = \beta$ and $\theta_4 = 0$).

3. SOLUTION PROCEDURE

For each part, a separate solution is assumed, which satisfies the classical simply-supported boundary conditions at $x = 0$ and L. This is done for the special construction for which there is no coupling between extension and shear, and between bending and twisting action. This means that for all parts

$$B_{ij} = A_{13} = A_{23} = D_{13} = D_{23} = 0 \tag{4}$$

The classical simply-supported boundary conditions[10] are

$$\text{SS-3: } w = M_{xx} = N_{xx} = v = 0 \tag{5}$$

The separated solution is

$$u(x, y) = U(y)\cos\frac{m\pi x}{L} \qquad v(x, y) = V(y)\sin\frac{m\pi x}{L} \qquad w(x, y) = W(y)\sin\frac{m\pi x}{L} \tag{6}$$

Substitution of eqns (6) into eqns (2), for the special construction yields

$$\begin{aligned} L_{11}U + L_{12}V + L_{13}W &= 0 \\ L_{21}U + L_{22}V + L_{23}W &= 0 \\ L_{31}U + L_{32}V + L_{33}W &= 0 \end{aligned} \tag{7}$$

where the L_{ij} are linear differential operators.

Elimination of U and V through the use of the first two of eqns (7) and through substitution into the third one yields a single higher-order ordinary differential equation in W alone. This higher-order equation assumes the form

$$F_8 \frac{d^8 W}{d\theta^8} + F_6 \frac{d^6 W}{d\theta^6} + F_4 \frac{d^4 W}{d\theta^4} + F_2 \frac{d^2 W}{d\theta^2} + F_0 = 0 \tag{8}$$

where the F's are constants that contain the external load q and structural geometric parameters (A_{ij}, D_{ij}, $\bar{h}$, R, etc.). Note that some of the F's change according to the case of load behavior during buckling.

The solution procedure is similar to the one described for rings and arches in Article 7.3 of Ref. 11. The number of equations is higher, the equations themselves are more complex, and a closed form solution is not expected as in Ref. 11. Nevertheless, the overall procedure can be followed and a numerical estimate can be achieved. Thus, first assume for $W(\theta)$ a solution of the form

$$W = C \exp(r\theta) \tag{9}$$

Since the order of the equation is eight, then substitution into eqn. (8) yields an eighth degree polynomial in r. Thus, eight roots are expected for each geometry and load level. If the eight roots are distinctly different, the general solution for $W(\theta)$ is given by

$$W(\theta) = \sum_{i=1}^{8} C_i \exp(r_i \theta) \tag{10}$$

If double roots occur, the form of eqn. (9) is modified accordingly.

The form of the solution for $U(\theta)$ and $V(\theta)$ is similar to that of eqn. (10), and the solution for U, V and W is given in terms of eight constants.

There exist eight unknowns, C_i's, for each part, and the total number of unknowns is 32. Use of the 32 boundary and auxiliary conditions leads to a system of 32 linear, homogeneous algebraic equations in 32 unknowns. For a nontrivial solution to exist the determinant of the coefficients must vanish. This yields the characteristic equation.

4. NUMERICAL EXAMPLES

Numerical results are presented graphically and in tabular form and they correspond to a special geometry.

TABLE 1
Critical load for a delaminated circular cylindrical shell

α	$\bar{h}=0·5$			$\bar{h}=0·3$ (0·7)		
	I	II	III	I	II	III
0·0	3·0000 (A)	4·0000 (A)	4·5000 (A)	3·0000 (A)	4·0000 (A)	4·5000 (A)
0·2	2·8846 (S)	3·8433 (S)	4·3215 (S)	2·8634 (S)	3·8141 (S)	4·2880 (S)
0·6	2·2705 (S)	2·9878 (S)	3·3290 (S)	2·4655 (S)	3·2638 (S)	3·6519 (S)
1·0	2·1974 (A)	2·8250 (A)	3·0958 (A)	2·4077 (S)	3·1905 (S)	3·4099 (S)
1·4	1·7689 (A)	2·2678 (A)	2·4817 (A)	2·1262 (A)	2·7258 (A)	2·9637 (A)
1·8	1·5880 (A)	2·0541 (A)	2·2610 (A)	1·7818 (A)	2·1175 (A)	2·1981 (A)
2·2	1·5460 (A)	2·0128 (A)	2·2239 (A)	1·3583 (A)	1·5327 (A)	1·5878 (A)
2·6	1·3871 (S)	1·7692 (S)	1·9284 (S)	1·0062 (A)	1·1413 (A)	1·2106 (A)
3·0	1·2186 (S)	1·5731 (S)	1·7290 (S)	0·7665 (A)	0·8798 (A)	0·9769 (A)
3·4	1·1339 (S)	1·4791 (S)	1·6372 (S)	0·6028 (A)	0·6944 (A)	0·8321 (A)
3·8	1·1088 (S)	1·4547 (S)	1·6160 (S)	0·4885 (A)	0·5536 (A)	0·7485 (A)
4·2	1·0373 (A)	1·3477 (A)	1·4877 (A)	0·4073 (A)	0·4511 (A)	0·7101 (A)
4·6	0·9219 (A)	1·2110 (A)	1·3479 (A)	0·3497 (A)	0·3868 (A)	0·6971 (A)
5·0	0·8375 (A)	1·1096 (A)	1·2428 (A)	0·3101 (A)	0·3746 (A)	0·6255 (A)
5·4	0·7831 (A)	1·0424 (A)	1·1715 (A)	0·2851 (A)	0·3701 (A)	0·5882 (A)
5·8	0·7562 (A)	1·0081 (A)	1·1340 (A)	0·2729 (A)	0·3685 (A)	0·5633 (A)
6·2	0·7500 (A)	1·0000 (A)	1·1250 (A)	0·2700 (A)	0·3667 (A)	0·5527 (A)

α	$\bar{h}=0·1$ (0·9)			$\bar{h}=0·01$ (0·99)		
	I	II	III	I	II	III
0·0	3·0000 (A)	4·0000 (A)	4·5000 (A)	3·0000 (A)	4·0000 (A)	4·5000 (A)
0·2	2·9360 (S)	3·9135 (S)	4·4019 (S)	0·2019 (A)	0·2021 (A)	0·2021 (A)
0·6	2·2377 (A)	2·2543 (A)	2·2545 (A)	0·0224 (A)	0·0226 (A)	0·0226 (A)
1·0	0·8036 (A)	0·8200 (A)	0·8210 (A)	0·0080 (A)	0·0082 (A)	0·0082 (A)
1·4	0·4086 (A)	0·4245 (A)	0·4265 (A)	0·0041 (A)	0·0042 (A)	0·0043 (A)
1·8	0·2461 (A)	0·2614 (A)	0·2649 (A)	0·0025 (A)	0·0026 (A)	0·0027 (A)
2·2	0·1641 (A)	0·1784 (A)	0·1841 (A)	0·0016 (A)	0·0018 (A)	0·0018 (A)
2·6	0·1170 (A)	0·1300 (A)	0·1388 (A)	0·0012 (A)	0·0013 (A)	0·0014 (A)
3·0	0·0877 (A)	0·0990 (A)	0·1118 (A)	0·0009 (A)	0·0011 (A)	0·0011 (A)
3·4	0·0683 (A)	0·0771 (A)	0·0960 (A)	0·0007 (A)	0·0008 (A)	0·0010 (A)
3·8	0·0550 (A)	0·0605 (A)	0·0877 (A)	0·0006 (A)	0·0006 (A)	0·0010 (A)
4·2	0·0457 (A)	0·0591 (A)	0·0853 (A)	0·0005 (A)	0·0006 (A)	0·0009 (A)
4·6	0·0391 (A)	0·0531 (A)	0·0792 (A)	0·0004 (A)	0·0005 (A)	0·0008 (A)
5·0	0·0346 (A)	0·0493 (A)	0·0722 (A)	0·0004 (A)	0·0005 (A)	0·0007 (A)
5·4	0·0317 (A)	0·0468 (A)	0·0680 (A)	0·0003 (A)	0·0005 (A)	0·0007 (A)
5·8	0·0303 (A)	0·0466 (A)	0·0663 (A)	0·0003 (A)	0·0005 (A)	0·0007 (A)
6·2	0·0300 (A)	0·0410 (A)	0·0654 (A)	0·0003 (A)	0·0004 (A)	0·0007 (A)

I, II and III represent load cases.

The geometry is a quasi-isotropic geometry (as in Ref. 4) with the delamination parallel to the shell reference surface and extending along the entire length of the panel. Moreover, the panel is assumed to be very long and thus the solution is not affected by the $x = \text{constant}$ boundary conditions.

The purpose of the generated results is to assess the effect of several parameters for the special geometry.

Three models of load behavior, during the buckling process, have been employed[6] in order to establish critical conditions for laterally loaded shells.

Moreover, the results are used to study the effect of delamination position through-the-thickness, $\bar{h}$, and from the edge of the panel, β. In addition, the effect of delamination size, α, is also assessed. In generating results $L/\pi R$ was taken to be extremely large (10^6). For these very long cylindrical shells and panels, the effect of R/h is extremely small provided that R/h is taken to be larger than 50. This is particularly true for load cases I and II. Results are presented for load case III only, for a panel with $\phi = \pi$ and several values of R/h. Finally, results are presented for both clamped and simply-supported panels, in terms of the critical load parameter $|\lambda|$ $(= |q|R^3/D)$.

Table 1 presents values of the critical load for several delamination positions, $\bar{h}$ corresponding to all three cases of load behavior. As expected, when there exists an extremely small delamination size ($\alpha \approx 0$), the critical load assumes the values corresponding to the perfect geometry.[6] As the delamination size increases the critical load decreases, and when the cylindrical shell is (virtually) completed delaminated ($\alpha = 2\pi$), the critical load approaches the value of $\bar{h}^2$ (or $(1 - \bar{h})^2$) times the value corresponding to the complete and perfect cylindrical shell. Note that because of the linearity of the analysis the critical loads for $\bar{h}$ and $1 - \bar{h}$ are identical. Finally, the letters 'A' and 'S' denote the character of the buckling mode. 'A' stands for antisymmetric and 'S' for symmetric, with respect to the midposition of the delaminated area. Note that when the mode is antisymmetric there exists contact between the upper and lower parts of the delaminated area in the postbuckling region. Part of the results shown in Table 1 are plotted in Fig. 2.

Similar results are shown on Table 2 for long delaminated thin cylindrical panels with clamped straight boundaries and three sizes ($\theta = \pi$, $\pi/2$ and $\pi/4$). The observations are similar, but there exist a few points that need special attention. First, the difference in critical values for the three load cases, corresponding to the same geometry, are not as far apart

TABLE 2

Critical loads for delaminated panels of angle π, π/2 and π/4 (rad.) (clamped boundaries)

φ	α	$\bar{h}=0·5$			$\bar{h}=0·1$ (0·9)		
		I	II	III	I	II	III
	0·0	8·0000	9·0590	10·6714	8·0000	9·0013	10·5973
	0·2	7·9029	8·8896	10·4365	7·9878	8·9863	10·5766
	0·6	5·8120	6·4523	7·1398	2·2379	2·2546	2·2548
	1·0	3·9435	4·3773	4·8031	0·8037	0·8201	0·8210
	1·4	3·4318	3·8328	4·2667	0·4086	0·4246	0·4266
π	1·8	3·3932	3·7924	4·2205	0·2462	0·2615	0·2649
	2·2	3·1340	3·4921	3·8211	0·1641	0·1783	0·1843
	2·6	2·6198	2·9294	3·2331	0·1171	0·1299	0·1391
	3·0	2·1424	2·4087	2·7706	0·0877	0·0986	0·1125
	3·141	2·0006	2·2506	2·6495	0·0800	0·0900	0·1060
		I	II	III	I	II	III
	0·0	32·5951	34·1610	34·4314	32·4337	33·9942	34·2595
	0·2	29·2116	20·5118	30·7211	20·1664	20·1833	20·1834
	0·4	18·6728	19·3474	19·4383	5·0395	5·0562	5·0563
	0·6	13·9258	14·4587	14·5315	2·2381	2·2548	2·2550
π/2	0·8	12·8737	13·4018	13·4780	1·2577	1·2741	1·2748
	1·0	12·7255	13·2380	13·3102	0·8039	0·8202	0·8212
	1·2	11·4625	11·9061	11·9650	0·5575	0·5735	0·5750
	1·4	9·5711	9·9728	10·0310	0·4089	0·4247	0·4268
	1·57	8·1139	8·5034	8·5704	0·3246	0·3402	0·3428
		I	II	III	I	II	III
	0·0	130·9623	132·6093	132·6703	130·6081	132·2657	132·3108
	0·1	117·6815	119·0511	119·0993	129·2294	130·8456	130·9050
	0·2	75·0366	75·7404	75·7616	20·1680	20·1847	20·1847
	0·3	55·5773	56·1281	56·1449	8·9624	8·9791	8·9791
π/4	0·4	51·0367	51·5811	51·5985	5·0404	5·0571	5·0573
	0·5	50·5280	51·0603	51·0771	3·2251	3·2417	3·2419
	0·6	45·8650	46·3277	46·3416	2·2391	2·2557	2·2560
	0·7	38·4775	38·8976	38·9110	1·6445	1·6611	1·6615
	0·785	32·6755	33·0862	33·1013	1·3074	1·3238	1·3244

I, II and III represent load cases.

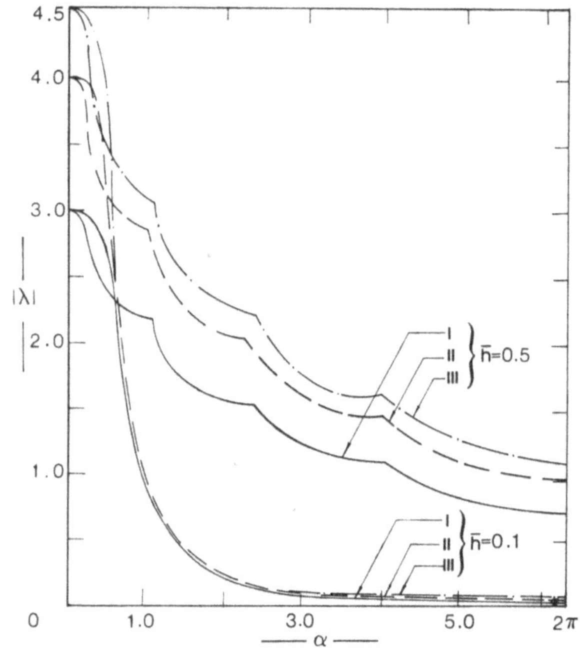

FIG. 2. Critical loads for a long, thin cylindrical shell.

percentage-wise as are those corresponding to the complete cylinder. Second, decrease in the panel width results in considerable increase in the critical load parameter. Moreover, the buckling mode for the panels is antisymmetric for all geometries for which results are presented. The implication of contact in the postbuckling region exists for these geometries too. The results corresponding to widths of ($\phi =$) π and $\pi/2$ are presented graphically in Figs. 3 and 4.

Table 3 clearly shows the small effect of R/h on the critical load for load Case III. Results are presented for $\phi = \pi$ only.

Table 4 presents partial results for simply-supported boundary conditions. The panel width corresponds to $\phi = \pi/2$ and only two values for $\bar{h}$ are shown (0·5 and 0·1 (0·9)). The trends are similar to those corresponding to the clamped case (see Table 2). The results of Table 4 are shown graphically in Fig. 5.

Finally, the effect of delamination position from the edge of the panel, β, on the critical load parameter is shown on Tables 5–8. The results in Table 5 correspond to a clamped panel with $\phi = \pi/2$ and for a delamination size (α)

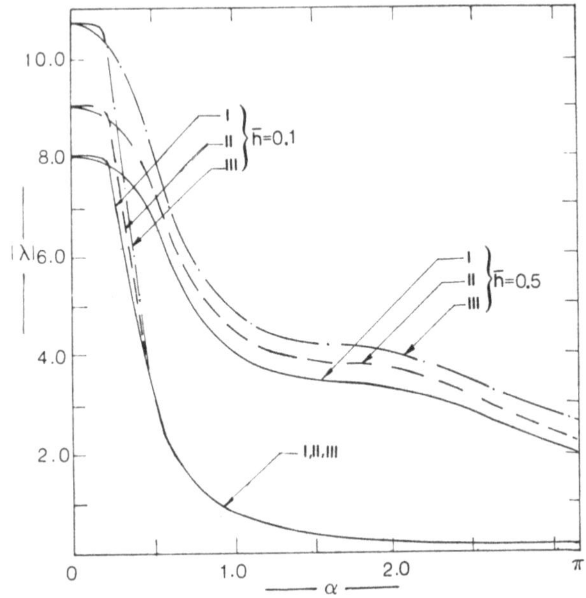

FIG. 3. Critical loads for a long, thin, clamped, cylindrical panel ($\phi = \pi$).

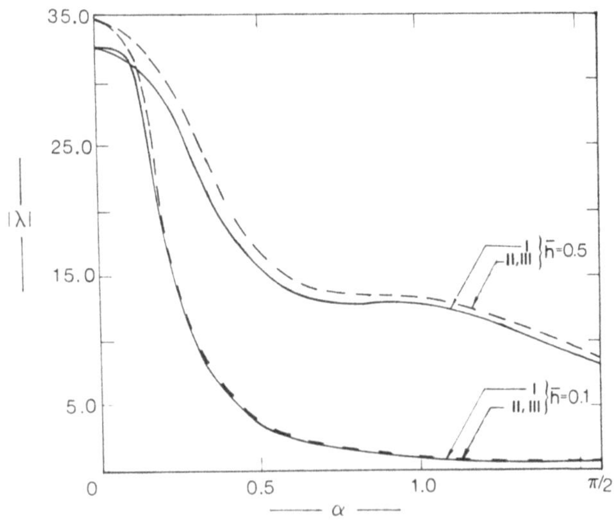

FIG. 4. Critical loads for a long, thin, clamped, cylindrical panel ($\phi = \pi/2$).

TABLE 3
Effect of R/h on critical load for a delaminated panel (clamped boundaries; $\phi = \pi$; load case III)

α	$\bar{h} = 0{\cdot}5$						
	$R/h = 50$	100	200	1 000	2 000	5 000	15 000
0·0	10·889 2	10·671 4	10·615 3	10·601 4	10·613 7	10·703 8	11·527 0
0·2	10·436 2	10·436 5	10·436 7	10·440 7	10·434 5	10·417 0	10·414 2
1·0	4·803 0	4·803 1	4·803 2	4·805 4	4·811 8	4·857 4	5·260 5
1·8	4·220 4	4·220 5	4·220 6	4·223 1	4·230 7	4·283 4	4·733 3
2·6	3·232 9	3·233 1	3·233 1	3·235 2	3·242 2	3·289 9	3·697 5
α	$\bar{h} = 0{\cdot}1$						
	$R/h = 50$	100	200	1 000	2 000	5 000	15 000
0·0	10·600 4	10·597 3	10·596 7	10·600 4	10·613 5	10·703 8	11·526 8
0·2	10·574 7	10·576 6	10·577 6	10·582 3	10·595 4	10·685 4	11·506 1
1·0	0·821 0	0·821 0	0·821 0	0·821 0	0·821 1	0·821 7	0·827 7
1·8	0·265 0	0·264 9	0·265 0	0·265 3	0·266 4	0·273 6	0·336 7
2·6	0·139 1	0·139 1	0·139 1	0·140 8	0·146 1	0·178 5	0·206 4

TABLE 4
Critical load for a delaminated panel of angle $\pi/2$ (rad.) (simply supported)

α	$\bar{h} = 0{\cdot}5$			$\bar{h} = 0{\cdot}1$ (0·9)		
	I	II	III	I	II	III
0·0	15·000 0	16·001 9	16·071 4	15·000 0	16·001 9	16·071 4
0·2	14·345 6	15·295 2	15·362 8	14·917 6	15·916 5	15·982 8
0·4	10·822 2	11·480 2	11·525 9	5·035 6	5·052 2	5·052 4
0·6	7·385 4	7·825 6	7·856 1	2·236 8	2·253 5	5·253 7
0·8	5·469 4	5·810 5	5·834 3	1·256 9	1·273 9	1·274 1
1·0	4·461 2	4·751 1	4·771 6	0·803 4	0·819 9	0·820 2
1·2	3·960 7	4·223 7	4·242 2	0·556 9	0·573 3	0·573 3
1·4	3·772 7	4·024 1	4·042 1	0·408 5	0·424 4	0·426 3
1·57	3·750 0	3·999 9	4·017 8	0·324 3	0·340 0	0·342 4

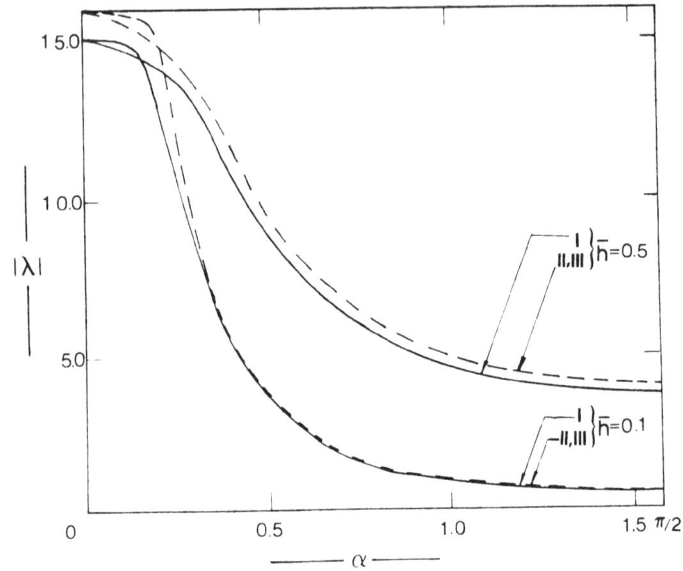

FIG. 5. Critical loads for a long, thin, simply-supported, cylindrical panel ($\phi = \pi/2$).

of 0·2 radians. Results are presented for various positions, β, and several through-the-thickness positions, $\bar{h}$. It is seen that this effect is relatively small for $\bar{h} = 0{\cdot}5$ and 0·3 (0·7), but negligibly small for $\bar{h} = 0{\cdot}1$ (0·9) and 0·01 (0·99). For the former case it appears that the strongest configuration corresponds to a symmetric delamination ($\beta = 0{\cdot}6854$), while the weakest configuration corresponds to $\beta = 0{\cdot}4$. Note also that when $\beta = 0{\cdot}1$ the critical load is almost as high as for the symmetric delamination. Table 6

TABLE 5

Effect of delamination location on the critical load (clamped boundaries; $\phi = \pi/2$; $\alpha = 0{\cdot}2$)

β \ $\bar{h}$	0·5	0·3	0·1	0·01
0·0	28·884 7	29·690 1	20·166 3	0·201 8
0·1	29·205 9	30·652 2	20·166 5	0·201 8
0·2	28·996 2	30·235 9	20·166 8	0·201 8
0·3	28·406 8	28·664 6	20·167 3	0·201 8
0·4	28·066 9	27·868 4	20·167 3	0·201 8
0·5	28·365 0	28·508 5	20·166 8	0·201 8
0·6	28·967 5	29·994 2	20·166 5	0·201 8
0·685 4	29·211 5	30·709 7	20·166 3	0·201 8

TABLE 6
Effect of delamination location on the critical load (clamped boundaries; $\phi = \pi$; $\alpha = 0{\cdot}4$)

β \ $\bar{h}$	0·5	0·3	0·1	0·01
0·0	6·6437	6·9669	5·0392	0·0504
0·2	7·1272	7·4631	5·0393	0·0504
0·4	6·9441	7·4144	5·0393	0·0504
0·6	6·0750	6·6369	5·0392	0·0504
0·8	5·7123	6·2632	5·0392	0·0504
1·0	5·9823	6·4943	5·0392	0·0504
1·2	6·6996	7·1487	5·0391	0·0504
1·3708	7·1897	7·5686	5·0390	0·0504

TABLE 7
Effect of delamination location on the critical load (simply supported; $\phi = \pi/2$; $\alpha = 0{\cdot}2$)

β \ $\bar{h}$	0·5	0·3	0·1	0·01
0·0	14·0409	13·7880	14·4631	0·2018
0·1	13·4714	13·2184	13·6704	0·2018
0·2	13·0469	12·8000	13·0147	0·2018
0·3	12·9875	12·7535	12·7999	0·2018
0·4	13·2688	13·0481	13·0864	0·2018
0·5	13·7477	13·5301	13·7893	0·2018
0·6	14·1944	13·9668	14·7922	0·2018
0·6854	14·3456	14·1121	14·9176	0·2018

TABLE 8
Effect of delamination location on the critical load (simply supported; $\phi = \pi/2$; $\alpha = 0{\cdot}6$)

β \ $\bar{h}$	0·5	0·3	0·1	0·01
0·0	8·2869	10·7108	2·2364	0·0224
0·1	8·0421	9·6941	2·2366	0·0224
0·2	7·7323	8·9581	2·2366	0·0224
0·3	7·5215	8·6057	2·2366	0·0224
0·4	7·4127	8·5228	2·2366	0·0224
0·4854	7·3854	9·4696	2·2368	0·0224

shows results for $\phi = \pi$ and $\alpha = 0{\cdot}4$ radians. The trends are as those shown in Table 5.

The results of Tables 7 and 8 are for simply-supported panels. The trends for $\phi = \pi/2$ and $\alpha = 0{\cdot}2$ are similar to those corresponding to Tables 5 and 6. On the other hand as the size of delamination is increased some differences in behavior are observed.

Needless to say, further and more detailed studies need be performed before we acquire complete and full understanding of delamination buckling and growth of laminated shells. This is indeed a very complex response.

ACKNOWLEDGEMENTS

The work is primarily supported by the Air Force Office of Scientific Research under AFOSR Grant No. AFOSR-86-0038. The Program Manager is Dr A. K. Amos of the Directorate of Aerospace Sciences. The financial support of AFOSR is greatly appreciated and gratefully acknowledged.

REFERENCES

1. SALLAM, S. and SIMITSES, G. J., Delamination buckling and growth of flat, cross-ply laminates, *Composite Structures*, **4** (1985), 361–381.
2. KULKARNI, S. V. and FREDERICK, D., Buckling of partially debonded layered cylindrical shells, *Proceedings of AIAA/ASME/SAE 14th SDM Conference*, Williamsburg, Virginia, March, 1973.
3. JONES, R. M., Buckling of stiffened two-layered shells of revolution with a circumferentially cracked unbonded layer, *AIAA J.*, **7** (1969), 1511–1517.
4. TROSHIN, V. P., Effect of longitudinal delamination in a laminar cylindrical shell on the critical external pressure, *J. Comp. Mater.*, **17**, No. 5 (1983), 563–567.
5. SALLAM, S. and SIMITSES, G. J., Delamination buckling of cylindrical shells under axial compression, *Composite Structures*, accepted for publication (to appear in early 1987).
6. SIMITSES, G. J. and ANSWANI, M., Buckling of thin cylinders under uniform lateral loading, *J. appl. Mech.*, **41**, No. 3 (1974), 827–829.
7. BUDIANSKY, B., Notes on nonlinear shell theory, *J. appl. Mech.*, **35**, 2 (1968), 393–401.
8. SANDERS, J. L., Nonlinear theories of thin shells, *Appl. Math.*, **21** (1963), 21–36.
9. JONES, R. M., *Mechanics of Composite Materials*, New York, McGraw-Hill, 1975.
10. HOFF, N. J., The perplexing behavior of thin circular cylindrical shells in axial compression: *Israel J. Technol.*, **4**, No. 1 (1966), 1–28.
11. SIMITSES, G. J., *Elastic Stability of Structures*, Englewood Cliffs, N.J., Prentice-Hall, 1976 (second printing R. E. Krieger Publishing Co., Melbourne, FL., 1985).

22

Composite Tension Members for Structural Applications

Y. C. T. Yeung and B. E. Parker
*British Ropes Ltd, Carr Hill,
Doncaster, South Yorkshire DN4 8DG, UK*

ABSTRACT

A pilot study has been carried out on 45/49 mm model composite strands with a nominal breaking strength of 1200–1500 kN, made from 5 mm round pultruded rods, unidirectionally reinforced respectively with E-glass, aramid Kevlar 49 and a PAN-based carbon fibre. This study has demonstrated the potential of such composite strands as structural tension members. High specific strength and stiffness, excellent fatigue lives and acceptable creep properties are obtainable. In particular, the Kevlar and carbon fibre strands were shown to have distinct performance advantages over conventional steel strand.

INTRODUCTION

Recent advances in the production and usage of high strength fibres have created a growing interest in their potential for advanced structures in the commercial sector, such as suspension and cable-stayed bridges. Although it is unlikely that high strength fibre-reinforced composites will ever totally replace traditional material such as steel and concrete in structural engineering, they are increasingly challenging the supremacy of traditional materials in numerous applications in view of their combination of enticing properties, increasing availability and diminishing cost. In addition, composites are presenting unique opportunities to constructional designers and engineers to project advanced structures such as very deep water offshore platforms[1] and longspan bridges[2,3] where design requirements can only be effectively met by materials possessing very high specific strength and stiffness.

Recognition of the potential of high strength fibres and composites has led the authors' company to explore actively the application of composite technology to the design, manufacture and evaluation of lightweight flexible tension members in long continuous lengths. Applications envisaged for such tension members include antenna, mast, bridge and wind energy convertor stays and mooring lines for offshore platforms operating in deep waters. An initial study has recently been carried out and given favourable results. Model strands consisting of 61 × 5 mm or 73 × 5 mm round pultruded rods assembled in a close-packed hexagonal array were manufactured with a slight twist (long helical pitch) using a newly developed method of manufacture.

Experimental lengths (30 m) of the strands were produced from pultruded rods, unidirectionally reinforced respectively with three types of high strength fibres: glass, aramid and carbon. These strands have been evaluated for breaking strength, axial modulus and tension–tension fatigue performance, and compared with an equivalent steel strand. The individual pultruded rods have also been tested for breaking strength, axial modulus and creep properties. This chapter is concerned with some of the findings of this exploratory study.

STRAND DESIGN AND MANUFACTURE

The design approach to the composite tension members has, to a large extent, been influenced by the traditional involvement of the authors' company in the manufacture of structural steel wire ropes and strands—tension members which combine high strength and flexibility and which are normally supplied to the customer in coiled or reeled form, with or without end fittings. Another salient feature of wire ropes/strands as load bearing members is the high level of redundancy present in the form of a multiplicity of parallel load paths.

The experimental composite tension members were essentially long continuous strands assembled from 5 mm pultruded rods, predisposed in a close-packed hexagonal array and the rods were unidirectionally reinforced respectively, with the three common types of high strength fibres: E-glass, aramid Kevlar 49 and a PAN-based carbon fibre. Pertinent details of the reinforcing fibres as given by the suppliers are listed in Table 1.

TABLE 1
Fibre details

	Glass	Aramid	Carbon
Type	E-glass	Kevlar 49	XA-S High performance
Supplier	Fibreglass	Du Pont	Hysol Grafil
Tensile strength (MPa)	1 700–2 400	2 760	3 300–3 600
Tensile modulus (GPa)	55–71	124	220–240
Ultimate elongation	3·1–3·4%	2·5%	1·4–1·5%
Density (g/cm³)	2·56	1·45	1·79

In view of the pultruded rods' limited resistance to bending compared to steel wire, the strands were made using a novel method of manufacture, a slight helical twist (long pitch) being introduced into the strand during manufacture. The slight twist, typically 2–3° on the outer rods, gives sufficient coherence to the strand for handling, coiling and transportation purposes, but is too small to affect significantly axial strength and stiffness. The glass composite strands comprised 73 rods and measured 49 mm (circumscribing circle diameter); the aramid and carbon strands were smaller at 45 mm and were assembled from 61 rods. Cross-sectional views of the 49 mm 1 × 73 glass and 45 mm 1 × 61 carbon strands are shown in Fig. 1. The comparator steel strand was also of 45 mm 1 × 61 construction. For added protection against mechanical damage during handling and installation, the strands were covered with a plastic jacket, as shown in Fig. 1. The jacket also served to preserve strand shape. Most of the mechanical testing was, however, conducted on unjacketed strand.

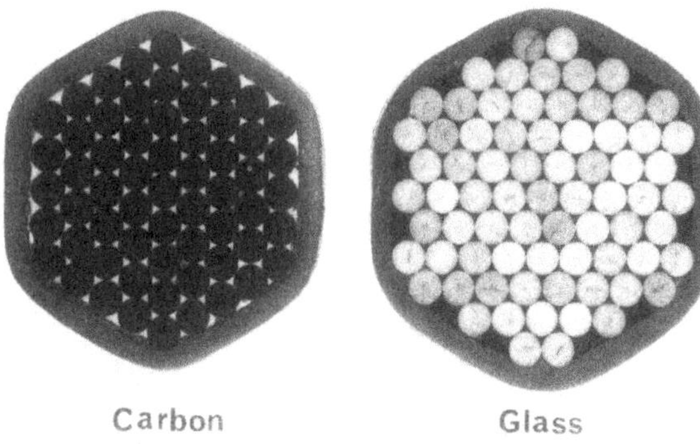

FIG. 1. Sectional views of 45 mm 1 × 61 carbon strand and 49 mm 1 × 73 glass strand.

ROD EVALUATION

The rod members were produced in long continuous lengths by the now well established pultrusion process. In this process, the reinforcing fibres in the form of tows or rovings, carefully tensioned, are pulled through a bath of resin and into a heated die, where actual forming and curing take place. The cured pultruded rod is used to transmit the pull provided by grip-pullers. The reinforcing fibres lie substantially in the axial direction resulting in high axial stiffness and strength. For the experimental strands, a vinyl ester resin matrix, Derakane 470-36, was selected, for good processability coupled with good mechanical properties. The fibre content was nominally 63% by volume.

Although the rods are strictly only semi-finished products, a comprehensive test programme has been undertaken to generate property data on the various composite rods. The rod data give an insight into how efficiently the reinforcing fibre is being utilised in the fibre–rod–strand manufacturing route. Also, strand properties are largely controlled by rod properties, and rod data are easier to acquire. The average rod properties, including physical data and short-term quasi-static test data, are given in Table 2.

TABLE 2
Rod properties

	E-Glass	Kevlar 49	XAS Carbon	Roping steel
Diameter (mm)	5·09	5·04	5·02	4·99
Fibre volume fraction	61%	63%	63%	—
Fibre weight fraction	76%	67%	72%	—
Density (g/cm^3)	2·02	1·36	1·57	7·86
Mass/length (g/m)	40·4	27·1	31·1	154
Calculated strength[a] (kN)	26·3	35·2	43·0	—
Breaking strength (kN)	21·8	31·7	30·3	35·7
Strength realisation	83%	90%	70%	—
Tensile strength (MPa)	1 060	1 590	1 610	1 820
Specific strength (km)	53	119	105	24
Fibre tensile strength[b] (MPa)	1 680	2 480	2 520	—
Elongation	2·3%	2·5%	1·2%	3·2%
Tensile modulus (GPa)	45·7	64·3	136	200
Specific modulus (GPa)	22·6	47·3	86·6	25·4
Bend ratio[c]	55	116	271	3

[a] Given by rule-of-mixtures, based on a mean fibre strength.
[b] Composite strength normalised to 100% fibre content.
[c] Ratio of bend diameter to rod diameter causing failure of rod by rupture.

Comparative data for a 5 mm steel (low relaxation) roping wire are listed. It should be noted that the mechanical properties given relate to the axial direction, the direction of prime interest.

A major problem encountered in the testing of unidirectionally-reinforced composite rods resides in the design of end fittings or grips (terminations) which can develop fully the strength of the rods. The low axial shear strength and transverse compressive strength of the composite rods preclude the use of split wedge-type action jaws traditionally used for steel wire and rod. Fittings for composite rods should be designed to transfer load gradually, reduce transverse forces and prevent crushing. A variety of gripping methods were tried, with varying degree of success. These included reinforcing the ends with overwrappings of glass fibre braid impregnated in resin and testing in wedge grips; splitting the rod ends and casting the split ends in resin-potted conical sockets; tab grip adaptors according to ASTM D3916-80 used in standard machine wedge grips. The most efficient grip for routine testing was a 150 mm long bolted split clamp, with the inner faces contoured and sandblasted to accept the rod. Of the three types of composite rods, the Kevlar-reinforced rod proved to be the most difficult to grip, and the 150 mm bolted clamp was the only reasonably effective method of gripping it. Analytical considerations indicate that the design of a fitting which develops 100% of the composite rod intrinsic strength may not be practically realisable.

Not unexpectedly, the data obtained from tests on the pultruded rods, in particular the breaking strength results, were subject to significant scatter. For instance, the breaking strength of 21·7 kN quoted in Table 2 for E-glass rod was averaged over 73 individual rod results ranging from 18·3 to 24·7 kN with a standard deviation of 1·32 kN. However, it is believed that a significant portion of the dispersion is associated with the less than perfect gripping technique used. More sophisticated gripping techniques currently being investigated should give improved strength values with less scatter and which reflect intrinsic rod strength. In interpreting the results in Table 2, it should be noted that the calculated rod strength was based on fibre strength data quoted by the suppliers, and as a consequence, undue significance should not be attached to the strength realisation figures obtained.

Notwithstanding the above remarks, the rod results show that the composite rods hold out the promise of considerable weight savings over steel wire from standpoints of both strength and axial stiffness. The only exception is the glass-reinforced rod whose specific modulus does not quite match that of steel.

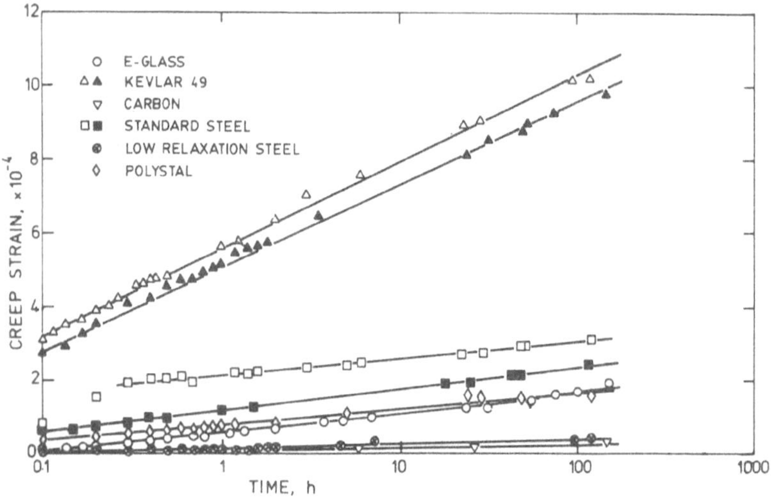

FIG. 2. Tensile creep properties of 5 mm composite rods at 50% breaking strength and 20°C. Data for two steel roping wires are also shown.

A property of concern in the application of the composite strands as structural members is the creep, creep rupture and stress-relaxation behaviour under uniaxial load. In view of the fact that rod tests are considerably cheaper to conduct and the strand creep behaviour should conform closely to the rod creep behaviour, a long-term creep test programme on the 5 mm composite rods has been initiated and is on-going. The preliminary 100 h creep data at 50% breaking strength for the various composite rods are presented in Fig. 2, along with data obtained on the low relaxation steel wire, a standard steel roping wire and a commercially available E-glass reinforced composite rod called Polystal.[4]

The following creep rates in diminishing order are indicated: Kevlar 49 rod ($0{\cdot}2 \times 10^{-3}$/decade); E-glass rod ($0{\cdot}053 \times 10^{-3}$/decade); standard steel wire ($0{\cdot}048 \times 10^{-3}$/decade); Polystal rod ($0{\cdot}041 \times 10^{-3}$/decade); low relaxation steel wire ($0{\cdot}011 \times 10^{-3}$/decade) and XAS carbon rod ($0{\cdot}006 \times 10^{-3}$/decade). The results show that with the exception of the Kevlar 49 rod, the composite rods show excellent creep resistance. The E-glass and Polystal rods have creep rates which are similar to that of a standard roping wire, and notably, XAS carbon appears to be even more creep resistant than a low relaxation steel wire intended for prestressing applications. Published data[5,6] confirm the excellent creep resistance of unidirectional carbon fibre reinforced composites. The creep rate obtained with the Kevlar 49 rod is rather higher than some published data[7] on the creep rate of a unidirectional Kevlar 49/epoxy composite in the 1–1000 min time period but is similar to that of the higher modulus aramid yarns.[8,9] Fatigue induced creep strains should follow the same trend as static creep strain but more importantly, the dynamic stiffness should be unaffected by creep. Although these preliminary creep tests have given some indication of the creep performance of the composite rods relative to more conventional materials, further testing to longer time durations and at different loads are required to establish a broader data base and this is currently being pursued.

STRAND EVALUATION

Property data acquired from single rod tests, such as those listed in Table 2, provide some useful guidelines as to likely strand behaviour. However, it was clear that tests on strand samples were essential to establish unequivocally certain properties, in particular the mechanical properties, for two main reasons. First, the strand can essentially be regarded as a bundle of rods, and its mechanical response will be dictated by the individual rod statistics. Secondly, it was recognised that the end terminations, which transmit the load to the structure, are a vital part of a strand assembly (strand plus termination), and in this respect tests on the assembly will reveal the overall system behaviour rather than the strand performance per se.

Strand specimens were subjected to two main categories of tests: (1) quasi-static tensile tests to failure and (2) constant amplitude tension–tension fatique tests. During the quasi-static tensile tests which were carried out on specimens with a test length in the range 1·5–3·7 m, the breaking strength (maximum load attained) was noted, along with the load–extension behaviour. A 100 kN hydraulic pulsator was employed for the fatigue tests, but the specimens had to be kept short (0·8–1·5 m) to achieve the necessary load amplitude levels. All the strand specimens were terminated with resin-potted cones. Used extensively in the wire rope industry for terminating wire ropes, resin cone terminations will develop the full strength of conventional steel wire rope products, with few exceptions. These resin cone terminations have proven to be reasonably successful for terminating the composite strands inasmuch as rod slip does not occur, but it is unlikely that they work as efficiently as with steel wire rope products.

TABLE 3
Strand properties (static)

	E-Glass	Kevlar 49	XAS Carbon	Steel
Construction	49 mm 1 × 73	45 mm 1 × 61	45 mm 1 × 61	45 mm 1 × 61
Mass/length (kg/m)	2·95 (2·46)[c]	1·65	1·89	9·40
Breaking strength (kN)	1 200 (1 000)[c]	1 520	1 400	2 100
Strength realisation[a]	76%	79%	72%	96%
Specific strength (N/m²)	41	94	75	23
Fibre stress at break[b] (MPa)	1 330	2 010	1 850	1 750
Elongation	1·80%	1·63%	0·88%	3·5%
Axial modulus (GPa)	44·3	77·3	132	197
Specific modulus (GPa)	21·9	56·8	84·1	25·1

[a] Rod to strand strength realisation.
[b] Strand strength normalised to 100% fibre content.
[c] Value in parentheses is breaking strength normalised to a 1 × 61 construction.

The tensile test results are recorded in Table 3, the values listed being averaged over several tests. Again, for comparison, equivalent steel performance values are given. Although the actual breaking strengths of the composite strands fall well short of the 2100 kN achievable in a steel strand, it is evident that they have a 2–4 fold advantage in specific strength. In long-laid steel strands, steel wire is very efficiently utilised, resulting in very high strength realisation (typically 96–98%) and an axial modulus close to the wire modulus. This efficient utilisation is attributable to the elasto-plastic stress–strain behaviour of steel rope wire and the uniformity in material properties. In contrast, the composite rods behave elastically to break, rendering the performance of composite strands more sensitive to even small manufacturing variations. In addition, the strength of the composite rods show a larger scatter than steel wire. Nonetheless, it is believed that the breaking strength results have, to some degree, been penalised by the less than perfect terminations, and improved terminations should give enhanced performance. Hence, the strength values given in Table 3 for the composite strands should be regarded as preliminary results.

The strand failure occurred typically as a rather sudden single event with rod breaks scattered over the specimen length, although in some instances, the main fracture was preceded by several distinct individual wire breaks. The rod fractures appeared to be generally clear of the terminations but closer inspection suggested that the rod breaks initiated at the strand/termination interface and propagated away in the form of longitudinal splitting followed by fibre break. The progression of this mode of failure was clearly observable in the specimens undergoing fatigue failure.

In contrast to the breaking strength, the axial modulus of the rod is closely reproduced in strand form. The Kevlar and carbon strands achieved a higher specific stiffness than steel: 0·7 (glass), 1·0 (steel), 2·2 (Kevlar), 3·3 (carbon). Hence in applications where stiffness is a key performance criterion, Kevlar and carbon strands are strong contenders. Figure 3 shows the fatigue performance of the composite strands versus steel. The solid line represents the fatigue curve for the steel strand. This fatigue curve is well supported by available test data (not shown) acquired on larger steel strands of similar construction. A considerable number of the earlier tests on the Kevlar 49 and carbon strands were carried out at relatively low stress ranges and were run-outs; these are not shown in Fig. 3. The comparative results indicate that on a stress basis, the fatique performance of the glass strands is rather inferior to that of the steel strand. On the other hand, the Kevlar and carbon strands appear to out-perform easily the steel strand with lives some 5–10 times that of steel. Available information[6,7,10–12] suggests that the tension fatigue behaviour of unidirectional carbon and Kevlar reinforced *laminates* is significantly better than that obtained with the composite strands. This discrepancy is primarily ascribed to

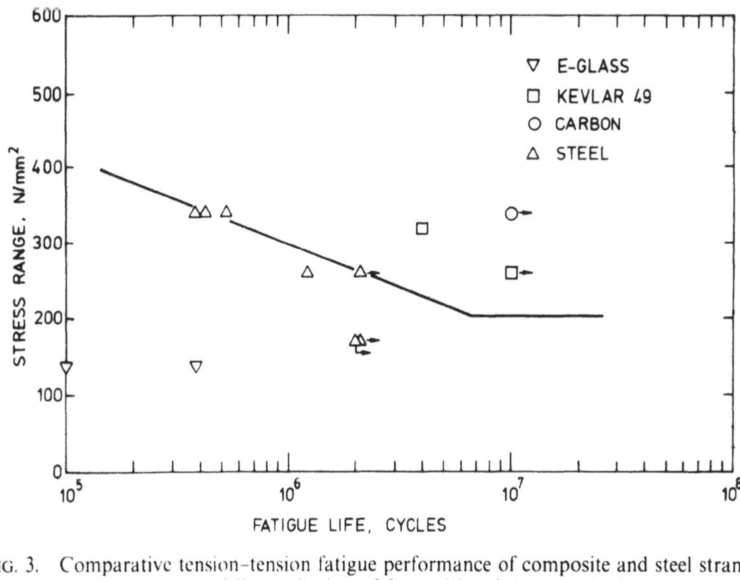

FIG. 3. Comparative tension–tension fatigue performance of composite and steel strands. → run-out (failure criterion of five rod breaks not reached).

imperfections in the terminations employed, thin laminates being considerably easier to grip.

Some comments on the mode of fatigue-induced failure are in order. On a macroscopic scale, the mode of strand failure occurred as a progressive failure of individual rods. Rod failure was typified by axial splitting followed by further degradation in the form of fibre break. Thus, although the individual rod failures may be rapid, the multiplicity of rods, which make up the strand, ensures that loss of strand integrity is not sudden and catastrophic.

CONCLUSIONS

The preliminary study reported here has demonstrated the potential of composite strands as tension members. Their performance advantages over conventional steel strand should given enhanced design capability to structural engineers concerned with tension members, and create new application possibilities where steel strands have perhaps never been considered. The preliminary study has yielded results which are sufficiently promising to justify a further and broader-based study with the main aims of consolidating the data base, maximising performance and identifying immediate applications for field experience acquisition. Specific aspects being currently addressed include improved end terminations; material optimisation; deterioration monitoring; long-term creep characteristics and long-term durability in a marine environment.

In view of the general lack of understanding in the effective use of composites, and the current high cost of composites versus traditional materials, large structures (very deep water moorings, long-span bridges) utilising composite strands should be regarded as developments of the future. In the meanwhile, field experience should be actively sought in secondary small-scale applications. Two such applications currently being explored can be cited. The first pertains to the usage of composite rods and strands as long continuous tension members for electro-mechanical and optical fibre cables. The second concerns the deployment of temporary bridges where lightness is of primary consideration. Field experience accruing from such applications, in conjunction with laboratory development work, will go a long way towards promoting acceptance of composite strands for future, large-scale utilisation.

The high cost of composite rods, currently in the order: steel, 1: glass, 5: Kevlar, 15: and carbon, 20 for small development quantities, inhibits their use. However, as demand and the scale of production of high strength fibres increase, so the cost differential will decrease. Cost considerations can also be misleading. Although the unit cost of the composite strands can be high, their deployment costs may be significantly reduced, considering their weight advantage. Where performance rather than cost is the dominant selection criterion, the composite strands offer major advantages over traditional steel strands in terms of:

—high strength-to-weight ratio;
—excellent tension fatigue performance;
—axial stiffness similar to steel;
—freedom from corrosion;
—good creep resistance;
—lightweight and easy to deploy.

REFERENCES

1. Salama, M. M., Lightweight materials for mooring lines of deepwater tension leg platforms, *Marine Technology*, **21** (1984), 234–241.
2. Meier, U., Proposal for a CFRP bridge across the Strait of Gibraltar at its narrowest site, *Proc. 14th Reinforced Plastics Congress 1984*, Brighton, England, 5–7 Nov. 1984, paper C46/86, The British Plastics Federation.
3. Meier, U., Future use of advanced composites in bridge construction engineering, *Proc. 2nd Int. Conf. on Fibre Reinforced Composites 1986*, Institute of Mechanical Engineers, University of Liverpool, April 1986.
4. Polystal high strength fibre composites, technical leaflet issued by Bayer AG.
5. Soliman, F. Y., Creep and rupture of graphite–epoxy composites, Composite Materials: Testing and Design, ASTM STP 460, American Society for Testing and Materials, 1969, pp. 254–270.
6. Sturgeon, J. B., Fatigue and creep testing of unidirectional carbon fibre reinforced plastics, *Proc. 28th Annual Technical Conference*, Reinforced Plastics/Composites Institute, Washington D.C., The Society of the Plastics Industry, Inc., Feb. 1973, sect. 12-B.
7. Miner, L. H., Wolffe, R. A. and Zweben, C. H., Fatigue, creep and impact resistance of aramid fiber reinforced composites, Composite Reliability, ASTM STP 580, American Society for Testing and Materials, 1975, pp. 549–559.
8. Riewald, P. G., Creep, stress-relaxation and stress-rupture of Kevlar aramid yarns, impregnated strands and ropes, Report No. KV-76-50, Wilmington, Delaware, E.I. du Pont de Nemours & Company, 1976.
9. Gerritse, A. and Schürhoff, H. J., Prestressing with aramid tendons, Technical Contribution to FIP 10th Congress, New Delhi 1986, Session 12: New Developments in Materials.

10. JONES, C. J., DICKINSON, R. F., ADAMS, T., REITER, H. and HARRIS, B., The environmental fatigue behaviour of reinforced plastics, *Proc. Roy. Soc. London*, **A396**, 1984, 315–338.
11. OWEN, M. J. and MORRIS, S., An assessment of the potential of carbon fibre reinforced plastics as fatigue resistance materials, *Proc. 25th Annual Technical Conference*, Reinforced Plastic/Composites Division, Washington DC, The Society of the Plastics Industry, Inc., 1970 sect. 8-E.
12. AGARWAL, B. D. and BROUTMAN, L. J., *Analysis and Performance of Fiber Composites*, New York, John Wiley, 1980, pp. 223–246.

23

Design of a Sofa for a Passenger Train

A. J. M. FERREIRA,* J. A. FIGUEIRAS† and A. T. MARQUES*

* *Departamento de Engenharia Mecânica,*
† *Departamento de Engenharia Civil,*
Faculdade de Engenharia da Universidade do Porto,
Rua dos Bragas, 4099 Porto Codex, Portugal

ABSTRACT

A broad description of a practical application of composite materials in a sofa for a passenger train is given. Considerations about the problem of design in general, i.e. selection of materials for design purposes, analysis of applied loads, application of design criteria, selection of materials for the best performance/cost relationship and some remarks about the fabrication processes are presented.

For design data purposes several types of composite materials were selected as well as various types of constructions (i.e. sandwich construction and single type of reinforcement construction).

The sofa was considered as a shell which is simulated by finite elements. An elasto-plastic analysis of the anisotropic shell considered was carried out by means of the finite element displacement methods, for types of laminates and constructions referred to above, using quadratic elements. A modified Huber–Mises yield criterion has been used to assess the structural integrity.

1. INTRODUCTION

An increasing number of structural applications for composite materials has been observed in recent years. This is due, not only to a better knowledge of the mechanical behaviour of composite materials traditionally used in commercial applications, but also to the use of advanced composite materials for such applications.

The high specific stiffness and specific strength of these materials are particularly applicable for satisfying the structural constraints imposed on the problem described in this chapter. The analysis of structures with complex geometric and support conditions is not an easy task and requires the use of numerical solutions, even for homogeneous materials. Hence, for the design of a sofa for a passenger train, it was necessary to work with finite element programmes which account for the anisotropy of material properties, not only along the surface of the laminates but also through the thickness.

As far as fire resistance of the composite materials is concerned, an increasing effort has been made to develop new materials or new additives which will accomplish the fire requirements imposed in transport services.

2. SERVICE CONDITIONS AND MATERIALS

2.1. Service Conditions

The sofa is intended for an inter-city train and its general shape, given by the railways company, is presented in Fig. 1. The main requirements imposed are the following:

—Light weight (320 N for the complete sofa; 210 N for the shell).
—Minimum fire risks (inflammable characteristics according to ORE B 106 Class A).
—No rupture or permanent strains should occur when the static loads (Fig. 1) are applied for given supporting conditions.

TABLE 1
Mechanical properties of materials used as skin in sandwich construction

Material	E_1 (MPa)	E_2 (MPa)	G (MPa)	$\bar{\sigma}$ (MPa)	$\bar{\tau}$ (MPa)	ρ (N/mm^3)	ν
Glass woven roving/resin (63% glass)	24 740	24 740	4 500	338	100	$0{\cdot}15 \times 10^{-4}$	0·18
Aramid woven/resin	42 000	42 000	2 500	455	49	$0{\cdot}24 \times 10^{-4}$	0·18
Carbon woven/resin	73 000	73 000	27 600	476	100	$0{\cdot}27 \times 10^{-4}$	0·18

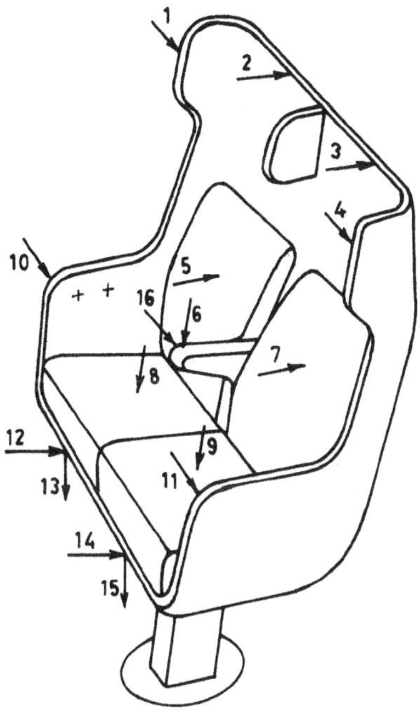

FIG. 1. Sofa model with loads to be applied.

2.2. Materials

A selection of materials has been made bearing in mind the conditions imposed above. The selected materials and their properties are presented in Tables 1 and 2. In Table 2 are presented the mechanical properties of core materials used in sandwich construction, and Table 1 presents the characteristics of materials used as a skin in sandwich construction or used as single type of reinforcement construction.

TABLE 2
Mechanical properties of materials used as core in sandwich construction

Material	E (MPa)	G (MPa)	$\bar{\sigma}$ (MPa)	$\bar{\tau}$ (MPa)	ρ (N/mm^3)	ν
PVC foam	72·6	19·3	4·08	2·16	$0{\cdot}08 \times 10^{-5}$	0·3
PVC foam	100	45	4·08	2·16	$0{\cdot}13 \times 10^{-5}$	0·3
PU foam	20	4	1	0·3	$0{\cdot}8 \times 10^{-6}$	0·3
PU foam	120	50	6	3	$0{\cdot}13 \times 10^{-5}$	0·3

3. STRUCTURAL BEHAVIOUR MODELLING

In the analysis of composite laminate structures the designer has to be able to account not only for a given loading, but also for the directional stiffness and strength properties of each laminae and the orientation of the various laminae. A powerful numerical model is necessary to perform this task. The main features of the finite element code developed for the analysis of thick and thin composite laminate shell structures[1,2] used in this work, is briefly described here.

3.1. Finite Element Modelling

Quadratic isoparametric curved shell elements derived from the degenerate shell concept are used. They are generally the best choice for the discretization of shells, as they accommodate complex geometries and different non-linear behaviours without having to use refined meshes.

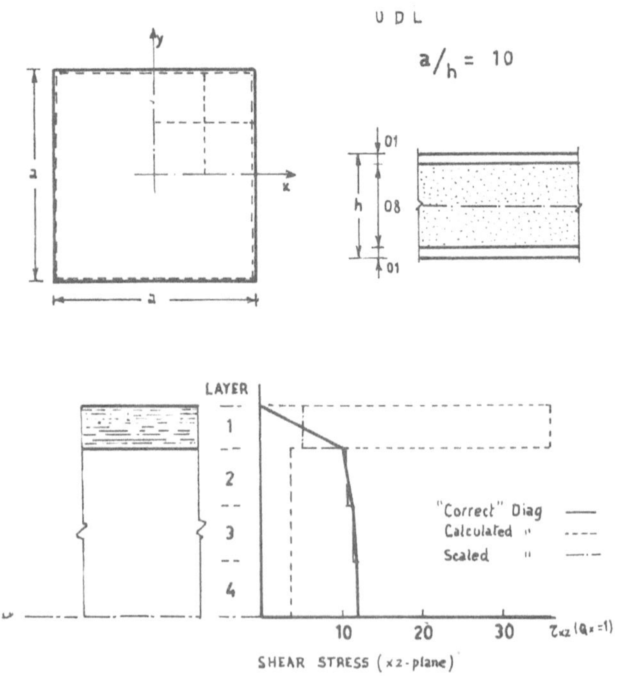

FIG. 2. Three-ply simply-supported square plate of symmetric section showing bending and shear stress distributions.

The transverse shear effects, which assume greater importance in anisotropic composite laminate analysis wherein the ratio of Young's modulus to shear modulus can be very large, are taken into account in the present formulation.

The 8-node Serendipity, the Heterosis and the 9-node Lagrangian elements are simultaneously considered. Reduced and selective integration rules are used to perform the numerical integration rules of the element matrices.

If the structure is built from a series of layers of different materials an appropriate discretization through the thickness has to be used. A layered approach based on a mid-point integration rule is employed to accommodate the variable material properties and to monitor the non-linear material behaviour across the thickness.

3.2. Material Modelling

To formulate the elastic constitutive relationships, a state of anisotropy with three mutually orthogonal planes of symmetry at each layer is assumed. Fibre-reinforced laminate structures can be considered as having, generally, this sort of symmetry.

The generalized Hook's law, referred to the principal material axes (1, 2, 3 of each layer, where 1 and 2 axes are the layer mid-surface) and assuming $\sigma_3 = 0$ (zero normal stress) can be written as:

$$\boldsymbol{\sigma} = \bar{\mathbf{D}} . \boldsymbol{\varepsilon}$$

where

$$\bar{\mathbf{D}} = \begin{bmatrix} \bar{D}_1 & \bar{D}_{12} & 0 & 0 & 0 \\ \bar{D}_{12} & \bar{D}_2 & 0 & 0 & 0 \\ 0 & 0 & \bar{D}_3 & 0 & 0 \\ 0 & 0 & 0 & \bar{D}_4 & 0 \\ 0 & 0 & 0 & 0 & \bar{D}_5 \end{bmatrix}$$

being

$$\bar{D}_1 = E_1/(1 - \nu_{12}\nu_{21})$$

$$\bar{D}_2 = E_2/(1 - \nu_{12}\nu_{21})$$

$$\bar{D}_{12} = E_2\nu_{12}/(1 - \nu_{12}\nu_{21})$$

$$\bar{D}_3 = G_{12}$$

$$\bar{D}_4 = k_1 . G_{13}$$

$$\bar{D}_5 = k_1 . G_{23}$$

If the principal axes of anisotropy 1, 2 do not coincide with the reference axes x, y but are rotated by a certain angle θ, the new elasticity matrix D is determined by:

$$D = T^{\mathrm{T}} \bar{D} T$$

with

$$\varepsilon_{(1,2,3)} = \mathbf{T}\varepsilon_{(x,y,z)}$$

The terms K_1 and K_2 are shear correction factors in the $\overline{13}$ and $\overline{23}$ planes respectively.

In the present finite element formulation a deformed 'normal' to the middle surface is assumed to remain straight but not necessarily normal to the middle surface.[3] This assumption makes the transverse shear strain constant through the thickness, which is a rough approximation to the actual variation, even for a homogeneous cross-section. Hence, the shear correction factors K_1 and K_2 must be introduced in order to approximate, on an average basis, the transverse shear strain energy. A function $g(z)$, which determines the shape of the shear stress diagram, can be obtained,[4] and a realistic value of the interlaminar shear stress can be evaluated according to:

$$Z_{xy} = \bar{G}_{13} \cdot \gamma_{xz} \frac{g(z)}{\bar{g}}$$

where $\bar{G}_{13}$ and $\bar{g}$ are mean values of G_{13} and $g(z)$ across the thickness.

4. DISCUSSION OF RESULTS

4.1. Structural Analysis

The structural analysis of the sofa has been made by using 8-node quadratic isoparametric elements to gain confidence in the finite element solution given by the 8-node element, the same finite element mesh has been appraised by using the Heterosis and 9-node elements. The choice has proved to be the correct one, as the 8-node elements give less computer time consuming solutions and the elastic results are pretty much the same for shell thicknesses over 15 mm. However, for shell thicknesses less than 15 mm, large differences in the results were observed. The finite-element mesh used to perform the structural analysis of the sofa is presented in Fig. 3. It corresponds to a coarse discretization which represents, with

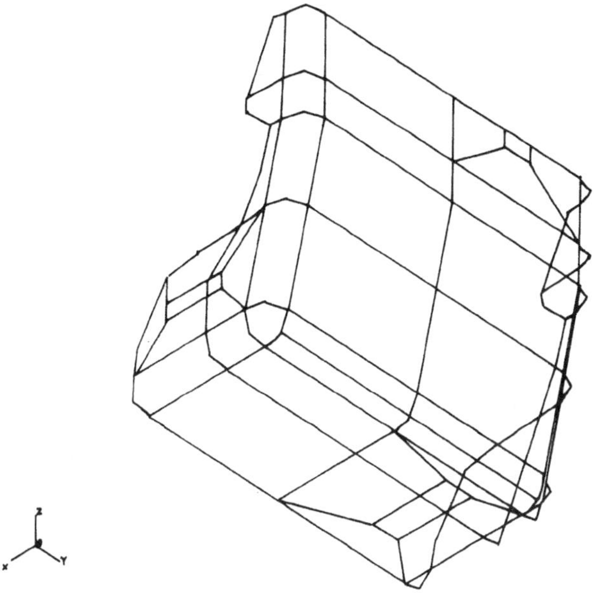

FIG. 3. Finite element mesh of the sofa.

enough accuracy, the geometric, loading and boundary conditions. In this analysis the loads are applied one at a time.

From Table 3 it can be seen that a determinant situation for designing occurs when load 3 is applied. Hence, the discussion is on results obtained by the application of load 3. Bearing in mind that the sofa is fastened to the wall (elements 3–6, Fig. 3), and has a leg-support to the floor (elements 40–41, Fig. 3), the influence on the maximum displacement of the distance from the support to the wall is presented in Table 4, for a certain type of construction. The following discussion will be based upon a sofa fastened to the wall and a leg-support apart from the wall 75% of the length of the sofa.

The influence of the skin material construction in the maximum displacement of the sofa is presented in Table 5, for a given core (19 mm of PVC foam: $E = 100$ MPa) and an overall thickness of 25 mm. It is quite evident, and as was expected, that a significant (50%) reduction in maximum displacement is obtained when carbon-fibre woven-roving is used as reinforcement, as compared with the use of glass-fibre woven-roving.

TABLE 3
Displacement versus individual loads

Load	Displacement (mm)	Thickness = 25 mm
1	4·25	
2	8	
3	16	
4	3·2	
5	6	
6	0·45	
7	17	
8	0·9	

TABLE 4
Maximum displacement × distance of floor support from wall

Sofa length %	Displacement (mm)
50	25
70	22·5
75	16
80	14

TABLE 5
Maximum displacement × skin material

Material	Displacement (mm)	Thickness = 25 mm
Glass woven + resin (63% fibre)	16	
Aramid woven + resin (50% fibre)	14·7	
Carbon woven + resin (50% fibre)	8	

Note: Core material used: PVC ($E = 100$ MPa).

The core material has a great influence on the maximum displacement (Table 6) obtained, and it can be concluded that a core material with a flexural modulus around 100 MPa must be applied to accomplish the maximum displacement (20 mm) and maximum weight (210 N) criteria imposed.

Another factor which greatly influences the maximum displacement is the core thickness. It was seen that by varying the core thickness from 0 to 25 mm, for the same type of skin, the maximum displacement was reduced from 200 mm to 15 mm.

TABLE 6
Maximum displacement × as function of core material

Material	Displacement (mm)	Thickness = 25 mm
PVC core ($E = 100$ MPa)	18·7	
PVC core ($E = 72{\cdot}6$ MPa)	25·2	
Polyurethane ($E = 20$ MPa)	30	
Polyurethane ($E = 120$ MPa)	17·5	

The remaining criteria to be used state that no permanent strains or rupture should occur when the static loads are applied individually. Those conditions were checked for one type of construction (i.e. 19 mm core PVC 100 or PU 120 and a 3 mm skin either side of woven roving glass fibre reinforced polyester resin laminate).

From the analysis of Fig. 4, one can conclude that no permanent strains occurred. In fact, the plastic behaviour begins for a load ten times bigger. In this figure it can also be seen that the non-linear geometric behaviour of the structure reduces, due to membrane effects, the maximum displacement occurring when high loads are applied.

The Huber–Mises criterion was fully satisfied in the structure, as the effective stress was observed to be ten times smaller than allowable stress. As far as normal and shear stresses are concerned, individual comparisons were made and it was seen that normal and shear stresses were about ten times smaller than the respective allowable stresses.

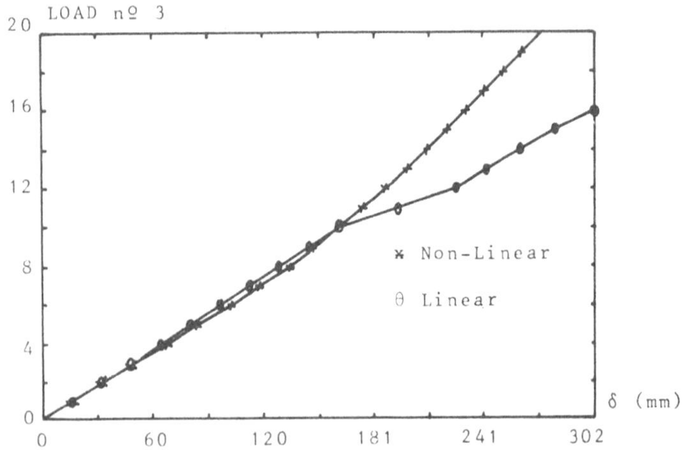

FIG. 4. Linear and non-linear behaviour of the 25 mm sandwich structure. Maximum displacement occurs in the loading point.

4.2. Selection of Materials and Processes

After selecting the materials from a structural point of view, other factors were considered as relevant for selecting the appropriate materials: fire resistance, price and availability.

The fire resistance data of epoxide resins, polyester resins, PVC foam and PU foam were checked. It could be concluded that certain types of each of these materials would be satisfactory for a class A material as far as flammability properties are concerned, when tested according to BS 476 (this gives classifications similar to the ORE flammability standard).

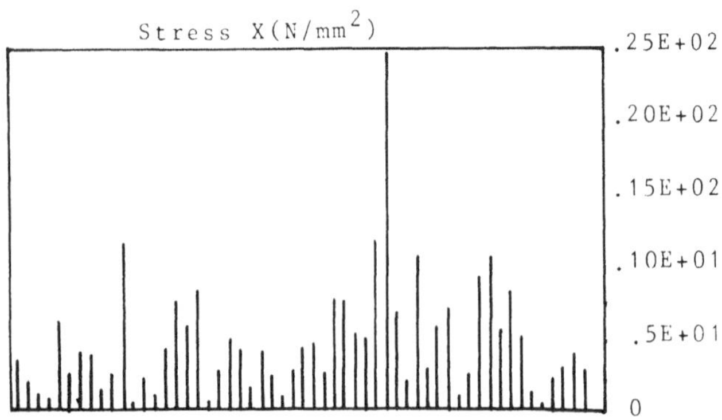

FIG. 5. Maximum σ_{xx} stresses over the elements.

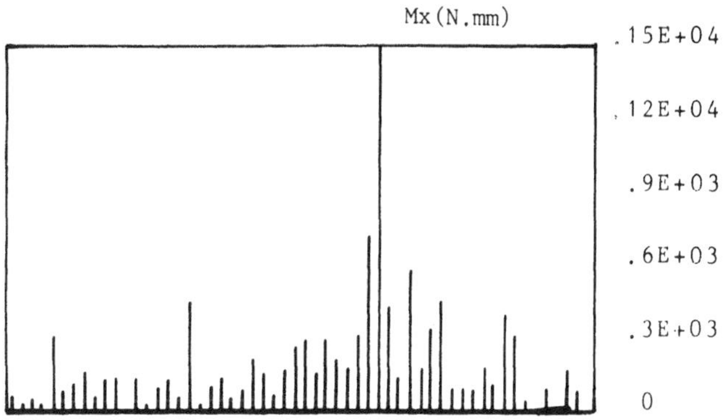

FIG. 6. Maximum M_x moments over the elements.

Considering the price and availability factors, mainly in Portugal, the choice was for a sandwich construction with a core in PVC or PU foam and skin of woven-roving glass fibre reinforced polyester laminates which give the best cost/performance relationship.

The choice of the fabrication technique was based on two main factors: number of pieces and quality.

As the number of sofas to be made was 1500 the obvious choice would go for contact moulding or vacuum bag moulding. As quality (weight fraction constant, no voids, ...) was the other predominant factor, the vacuum bag moulding process has been selected as the most suitable for this particular situation.

5. CONCLUSIONS

(i) A structural analysis by a finite element technique, accounting for variable material properties through the thickness, has defined a sandwich construction suitable for a sofa for a passenger train.

(ii) The 8-node quadratic isoparametric elements give adequate solutions for the finite element mesh adopted and thickness greater than 15 mm.

(iii) For thinner shells, 9-node Lagrangian or Heterosis elements must be used.

(iv) The materials used as skin, or as core, the laminate thickness, the type and the position of the support have great influence on the deformation characteristics of the sofa.

(v) Using a vacuum bag moulding technique, a sandwich construction of PVC or PU foam ($\sim$ 100 MPa) as the core, and woven-roving glass fibre reinforced polyester resin as the skin has been chosen, giving the best/cost performance relationship.

(vi) Further developments have to be considered to take into account the possibility of using reinforcing ribs.

ACKNOWLEDGEMENTS

The authors wish to thank Industrias Molaflex SARL, and, in particular, the president of the board of directors Dr Pereira da Costa, for their support to carry out this work.

REFERENCES

1. Owen, D. R. J. and Figueiras, J. A., Anisotropic elasto-plastic finite element analysis of thick and thin plates and shells, *Int. Journal for Numerical Methods in Engineering*, **19** (1983), 541–566.
2. Figueiras, J. A. and Owen, D. R. J., Analysis of elasto-plastic and geometric nonlinear anisotropic plates and shells, in: *Finite Element Software for Plates and Shells* (E. Hinton and D. R. J. Owen eds), Swansea, UK, Pineridge Press, 1984, pp. 235–326.
3. Mindlin, R. D., Influence of rotary inertia and shear on flexural motions of isotropic, elastic plates, *J. appl. Mech.*, **18** (1), *Trans. ASME*, **73** (1951), 31–38.
4. Zienkiewicz, O. C., *The Finite Element Method*, 3rd edn, London, McGraw-Hill, 1977.
5. Srinivas, S., A refined analysis of composite laminates, *Journal of Sound and Vibration*, **30** (4) (1973), 495–507.

24

Innovative Use of Syntactic Foam in GRP Sandwich Construction for Wind Generator Nacelles

M. J. Seamark

Balmoral Glassfibre, Hareness Road, Altens Industrial Estate, Aberdeen AB1 4LE, Scotland, UK

ABSTRACT

Wind energy is increasingly being harnessed as an alternative to finite sources of energy, and several UK companies have been in the forefront of the development of wind turbine technology.

Whilst the use of glass reinforced plastics (GRP) in the fabrication of blades is well known and many papers have been published, the use of these materials in the construction of nacelles is a recent innovation.

This paper covers some of the background of what is believed to be the unique choice of a GRP/syntactic foam sandwich construction, over more traditional building materials. Aspects of previous use of this combination of materials in building cladding panels, specification requirements and production are also referred to.

1. INTRODUCTION

James Howden & Company of Glasgow and the Wind Energy Group, comprising British Aerospace and Taylor Woodrow Construction, are recognised as leaders in the field of harnessing of wind energy through the development of wind turbines, and both have many installations operating in wind farms in California.

Both groups conducted extensive studies into the evaluation of materials available for the construction of nacelles. They both finally decided upon a very special form of glass reinforced plastic (GRP) sandwich construction and approached Balmoral Glassfibre of Aberdeen, who are acknowledged leaders in the design and fabrication of large structures in these materials.

This chapter is intended to provide a brief background to the choice of GRP, what are believed to be certain unique aspects of its sandwich construction, where these techniques have been used extensively and successfully prior to their use for nacelles, and the specification and performance requirements.

2. BACKGROUND

When the initial detailed feasibility studies were undertaken on the construction of the nacelle, which houses the tower top machinery on each wind generator, the material selection was a crucial consideration to their successful entries into this market. Fabricated steel, aluminium and GRP were the main material contenders.

Taking into account the major requirement of a high strength to weight ratio, outstanding stiffness to withstand the high wind loadings anticipated, minimum maintenance and low tooling investment, GRP was considered the ideal choice material.

3. MATERIAL SELECTION—GRP/SYNTACTIC FOAM

GRP has been used extensively for over 30 years in large structures, and one-piece mouldings over 60 m in length have been in use for many years, in such things as naval minehunters and fishing trawlers. Predominantly these have been based upon single skin laminates, although sandwich construction has been employed on some hull designs. Single skin laminates can be up to 60 mm in thickness, but clearly their use in this application to provide the necessary stiffness would have proven uneconomical.

The use of sandwich construction methods, utilising GRP skins, enables more lightweight structures to be produced without sacrificing stiffness, having the core material in the neutral axis.

Typical core materials which have been used for GRP sandwich construction are:

rigid polyurethane (PU) foam;
PVC foam;
phenolic foam;

polyester foam;
end grain balsa;
Firet Coremat—a non-woven synthetic fibre web.

All have certain advantages and disadvantages; the limitations can be related to:

too low a shear strength—foams are friable;
too high cost;
too high density;
labour intensive.

Indeed, in the case of rigid PU foam, it has been found necessary to tie the GRP skins together through the PU foam to avoid the possibility of delamination at the GRP/foam interface.

An alternative core material which does not exhibit these limitations is syntactic foam. Unlike all other foams whose cell structure is formed by blowing agents, gases, etc., syntactic foams have a predictable cell size and structure, achieved by combining low density microspheres with a thermosetting resin, such as epoxy or unsaturated polyester resins. The range of microspheres is extensive and includes the following:

cenospheres—power station flyash;
phenolic spheres;
polyvinylidene chloride expandable spheres;
glass spheres.

Balmoral Hitec Buoyancy have developed an extensive range of syntactic foams, based on glass microspheres, for their subsea buoyancy down to a working depth of greater than 6000 m. The group's innovative approach to both new materials and processing led the Glassfibre Division to explore the possibility of spraying syntactic foam for cladding panels; hitherto it had been processed only by 'in air' mixing, vacuum mixing or vacuum impregnation. Figures 1 and 2 show a grade of glass microsphere under magnification and how they appear to the naked eye.

It was soon apparent, following an intensive programme of development, that the approach was entirely feasible, and Balmoral's process engineers designed their own 'in-house' equipment for spraying chopped glass fibres, polyester resin and syntactic foam through the same equipment and by careful control and monitoring of the air supply for the glass microspheres from the hopper feed, syntactic foam could be sprayed consistently at a density of 0·50 g/cm^3.

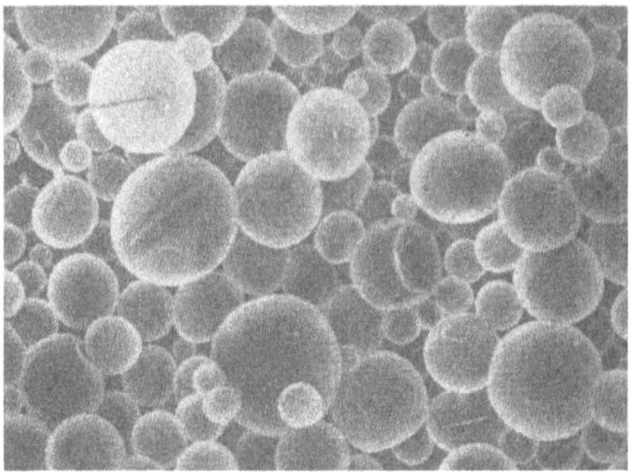

FIG. 1. Glass microspheres magnified 200 times.

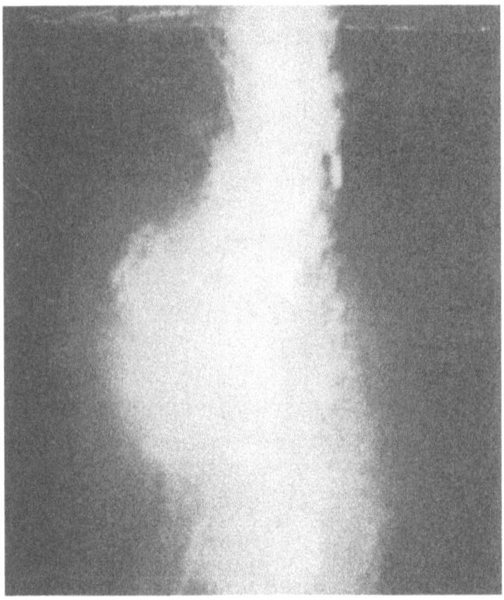

FIG. 2. Glass microspheres as seen by the naked eye.

4. BUILDING CLADDING PANELS

Once the technique had been established and mastered by the spray operators, the company confidently quoted for contracts for GRP cladding involving extensive use of this method of sandwich construction. Figures 3 and 4 illustrate examples.

FIG. 3. Cladding in GRP/syntactic—maltings plant at Elgin.

FIG. 4. Cladding in GRP/syntactic—Warwick University.

Due to applying simultaneously resin and chopped glass fibres followed immediately after consolidation by the syntactic foam core and then finally resin and chopped glass fibres, the laminate achieves a highly advanced stage of cure whilst still in the mould, the foam core when curing assisting the laminate cure. This lay-up and cure procedure has resulted in totally distortion-free mouldings, which due to their highly stable condition have exhibited no evidence of thermal movement in service, and which has been shown to be a problem with some other forms of sandwich GRP construction.

5. SPECIFICATION

Panels for the nacelles were designed to withstand a wind loading of 120 mph, to operate in a desert environment of -15 to $+35°C$ and to show no significant deterioration over a period of 25 years.

The gel coat or resin-rich outer coating was pigmented white to BS 4800 10B15, and the polyester resin selected provided optimum weathering properties combined with acceptable fire resistance.

A special conductive inner surface coating was incorporated to ensure a suitable earth in case of lightning.

The insulation properties of the core material needed to ensure the temperature inside the nacelle never rose to an unacceptable level.

6. PRODUCTION

Altogether over 100 nacelles have now been produced for the two groups, and Fig. 5 illustrates a typical wind farm on which they are installed. Figure 6 shows a turbine being hoisted into position.

Whilst specific epoxy resin systems are used predominantly for subsea buoyancy foam, an unsaturated polyester resin was quite satisfactory for these components, providing excellent processing and curing characteristics. Moulds were designed and split lines incorporated to assist moulding; the flanges also served to provide additional stiffness. Figures 7 and 8 show interior and exterior views of the nacelle.

Syntactic foam based on a specific grade of glass microsphere and a special unsaturated polyester resin were sprayed up to the required thickness. The only limitation of the spray technique adopted is that care should be taken to avoid deep complex shapes, not a problem with the design of these nacelles.

FIG. 5. Wind energy farm in California.

FIG. 6. Lifting the Howden HWP-300 into position at Burgar Hill, Orkney.

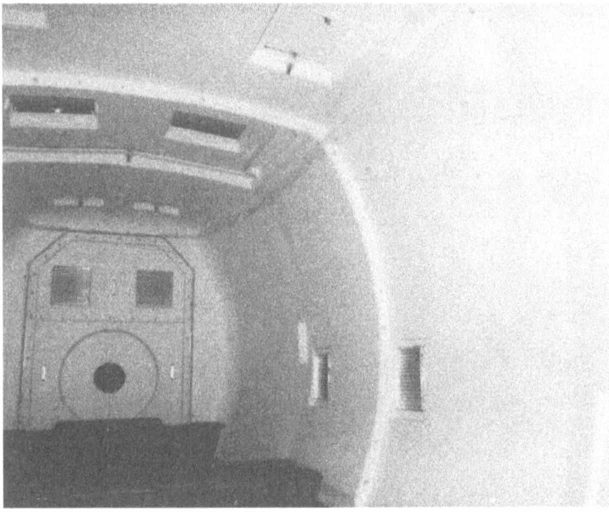

FIG. 7. Interior view of Howden nacelle.

FIG. 8. Exterior view of Howden nacelle.

FIG. 9. Three nacelles for Wind Energy Group.

7. SUMMARY

Balmoral believes this innovative use of syntactic foam in GRP sandwich construction to be unique in the world today and serves to highlight the great versatility of these materials, to produce high performance composites for an application with such an exciting growth potential. We have been proud to be associated with both Howden and WEG in this exciting development.

This is an outstanding success story, with mouldings having a predicted life of 25 years, which could well turn out to be 100 years, judged by the case studies of the successful use of GRP over the past 35 years, coupled with the continuing improvements in both materials and design.

25

A Design Study of Material Production for Composite Coil Springs

W. A. LEE

National Engineering Laboratory, East Kilbride, Glasgow G75 0QU, Scotland, UK

ABSTRACT

The need to produce hollow, continuous, multi-layered stock material of highly aligned continuous fibres at the high production rates required for automotive springs led to the design of a specialised winding facility. The facility described incorporates the more desirable features of both braiding and filament winding.

1. INTRODUCTION

The potential of fibre reinforced plastics (FRP) for primary load-bearing components in vehicle construction led to the formation of an industrial research consortium at the National Engineering Laboratory (NEL). The members of this Consortium, who contributed both financially and technically, are given in Table 1 and included component suppliers, material suppliers and car companies. Although the project management and most of the investigations were carried out by NEL, assistance in specific areas was undertaken by AERE, Harwell. The overall research programme was aimed at the practical assessment of FRP materials by the design, manufacture and testing of three typical components: a road wheel, a suspension arm and a coil spring.[1]

The coil spring was selected for investigation to assess the feasibility of an FRP design replacing a high-specification steel spring for an existing car application. On a more general level, the spring was considered as representative of other torsionally loaded components.

TABLE 1
Industrial members of the Consortium

Car Companies	Component Suppliers
BL Technology	Dunlop
Ford	GKN
	Jonas Woodhead
Material Suppliers	Lucas Industries
BP Research	AE
Ciba-Geigy	Mintex
Hysol Grafil	
Shell	
Pilkingtons	
Turner and Newall	

This chapter outlines the development work on the carbon fibre coil spring, with specific emphasis on a design study which was undertaken as part of the project to identify a possible production route for the manufacture of the carbon fibre reinforced plastic (CFRP) stock material. It is important, when considering the possibility of composite components in industries such as the car industry, that they can be produced successfully on a mass production basis. The relatively low production rates currently achieved with conventional FRP processes inhibit their industrial acceptance. Two features contribute to this: fibre placement and resin-curing speed. The coil spring for a car was a good example of a component which would be required to be produced in large quantities annually and thus provided the Consortium with a component for an investigation into the process-dependence of production rates.

2. COMPONENT SPECIFICATION

The spring selected for development was from the rear suspension of a BL Rover saloon car. The specification of the spring is given in Table 2, as are the details of the CFRP replacement spring. Well-documented design formulae and procedures existed for conventional spring material[2] which provided a starting point for the composite replacement design. From the consideration of the basic theory of helical springs, the FRP stock material would require a maximised torsional stiffness which would be obtained from orientation of the continuous reinforcing fibres at $\pm 45°$ circumferentially in the spring stock material in order to maximise the shear modulus[3] as shown in Fig. 1. The use of hollow stock was feasible for

TABLE 2
Spring specifications

	Steel spring	CFRP spring
Spring rate (kN/m) at bump	13·73	13·73
Free height (mm)	350	350
Wire diameter (mm)	11	
Stock diameter (mm)		8 mm i.d. 16 mm o.d.
Mean coil diameter (mm)	127·5	140
Fitted height (mm) (load N)	207 (1963)	207 (1963)
Bump height (mm) (load N)	132 (2993)	132 (2993)
Rebound height (mm) (load N)	327 (316)	327 (316)

CFRP springs, whereas this was generally not practical in steel. This offered a more efficient design of cross-section for the stock material and had an influence on the manufacturing route chosen.

After consultation with industry, a target production figure of 500 000 springs per annum was set for the design study.

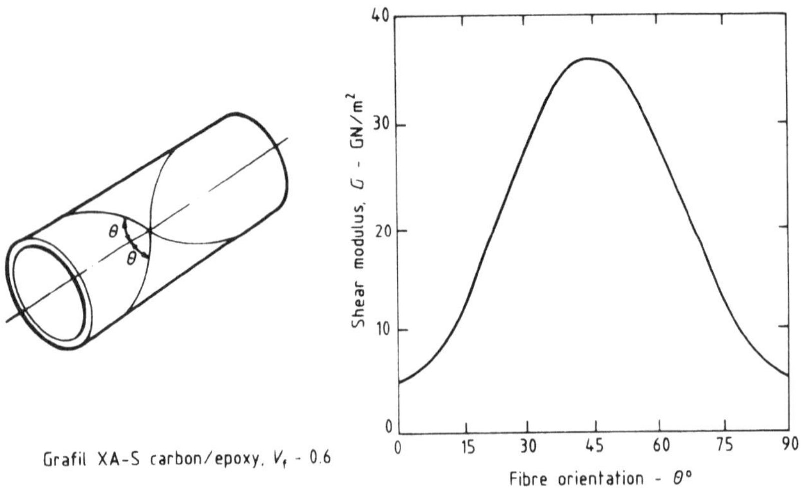

FIG. 1. Theoretical variation of shear modulus with braid angle.

3. MANUFACTURE OF SPRING

A major difference between FRP coil springs and conventional metal springs was the method of manufacture. Conventional spring manufacture utilises the metal's ability to deform plastically to produce the helical coils.

This type of deformation is not possible with thermosetting plastics, therefore, both the manufacture and the coiling of the FRP stock required to be completed prior to the resin being cured.

In general, the requirement was to lay down carbon fibre filament bundles, called tows, onto a flexible tube or core material. The resin impregnation of the tows was carried out either during the fibre lay-down or by vacuum impregnation on completion of dry winding.

For the manufacture of a spring, the stock material, comprising the core and carbon fibres impregnated with the uncured resin, was wound onto a segmented spring former or mandrel of the required form. The stock material was secured with a clamp onto the segmented former as shown in Fig. 2. To complete the manufacturing process the assembly was subjected

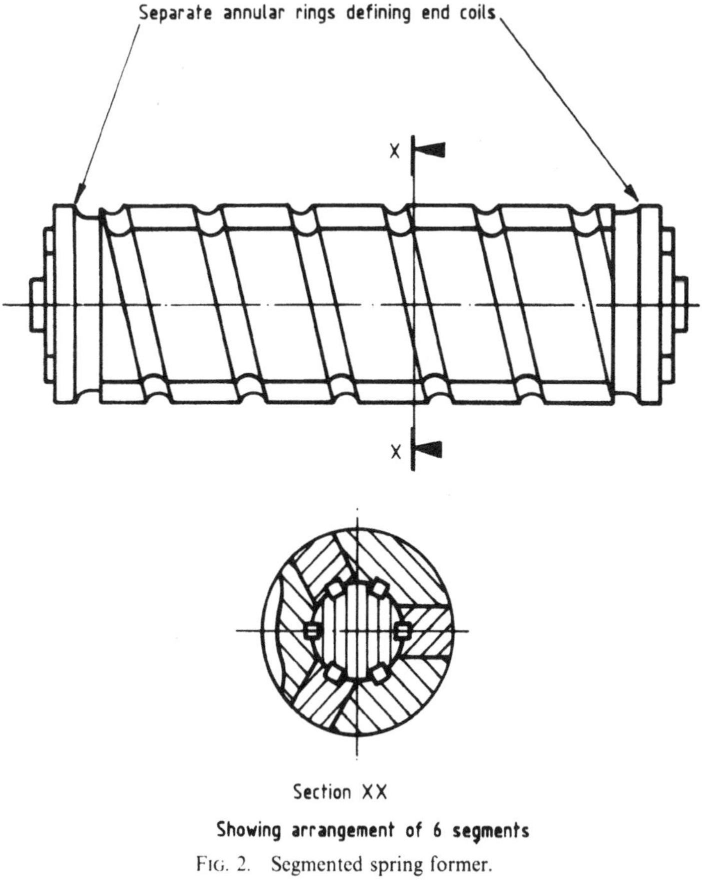

FIG. 2. Segmented spring former.

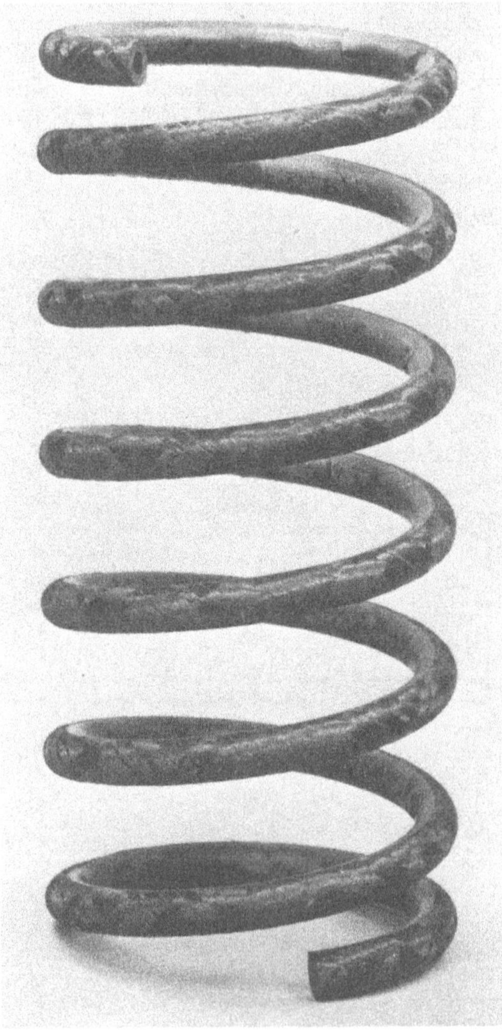

FIG. 3. CFRP spring.

to a heating cycle to cure the resin system. The spring was removed from the former by removing the central core in the former, thereby allowing the segments to collapse and fall away from the spring helix. Figure 3 shows a CFRP spring produced following this procedure.

4. MANUFACTURE OF STOCK MATERIAL

Initially two methods were investigated for the production of the CFRP stock for spring manufacture. The first employed was a single head, multi-carrier, braiding machine modified to operate horizontally. The head ran at a fixed speed of 14 carrier circuits round the head per minute and, in conjunction with a variable speed drive to the haul-off unit, this allowed the braid angle to be selected. When the core tube was passed through the head, carbon fibre tows from the carriers were laid down in an interwoven braid pattern on the tube. Successive passes through the head allowed the stock material to be built up to the desired diameter. The fibres were braided dry and post impregnated by either passing each successive layer through a resin bath or by vacuum impregnation on completion of all the braiding.

The second method of production was by filament winding at AERE Harwell. In this process the carbon fibre was laid down as a single tow onto the core which was rotated between two fixed stocks. The tow carrier traversed back and forth along the length of the core, until the desired diameter was produced. The tow carrier incorporated sophisticated resin impregnation rollers allowing impregnation of the two prior to laydown.

When considered as methods of spring stock production, both routes presented advantages and disadvantages. Braiding, while having the advantage of being an in-line process capable of producing long lengths of continuous stock, did however produce stock with an interlocking weave pattern; which compounded to fibre damage during laydown, since the tight bends the tow required to negotiate resulted in an unacceptable reduction of modulus and fatigue life in comparison to filament wound stock springs. While the fibre interlocking had a detrimental effect on the mechanical properties of the stock, the interlocking construction did however give good post-formability. Filament winding, while producing stock with a relatively high modulus, due to low fibre damage and higher volume fractions, resulting from the highly developed resin impregnation methods, is a process only capable of producing discrete lengths. This rendered it unsuitable for the mass production of continuous stock which would be required for a production facility.

5. SPIRAL WINDING

Based upon the experimental work carried out on the braided and filament wound springs, a design study was undertaken to identify a possible method for the mass production of the CFRP stock, which would combine the high mechanical properties obtained from the filament winding process with the continuous stock thoughput of the braiding process. Figure 4 shows the required production flow from the raw materials to the finished spring. The process finally decided upon to combine these benefits was spiral winding (also known as wave winding or shell winding).

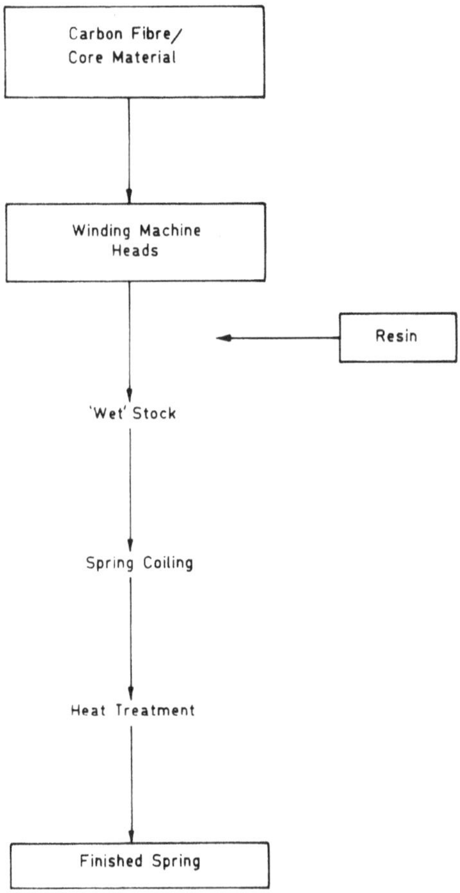

FIG. 4. Flow chart of spring production.

Spiral winding employs a rotating head mounted on which are a number of spools loaded with carbon fibre tows, as shown in Fig. 5. Winding is achieved by rotating the head while pulling the core tube through a hole in its centre; the tow from each spool is fed onto the core tube via a guide ring which maintains the lay-down angle and spreads the fibre. The angle at which the tow is wound depends upon the head rotational speed, the core diameter and the axial pull-through speed of the core.

After it was decided to use a spiral-winding technique for stock production, a design layout of the machinery was undertaken. The required number of winding heads, the number of spools per head, rotational speeds, etc., were calculated based on a production rate of 3 m/min. This represented sufficient stock for one spring to the specification and dimensions given in Table 2. The quantity of tow on each spool ideally would be sufficient for the completion of a working shift thereby avoiding the need to stop winding for the replenishment of the spools.

The main criteria in determining the number of heads and spools per head required to produce the desired stock design were based upon the spread which can be obtained from the tow when wrapped over the core and the associated packability of the fibre. Both 6 k and 12 k filament carbon fibre tow were tested to determine spread and packability, resulting in a decision to design the machine for 6 k tow on account of its better lay-down on the 8 mm diameter core tube.

To produce the stock material for a spring having inside and outside diameters of 8 and 16 mm respectively required a machine having 24 winding heads. Resin impregnation of the fibres was achieved by passing the stock through a resin bath incorporating sizing orifices between each head. To take account of the increasing diameter of the stock as it passed through each successive head, an increase in the number of tows being laid down would be required to maintain total coverage. A reduction in the head rotational speed would also be required to maintain a constant winding angle. Figure 5 shows, schematically, a possible 24-head winding machine with an indication of the number of spools on each head with its associated rotational speed. For the production of a balanced stock, each head would rotate in the opposite direction to the one on either side giving a $\pm 45°$ winding, the head being driven through a belt or chain from one or two contra-rotating lay shafts; the actual head speed relative to the lay shaft would be governed by the connecting drive ratios.

In order to support the stock material during production, a haul-off unit, as detailed in Fig. 6, would have to be spaced between every six heads to pull the stock through the machine. The haul-off force required would have

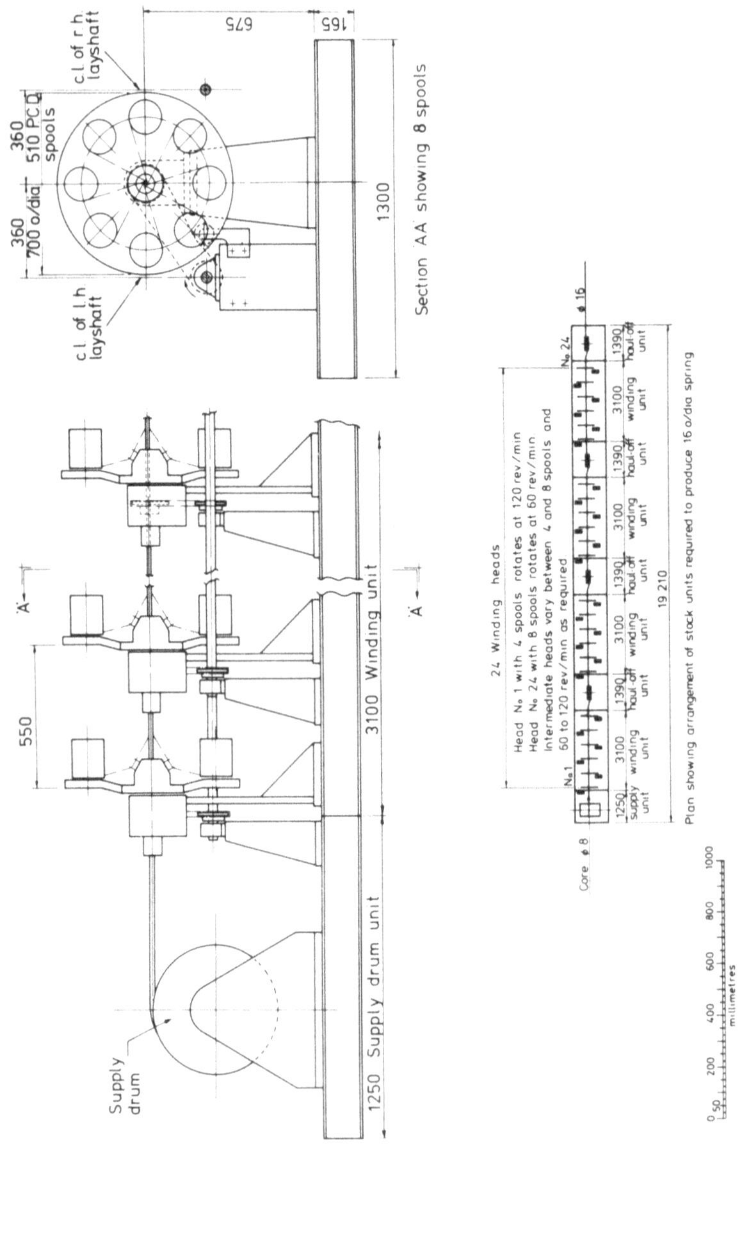

FIG. 5. Multi-head spiral winding machine (dimensions in mm).

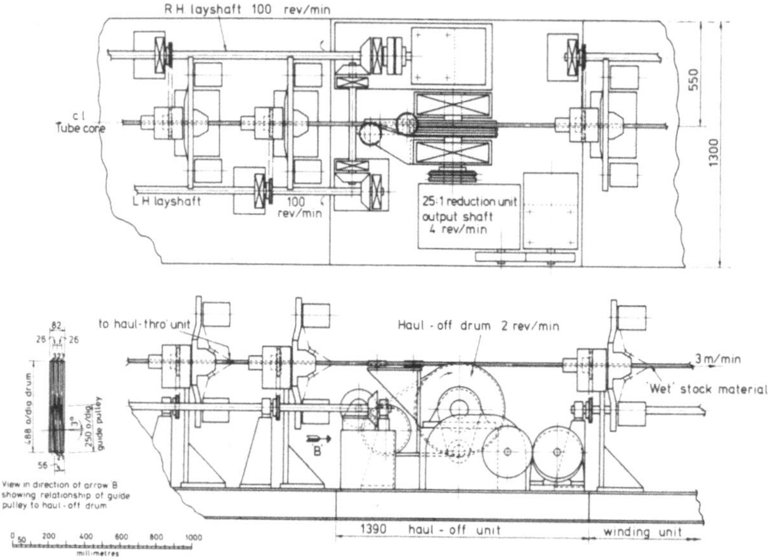

FIG. 6. Stock haul-off unit (dimensions in mm).

to overcome the tow tension, the impregnation and sizing orifices' resistance, and friction. The proposed haul-off units would work in a similar manner to a capstan. Firstly the stock would be guided onto the drum by a pair of rollers displacing the material off-centre. After winding over the drum and angled guide pulley, the stock returns to the centre of the machine ready to enter the next set of heads. Synchronisation between the haul-off units and winding heads would require to be very carefully controlled. There are many aspects of the machine design just outlined which are obvious areas for practical development. For example, the method of tow tension during winding which is important in obtaining constant fibre spread, resin impregnation during winding and quality control. The study indicated that these areas were not considered to pose unsurmountable problems in the design of a prototype machine.

A costing exercise was undertaken for the mass production of CFRP springs using stock material manufactured by this process. The exercise was based upon starting from a greenfield site and included: building, machinery, raw material, labour, power and depreciation costs. From the figures, the machinery cost accounted for approximately 10% of the overall cost. The major cost being that of the carbon fibre tow required for the springs' production.

6. EXPERIMENTAL TEST FACILITY

To test and demonstrate the spiral winding principle for the manufacture of CFRP stock, a test facility was designed and manufactured at NEL. Figure 7 is a schematic diagram of the test rig based upon a single winding head that was capable of winding up to eight tows at once. The stock diameter was increased by pulling the core tube back and forward through the head, using a cable attached to either end of the core tube. The cable which formed a continuous loop with the core, was wrapped several times around a helical grooved drum, which was connected to a variable speed drive unit. Rotation of the helical drum resulted in the cable being taken in by the drum at one side, paid out at the other and thus pulled the core through the head. Reversal of the drum rotation resulted in the core travelling in the opposite direction.

By maintaining the same direction of rotation of the winding head during the back and forward motion of the core, each successive layer of fibres was laid down in the opposite helical path to the previous one. The

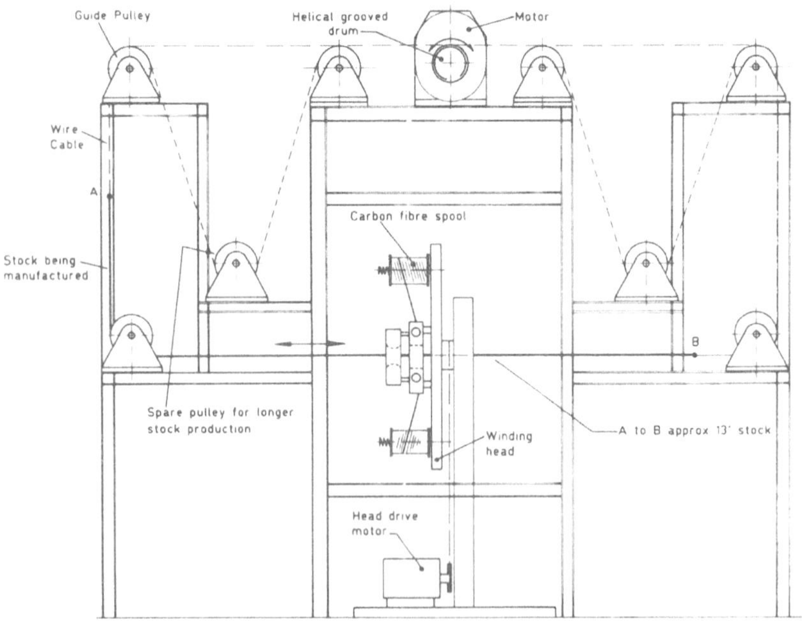

FIG. 7. Schematic elevation of spiral winding test facility.

angle at which the tow was wound onto the core depended upon the rotational speed of the winding head and the through-speed of the core. Fine adjustment of the laydown angle was controlled by adjusting the electronically variable speed unit of the head in preference to the mechanically adjustable unit on the core speed.

Figure 8 gives the relationship between the winding head speed and the stock diameter for a range of core-through speeds. This graph was used during winding to maintain the correct head speed for a $\pm 45°$ winding.

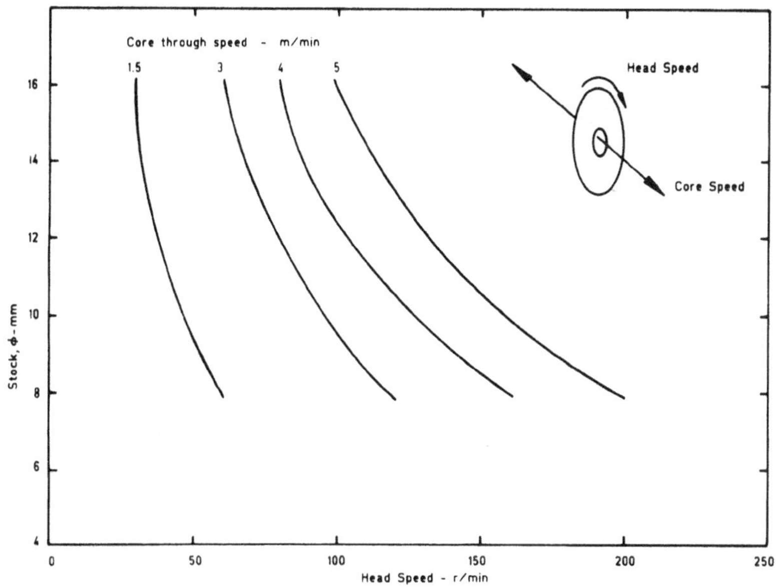

FIG. 8. Winding head speed versus core through speed for 45° lay-down.

Figure 9 illustrates the test winding facility showing the core passing through the head on which was mounted the spools of carbon fibre tow. After each layer of fibres was wound it was resin impregnated by passing the stock through an in-line resin bath and volume fraction/consolidation orifice positioned either side of the head. One of these units can be seen to the left-hand side of the head on Fig. 9. The direction change of the stock through the machine was controlled by limit switches positioned on the pull-through cable activating the drive controls.

Figure 10 shows in greater detail the tows being wound onto the core via a stationary guide and fibre-spreading unit. The tow tension was controlled using a simple spring-loaded friction pad. Initially four tows were required to be wound for complete coverage of the core. As the stock diameter increased, further fibre spools were incorporated into the winding, until all eight spools loaded with 6 k tow were being wound, bringing the diameter up to 16 mm.

FIG. 9. Spiral winding test facility.

Having completed the final pass, the impregnated stock was removed from the machine to be coiled onto the mandrel as outlined earlier in the chapter. The quality of the stock material produced, while not as good as had been obtained by filament winding, was encouraging. The reasons for the lower quality of stock produced from the test facility was due to its relatively simple tow tensioning and fibre spreading devices, which did not give sufficient control of the fibre during lay-down. The test facility was also built to demonstrate the process and was not expected to produce stock comparable to the highly developed filament winding process.

FIG. 10. Winding head.

7. CONCLUSION

The experimental facility demonstrated the principle of spiral winding as a possible method of producing stock material for CFRP springs. This work also illustrated the general need when considering component production in reinforced plastics, to examine, at length, existing processes and equipment, to achieve the desired material form, mechanical properties and production rates.

ACKNOWLEDGEMENTS

The author wishes to thank his immediate colleagues in the Materials Engineering Division, members from other assisting divisions within NEL and the collaboration and supportive work undertaken at AERE, Harwell. The structure of the Consortium allowed individuals and groups to make valuable contributions which are gratefully acknowledged.

The work was undertaken with part funding from the Mechanical Engineering and Machine Tool Requirements Board and this chapter is published by permission of the Director, National Engineering Laboratory, Department of Trade and Industry. This chapter is Crown copyright.

REFERENCES

1. WOOTTON, A. J., HENDRY, J. C., CRUDEN, A. K. and HUGHES, J. D. A., Structural automotive components in fibre reinforced plastics. *Composite Structures—3* (Marshall, I. H. ed.), Elsevier Applied Science, London, 1985, pp. 19–42.
2. WAHL, A. M., *Mechanical Springs*, Cleveland, Ohio, Penton Publishing, 1944.
3. HENDRY, J. C. and PROBERT, C., Carbon fibre springs. *Third International Conference on Carbon Fibres*, London, October 1985. Plastics and Rubber Institute (also Materials and Design, **7**, (6) (1986).

26

Multilevel Optimization Procedure of Composite Structure

LI WEIJI and SUN BAOHUA

Department of Aircraft Engineering, Northwestern Polytechnical University, Xian, Shaanxi Province, People's Republic of China

ABSTRACT

We present a multilevel optimization procedure for minimum weight design of composite structures. The optimization process is divided into two levels: system level and component level. In system level the thicknesses of laminated composite plates represent the design variables. The well-known optimality criterion method based on the Kuhn–Tucker condition is used to modify the design variables. The detail design of laminated plates, i.e. to determine the optimal ply thicknesses, is handled in component level. In component level, keeping the thickness of laminated composite plates and the nodal displacements invariant, the ply thicknesses are adjusted according to the strain energy possessed by each ply. As an illustrative example, a cantilever composite beam, which is subjected to several load conditions and multiple constraints, is studied. All the five examples given in this chapter present very satisfactory results. In general, the optimal design can be obtained after 3–5 iterations.

OPTIMIZATION PROCEDURE

1. System Level Optimization

For the structure composed of laminated composite plates, metallic plates and rods, the thicknesses of composite plates and metallic plates and the cross-sectional areas of rods are considered design variables. The well-known optimality criterion method based on the Kuhn–Tucker condition is used to modify the design variables. It is necessary for optimality criterion design to identify the active constraints. References 5 and 6 point out that it is necessary to only involve the most critical constraint in optimality criterion design. Because the displacement constraints influence significantly the design of every structural element, the most critical displacement constraint is involved in the optimality criterion design in this paper. The recurrence formula for modifying the design variables to minimize the structural weight is derived as follows.

The mathematical model of this problem can be stated as

$$\text{Min } W = \sum_{i=1}^{n} P_i l_i T_i \tag{1}$$

$$\text{S.T. } g_j(T_i) = \Delta_j - \Delta_j^* \le 0 \qquad j = 1, \ldots, m \tag{2}$$

where T_i is design variable; P_i is specific weight of material; l_i is surface area of plate or length of bar; n is number of elements; m is number of nodal displacement degrees of freedom; Δ_j is actual displacement of jth displacement degree of freedom; W is structural weight; Δ_j^* is allowed displacement of jth displacement degree of freedom; $g_j(T_i)$ is jth displacement constraint function; and subscript i indicates ith element.

Assume that the displacement of the pth degree of freedom is most critical. The Lagrange function can be formed by formulae (1) and (2):

$$L = \sum_{i=1}^{n} P_i l_i T_i + \lambda(\Delta_p - \Delta_p^*) \tag{3}$$

where λ is Lagrange multiplier.

From the optimality condition we have

$$P_i l_i + \lambda \frac{\partial \Delta_p}{\partial T_i} = 0 \qquad i = 1, \ldots, n \tag{4}$$

$$\Delta_p - \Delta_p^* = 0 \tag{5}$$

Suppose that the nodal displacement vector

$$\{\mathbf{r}\} = \{r_1, r_2, \ldots, r_m\}$$

and the unit load vector

$$\{\mathbf{S}\} = \{0, 0, \ldots, 1, \ldots, 0\}$$

Where all elements are zero except the pth element, which equals unity.

From the unit load method we learn

$$\Delta_p = \{\mathbf{r}\}^{\mathrm{T}}\{S\} \tag{6}$$

$$\{\mathbf{S}\} = [k]\{\mathbf{d}\} \tag{7}$$

where $\{\mathbf{d}\}$ is nodal displacement vector under unit load and $[k]$ is structural stiffness matrix. From the equilibrium equation

$$[k]\{\mathbf{r}\} = \{\mathbf{p}\} \tag{8}$$

it is derived that

$$\frac{\partial}{\partial T_i}\{\mathbf{r}\} = -[k]^{-1}\left(\frac{\partial}{\partial T_i}[k]\right)\{\mathbf{r}\} \tag{9}$$

$$\frac{\partial}{\partial T_i}\{\mathbf{r}\}^{\mathrm{T}} = -\{\mathbf{r}\}^{\mathrm{T}}\left(\frac{\partial}{\partial T_i}[k]\right)[k]^{-1} \tag{10}$$

Substituting (6) and (7) in (10) we find

$$\frac{\partial}{\partial T_i}\Delta_p = -\{\mathbf{r}\}^{\mathrm{T}}\left(\frac{\partial}{\partial T_i}[k]\right)\{\mathbf{d}\} \tag{11}$$

If the relationship between element stiffness and design variable is linear, the following formula is true:

$$\frac{\partial}{\partial T_i}[k] = \frac{[k]_i}{T_i} \tag{12}$$

Formula (11) can be rewritten as

$$\frac{\partial}{\partial T_i}\Delta_p = -\{\mathbf{r}\}_i^{\mathrm{T}}[k]_i\{\mathbf{d}\}_i/T_i \tag{13}$$

Substituting (13) in (4) the optimality criterion can be derived:

$$P_i l_i T_i - \lambda\{\mathbf{r}\}_i^{\mathrm{T}}[k]_i\{\mathbf{d}\}_i = 0 \tag{14}$$

$$1 = \lambda(\{\mathbf{r}\}_i^{\mathrm{T}}[k]_i\{\mathbf{d}\}_i)/(P_i l_i T_i) \tag{15}$$

The Lagrange multiplier is calculated by means of summation for every element in (14):

$$\sum_{i=1}^{n} P_i l_i T_i - \lambda\sum_{i=1}^{n}\{\mathbf{r}\}_i^{\mathrm{T}}[k]_i\{\mathbf{d}\}_i = 0 \tag{16}$$

$$W - \lambda\Delta_p = 0 \tag{17}$$

Referring to (5)

$$\lambda = W/\Delta_p^* \tag{18}$$

From the optimality criterion (15) the recurrence formula for modifying the design variables is as follows:

$$T_i^{\nu+1} = [(W^\nu/\Delta_p^*)(\{\mathbf{r}\}_i^T[k]_i\{\mathbf{d}\}_i)/(P_i l_i T_i)]^\eta T_i^\nu \tag{19}$$

where η is a coefficient of step length control and ν is iteration number.

Coefficient η can be changed in the iteration process. Decreasing η may cause the iteration process to become unstable and increase the iteration number. However, the extremely large η may cause the oscillation of the iteration process. In this chapter we suggest that η be 0·1–0·5. In addition, all of the elements should be divided into active elements and passive elements. The elements in which $\partial\Delta_p/\partial T_i \geq 0$ are passive, otherwise they are active. For active elements the design variables are modified by (19). For passive elements the design variables should keep the same value as in the last iteration.

2. Component Level Optimization

Component level optimization is performed on the basis of system level optimization. Keeping the thickness of laminated composite plate and the nodal displacements invariant, the ply thicknesses of a laminated composite plate are adjusted to make the material distribution among plies more reasonable. The adjustment of ply thickness is based on the maximum strain energy criterion. It means that for certain load conditions, structural configuration and materials the whole structural flexibility is adjusted to meet constraints and maximize the structural strain energy, thus obtaining a minimum weight design. The structure is deformed by the action of applied load, storing certain strain energies in the elements of the structure. The stored strain energy in an element is a measure of its ability to resist applied load. It is reasonable that the material distribution among the structural parts should be proportional to the stored strain energy in it. It is considered in this chapter that the ply thickness of a laminated composite plate should be proportional to the possessed strain energy in that ply. Increasing the ply thickness, which means more strain energy, should be a benefit for decreasing structural weight. Based on the above-mentioned idea we introduce the recurrence formula for adjusting the ply thickness:

$$t_{ij} = \left(\frac{e_{ij}}{\sum_{j=1}^{NL} e_{ij}} \right) T_i \tag{20}$$

$$e_{ij} = \tfrac{1}{2}\{\mathbf{r}\}_i^T[k]_{ij}\{\mathbf{r}\}_i \tag{21}$$

where t_{ij} is thickness of jth ply in ith composite plate; e_{ij} is strain energy possessed by jth ply in ith composite plate; and NL is number of plies in a composite plate.

3. Uniformly Scaling Design

For feasibility of the optimal design it is necessary to perform the uniformly scaling design, i.e. every design variable is multiplied by a common factor. At this stage the element strength should be checked. The Tasi–Hill criterion is taken as the strength criterion for composite plates in this chapter:

$$H_{ij} = (\sigma_{ij}^x/F_x)^2 + (\sigma_{ij}^y/F_y)^2 - (\sigma_{ij}^x\sigma_{ij}^y/F_x^2) + (\tau_{ij}/F_{xy})^2 \tag{22}$$

where σ_{ij}^x is stress in fibre direction for jth ply of ith composite plate; σ_{ij}^y is stress normal to fibre direction for jth ply of ith composite plate; τ_{ij} is shear stress for jth ply of ith composite plate; F_x is axial strength of unidirectional composite; if $\sigma_{ij}^x > 0$, F_x is axial tension strength; if $\sigma_{ij}^x < 0$, F_x is axial compression strength; F_y is transverse strength of unidirectional composite; if $\sigma_{ij}^y > 0$, F_y is transverse tension strength; if $\sigma_{ij}^y < 0$, F_y is transverse compression strength; F_{xy} is shear strength; H_{ij} is Tsai–Hill number of jth ply of ith composite plate; and if $H_{ij} \leq 1$, the strength requirement is considered to be satisfied for the jth ply of the ith composite plate and vice versa.

The square root of the Tsai–Hill number of each ply, $\sqrt{H_{ij}}$, is taken as the strength index. Compute the maximum strength index:

$$a = \max(\sqrt{H_{1,1}}\,\sqrt{H_{1,2}}\cdots\sqrt{H_{n_1,\mathrm{NL}}}) \tag{23}$$

where n_1 is number of composite plates.

For metallic elements compute the maximum ratio of actual stress to allowable stress:

$$b = \max(\sigma_1/[\sigma_1], \sigma_2/[\sigma_2], \ldots, \sigma_{n_2}/[\sigma_{n_2}]) \tag{24}$$

where n_2 is number of metallic elements.

Meanwhile, the displacement constraint of structure should also be checked. Computing the ratio of the most critical displacement to allowable displacement.

$$c = \Delta_p/\Delta_p^* \tag{25}$$

The maximum value among a, b, c is the common multiplier to modify design variables in a uniformly scaling design.

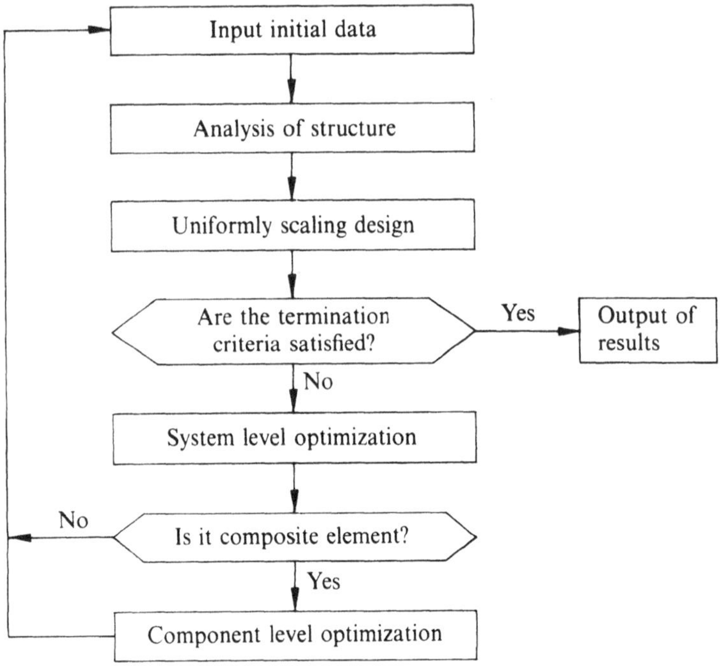

FIG. 1. Flow diagram for the algorithm.

4. Termination Criterion of Iteration Process

The iteration process is terminated whenever one of the following criteria is met. (1) The number of redesign cycle exceeds a prescribed value. (2) The change of structural weight between two adjacent iterations is less than a specified value.

5. Flow Diagram

To visualize the above-mentioned algorithm a flow diagram is given (see Fig. 1).

ILLUSTRATIVE EXAMPLES

In this chapter we study a cantilever beam composed of aluminium alloy plates, rods and boron/epoxy composite plates having midplane symmetry and fixed fibre orientations. The upper and lower covers are made of boron/epoxy composite plates. The shear panels, rods and ribs are made of

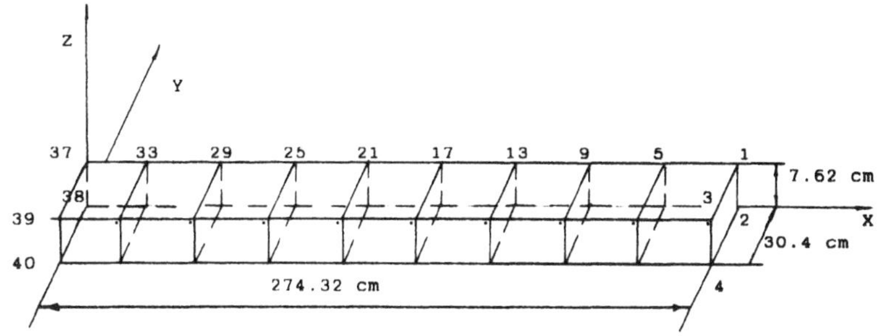

FIG. 2. Cantilever beam.

aluminium alloy. The mechanical model of finite element analysis of this structure consists of 18 plane stress composite membranes, 27 aluminium shear panels and 18 uniform axial force rods (see Fig. 2). The material properties and allowable stresses for boron/epoxy composite, the material properties of aluminium alloy and the initial and lower limit sizes are shown in Tables 1–4, respectively.

Example 1

The beam is under the action of bending moment and subjected to strength and tip section displacement constraints. The upward load, 4·448 kN, is acting on the nodes 1, 2, 3, 4. The Z-direction displacement of the tip section nodes is not allowed to be greater than 25·4 cm. After two

TABLE 1
Material properties of boron/epoxy composite

E_{11} (GN/m^2)	E_{22} (GN/m^2)	G_{12} (GN/m^2)	ρ (kg/m^3)	ν_{12}
$2{\cdot}069 \times 10^2$	$1{\cdot}862 \times 10$	4·482	$2{\cdot}007 \times 10^3$	0·21

TABLE 2
Allowable stresses of boron/epoxy composite

F_x (GN/m^2)		F_y (GN/m^2)		F_{xy} (GN/m^2)
$\sigma_x > 0$	$\sigma_x < 0$	$\sigma_y > 0$	$\sigma_y < 0$	
1·214	2·689	0·078 6	0·308	0·014 5

TABLE 3
Material properties of aluminium alloy

E (GN/m^2)	G (GN/m^2)	ρ (kg/m^3)	ν	Allowable stress (GN/m^2)	
				$[\sigma]$	$[\tau]$
72·397	27·842	2·768 × 10^3	0·3	0·462	0·267

TABLE 4
Initial and lower limit sizes

	Initial size	Lower limit size
Ply thickness of composite laminate (cm)	0·127	0·0127
Thickness of metal plate (cm)	0·254	0·002 54
Cross-sectional area of metal rod (cm^2)	0·645	0·006 45

iterations the structural weight is decreased from 41·58 to 9·05 kg and the tip section displacement is equal to 25·4 cm. The maximum Tsai–Hill number of a composite plate is 0·9795 and the ratio of maximum stress to allowable stress for aluminium alloy elements is 0·738. This design has satisfied strength and displacement constraints and is considered optimal.

Example 2

The beam is under the action of torque moment and subjected to strength constraint and tip section twist angle constraint. The upward load, 9·341 kN, is acting on nodes 1 and 2, and the downward load, 9·341 kN, is acting on nodes 3 and 4. The tip section twist angle is limited to be equal to or less than 5°. After two iterations the structural weight is decreased from 11·34 to 7·18 kg. The tip section twist angle is equal to 5°. The maximum Tsai–Hill number of a composite plate is 0·2672. The ratio of maximum stress to allowable stress for aluminium alloy elements is 0·3034. This design has satisfied strength and twist angle constraints.

Example 3

The beam is under the action of bending and torque moments and subjected to strength, tip section displacement and twist angle constraints. The upward load, 15·568 kN, is acting on nodes 1 and 2, and the downward load, 6·672 kN, is acting on nodes 3 and 4. In the tip section the Z-direction displacement is not allowed to be greater than 25·4 cm and the twist angle is limited to be equal to or less than 5°. After three iterations the structural weight is decreased from 41·87 to 18·81 kg. In the tip section the maximum Z-direction displacement is 17·87 cm and the twist angle is equal to 5°. The maximum Tsai–Hill number of a composite plate is 0·7978. The ratio of maximum stress to allowable stress for aluminium alloy elements is 0·6265. This design has satisfied strength constraints and tip section displacements and twist angle constraints.

Example 4

The beam is under the action of bending moment and subjected to special constraint on the deflection distribution along the structural axis. The applied load is the same as for Example 1. In order to emphasize displacement constraints the higher strength data of material have been selected (see Table 5). After four iterations the deflection distribution along the structural axis is in good agreement with the given deflection distribution (see Table 6). The structural weight is decreased from 62·87 to 14·39 kg. The change of structural weight obtained in this chapter is compared with other references (see Table 7). It is obvious from Table 7 that the method presented here is much more effective.

Example 5

The beam is under the action of torque moment and subjected to a special constraint on twist angle distribution along the structural axis. The

TABLE 5

Boron/epoxy composite material (GN/m^2)					Metal material	
F_x ($\sigma_x > 0$)	F_x ($\sigma_x < 0$)	F_y ($\sigma_y > 0$)	F_y ($\sigma_y < 0$)	F_{xy}	$[\sigma]$	$[\tau]$
0·8826	0·8826	0·04778	0·1834	0·07033	0·6895	0·6895

TABLE 6

Node	1	5	9	13	17	21	25	29	33
Computed deflection (cm)	17·78	13·61	9·89	7·25	4·98	3·09	1·68	0·74	0·21
Given deflection (cm)	17·78	15·24	10·16	7·62	5·08	3·81	2·03	0·76	0·25

TABLE 7

Weight (kg) \ Iteration number	1	2	3	4	5	...	10
Ref. 3	84·41	22·95	18·15	15·72	14·71		14·17
Ref. 4	87·68	43·44	28·36	20·65	14·02		13·19
This chapter	62·87	20·08	14·42	13·76	14·39		

TABLE 8

Station	1–4	5–8	9–12	13–16	17–20	21–24	25–28	29–32	33–36
Computed twist angle (°)	4·5	3·92	3·41	2·92	2·44	1·95	1·47	0·97	0·47
Given twist angle (°)	4·5	4·00	3·50	3·00	2·50	2·00	1·50	1·00	0·50

material properties are the same as in Example 4, and the applied load is as in Example 2. After two iterations the structural weight is decreased from 12·58 to 7·89 kg. The twist angle distribution along the structural axis is also in good agreement with the given twist angle distribution (see Table 8).

CONCLUSION

It is shown from theoretical analysis and illustrative examples that the multilevel optimization procedure presented in this chapter can effectively be used to solve the optimization problem of composite structures. The design variables are reduced to a great extent in each level of optimization. The component level optimization based on strain energy concepts is more visualized and easy to implement. The satisfactory results are given for optimal design of composite structures which are under the action of multiple load conditions and subjected to strength, displacement constraints and some special constraints on deflection and twist angle distribution along the structural axis. As compared with some related references, the proposed method may be more readily accepted by engineers because of the simplicity in implementation and the clarity in concept. The illustrative examples also confirm the efficiency of the method. In general, the optimal design can be obtained after 3–5 iterations.

REFERENCES

1. SCHMIT, L. A. *et al.*, A multilevel method for structural synthesis, AIAA Paper No. 84-0853-cp.
2. WANG GUANGYUAN *et al.*, Two-phase optimization method for structural design, *Acta Mechanica Sinica*, **15** (1983), 376–387.
3. KHOT, N. S. *et al.*, Optimum design of composite structures with stress and displacement constraints, AIAA Paper No. 75-141.
4. ZHU JIANSEN *et al.*, Optimum design of composite structures based on sequential quadratic programming, Technical Report NHJB-85-2492, Nanjing Institute of Aeronautics.
5. LI WEIJI, A rapid optimization procedure for laminated composite plates, *Proceedings of The Second International Conference on Civil and Structural Engineering Computing*, Vol. 1, 1985, pp. 383–390, London.
6. TSAI, S. W. *et al.*, *Introduction to Composite Materials*, Technical Publishing Company, 1980, pp. 115–164.

27

Optimum Design of Hybrid Fibrous Laminated Composite Plates Subject to Axial Compression

MITSUNORI MIKI and KOUJI TONOMURA

Materials System Research Laboratory, Kanazawa Institute of Technology, 7-1 Ogigaoka, Nonoichi, Ishikawa 921, Japan

ABSTRACT

A flexural lamination parameter diagram of hybrid fibrous laminates is constructed and a new analytical design method is proposed to determine the optimum lay-up of hybrid laminated plates subject to axial compression. The object of the design problem is to minimize the cost of the hybrid plates which have a composite surface with high performance and high cost, and a composite core with low cost, under the constraint that the plate has a given buckling strength. The optimum lay-up can be obtained from the material constants of the two kinds of unidirectional composites used, the plate aspect ratio, and the constraint on the buckling load.

INTRODUCTION

The design criterion for laminated plates subject to compressive load is the buckling strength. The buckling strength of fibrous laminated plates varies with the change of materials, stacking sequence and the orientation angles. Consequently, the maximum buckling strength can be attained by an optimization in the stacking sequence and the orientation angles. For this design problem, Hirano,[1] Wang,[2] Fukunaga[3] and Miki[4] respectively proposed the optimum design method based on different approaches. However, the analysis for hybrid composites has not yet been performed.

The authors have proposed a new analytical design method of determining the optimum lay-ups of fibrous laminated plates with the maximum buckling strength.[4] The proposed method in this chapter is based on the previous method, but is extended to obtain the cost-minimum solution instead of the maximum buckling strength.

The aspect ratio of the plate is assumed to be equal or less than unity because the analytical complexity increases when the aspect ratio is greater than unity. This type of analysis will be dealt with in another report.

PROBLEM DEFINITION

The laminated hybrid plates intended for this study are multiple balanced angle-ply symmetric sandwich-type laminates with surface and core of different composites. It is assumed that the surface material has a high stiffness, but is expensive, and the core material does not have a high stiffness, but is cheap. Each ply has the fiber orientation of $\pm\theta_k$. They are considered to be orthotropic plates, of which flexural stiffness elements D_{16} and D_{26} are equal to zero. The core thickness ratio R_c is defined as the ratio of the core thickness h_c to the total thickness h of the plate, as follows:

$$R_c = h_c/h \tag{1}$$

Consider the general class of laminated rectangular plates that are simply supported along edges $x = 0$, $x = a$, $y - 0$ and $y = b$, and subjected to a uniform in-plane force in the x-direction. The load per unit width is denoted by $\bar{N}_X$ and the plate aspect ratio r is defined by the following:

$$r = a/b \leq 1 \tag{2}$$

The aspect ratio is assumed to be equal or less than unity in this chapter.

The constraint condition assumed is as follows. That is, the total thickness h should be less than a given upper bound and the buckling load per unit width $(\bar{N}_X)_{cr}$ should be greater than a given lower bound.

In this case, however, the total thickness h always becomes the upper bound because the buckling load invariably increases when the thickness increases if the lay-up is fixed. Then the above constraints can be rewritten by using the reduced buckling stress $(\bar{N}_X)^*_{cr}$ as follows:

$$(\bar{N}_X)^*_{cr} \geq \tilde{N}^*_X \qquad (\tilde{N}^*_X\text{: constant}) \tag{3}$$

where

$$(\bar{N}_X)^*_{cr} = 12b^2(\bar{N}_X)_{cr}/\pi^2 h^3 \tag{4}$$

and $(\bar{N}_X)_{cr}$ is the buckling load per unit width.

The objective of the problem considered here is to minimize the cost of the hybrid plate, and the design variables are the fiber orientation angles $\pm\theta_k$ of constituent plies. The minimization of the cost corresponds to the maximization of the core thickness ratio R_c when the thickness is constant, because the surface material is assumed to be more expensive than the core material.

ANALYTICAL

Flexural Lamination Parameter Diagram of Hybrid Composites

The stacking sequence of the hybrid laminate under consideration is represented by

$$[(\pm\theta_n)/\cdots/\pm\theta_{N_c+1})_{\mathrm{FRP_s}}/(\pm\theta_{N_c}/\cdots/\pm\theta_2/\pm\theta_1)_{\mathrm{FRP_c}}]_{\mathrm{S}} \tag{5}$$

where $\mathrm{FRP_c}$ and $\mathrm{FRP_s}$ represent core and surface materials respectively, and n and N_c represent the half number of the total plies and the half number of plies of the core. The equation of the normalized flexural stiffness D^*_{ij} $(=12D_{ij}/h^3)$ of this type of hybrid laminate can be derived as follows from the stiffness equation given by Tsai and Hahn:[5]

$$\begin{bmatrix} D^*_{11} \\ D^*_{22} \\ D^*_{12} \\ D^*_{66} \end{bmatrix} = \begin{bmatrix} \bar{U}_1 & W^*_{1c} & W^*_{2c} & W^*_{1s} & W^*_{2c} \\ \bar{U}_1 & -W^*_{1c} & W^*_{2c} & -W^*_{1c} & W^*_{2s} \\ \bar{U}_4 & 0 & -W^*_{2c} & 0 & -W^*_{2s} \\ \bar{U}_5 & 0 & -W^*_{2c} & 0 & -W^*_{2s} \end{bmatrix} \begin{bmatrix} 1 \\ U_{2c} \\ U_{3c} \\ U_{2s} \\ U_{3s} \end{bmatrix} \tag{6}$$

where $\bar{U}_m$ $(m=1,4,5)$ is represented as

$$\bar{U}_m = R_c^3 U_{mc} + (1-R_c^3)U_{ms} \tag{7}$$

and U_{mc} and U_{ms} denote the material constants of the core and the surface. The factor W^*_{ij} $(i=1,2;\ j=\mathrm{c},\mathrm{s})$ is referred to as a flexural lamination parameter of hybrid composites and can be obtained by the following definitions:

$$W^*_{1c} = \sum_{k=1}^{N_c} \xi_k \cos 2\theta_k \qquad W^*_{2c} = \sum_{k=1}^{N_c} \xi_k \cos 4\theta_k$$
$$W^*_{1s} = \sum_{k=N_c+1}^{n} \xi_k \cos 2\theta_k \qquad W^*_{2s} = \sum_{k=N_c+1}^{n} \xi_k \cos 4\theta_k \tag{8}$$

where θ_k represents the fiber orientation angle of the kth ply, and ξ_k means a weighting factor defined by

$$\xi_k = (2Z_k/h)^3 - (2Z_{k-1}/h)^3 \tag{9}$$

where Z_k represents the distance from the mid plane to the upper surface of the kth ply.

In order to investigate the effect on flexural stiffness of hybrid laminates, the allowable region of the flexural lamination parameters should be clarified.

When the number of the kinds of fiber orientation angles in the surface and the core is one in each portion, the lay-up is represented as

$$[\{(\pm\theta_s)_{N_s}\}_{\mathrm{FRP_s}}/\{(\pm\theta_c)_{N_c}\}_{\mathrm{FRP_c}}]_{\mathrm{S}} \tag{10}$$

Then eqn. (8) is reduced to

$$\begin{aligned} W_{1c}^* &= R_c^3 \cos 2\theta_c & W_{2c}^* &= R_c^3 \cos 4\theta_c \\ W_{1s}^* &= (1 - R_c^3)\cos 2\theta_s & W_{2s}^* &= (1 - R_c^3)\cos 4\theta_s \end{aligned} \tag{11}$$

In this case, the following relations hold from a trigonometric identity:

$$W_{2c}^* = 2W_{1c}^{*2}/R_c^3 - R_c^3 \qquad W_{2s}^* = 2W_{1s}^{*2}/(1 - R_c^3) - (1 - R_c^3) \tag{12}$$

When the number of the kinds of fiber orientation angles in the surface and the core is more than one, the allowable regions of the flexural lamination parameters can be obtained from a similar procedure to that for the flexural lamination parameters of non-hybrid laminates,[6] as follows:

$$2W_{1c}^{*2}/R_c^3 - R_c^3 \le W_{2c}^* \le R_c^3 \tag{13}$$

$$2W_{1s}^{*2}/(1 - R_c^3) - (1 - R_c^3) \le W_{2s}^* \le 1 - R_c^3 \tag{14}$$

Figure 1 shows the allowable region of the flexural lamination parameters of hybrid laminates. Any lay-up can be represented by a point in each allowable region.

Constraint Region

The constraint on the buckling load can be considered by using the reduced buckling stress defined in eqn. (4). The reduced buckling stress can be written as follows:[7]

$$(\bar{N}_X)_{cr}^* = (m/r)^2 D_{11}^* + 2(D_{12}^* + 2D_{66}^*) + (r/m)^2 D_{22}^* \tag{15}$$

where m is the half wave number. Substitution of eqn. (6) into eqn. (15)

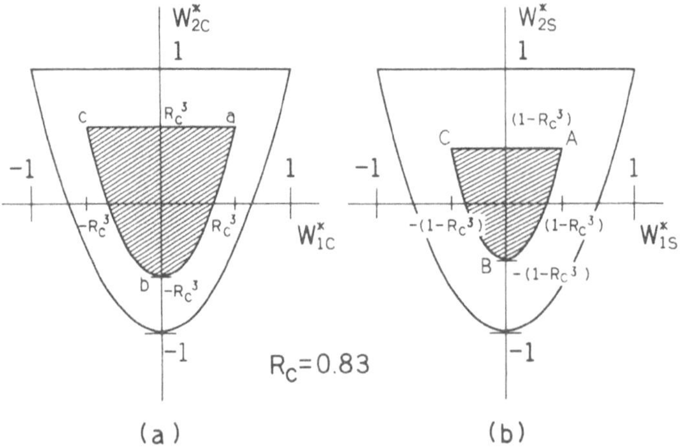

FIG. 1. Feasible regions of hybrid lamination parameters (left: core material; right: surface material).

yields the equation of the contour line of the buckling stress on the flexural lamination parameter diagram of the surface, as follows:

$$W_{2s}^{*} = \{AU_{2s}W_{1s}^{*} + AU_{2c}W_{1c}^{*} - BU_{3c}W_{2c}^{*} + C\bar{U}_1 + 2\bar{U}_4 + 4\bar{U}_5 - (\bar{N}_X)_{cr}^{*}\}/BU_{3s} \quad (16)$$

where

$$A=(m/r)^2-(r/m)^2 \qquad B=6-(m/r)^2-(r/m)^2 \qquad C=(m/r)^2+(r/m)^2 \quad (17)$$

Then the slope of the contour line is obtained as

$$S_s = AU_{2s}/BU_{3s} \quad (18)$$

The half wave number can be assumed to be one because the aspect ratio is assumed to be equal or less than unity:

$$m=1 \quad (19)$$

The constraint or feasible region can be obtained on the flexural lamination parameter diagram by taking the slope of the contour line of the buckling stress and the sign of the factor B in eqn. (16) into account.

The contour line of the normalized buckling stress on the flexural lamination parameter diagram of the surface can be drawn by using eqn. (16) when the values of the core lamination parameters W_{1c}^{*} and W_{2c}^{*} are fixed in their allowable region. Figure 2 shows the feasible region (shaded area) of the flexural lamination parameters of the surface.

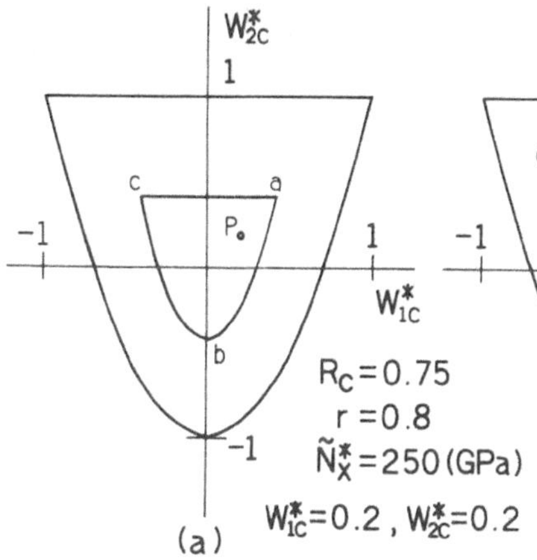

FIG. 2. Feasible design region.

Optimum Lay-up of the Surface

The effect of the core thickness ratio to the feasible region was investigated. It is found that the area of the feasible region decreases when R_c increases, and the region is reduced to a point when the core thickness ratio R_c is maximized, as shown in Fig. 3. The optimum lay-up of the surface can be obtained under this condition.

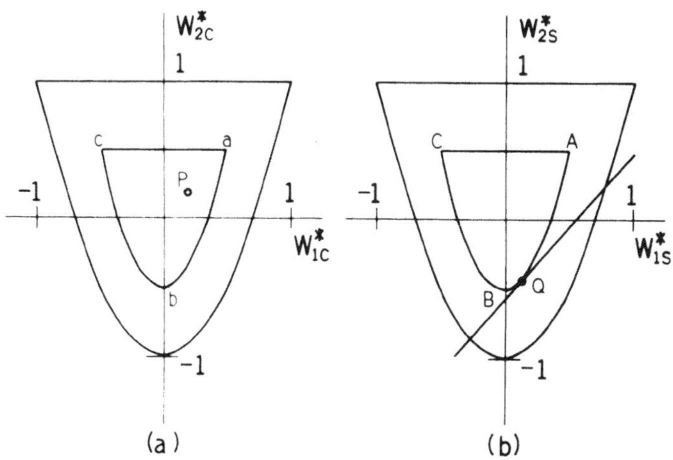

FIG. 3. The state in which the optimum condition is satisfied.

The optimum lay-up of the surface becomes angle-ply because the optimum flexural lamination parameters of the surface are represented by a point on the boundary curve ABC in Fig. 3 and the latter equation in eqn. (12) holds. Then the optimum fiber orientation angle can be obtained as follows.

(i) When $0 < r \leq \sqrt{2} - 1$, the reduced feasible region is represented by point A in Fig. 3. The flexural lamination parameters are obtained from eqn. (14) by

$$W_{1s}^* = 1 - R_c^3 \qquad W_{2s}^* = 1 - R_c^3 \tag{20}$$

The substitution of eqn. (20) into eqn. (11) yields the optimum fiber orientation of the surface as

$$\theta_s = 0 \tag{21}$$

(ii) When $\sqrt{2} - 1 < r \leq 1$, then the following two cases should be considered.

(a) In case $S_s \geq 4$, the reduced feasible region with the maximum value of R_c is represented by point A in Fig. 3. In this case, the optimum orientation is given by eqn. (21).

(b) In case $0 \leq S_s < 4$, the reduced feasible region is represented by a point on the curve AB in Fig. 3. In this case, the lamination parameters are obtained from the condition where the contour line of the buckling stress becomes a tangent line of the boundary curve ABC, as

$$W_{1s}^* = (1 - R_c^3)S_s/4 \qquad W_{2s}^* = (1 - R_c^3)(S_s^2/8 - 1) \tag{22}$$

Consequently, the optimum fiber orientation can be obtained from eqns (11) and (22) as

$$\theta_s = [\cos^{-1}(S_s/4)]/2 \tag{23}$$

From eqns (22) and (23), it is found that the optimum fiber orientation angle of the surface is determined only with the aspect ratio and the material constants of the surface and it does not depend on the core lamination parameters W_{1c}^* and W_{2c}^*.

Optimum Lay-up of the Core and the Core Thickness Ratio

The feasible region on the core lamination parameter diagram can be obtained when the surface lamination parameters W_{1s}^* and W_{2s}^* are fixed in their allowable region in the same way as the case for the surface. When the core thickness ratio R_c increases, the allowable region of the core lamination parameters shown in Fig. 1 expands with a similar shape, but the contour line moves towards the outside of the allowable region. As a result, the feasible region is reduced to a point on the boundary curve of the core lamination parameters when the core thickness ratio R_c is maximized. The optimum lay-up of the core becomes angle-ply in the same way as the surface.

It is found that the optimum fiber orientation angle of the core is determined only with the aspect ratio and the material constants of the surface and it does not depend on the surface lamination parameters W_{1s}^* and W_{2s}^*.

The core thickness ratio of the optimum hybrid laminates can be obtained from the optimum fiber orientations of the surface and the core.

CALCULATED RESULTS

It is assumed that the surface and the core materials are CFRP (T300/5208) and GFRP (Scotchply 1002) respectively. Their material constants are as follows (after Tsai and Hahn[5]):

CFRP (T300/5208)

$$U_1 = 76{\cdot}37 \quad U_2 = 85{\cdot}73 \quad U_3 = 19{\cdot}71 \quad U_4 = 22{\cdot}61 \quad U_5 = 26{\cdot}88 \quad \text{(GPa)}$$

(24)

GFRP (Scotchply 1002)

$$U_1 = 20{\cdot}47 \quad U_2 = 15{\cdot}40 \quad U_3 = 3{\cdot}33 \quad U_4 = 5{\cdot}53 \quad U_5 = 7{\cdot}47 \quad \text{(GPa)}$$

The constraint condition is assumed as

$$(\bar{N}_X)_{cr}^* \geq \tilde{N}_X^* \qquad \tilde{N}_X^* = 250,\ 300,\ 350,\ 400,\ 450 \quad \text{(GPa)} \tag{25}$$

Figure 4 and Table 1 show the optimum lay-ups as a function of the aspect ratio. From these results the optimum fiber orientations of the surface and the core increase rapidly from 0° at the aspect ratio, being about 0·6. It can be seen that the optimum fiber orientation does not depend on the constraint value of the buckling stress. The core thickness ratio decreases when the aspect ratio increases, and decreases when the lower bound of the buckling stress increases. The hybrid laminate cannot be designed for $R_c < 0$.

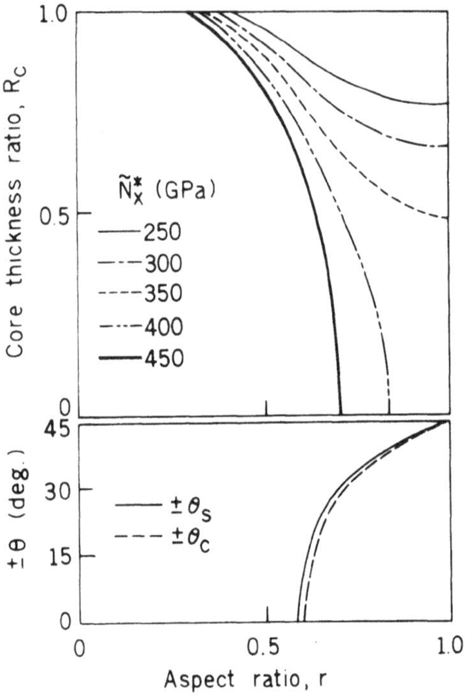

FIG. 4. Results of optimum design.

TABLE 1
Results of optimum design

Aspect ratio (r)	Core thickness ratio (R_c)	Orientation angle in surface $\pm\theta_s$ (deg)	Orientation angle in core $\pm\theta_c$ (deg)
0·1	1·000	0·00	0·00
0·2	1·000	0·00	0·00
0·3	1·000	0·00	0·00
0·4	0·968	0·00	0·00
0·5	0·892	0·00	0·00
0·6	0·778	11·64	6·28
0·7	0·659	30·46	29·45
0·8	0·569	37·34	36·85
0·9	0·511	41·65	41·44
1·0	0·491	45·00	45·00

Material: CFRP (T300/5208)/GFRP (Scotchply 1002)—$\tilde{N}_X^* = 350$ (GPa).

CONCLUSIONS

A new analytical method for designing minimum-cost hybrid laminates subject to axial compression has been investigated and the following conclusions were obtained.

(1) The core thickness ratio R_c is maximized when the contour lines of the normalized buckling stress become tangent lines of the allowable regions of the lamination parameters of the surface and the core. The optimum lay-ups can be obtained from this condition.

(2) The optimum lay-ups for the surface and the core become angle-plied and respective fiber orientations are obtained from the respective material constants and the plate aspect ratios. They do not depend on the constraint value of the buckling stress and the lay-up of the other material.

(3) The core thickness ratio of the optimum hybrid plate can be obtained from the lower bound of the buckling stress, the aspect ratio and the material constant of the surface and the core.

REFERENCES

1. HIRANO, Y., Optimum design of laminated plates under axial compression, *AIAA J.*, **17** (1979), 1017–1019.
2. WANG, J. T. S., Best angles against buckling for rectangular laminates, *Progress in Science and Engineering of Composites* (Hayashi, T. *et al.* eds), Tokyo, Japan Society for Composite Materials 4th Int. Conf. Comp. Mater., 1982, pp. 575–582.
3. FUKUNAGA, H., Compressive buckling characteristics and optimization of laminated plates, *Journal of the Japan Society for Aeronautical and Space Sciences*, **30** (1982), 482–489 (in Japanese).
4. MIKI, M., Optimum design of fibrous laminated composite plates subject to axial compression, *Composites '86: Recent Advances in Japan and the United States* (Kawata, K. *et al.* eds), Tokyo, Japan Society for Composite Materials, Proc. 3rd US–Japan Conf. on Composite Materials, 1986, pp. 673–680.
5. TSAI, S. W. and HAHN, H. T., *Introduction to Composite Materials*, Westport, Connecticut, Technomic, 1980, p. 108.
6. MIKI, M., *Recent Advances in Composites in the United States and Japan*, ASTM STP 864 (Vinson, J. R. *et al.* eds), Philadelphia, PA, American Society for Testing and Materials, 1985, pp. 387–400.
7. JONES, R. M., *Mechanics of Composite Materials*, New York, McGraw-Hill, 1975, p. 261.

28

Optimum Design Based on Reliability for Composite Laminates

YANG LING and MA ZUKANG

Department of Aircraft Engineering, Northwestern Polytechnical University, Xian, People's Republic of China

ABSTRACT

In aircraft structures the loads are mainly axial force and shearing force for panels. However, between the two forces there is a definite ratio for a concrete problem, then they can be seen as a general load. For the skin of a composite wing with multiple wall structure the optimum design based on reliability of laminates under the different values of general loading have been respectively computed. Gradient methods with accelerating convergence are adopted to obtain optimal ratios of layers for laminates under general loading. This optimization problem is subject to main structural system reliability constraint and some secondary constraints (i.e. minimum ratio of thickness, balance and symmetry). The method presented in this chapter is a very efficient one for obtaining optimum design based on reliability of composite structures. It is verified by some illustrative examples.

INTRODUCTION

In aircraft structures, fibre reinforced composites are adopted to decrease structural weight. Because the standard deviations of mechanical properties of composite materials are relatively large and the strength of composite products also depends on technological processes and levels of workers, the research for reliability of composite structures possesses special meaning.

In aircraft structures the loads are mainly axial force and shearing force for panels. However, between these two loads there is a definite ratio for a concrete problem, then they can be seen as a general load. For composite laminates it is important to study optimum design problems based on reliability under a general load.

OPTIMALITY CRITERION FOR LAYING-UP OF LAMINATES BASED ON RELIABILITY

In aircraft composite structures a balanced, symmetrical laminate is used. In general, it consists of four kinds of layers, the orientations of which are $0°$, $90°$ and $\pm 45°$ (a layer includes some plies of same direction). It is known from qualitative analysis that laying-up $0°$ and $\pm 45°$ are useful under axial force (along the $0°$ orientation) and shearing force. However, it is also necessary to consider the requirement of Poisson's ratio and expansion coefficient among composite components; therefore, the $90°$ layer has an additional constraint of limit of minimum thickness ratio which is $6 \sim 10\%$ of the total thickness. For optimum design of laminate based on reliability the number of layers of each orientation, n_i, is taken as a variable; therefore, there are four variables, $n_1, \ldots, n_4$ (corresponding respectively to $0°$, $90°$, $+45°$ and $-45°$).

For reliability optimization the weight of the laminate is taken as the design objective. The objective function to be minimized can be expressed as

$$W = \sum_{i=1}^{4} w_i = \sum_{i=1}^{4} \gamma A t_i \tag{1}$$

where

W = the total weight of the laminate;

w_i = the weight of the i orientation layers;

γ = the density of composite;

A = the area of the laminate;

t_i = the thickness of the i orientation layers.

The only main constraint needed is the laminate structural system reliability constraint. It can be expressed as

$$R_s^* - R_s \leq 0 \tag{2}$$

where

R_s = laminate structural system reliability;

R_s^* = the allowable system reliability.

Here R_s^* is assumed to be known, although it has to be determined in more difficult tasks. Therefore, the Lagrange function is

$$\phi(w_i, \lambda) = W + \lambda(R_s^* - R_s) = \sum_{i=1}^{4} w_i + \lambda(R_s^* - R_s) \tag{3}$$

From $\partial\phi/\partial w_i = 0$, we obtain

$$1 - \lambda \frac{\partial R}{\partial w_i} = 0$$

i.e.

$$\frac{\partial R_s}{\partial w_i} = \frac{1}{\lambda} = \text{constant} \tag{4}$$

This formula (4) is the optimality criterion for the composite laminate. When this criterion is satisfied, it means that the increments of structural system reliability are the same due to increasing unit weight in any orientation layers, and if the optimal ratio of each orientation layer has to be approached, the condition $R_s \approx R_s^*$ has to be satisfied.

For the laminate the number of layers increases or decreases by one ply at least, so the formula (4) corresponds to the same value of reliability increments due to an increase of one ply in each orientation layer.

Let $\partial R_s/\partial n_i$ be the increment of structural system reliability due to increasing one ply in the i orientation layers. Because the weight of each ply is the same, $\partial R_s/\partial n_i$ can be substituted by $\partial R_s/\partial w_i$, and here $\partial R_s/\partial w_i$ is the increase of structural system reliability due to increasing the weight of one ply in the i orientation layers.

ITERATIVE FORMULA OF LAMINATE RELIABILITY OPTIMIZATION

Referring to the criterion method (in physical concept it also is a gradient method with accelerating convergence) presented by Refs 1 and 2, the author of this paper develops the following iterative formula.

For each iteration cycle (k), $w_i^{(k)}$ is taken as the weight of i orientation layers. If i layers are effective in improving structural system reliability, then the weight of i layers is increased at the next step, that is, $w_i^{(k+1)}$ should be directly proportional to $\partial R_s/\partial w_i$; the term $\Delta R_s/\Delta w_i$ can be approximately substituted for $\partial R_s/\partial w_i$. Similarly, $w_i^{(k+1)}$ should be directly proportional to $R_s^*/R_s^{(k)}$, where $R_s^{(k)}$ is system reliability at the kth iteration cycle. Then the iterative formula can be expressed as

$$w_i^{(k+1)} = w_i^{(k)} \left(\frac{\Delta R_{s,\Delta w,i}^{(k)}}{\overline{\Delta R_{si}^{(k)}}} \right)^{\gamma_1^{(k)}} \left(\frac{R_s^*}{R_s^{(k)}} \right)^{\gamma_2^{(k)}} \tag{5}$$

where subscript $\Delta w,i$ is the weight increased one ply in i orientation layers, and $\Delta R_{s,\Delta w,i}^{(k)}$ is the increment of structural system reliability in increasing one ply in i orientation layers at the kth step (i.e. $\Delta R_{s,\Delta w,i} = \Delta R_s/\Delta w_i$). The $\overline{\Delta R_{si}^{(k)}}$ is the mean value of increments of structural system reliability due to increasing each orientation layer respectively at the kth step, thus

$$\overline{\Delta R_{si}^{(k)}} = \frac{1}{4} \sum_{i=1}^{4} \Delta R_{s,\Delta w,i}^{(k)} = \overline{\left(\frac{\Delta R_s}{\Delta w_i} \right)}$$

and $\Delta R_{s,\Delta w,i}^{(k)}/\overline{\Delta R_{si}^{(k)}}$ is the sensitivity coefficient of i orientation layers. The $\gamma_1^{(k)}, \gamma_2^{(k)}$ are exponents to assist in controlling the convergence.

As mentioned before for 90° layers, constraint of limit of minimum thickness ratio needs to be satisfied. It can be expressed as

$$\frac{n_2}{\sum_{i=1}^{4} n_i} \geq 6 \sim 10\% \tag{6}$$

In addition, if it is necessary to lay-up ±45° layers in the form of balance and symmetry to avoid the coupling effect, a reasonable quantitative difference between +45° and −45° layers is determined less than or equal to 2 in the process of iteration and the number of +45° and −45° layers are even numbered respectively. It can be expressed as

$$|n_3^{(k)} - n_4^{(k)}| = 0 \text{ or } 2 \tag{7}$$

Formulae (6) and (7) are called secondary constraints (see Figs 1 and 2).

It is obvious that the ratio of reliability and sensitivity coefficients is considered as the standard for terminating the iteration calculation. The ratio of total weight change also can be considered as a parameter.

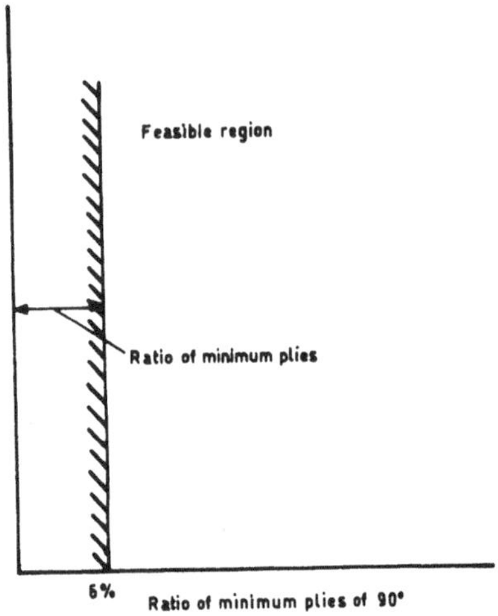

FIG. 1. The constraint of ratio of minimum plies.

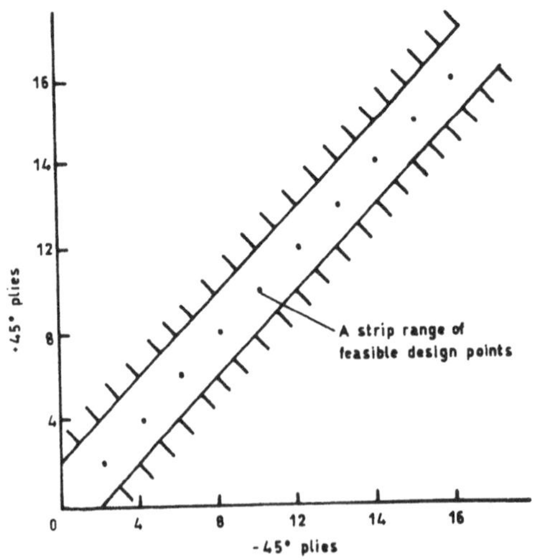

FIG. 2. The constraint of balance and symmetry for ±45° layers.

Therefore, one obtains

$$0{\cdot}9995 \leq R_s^*/R_s \leq 1{\cdot}0005 \tag{8a}$$

$$0{\cdot}95 \leq \frac{\Delta R_s}{\Delta w_i} \Bigg/ \left(\overline{\frac{\Delta R_s}{\Delta w_i}}\right) \leq 1{\cdot}05 \tag{8b}$$

$$d^{(k+1)} = \left| \frac{W^{(k+1)} - W^{(k)}}{W^{(k)}} \right| \leq 5\% \tag{9}$$

Because in the following examples the convergence is very fast, the weight of each orientation layer is taken as a round number at each iterative cycle. Their sum is total weight, $W^{(k)}$.

In general, the above treatment is used when the total layer numbers of the laminate are large. If total numbers are not so large, then it used to be taken in round numbers after three or four iterations, or after the terminating iteration.

Finally, after the iteration is terminated, in the light of specific conditions it is decided whether there is a need to exactly check the reliability R_s.

ILLUSTRATIVE EXAMPLES FOR COMPOSITE LAMINATE LAYING-UP

Some illustrative examples are given for explaining the method of optimum design based on reliability of composite laminates.

For the skin at a lower panel of composite wing with multiple wall structures under the different values of general loading, the optimization procedures have been respectively computed. From statistical data two groups of general loading, i.e. the different ratios between axial force and shearing force, are used: $\{1, 0, 0{\cdot}2214\}$ and $\{1, 0, 0{\cdot}1771\}$. Though these two groups of loading are not loaded at the same time, they are applied to the same wing. Therefore, the optimum design of the laminate is calculated under both these two single groups of loading individually applied and under the combined two groups of loads.

In this chapter the composite laminate is made of Graphite Epoxy T300/648. The material properties are as follows:

$E_1 = 12{\cdot}3 \times 10^5$ kg/cm^2 $\qquad E_2 = 0{\cdot}8045 \times 10^5$ kg/cm^2

$G_{12} = 0{\cdot}375 \times 10^5$ kg/cm^2 $\qquad \nu = 0{\cdot}33$

$X_t = 14\,240$ kg/cm^2 $\qquad X_c = 10\,500$ kg/cm^2

$Y_t = 346$ kg/cm^2 $\qquad Y_c = 1700$ kg/cm^2

$S = 668$ kg/cm^2 $\qquad t = 0{\cdot}0125$ cm (a ply thickness)

The composite laminate is of anisotropic, brittle laminated construction. The authors of this chapter, according to this feature, present a method of full quantity loading for computing the reliability of composites. For the composite laminate and other brittle materials, the advantage of this method is that computing is more effective and its concept is more clear than the present incremental loading method.[3]

For computing structural system reliability of laminates, the normal distribution is used to simulate actual load and strength probability density distributions. Let the coefficient of variation of load and strength be 0·2 and 0·1, respectively. The allowable system reliability R_s^* is taken as 0·999.

For the calculation method of the system failure probability P_{sf}, refer to Refs 4–6.

Example 1

This shows the reliability optimization design of laminates under a general loading {1, 0, 0·2214}.

The 0° direction is taken as the X-axis of the laminate. {1, 0, 0·2214} means that the ratio between axial tensile load N_x and shearing load N_{xy} is 1 to 0·2214, and the load in the Y direction equals zero.

As a matter of experience, the exponents $\gamma_1^{(k)}$ and $\gamma_2^{(k)}$ are taken as 1/2 and 1·0, respectively.

From Table 1 the optimum number of plies of each orientation layer and ratio of thickness of each layer can be obtained (see second column in brackets in Table 1).

There is a definite relationship between +45° and −45° so that the changes of numbers of plies for 0° and +45° orientation layers are illustrated (Fig. 3). The changes of weight of laminate, sensitivity coefficient of 0° and +45° layers, and R_s/R_s^* with iteration number are respectively illustrated in Figs 4–6.

Example 2

This gives the reliability optimization design of laminates under a general loading {1, 0, 0·1771}.

The process of iteration and the optimal solution are shown in Table 2 and Figs 7–10.

Example 3

This shows the reliability optimization design of laminates under the combined two groups of loads.

When there are two groups of loads applied, not simultaneously, the

TABLE 1

Calculation for Example 1

Term	Iteration number		
	0	1	2
$n_1^{(k)}$ (plies)	10	20	21 (41·18%)
$n_2^{(k)}$	10	4	4 (7·84%)
$n_3^{(k)}$	10	14	14 (27·45%)
$n_4^{(k)}$	10	14	12 (23·53%)
$t^{(k)}$ (cm)	0·500 0	0·650 0	0·637 5
$W^{(k)}$ (kg)	0·080 0	0·104 0	0·102 0
$R_s^{(k)}$	0·611 392	0·999 257	0·999 258
$(\Delta R_s/\Delta w_1)^{(k)}/\overline{(\Delta R_s/\Delta w_i)}^{(k)}$	1·552 23	1·091 84	1·027 30
$(\Delta R_s/\Delta w_2)^{(k)}/\overline{(\Delta R_s/\Delta w_i)}^{(k)}$	0	0	0
$(\Delta R_s/\Delta w_3)^{(k)}/\overline{(\Delta R_s/\Delta w_i)}^{(k)}$	1·310 8	0·908 16	0·972 71
$(\Delta R_s/\Delta w_4)^{(k)}/\overline{(\Delta R_s/\Delta w_i)}^{(k)}$	0·137 38	~0	0
R_s^*/R_s	1·633 976	0·999 743	0·999 742
$\sum_{i=1}^{4} n_i$ (plies)	40	52	51
$d^{(k)}$		30%	1·923%

TABLE 2

Calculation for Example 2

Term	Iteration number				
	0	1	2	3	4
$n_1^{(k)}$ (plies)	10	17	21	21	21
$n_2^{(k)}$	10	4	4	4	4
$n_3^{(k)}$	10	8	10	12	12
$n_4^{(k)}$	10	8	10	10	12
$t^{(k)}$ (cm)	0·500 0	0·462 5	0·562 5	0·587 5	0·612 5
$W^{(k)}$ (kg)	0·080	0·074	0·090	0·094	0·098
$R_s^{(k)}$	0·661 02	0·944 44	0·997 50	0·999 21	0·999 62
$(\Delta R_s/\Delta w_1)^{(k)}/\overline{(\Delta R_s/\Delta w_i)}^{(k)}$	1·250 87	1·275 10	0·929 00	0·989 00	1·053 30
$(\Delta R_s/\Delta w_2)^{(k)}/\overline{(\Delta R_s/\Delta w_i)}^{(k)}$	0	0	0	0	0
$(\Delta R_s/\Delta w_3)^{(k)}/\overline{(\Delta R_s/\Delta w_i)}^{(k)}$	0·749 14	1·408 64	1·025 68	0·899 70	0·946 70
$(\Delta R_s/\Delta w_4)^{(k)}/\overline{(\Delta R_s/\Delta w_i)}^{(k)}$	~0	0·031 61	1·045 23	1·110 00	
R_s^*/R_s	1·511 31	1·057 77	1·001 50	0·999 79	0·999 38
$\sum_{i=1}^{4} n_i$	40	37	45	47	49
$d^{(k)}$		7·5%	21·6%	4·44%	4·26%

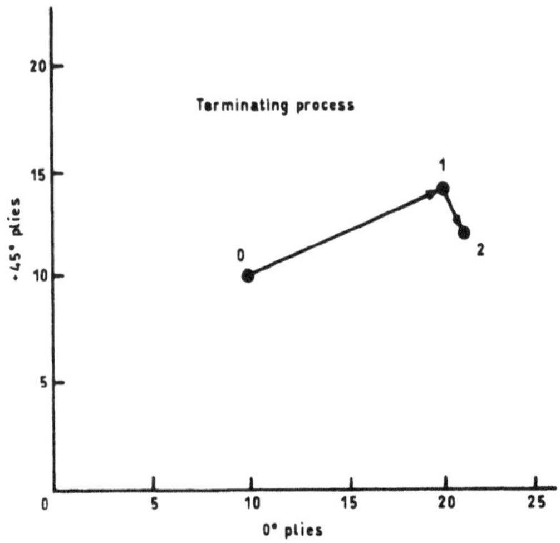

FIG. 3. The changes of plies for 0° and +45° layers for Example 1.

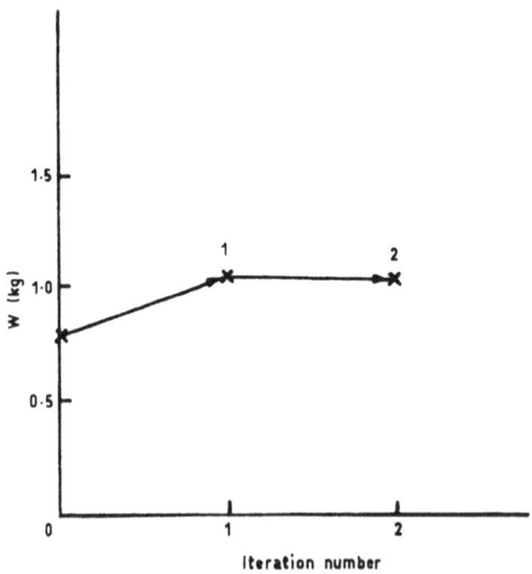

FIG. 4. The changes of weight of laminate for Example 1.

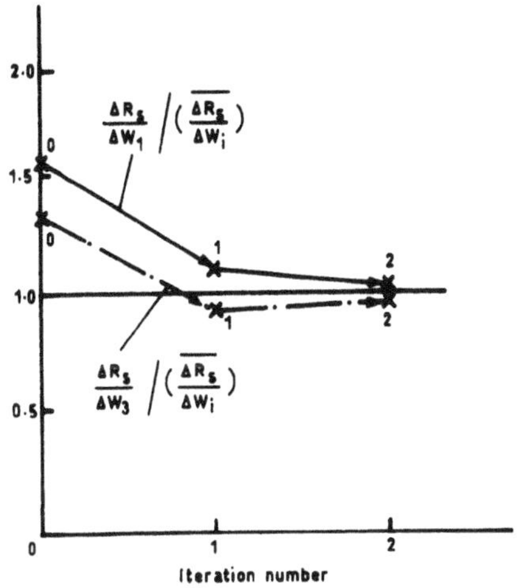

FIG. 5. The sensitivity coefficient of 0° and +45° layers for Example 1.

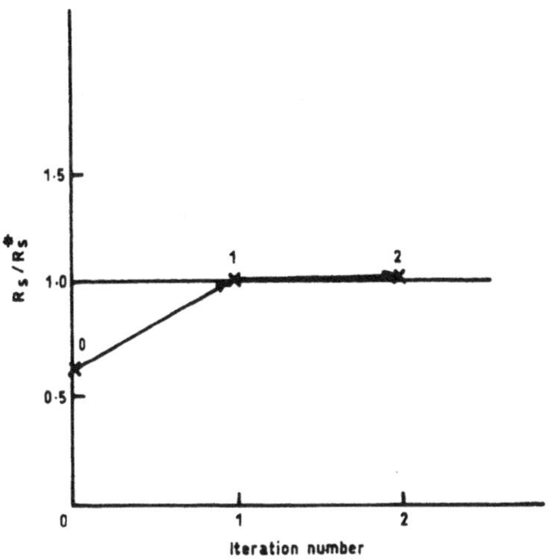

FIG. 6. The changes of R_s/R_s^* for Example 1.

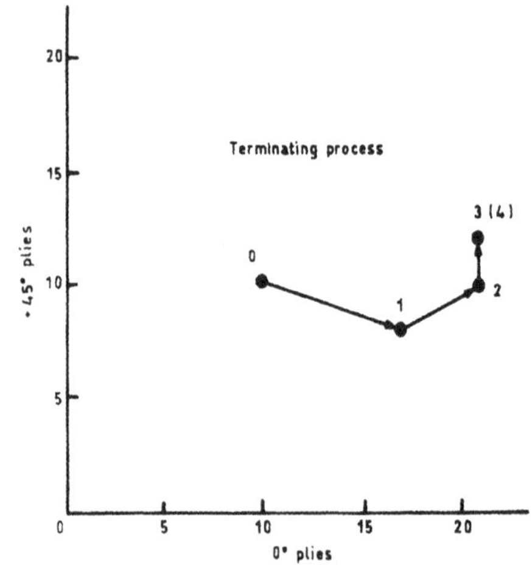

FIG. 7. The changes of plies for 0° and +45° layers for Example 2.

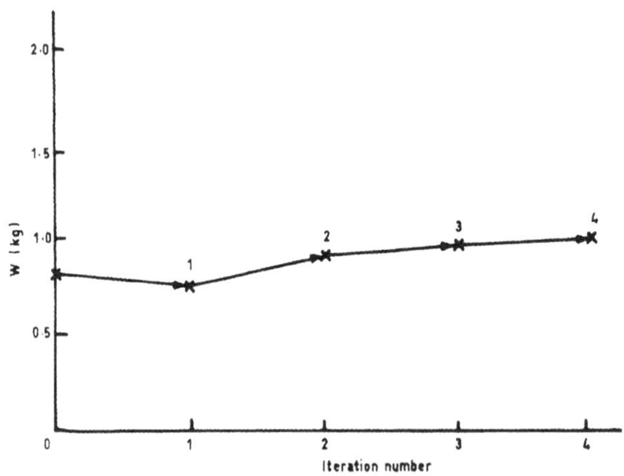

FIG. 8. The changes of weight of laminate for Example 2.

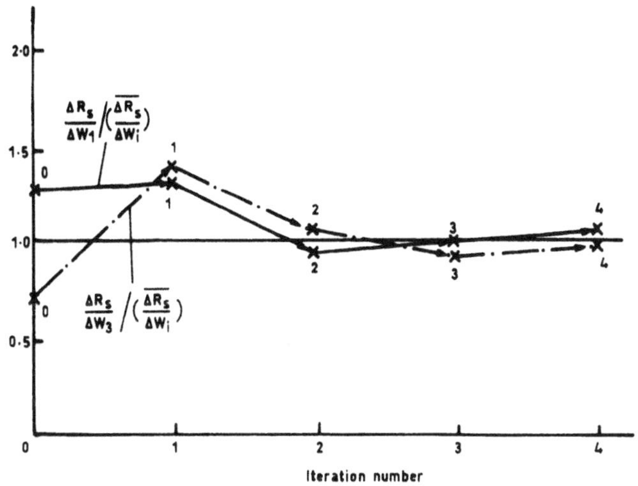

FIG. 9. The sensitivity coefficient of 0° and +45° layers for Example 2.

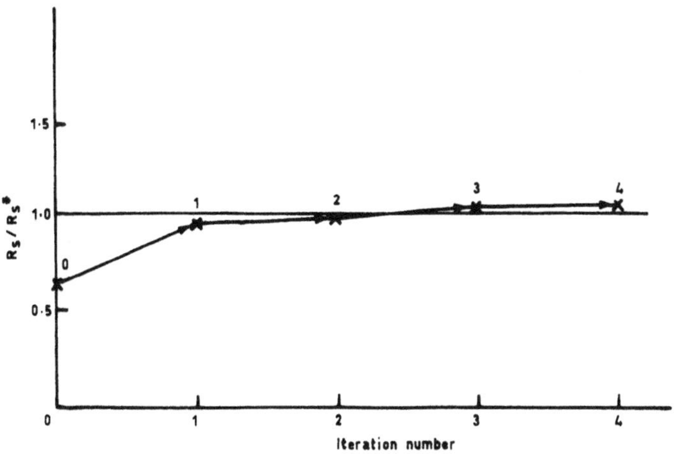

FIG. 10. The changes of R_s/R_s^* for Example 2.

values of system reliability R_s/R_s^* are needed to check their magnitudes under each group of loading. When the values of R_s/R_s^* deviate from 1·0 too much, the iteration for different groups of loading may be individually carried out. When the values of R_s/R_s^* are near to 1·0 at the kth iteration cycle, the method of combined solution is adopted subsequently.

For the combined solution, refer to Refs 1 and 2. Let general variables n_{i1} and n_{i2} be formed by each group of solutions individually (the ratios of each orientation layer are known), then the combined solution n_i is a linear combination of them, i.e.

$$n_i^{(k)} = c_1^{(k)} n_{i1}^{(k)} + c_2^{(k)} n_{i2}^{(k)} \tag{10}$$

and in order to keep to a suitable weight, let

$$c_1^{(k)} + c_2^{(k)} = 1 \quad \text{or} \quad c_1^{(k)} = 1 - c_2^{(k)} \tag{11}$$

Using the symbol c instead of c_1, then

$$n_i^{(k)} = c^{(k)} n_{i1}^{(k)} + (1 - c^{(k)}) n_{i2}^{(k)} \tag{12}$$

By the aid of the unidimensional search method a reasonable value of $c^{(k)}$ can be determined.

In this paper three conditions are adopted to estimate a reasonable initial point value of $c^{(k)}$.

(1) The total thickness of laminate $n_t^{(k)}$ ($n_t^{(k)} = \sum_{i=1}^{4} n_i^{(k)}$) approaches the same value at the kth iteration cycle under two groups of loads individually. The plies of each layer in these two conditions, either large or small, thereafter combine according to the plies of each layer. Thus, we have

$$n_i^{(k+1)} = 0{\cdot}5 n_{i1}^{(k)} + 0{\cdot}5 n_{i2}^{(k)} \tag{13}$$

that is, $c^{(k)} = 0{\cdot}5$.

There is a need for continuing to the $(k+1)$th iteration under different single groups of loads individually. The laminate consists of each orientation layer as obtained above, again and again, until the standard of termination is satisfactory.

(2) If the plies of each layer under a group of loads are overall larger than or equal to one another under another group of loads at the kth iteration, then $c^{(k)}$ is taken as 1·0:

$$n_i^{(k+1)} = n_{i1}^{(k)} \qquad (\text{if } n_{i1} \geq n_{i2}) \tag{14}$$

(3) If the plies of some layers under a group of loads are approximately equal to one another under another group of loads but the plies of the

remaining layers are larger than others at the kth iteration, then $c^{(k)}$ is close to 1·0.

Because the difference between these two groups of loads, $\{1, 0, 0{\cdot}2214\}$ and $\{1, 0, 0{\cdot}1771\}$, is not large at the kth iteration, the plies of each layer are overall larger than or equal to one another. This is the second condition mentioned above, that is, $c = 1{\cdot}0$. Then laminate $\{21, 4, 14, 12\}$ is the result of reliability optimization under both groups of loads not applied simultaneously. The reliability indexes R_s of this laminate are as follows:

$$R_s^{(1)} = 0{\cdot}999\,26 \qquad R_s^*/R_s^{(1)} = 0{\cdot}999\,74$$

$$R_s^{(2)} = 0{\cdot}999\,73 \qquad R_s^*/R_s^{(2)} = 0{\cdot}999\,27$$

where superscripts (1) and (2) represent the two groups of loads, respectively.

CONCLUSIONS

(1) When the shearing load applied to the laminate is positive, the total failure of the laminate is caused by $-45°$ layers. And this feature is also shown from the sensitivity coefficient of each orientation layer. Hence, from the increasing system reliability point of view the $-45°$ layers should be possibly reduced. However, because there is a requirement for a balanced and symmetrical laminate, $+45°$ and $-45°$ layers should ensure matching. The requirement for match is shown by formula (7) (under this condition the laminates are symmetrical but not entirely balanced). According to this additional constraint, the ratio of sensitivity coefficient $(\Delta R_s/\Delta w_i)/(\Delta R_s/\Delta \bar{w}_i)$ will be close to but not exactly equal to 1·0.

(2) The plies of the $\pm 45°$ layers fall in a strip range of design points, as shown in Fig. 2. After terminating the iterative calculation, there are still other optimal laminate solutions which satisfy the requirement of reliability index near the selected one, e.g. in Example 1 $\{21, 4, 12, 12\}$ (49 plies) and $\{20, 4, 14, 12\}$ (50 plies) are the choice. The thickness of each layer may be adjusted in 1–2 plies; finally, there is a need to check the reliability if the index is met. During the iteration process the results of the kth iteration may be also regulated in 1–2 plies; after analysis a better laminate is obtained to iterate for the $(k+1)$th cycle. The iterative cycle can be reduced in this way.

(3) As before, the minimum thickness of the 90° layers should be ensured; from computing results it is known that increasing the plies of 90° layers is not useful for improving structural system reliability, i.e. the

sensitivity coefficient $(\Delta R_s/\Delta w_2)^{(k)}/(\overline{\Delta R_s/\Delta w_i}) \approx 0$. Therefore, they need not join iteration but need only check if the requirement of minimum volume ratio is satisfied.

(4) From these examples it is shown that the terminating condition can be satisfied by 3–4 cycles, and no requirements are presented for initial value selected, so this method can be used by inexperienced people. These advantages show that the iterative formula for the gradient method with accelerating convergence is very efficient and the values of exponents γ_1 and γ_2 selected are also reasonable.

REFERENCES

1. FENG, Y. S. and MOSES, F., A method of structural optimization based on structural system reliability, *J. Struct. Mech.*, **14**, No. 4 (1986).
2. FENG, Y. S., A new method of configuration optimization for large-scale structures, 16th ICTAM, 1984.
3. MOSES, F. and RASHEDI, M. R., The application of system reliability to structural safety, Fourth Int. Conf. on Application of Statistics and Probability in Soil and Structural Engineering, Università di Firenze, Italy, 1983, Pitagora Editrice.
4. DITLEVSON, O., Narrow reliability bounds for structural systems, *J. Struct. Mech.*, **I**(4) (1979).
5. STEVENSON, J. D., Reliability analysis and optimum design of structural systems with applications to rigid frames, Division of Solid Mechanics and Structures, Report No. 14, Case Western Reserve University, Cleveland, Ohio, 1967.
6. THOFT-CHRISTENSEN, PALLE and BAKER, MICHAEL J., in: *Structural Reliability Theory and its Application*, Berlin, Heidelberg, New York, Springer-Verlag, 1982.

29

Multilevel Optimization of Composite Structures

R. I. WATKINS

National Institute for Aeronautics and Systems Technology,
CSIR, PO Box 395, Pretoria 0001, South Africa

ABSTRACT

A multilevel optimization scheme for large laminated composite structures, which allows the use of ply thickness and orientation as variables, is described. The lower level of optimization uses a multi-criteria objective function which combines a weight function and a strain energy change function into a utility function which is minimized and in which the relative importance of each part is reflected by weighting coefficients. Minimizing the change in strain energy ensures load path continuity in the overall structures when switching between upper and lower levels of optimization, and so decouples the problems at the two levels. This system has been used to optimize various aircraft-type structures using strain, buckling, displacement and gauge constraints, and an example is given to illustrate the effectiveness of the multilevel optimization scheme.

1. INTRODUCTION

Laminated composite structural elements may be constructed from multiple layers of orthotropic materials of different thicknesses and orientated at various angles. Even a relatively simple composite plate therefore has many more design variables associated with it (two per layer) than a comparable isotropic plate which has total thickness as its only design variable. As a result, the magnitude of structural design problems can increase dramatically with the use of composite materials. This, combined with the highly nonlinear way in which strains and deflections vary with ply orientation, makes the design problem so complex that normal design intuition cannot always be considered to be a reliable guide and simple sizing formulae become inaccurate and ineffective.

The use of some modern mathematical optimization technique combined with a composite structural analysis method therefore appears to be an attractive solution to the problem. The inordinately large number of design variables and related constraints associated with composite structures, however, may lead to problems that cannot readily be solved using existing structural optimization algorithms and software (developed with isotropic structures in mind). These difficulties are the most probable reasons for relatively few researchers having been attracted to the problem of optimal design of large multi-element laminated composite structures. A few large-scale composite structural synthesis systems have been described by Khot,[1] Starnes and Haftka,[2] Schmit and Mehrinfar[3] and Sobieszczanski-Sobieski.[4] All these systems, however, suffer from limited design flexibility and capability in that full design variable freedom is not allowed (only layer thickness and not orientation is varied[1-3]) or the interaction of all the design variables and constraints is not properly taken into account.[4] Computational and numerical restrictions are probably at least partially responsible for these limitations. A multilevel optimization scheme which overcomes these problems is described in this chapter.

2. PROBLEM FORMULATION

In order to avoid the problem of size and potential instability of the optimization problem (due to the many nonlinear functions) an approach needs to be adopted where the number of design variables under consideration at any point should be kept as low as possible. Furthermore, freedom of all design variables should only be allowed when optimizing small substructures which are well defined in terms of constraints and loading.

The use of multilevel optimization is particularly suitable in this case as it can be used to satisfy both of the above requirements. The upper level (or system level) of optimization is performed considering the entire structure and uses only the layer thicknesses of the laminated structural components as the design variables, i.e. the ply angles and the number of layers is kept fixed.

The lower level (or element level) of optimization considers only the

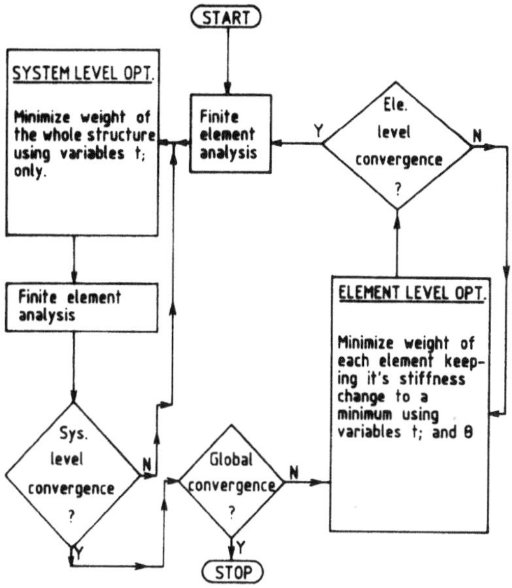

FIG. 1. Multilevel design logic.

individual elements where the major constraint is that the change of stiffness of the element should be kept to a minimum as weight is reduced. This ensures that the stiffness, and hence the load paths, in the overall structure do not change substantially, so preserving the continuity when switching back to the upper level of optimization. The system and element level problems are effectively decoupled in this manner, and the loads in any given element can thus be assumed to remain the same irrespective of the change in the design variables during the lower level of optimization. This constraint, together with strain, buckling and gauge constraints, lead to a well-defined problem at the element level and so full freedom of the design variables (layer thickness and ply angle) can be allowed.

A schematic description of the multilevel approach is given in Fig. 1. Results are obtained by iterating between system and element level problems.

Expressed more formally, the composite structure optimization problem can be written in general terms as:

Min $W(t)$

subject to: (i) $c(\theta, t) \leq 0$

(ii) $\theta^{l} \leq \theta \leq \theta^{u}$

(iii) $t^{l} \leq t \leq t^{u}$

where $W(t)$ is the total structural weight; $c(\theta, t)$ are the constraints; θ, t are the ply angle and thickness variables respectively; and l, u are superscripts indicating the lower and upper bounds respectively.

When expressed as a multilevel optimization problem it can be written as:

System level:
Find t such that
Min $W(t)$
subject to: (i) $G(t) \leq 0$
(ii) $t^{\text{l}} \leq t \leq t^{\text{u}}$

Element level:
Find t and θ such that
Min $w(t)$
subject to: (i) $\Delta k \rightarrow 0$
(ii) $g(\theta, t) \leq 0$
(iii) $\theta^{\text{l}} \leq \theta \leq \theta^{\text{u}}$
(iv) $t^{\text{l}} \leq t \leq t^{\text{u}}$

where $G(t)$ and $g(\theta, t)$ are the constraints applicable at the system and element level respectively; $w(t)$ is the element weight; and Δk is the change in element stiffness.

The two levels of optimization are described in more detail below.

3. ELEMENT LEVEL OPTIMIZATION

In order to obtain stable convergence of the overall design procedure, the behaviour of the structure, and the load paths within it, should not be altered significantly when switching from one optimization level to another. This can be achieved by requiring that the stiffness change at the element level optimization be kept to a minimum.

The stiffness change of the element can thus be used as a constraint (with relatively tight move limits) at the lower level of optimization or alternatively can be used as the objective function to be minimized.

The use of stiffness change alone, however, as the objective function will not generally be sufficient to drive the design to an optimum. This is particularly well illustrated in the case where the element satisfies all the constraints as there is then no incentive to change the design (and so reduce the volume of material), since the stiffness change will then be zero, i.e. a minimum. Stiffness change and weight are thus combined in a multi-criteria objective function that is minimized. The inclusion of weight as a part of the objective function ensures that both layer thicknesses and lay-up angles will be used such that a design will be obtained that achieves a good compromise of minimized weight and stiffness change, while satisfying all the constraints.

A number of ways of quantifying the change in element stiffness were evaluated[3,5] but using the change in element strain energy seemed to produce the most satisfactory results. This was also thought to provide a more accurate gauge of the load continuity at the upper optimization level, since both element strains and stiffness are taken into account. The element strain energy is evaluated as

$$U = \{\varepsilon_{p1}\}^T[R]\{\varepsilon_{p1}\} \tag{1}$$

and hence the change in strain energy is given by

$$\Delta U = \{\varepsilon_{p1}\}^T[R]\{\varepsilon_{p1}\} - U^* \tag{2}$$

where $\{\varepsilon_{p1}\}$ and $[R]$ are the element strains and laminate rigidity matrix respectively, as defined in classical lamination theory, and U^* is the strain energy of the element on entry to this level of optimization.

The multi-criteria objective function was formed using the basic weighted objectives method[6] of combining objective functions since this method does not intrinsically require information on the individual functions' optima. All the objective functions are added together using different weighting coefficients for each. The multi-criteria optimization problem is thereby transformed to a scalar optimization problem by creating one function of the form

$$f(x) = \sum_i q_i f_i(x) c_i \tag{3}$$

where $f_i(x)$ are the original objective functions, x are the variables, q_i are the weighting coefficients representing the relative importance of the criteria ($q_i \geq 0$ and $\sum_i q_i = 1$), and c_i are constant multipliers.

Note that the weighting coefficients q_i do not reflect proportionally the relative importance of the objectives but are only factors which, when varied, would locate different points in the design space. The q_i can, however, be made to reflect closely the relative importance of the objective functions if the factors c_i are used to ensure that the objective functions have numerical values of the same order. It has been suggested[6] that the best results are usually obtained if $c_i = 1/f_i^o$, where f_i^o is the ideal optimum of the objective function f_i within the bounds of the prescribed constraints.

In order to avoid the additional computational effort associated with a separate evaluation of the f_i^o value for the weight function and the problem of $c_i = \infty$ (since $f_i^o = 0$) for the strain energy change part of the objective function, certain assumptions are made. The f_i^o value for the weight function is assumed to be the element weight obtained at the system level optimization iteration immediately prior to the element level optimization. The f_i° value for the strain energy change function is taken to be the strain energy of the element upon entry to that level of optimization.

The total objective function thus has the form

$$f(x) = q_1 f_1(x)/W + q_2 f_2(x)/\Delta U \tag{4}$$

where W and ΔU are the f_i° values for the weight and strain energy change as described above. This form of eqn. (3) ensures that the separate objective functions are dimensionless and will have similar numerical values and thus neither should dominate the solution, unless vastly different weighting coefficients are assigned to the two parts.

The constraints considered in the element level optimization are higher mode buckling (to be prevented), strain (with upper and lower bounds), minimum lamina thickness and bounds on the lay-up angles.

4. RESULTS OF ELEMENT LEVEL OPTIMIZATION

A rectangular box section, which can be considered to be representative of a wing box, is used to demonstrate the suitability of the form of the multi-criteria objective function (given above) and the effect on the optimum of varying the weighting coefficients in the objective function. The layout and dimensions of the box section are shown in Fig. 2.

The construction of all the elements in the lower skin was defined to be of laminate type 1 (i.e. design variable linking was used), all the elements in the upper skin of laminate type 2, and all the elements in the webs and ribs of laminate type 3. The various laminate types are defined in Table 1, with

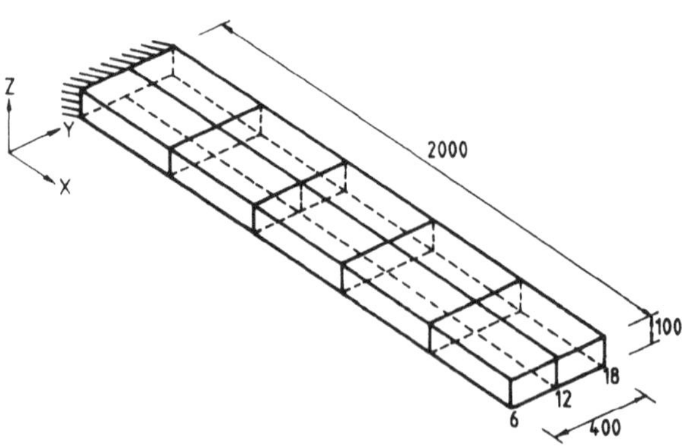

FIG. 2. Rectangular box beam.

TABLE 1

Laminate type	Layer number	Thickness (mm)	Ply angle (degrees)
1	1	2·5	0·0
	2	1·25	−45·0
	3	1·25	45·0
2	1	2·0	0·0
	2	2·0	−45·0
	3	2·0	45·0
3	1	1·0	−45·0
	2	1·0	45·0

TABLE 2

	Load 1			Load 2		
Node number	6	12	18	6	12	18
Force (N)	10 000	20 000	10 000	10 000	15 000	20 000

only the upper half of the symmetric lay-up being given (layer 1 is uppermost). All the layers were assigned the following material properties: $E_L = 130\,\text{GPa}$, $E_T = 9\,\text{GPa}$, $G_{LT} = 4{\cdot}8\,\text{GPa}$, $\nu = 0{\cdot}28$, and the longitudinal and transverse strains were limited to 0·004 while the shear strain was limited to 0·0055. The structure was subjected to the loads applied at the tip nodes, as given in Table 2, and was optimized using varying weighting coefficients in the objective function. The results are presented in Fig. 3.

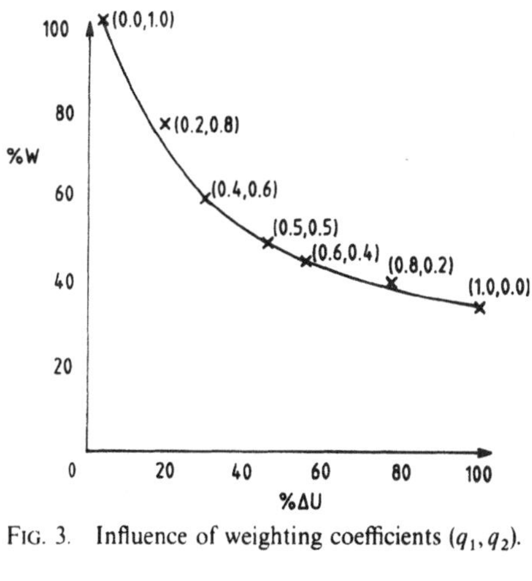

FIG. 3. Influence of weighting coefficients (q_1, q_2).

The results show how effectively the stiffness change of the structure can be controlled by minimizing the strain energy change in the various laminate types. This capability combined with the smooth, progressive transition to the other extreme (pure weight minimization) for all the intermediate weighting coefficients demonstrates the viability, and flexibility, of the form of objective function developed above and also the effectiveness of the weighting coefficients in representing the relative importance of the two parts of the multi-criteria objective function.

5. SYSTEM LEVEL OPTIMIZATION

A single criteria objective function, that of minimizing total structural weight, is used at the system level (as opposed to the multi-criteria objective function at the element level), and is written as

$$W(t) = \sum_{j}^{NEL} \sum_{i}^{L} (\rho A t)_{ij}$$

where NEL is the total number of elements in the finite element model, L is the number of layers in element j, A is the surface area of element j, and ρ and t are the density and thickness of layer i in element j respectively.

The constraints considered at the system level of optimization were primary mode panel buckling, strain and displacement limits on the structure and bounds on the lamina thicknesses.

6. RESULTS OF MULTILEVEL OPTIMIZATION

6.1. Influence of Weighting Coefficients Used at the Element Level

Numerous test cases were run using structures representative of delta wings, forward swept wings and straight rectangular wings and these all showed that the final results (in terms of weight and design variable values) and the speed of convergence of the multilevel optimization are relatively insensitive to the weighting coefficients used in the element level objective function. In the few cases where there were relatively large differences in the final ply angles chosen the related thicknesses were generally small relative to the entire laminate thickness and so their effect on the overall laminate stiffness (rigidity) was very small indeed. Since the final design variables for each test case were similar, the displacements of the loaded structure were, as could be expected, very similar as well.

6.2. Example—Multilaminate Rectangular Box Beam

The results obtained using the multilevel system of optimization have been very satisfactory and have proved to be both feasible and realistic. The example of a multilaminate rectangular box beam (representative of an aircraft wing box) given below demonstrates this. This example was chosen as the results are readily explained using simple design logic.

The general layout of the structure is the same as that shown in Fig. 2 and the material properties, strain limits and loads applied are the same as that given in the example on element level optimization. In this example, however, a greater number of laminate types (i.e. design variables) were considered and these are defined in Table 3. Design variable linking was used to define that the four root end elements in the top skin were of laminate type 1, the four root end elements of the top skin were of laminate type 2, the remaining bottom skin of laminate type 3 and the remaining top skin of laminate type 4. The shear webs (spars) and ribs were all defined to

TABLE 3
Multilaminate rectangular box results

Laminate type	Initial design		Final design	
	t (mm)	θ (degrees)	t (mm)	θ (degrees)
1. Layer 1	2·5	0·0	1·98	−4·2
2	1·25	−45·0	0·83	−50·1
3	1·25	45·0	0·02	47·1
2. Layer 1	2·5	0·0	2·07	−1·9
2	1·25	−45·0	1·31	−40·4
3	1·25	45·0	0·05	59·7
3. Layer 1	2·0	0·0	0·60	21·2
2	2·0	−45·0	1·71	−55·0
3	2·0	45·0	0·01	32·2
4. Layer 1	2·0	0·0	0·29	30·9
2	2·0	−45·0	1·95	−53·8
3	2·0	45·0	0·01	27·8
5. Layer 1	1·0	−45·0	1·41	−47·4
2	1·0	45·0	1·30	41·4
Weight (kg)	329·7		183·6	

be of laminate type 5. Linking was also used to ensure laminate symmetry about the midplane and the lay-ups given in Table 3 thus only define the one half of the laminate (layer 1 is the furthest from the midplane). Finally, a displacement limit of ± 120 mm was placed on all the tip nodes and all forms of buckling were constrained.

The initial and final converged designs are given in Table 1. The convergence criteria used was a weight change of less than 2% and this was achieved after three iterations between the two levels of optimization.

The ply orientations are given with respect to the element local x-axes which all lie parallel to the spanwise lines shown in the finite element grid (Fig. 2).

Laminates 1 and 2 (root end bottom and top skins respectively) show similar tendencies in their material distribution having the largest component in approximately the 0° direction to offer the necessary bending stiffness. Both have significant components in the region of −45° offering resistance to the torsion component of the second load case. However, the top skin, which is in compression, has distinctly more material in this region than the bottom skin so that the panels do not buckle (45° material offers optimum resistance to buckling for square plates).

Laminates 3 and 4 (tip elements in the bottom and top skin respectively) both have large components of material in approximately the 54° direction, with proportionately more being found in laminate 4 (top skin—under compression). Although not quite at the optimum angle, this material provides near optimal torsional stiffness to resist the torsion load component of the second load case, and also provides very good resistance to buckling (for laminate type 4). The material in layer 1 in both laminates is orientated at 20–30° where it provides a good compromise between additional torsional rigidity and bending stiffness.

7. CONCLUSIONS

This work has shown that general optimization methods for composite structures, capable of using full design variable freedom, can be developed for, and successfully implemented on, the present generation of computers. The multilevel optimization scheme is an effective way of reducing the number of design variables under consideration at any given stage in the optimization process and this makes it feasible to optimize large multi-element laminated composite structures. Finally, this multilevel system also has the appeal of being able to add many more constraint types at the element level to cater for effects which affect only the local element design, such as interlaminar stresses and edge effects if suitable mathematical expressions (even if empirically derived) can be found to quantify them. This potential for growth is one of the system's most attractive features.

REFERENCES

1. Khot, N. S., Computer program (OPTCOMP) for optimization of composite structures for minimum weight design, AFFDL TR-76-149, 1977.
2. Starnes, J. H. and Haftka, R. T., Preliminary design of composite wings for buckling, strength and displacement constraints, 19th AIAA, ASME Structures, Structural Dynamics and Materials Conference, Bethesda, MD, 1978.
3. Schmit, L. A. and Mehrinfar, M., Multilevel optimum design of structures with fibre-composite stiffened panel components, *AIAA J.*, **20** (1982), 138–147.
4. Sobieszczanski-Sobieski, J., An integrated computer procedure for sizing composite airframe structures, NASA TP 1300, 1979.
5. Schmit, L. A. and Ramanathan, R. K., Multilevel approach to minimum weight design including buckling constraints, *AIAA J.*, **16** (1978), 97–104.
6. Osyczska, A., *Multicriterion Optimization in Engineering: With Fortran Programs*, Chichester, Ellis Horwood Ltd, 1984.
7. Watkins, R. I. and Morris, A. J., A multicriteria objective function optimization scheme for laminated composites for use in multilevel structural optimization schemes, *Comp. Meth. appl. Mech. Engng* (to be published).
8. Watkins, R. I., Multilevel optimum design of large laminated composite structures, Ph.D. thesis, Cranfield Institute of Technology, 1986.

30

Thermal Expansion of Steel Fibre Reinforced Concrete up to 800°C

J. A. PURKISS

Department of Civil Engineering, Aston University, Aston Triangle, Birmingham B4 7ET, UK

ABSTRACT

The chapter presents the results on the thermal expansion of steel fibre reinforced concrete for three different volume fractions of plain steel fibres (0·75, 1·5 and 2·5%) and shows that below 300°C steel fibres inhibit expansion but that at higher temperatures there is little difference between steel fibre reinforced concrete and plain concrete. However, the residual strains obtained for steel fibre reinforced concrete are much higher than for plain concrete.

NOTATION

a_t	Thermal diffusivity
b	Quarter width of specimen
k	Temperature variation parameter
T	Temperature
t	Time
x	Distance
Δ	Incremental change
ψ	Rate of temperature rise

Subscripts

c	Centre of specimen
i	At time $t = t_i$
m	Mean

s Surface
x At point x
1 Quarter depth

A dot above a symbol indicates differentiation with respect to time.

INTRODUCTION

Recently RILEM have published a report[1] on the properties of concrete at high temperatures. It will be found that there is in this document little data on steel fibre reinforced concrete. It was decided, therefore, to attempt to fill this need and thus the present project was devised—that is, a series of tests on the thermal expansion of steel fibre reinforced concrete.

TEST PROCEDURE

The concrete mix was 2·13:1·75:1, 10 mm siliceous aggregate, medium sand and OPC. The water cement ratio was 0·57. The mix proportions are in line with but not identical to some earlier published work,[2] owing to the use of higher volume fraction of fibres.

The specimen size used was 500 × 50 × 50 mm prisms. A relatively small cross-section was chosen in order to attempt to reduce the effect of thermal gradients in the specimens on the results. Each specimen has a Chromel–Alumel thermocouple cast in at the geometric centre of the specimen since it would be necessary to relate the thermal movement to the volumetric mean temperature in the specimen.

Four different volume fractions of plain brass-coated steel fibres 20 mm long by 0·25 mm diameter were considered: zero, 0·75, 1·5 and 2·5%. It should be noted that some difficulty was experienced in mixing the concrete with 2·5% fibres.

Some of the specimens were cast vertically and the remainder horizontally. All specimens were air-cured in the laboratory for two months after demoulding, prior to testing.

Three different nominal test temperatures were studied: 300, 500 and 800°C. In each case the specimens were heated to the test temperature, at around 2·5°C/min in a proprietary kiln-type furnace, allowed to soak at temperature for 1 h and then allowed to cool to ambient conditions, during which time (on the heating and soaking parts of the cycle only) readings were taken of the specimen mid-point temperatures using a DVM and zero degrees Centigrade reference, the kiln temperature using the built-in temperature gauge and the thermal movements of the specimens which were transmitted to dial gauges outside the furnace by a system of quartz rodding. A maximum of three specimens could be tested at any one time. For further details Ref. 3 should be consulted.

In order to determine the volumetric mean temperature, one plain concrete specimen was instrumented with thermocouples on the surface, at quarter depth and at half depth into the specimen.

DETERMINATION OF VOLUMETRIC MEAN TEMPERATURE

For a square specimen heated equally on all four sides, the isotherms will be as shown in Fig. 1a. If the rounding at the corners is ignored (introducing a

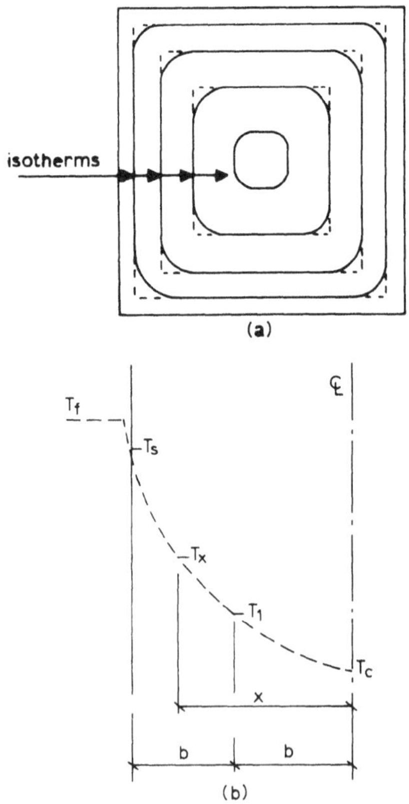

FIG. 1. Temperature distribution within the specimen.

small but not significant error) the temperature distribution can then be given by

$$T_x = T_c + k(x/2b)^2 \tag{1}$$

where the symbols are defined in Fig. 1b.

For the specimen fully instrumented with thermocouples, the temperature at the surface, T_s, is given by

$$T_s = T_c + K(2b/2b)^2$$

or

$$\Delta T_s = T_s - T_c = k \tag{2}$$

and the temperature at quarter depth, T_1, by

$$T_1 = T_c + k(b/2b)^2$$

or

$$\Delta T_1 = T_1 - T_c = k/4 \tag{3}$$

The best fit value of k for eqns (2) and (3) is

$$k = 4(\Delta T_1 + 4\Delta T_s)/17 \tag{4}$$

The mean volumetric temperature difference, ΔT_m, is given by

$$\Delta T_m = \frac{1}{8b^2/2}\int_0^{2b} (2x)k(x/2b)^2\,dx$$

or

$$\Delta T_m = k/2$$

or

$$\Delta T_m = 2(\Delta T_1 + 4\Delta T_s)/17 \tag{5}$$

but

$$\Delta T_m = T_m - T_c$$

so

$$T_m = (7T_c + 2T_1 + 8T_s)/17 \tag{6}$$

The value of T_m calculated from eqn. (6) for the reference specimen was plotted against both T_c and T_f (the furnace temperature). The results are given in Fig. 2.

In the thermal expansion tests only T_c and T_f were measured and the mean value of T_m calculated from T_c and T_f separately was used to plot the thermal expansion results. Note this procedure will give slight errors for the fibre reinforced concrete as the thermal conductivity is lower, as will be the specific heat but the errors are not likely to be significant. It was also noted that in the early stages of heating the kiln temperature measured by the built-in meter appeared lower than the mean temperature. On carrying out a calibration between a Chromel–Alumel thermocouple placed at the volumetric centre of the kiln and the built-in meter, the meter was discovered to be reading low up to about 300°C.

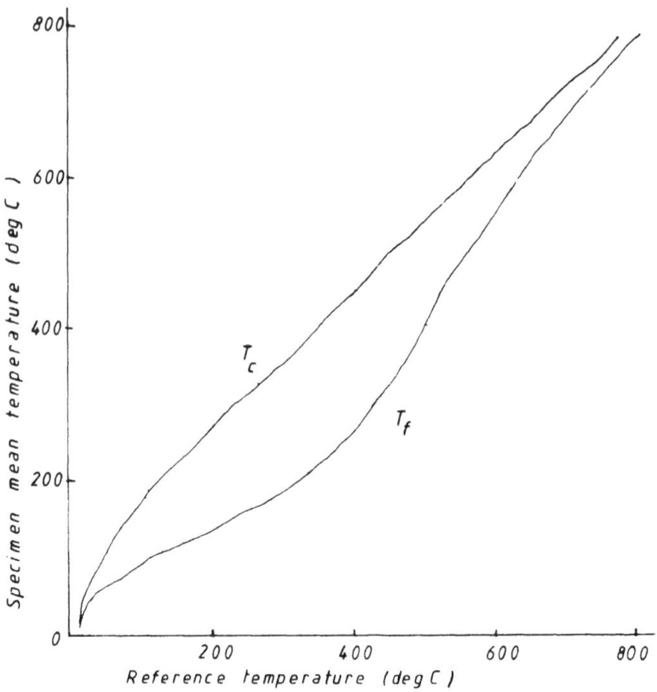

FIG. 2. Variation of the specimen temperature with both the furnace and central temperature.

The value of k calculated in eqn. (4) may also be used to produce values of the thermal diffusivity, a_t.

Considering unidirectional heat flow, then

$$a_t \frac{d^2T}{dx^2} = \dot{T} \tag{7}$$

but from eqn. (1)

$$\frac{d^2T}{dx^2} = 2k/4b^2 \tag{8}$$

If $\dot{T}$ is replaced by its finite difference approximation calculated considering the points either side of the current point, i.e.

$$\dot{T}_i = (T_{m,i+1} - T_{m,i-1})/\Delta t_{i+1,i-1} = \psi_i \tag{9}$$

then

$$a_t = (2b^2/k)\psi_i \tag{10}$$

The results from eqn. (10) are plotted in Fig. 3. The results are in agreement with existing data.[4]

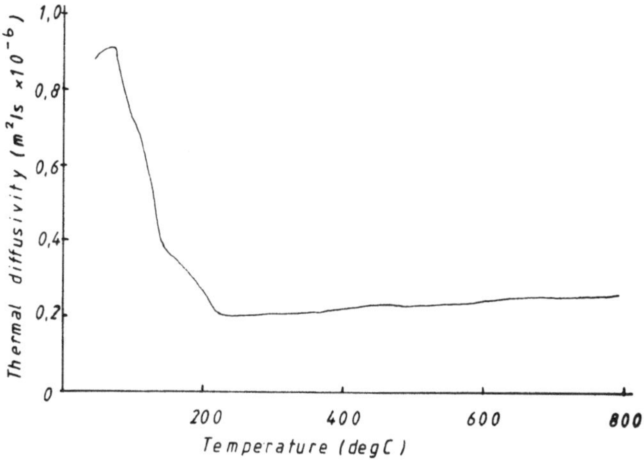

FIG. 3. Variation of thermal diffusivity with temperature.

THERMAL EXPANSION MEASUREMENTS

It should be noted that the thermal expansion measurements quoted in this chapter are inclusive of any shrinkage that may occur and thus are net values.

Table 1 provides a summary of the results, giving the number of specimens, maximum nominal temperature, maximum thermal strain attained and final residual strain.

All graphs have been plotted as mean results for all the tests at a particular temperature level, fibre percentage and direction of casting.

At each temperature level the vertically cast and horizontally cast specimens have been plotted separately.

(i) Series One (Figs 4 and 5)

In general, the addition of fibres appears to reduce the net expansion except for the 1·5% horizontally cast specimen which appears to give anomalous results.

(ii) Series Two (Figs 6 and 7)

Similar trends are observed as in Series One, except that the expansion is starting to increase rapidly at around 570°C due to the $\alpha \rightarrow \beta$ phase transformation in the quartz.

TABLE 1
Test results

Specimen reference	Nominal test temperature (°C)	Percentage steel fibres	Actual maximum temperature reached (°C)	Direction cast[a]	Maximum thermal strain (microstrain)	Residual strain[b] (microstrain)
3000V/1	300	None	277	VERT	4 060	−880
3000H/1	300	None	277	HOR	4 200	−600
3000H/2	300	None	251	HOR	3 420	−20
3000H/3	300	None	278	HOR	4 300	−580
3075V/1	300	0·75	258	VERT	3 660	−40
3075V/2	300	0·75	267	VERT	3 720	−180
3075H/1	300	0·75	250	HOR	3 500	60
3150H/1	300	1·50	229	HOR	3 220	20
3150H/2	300	1·50	262	HOR	4 380	160
3250V/1	300	2·50	268	VERT	3 880	−500
3250H/1	300	2·50	268	HOR	3 800	−60
6000V/1	600	None	550	VERT	10 680	2 460
6000V/2	600	None	578	VERT	14 300	2 200
6000H/1	600	None	572	HOR	17 260	5 000
6000H/2	600	None	561	HOR	14 380	2 240
6075V/1	600	0·75	575	VERT	11 840	1 260
6075V/2	600	0·75	580	VERT	13 320	2 620
6075H/1	600	0·75	569	HOR	12 480	2 400
6075H/2	600	0·75	584	HOR	15 260	2 720
6150V/1	600	1·50	581	VERT	14 300	2 640
6150V/2	600	1·50	583	VERT	14 160	3 620
6250V/1	600	2·50	578	VERT	12 540	3 020
6250H/1	600	2·50	584	HOR	14 420	3 320
8000V/1	800	None	784	VERT	16 020	2 900
8000V/2	800	None	797	VERT	16 520	3 680
8000V/3	800	None	786	VERT	16 240	3 020
8075V/1	800	0·75	797	VERT	15 880	4 160
8075V/2	800	0·75	789	VERT	16 340	4 700
8075H/1	800	0·75	789	HOR	17 840	6 480
8075H/2	800	0·75	793	HOR	16 600	6 040
8150V/1	800	1·50	798	VERT	17 520	6 180
8150V/2	800	1·50	796	VERT	18 060	6 800
8150H/1	800	1·50	782	HOR	16 620	7 020
8150H/2	800	1·50	790	HOR	17 480	6 560
8250V/1	800	2·50	793	VERT	17 940	6 820
8250V/2	800	2·50	798	VERT	19 720	7 000
8250H/1	800	2·50	798	HOR	17 220	5 800
8250H/2	800	2·50	800	HOR	16 700	5 080

[a] VERT indicates cast vertically as a column, HOR indicates horizontally cast as a beam.
[b] A negative sign indicates net contraction.

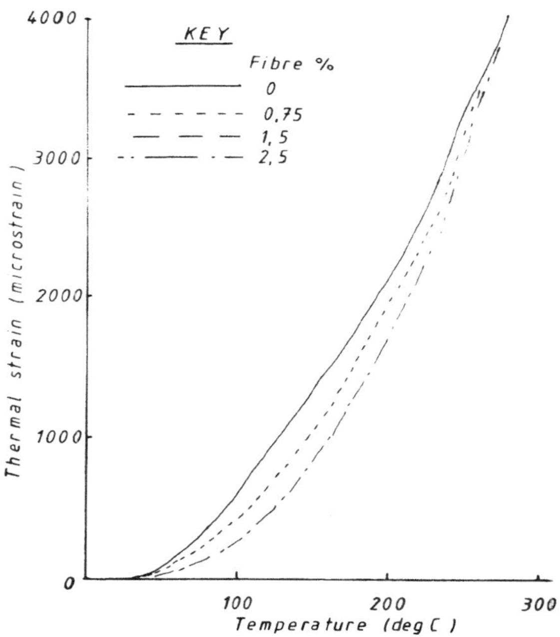

FIG. 4. Thermal expansion curves for vertically cast specimens up to 300°C.

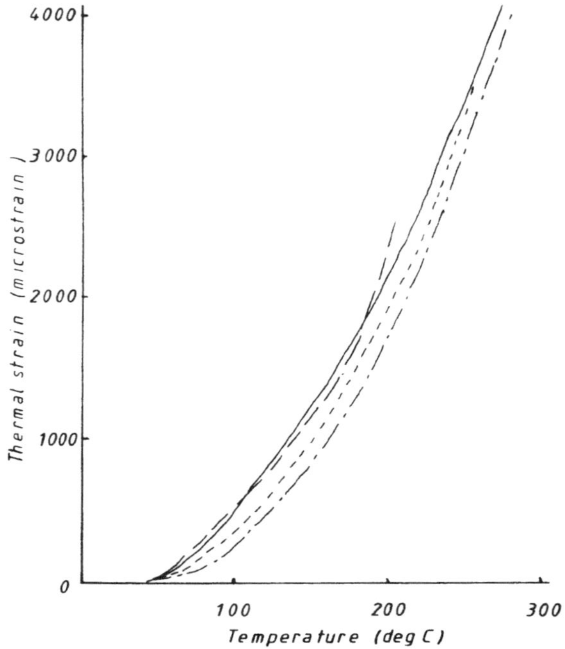

FIG. 5. Thermal expansion curves for horizontally cast specimens up to 300°C.

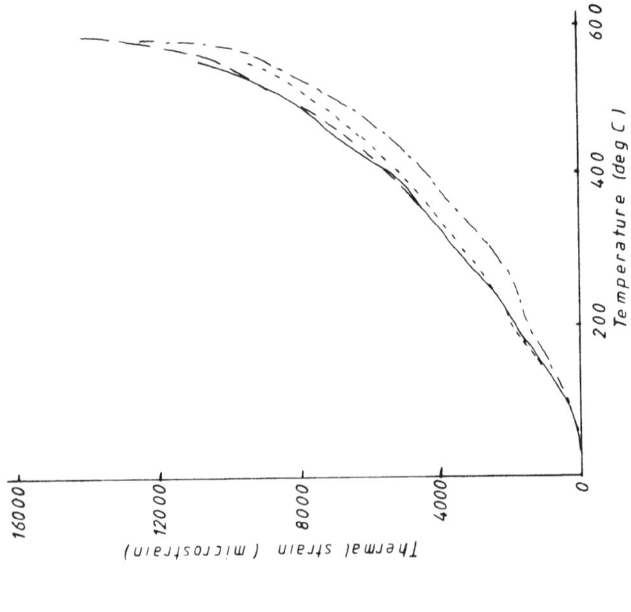

FIG. 6. Thermal expansion curves for vertically cast specimens up to 600°C.

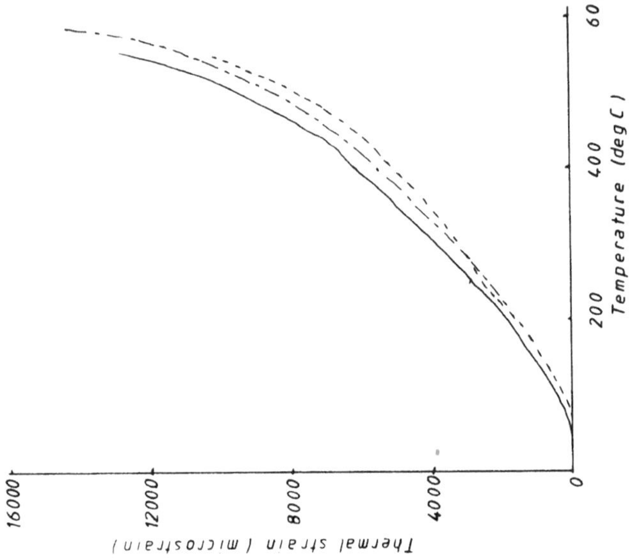

FIG. 7. Thermal expansion curves for horizontally cast specimens up to 600°C.

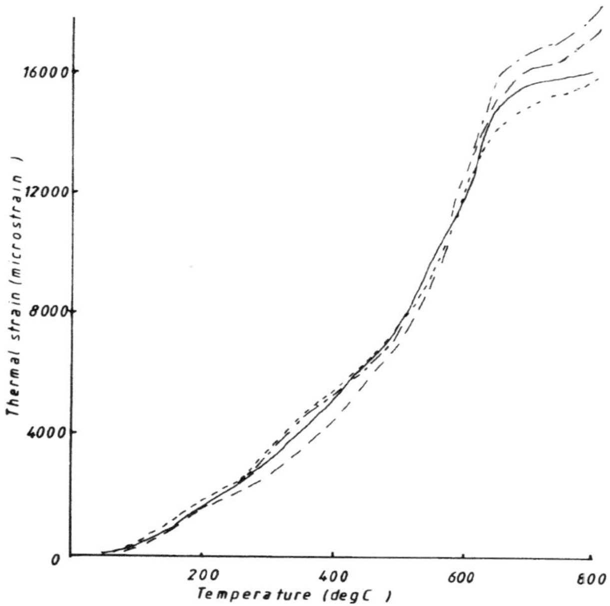

FIG. 8. Thermal expansion curves for vertically cast specimens up to 800°C.

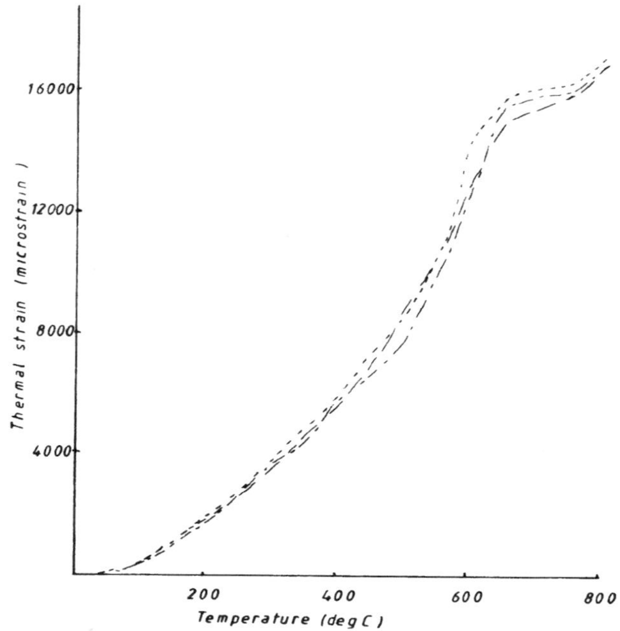

FIG. 9. Thermal expansion curves for horizontally cast specimens up to 800°C.

(iii) Series Three (Figs 8 and 9)

Above 700°C a severe reduction in the rate of expansion is observed. At around 550–600°C a steep increase in the rate appears owing to the phase transformation mentioned above.

In general, the results for plain concrete are as previously reported,[1,4] and the presence of fibres appearing to reduce the expansion of concrete.

A comparison of the results for vertically and horizontally cast specimens seems to suggest that there is little significant difference in the results.

(iv) Residual strains

Only the data for which a complete set of results exists are plotted (Fig. 10), where it is observed that a net residual contraction (or low expansion) for the Series One tests is changed into a large net expansion for the other two series, and that fibre reinforcement produces higher residual strains.

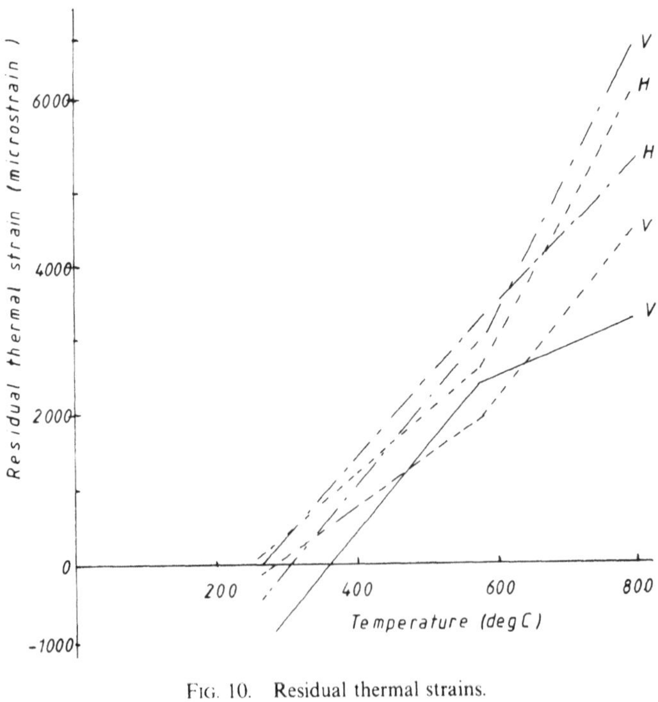

FIG. 10. Residual thermal strains.

CONCLUSIONS

At temperatures below 300°C the addition of fibres will cause a decrease in the thermal expansion of concrete, although the volume fraction appears to have only a secondary effect. At high temperatures the effect of steel fibres is not so marked and even appears to cause an increase in thermal expansion at higher volume fractions (thermal strain in steel at a temperature of 800°C is approximately 11 200 microstrain).

The interaction between the fibres and the concrete matrix is complex since both materials will reduce in stiffness and strength as the temperature increases.

The residual strains appear to increase with the addition of fibres, and the results seem to indicate that greater residual strains are attained with increasing volume fraction.

ACKNOWLEDGEMENT

The author wishes to thank Alan Sharpe and Graham Chidgey for carrying out the experimental work as part of a final year project.

REFERENCES

1. SCHNEIDER, U. (Ed.), *Rilem Report—Properties of Materials at Elevated Temperatures—Concrete*, Kassel Universität, 1985.
2. PURKISS, J. A., Steel fibre reinforced concrete at elevated temperatures, *International Journal of Cement Composites and Lightweight Concrete*, **6** (1986), 179–184.
3. SHARPE, A. and CHIDGEY, G., Fibre Reinforced Concrete at Elevated Temperatures, Final Year Project Report, Department of Civil Engineering, Aston University, 1986.
4. SCHNEIDER, U., Verhalten von Beton bei höhen Temperaturen, *Deutscher Ausschuss für Stahlbeton*, **337** (1982), 1–222.

31

Effects of Temperature Changes and Large Amplitude on Vibrations of Composite Triangular Plates

M. SATHYAMOORTHY

Department of Mechanical and Industrial Engineering, Clarkson University, Potsdam, New York 13676, USA

ABSTRACT

This paper is concerned with the nonlinear static and dynamic analysis of composite triangular plates subjected to thermal loading. The nonlinearities considered here arise due to large deformation resulting in nonlinear strain–displacement relations. Equations to represent the nonlinear behavior of composite plates are derived and solutions to these coupled set of governing nonlinear equations are reported. Numerical results indicate some interesting influences of temperature, amplitude of vibration, and geometric and material parameters of the composite plates. These parameters also have considerable effect on the nonlinear load–deflection behavior.

INTRODUCTION

Geometrically nonlinear static and dynamic analysis of composite plates of various geometries has received considerable attention in the literature.[1,2] In particular, vibrations of composite plates subjected to a given change in temperature are of concern to many design engineers when checking the aeroelastic performance of high-speed flying vehicles. Sunakawa and Uemura,[3] Pal[4,5] and Biswas[6] studied the nonlinear static behavior of plates subjected to thermal loads. Sunakawa and Uemura[3] considered the deformations and thermal stresses in a simply supported rectangular plate subjected to an arbitrary symmetrical temperature distribution. Pal[4,5] investigated the nonlinear static as well as dynamic behavior of isotropic and orthotropic heated circular plates using the Berger-type approximation. For rectangular plates with hinged and immovable edges Sunakawa[7] used the method of successive approximation to study the nonlinear vibration behavior of plates subjected to certain temperature changes. Pal[5,8] considered the large amplitude free vibration of rectangular and circular plates subjected to aerodynamic heating. Jones *et al.*[9] used the Berger-type approximation to investigate the large amplitude vibration behavior of both elastic and viscoelastic plates subjected to a parabolic temperature distribution. They have derived a formula for the fundamental frequency in terms of the critical buckling temperature and the corresponding fundamental frequency of the unheated plate. All these investigations clearly indicate the effects of temperature change and large amplitude on the static and dynamic behaviors of plates of various geometries. It can be seen that these effects cannot be ignored even when the temperature change is small.

In this paper, an analytical investigation is presented for the large amplitude vibration behavior of triangular plates with clamped stress-free boundary conditions. Large amplitude, free flexural vibrations of thin composite triangular plates subjected to certain types of temperature distribution are considered in this investigation. The nonlinearities investigated here arise due to large deformation or large amplitude vibration and are included in the nonlinear strain–displacement relations. Von Kármán-type nonlinear equations are presented for rectilinearly orthotropic thin plates in terms of stress function, F, and the lateral displacement, w. For a particular choice of temperature distribution, a solution for w is assumed in a separable form consisting of a product of a time function and a space function. Due to the complex nature of the governing equations, only solutions based on a single mode are presented with the aid of Galerkin's method. Using the appropriate mode shape for w and a suitable expression for the stress function, F, the coupled set of nonlinear governing equations are reduced to one nonlinear time-differential equation whose solutions are obtained by the elliptic integral method. Numerical results for the frequency ratios are presented for different isotropic and orthotropic triangular plates, including the effects of temperature distribution. Since the frequencies are amplitude dependent, these values are listed at various amplitudes of vibration of the plate. Results for some nonlinear static cases are also presented in terms of load deflation values. Present results are in good agreement with those in the literature for all the special cases.

GOVERNING EQUATIONS

For a rectilinearly orthotropic thin elastic plate of uniform thickness, h, the governing nonlinear equations in terms of the stress function, F, and lateral displacement, w, are

$$F_{,xxxx} + m^2 F_{,xxyy} + k^2 F_{,yyyy} = E_y(w^2_{,xy} - w_{,xx}w_{,yy}) + (1/h)(N_{2,xx} - q^2 N_{1,xx} + K^2 N_{1,yy} - q^2 N_{2,yy}) \quad (1)$$

$$D_1[w_{,xxxx} + 2(q^2 + 2p^2)w_{,xxyy} + K^2 w_{,yyyy}] = q(x, y) - \rho h w_{,tt} + h(F_{,yy}w_{,xx} + F_{,xx}w_{,yy} - 2F_{,xy}w_{,xy}) + M_{1,xx} + M_{2,yy} \quad (2)$$

In eqns (1) and (2), m^2, K^2, q^2, p^2 and D_1 are orthotropic material parameters which are defined below in terms of the elastic moduli E_x, E_y, the Poisson ratios ν_{xy}, ν_{yx}, and the shear modulus G_{xy}. It must be noted that a comma has been used to denote partial differentiation with respect to the indicated variable x, y or t:

$$K^2 = E_y/E_x \qquad q^2 = \nu_{yx} \qquad p^2 = \mu G_{xy}/E_x \qquad \nu^1 = 1 - \nu_{xy}\nu_{yx}$$
$$m^2 = (K^2 - 2p^2q^2 - q^4)/p^2 \qquad D_1 = B_{11}h^3/12 \qquad B_{11} = E_x/\nu^1 \quad (3)$$

In eqns (1) and (2), the effects of temperature change have been included by means of N_1, N_2 and M_1, M_2, which are defined as

$$N_1; N_2 = \int_{-h/2}^{h/2} (\bar{\alpha}_{11}; \bar{\alpha}_{22})T \,\mathrm{d}z \quad (4)$$

$$M_1; M_2 = \int_{-h/2}^{h/2} (\bar{\alpha}_{11}; \bar{\alpha}_{22})Tz \,\mathrm{d}z \quad (5)$$

Assuming the coefficients of thermal expansion to be α_{11} and α_{22}, the expressions for the two normal strains and the shear strain can be expressed in terms of stresses and temperature T as

$$\begin{aligned} \varepsilon_x &= b_{11}\sigma_x + b_{12}\sigma_y + \alpha_{11}T \\ \varepsilon_y &= b_{12}\sigma_x + b_{22}\sigma_y + \alpha_{22}T \\ \gamma_{xy} &= b_{44}\tau_{xy} \end{aligned} \quad (6)$$

Solution of the system of eqns (6) will result in the following equations for the stresses which are written in terms of the strains and the temperature T:

$$\begin{aligned} \sigma_x &= B_{11}\varepsilon_x + B_{12}\varepsilon_y + \bar{\alpha}_{11}T \\ \sigma_y &= B_{12}\varepsilon_x + B_{22}\varepsilon_y + \bar{\alpha}_{22}T \\ \tau_{xy} &= B_{44}\gamma_{xy} \end{aligned} \quad (7)$$

In eqn. (7) the coefficients are

$$B_{11} = -b_{22}\bar{b} \qquad B_{12} = b_{12}\bar{b} \qquad B_{22} = -b_{11}\bar{b}$$
$$\bar{\alpha}_{11} = (b_{22}\alpha_{11} - b_{12}\alpha_{22})\bar{b} \qquad \bar{\alpha}_{22} = (b_{11}\alpha_{22} - b_{12}\alpha_{11})\bar{b} \tag{8}$$
$$\bar{b} = 1/(b_{12}^2 - b_{11}b_{22})$$

In the case of isotropic plates, the coefficients b_{ij} and B_{ij} in eqn. (8) simplify to the following:

$$b_{11} = b_{22} = 1/E \qquad b_{12} = -\nu/E \qquad \alpha_{11} = \alpha_{22} = \alpha$$
$$B_{11} = B_{22} = E/(1-\nu^2) \qquad B_{12} = \nu B_{11}$$
$$\bar{\alpha}_{11} = \bar{\alpha}_{22} = -E\alpha/(1-\nu) \quad \text{and} \quad K^2 = 1 \qquad m^2 = 2 \tag{9}$$
$$q^2 = \nu \qquad p^2 = (1-\nu)/2 \quad \text{and} \quad E_y = E$$

Equations (1) and (2) are the two coupled nonlinear governing equations applicable to thin, elastic, rectilinearly orthotropic plates of constant thickness. The effects of temperature distribution are included in these equations by means of N and M. $q(x, y)$ is an external load acting on the plate (load per unit area of the plate), ρ is the mass per unit volume of the plate material and h is the uniform thickness of the plate. In the next section, approximate solutions to eqns (1) and (2) are presented considering orthotropic triangular plates subjected to certain temperature distributions.

METHOD OF SOLUTION

For a right-triangular plate of dimensions a (along x) and b (along y) with clamped stress-free boundaries, the following boundary conditions must be satisfied:

$$w = w_{,x} = F_{,yy} = F_{,xy} = 0 \quad \text{at } x = 0$$
$$w = w_{,y} = F_{,xx} = F_{,xy} = 0 \quad \text{at } y = 0 \tag{10}$$
$$w = w_{,n} = F_{,nn} = F_{,nt} = 0 \quad \text{at } 1 - (x/a) - (y/b) = 0$$

In eqn. (10), n and t denote the outward normal and tangential directions of the inclined edge of the right-triangular plate. Quantities $F_{,nn}$ and $F_{,tt}$ can be readily written in terms of $F_{,xx}$, $F_{,yy}$ and $F_{,xy}$.

The appropriate functions for w and F that will satisfy all the in-plane and out-of-plane boundary conditions given in eqn. (10) may be chosen as

$$w = w_0(\tau)\phi(x, y) \qquad F = F_0(\tau)\phi(x, y) \tag{11}$$

In eqn. (11), $w_0(\tau)$ and $F_0(\tau)$ are functions of nondimensional time τ defined as

$$\tau = t(E_x/\rho a^2)^{1/2} \tag{12}$$

Also $\phi(x, y) = x^2y^2[1-(x/a)-(y/b)]^2$, which are the same as the ones used in Refs 10 and 11. Assuming that the temperature-dependent changes of orthotropic material properties are negligible, the triangular plate is subjected to a two-dimensional temperature distribution of the form

$$T = T_0[1-(x/a)-(y/b)]^2 \tag{13}$$

with T_0 taken as the reference temperature.

Equations (11) and (13) are substituted into eqns (1) and (2) after deriving N_1, N_2 and M_1, M_2 from eqns (4), (5) and (13). The resulting error functions $E_1(w, F)$ and $E_2(w, F)$ from eqns (1) and (2) are integrated in the following manner:

$$\int_{y=0}^{(a-x)/r}\int_{x=0}^{a} E_1(w, F)\phi(x, y)\,\mathrm{d}x\,\mathrm{d}y = 0 \tag{14}$$

$$\int_{y=0}^{(a-x)/r}\int_{x=0}^{a} E_2(w, F)\phi(x, y)\,\mathrm{d}x\,\mathrm{d}y = 0 \tag{15}$$

The result of the integration shown in eqn. (14) will give F_0 in terms of w_0^2 used in eqn. (11). For the sake of saving space, the details of the integration and the precise definition of the coefficients involved are omitted. After performing the integration shown in eqn. (15), F_0 is eliminated by means of the relationship for F_0 found already in terms of w_0. For thin plates, this procedure will result in a single second-order time differential equation of the Duffing type in $w_0(\tau)$. Since the maximum deflection, w_{max}, occurs at $x/a = 1/3$ and $y/b = 1/3$ in a triangular plate, the magnitude of w_{max} becomes $w_0a^2b^2/729$. Representing the nondimensional maximum deflection by $A(w_{max}/h)$, the time-differential equation becomes

$$C_1A_{,\tau\tau} + C_2A + C_3A^3 = C_4q_0^* \tag{16}$$

where C_1, C_2, C_3 and C_4 are coefficients which depend upon the geometric and orthotropic material parameters of the plate as well as the coefficients of thermal expansion. In eqn. (16), q_0^* is the nondimensional uniformly distributed load acting on the plate given by $q_0^* = q_0a^4/E_xh^4$. r is the aspect ratio which is the ratio of the sides a/b. Solutions of eqn. (16) can be readily obtained by using the elliptic integral method when $q_0^* = 0$.[12] In the case of nonlinear static problems, A is independent of time and therefore eqn. (16)

will reduce to a nonlinear algebraic equation. The solutions to this equation will give the load–deflection values in the nonlinear range for triangular plates subjected to uniformly distributed loads.

RESULTS AND DISCUSSION

Numerical results are presented for both static and dynamic problems dealing with rectilinearly orthotropic and isotropic right triangular thin plates. In the case of orthotropic plates, the material constants for glass-epoxy (GE) and boron-epoxy (BE) plates are taken as $(E_x/E_y) = 3{\cdot}0$, $10{\cdot}0$; $\nu_{xy} = 0{\cdot}25$, $0{\cdot}22$; $(G_{xy}/E_x) = 0{\cdot}2$, $0{\cdot}033$. In the case of isotropic plates, the corresponding values are 1·0, 0·3, 0·3846. In the case of static problems, the nonlinear load–deflection values can be readily obtained by solving the nonlinear algebraic equation obtained by dropping the time-dependent term in eqn. (16). The variation of nondimensional maximum deflection, w, with the nondimensional uniformly distributed load, q_0^*, is shown in Tables 1 and 2 for isotropic and glass-epoxy plates at various values of T^*. The temperature parameter T^* is defined as $T^* = T_c a^2/(B_{11}h^2)$. It is clear that the load–deflection values exhibit the hardening type behavior. As the value

TABLE 1
Values of nondimensional load q_0^* for isotropic triangular plates ($r = 1{\cdot}0$)

A	T^*					
	0	50	100	150	200	250
0	0	0	0	0	0	0
0·5	321·52	322·91	324·29	325·65	327·03	328·40
1·0	649·57	652·34	655·09	657·83	660·57	663·31
1·5	990·64	994·79	998·93	1 003·04	1 007·15	1 011·25

TABLE 2
Values of nondimensional load q_0^* for glass-epoxy (GE) triangular plates ($r = 2{\cdot}0$)

A	T^*					
	0	50	100	150	200	250
0	0	0	0	0	0	0
0·5	819·01	820·64	822·28	823·91	825·55	827·18
1·0	1 649·18	1 652·45	1 655·72	1 658·99	1 662·26	1 665·53
1·5	2 501·68	2 506·59	2 511·49	2 516·40	2 521·31	2 562·21

of the temperature increases, there is a slight increase in the rigidity of the plate.

In the case of dynamic problems, the ratios of the nonlinear frequencies to the corresponding linear frequencies are presented at different nondimensional amplitudes considering different plate, material and temperature parameters. Such numerical results are presented in Tables 3, 4 and 5. Both isotropic as well as some orthotropic material plates have been

TABLE 3

Values of nondimensional frequency ratio $(\omega/\omega_0)10^4$ *for isotropic triangular plates* $(r = 1{\cdot}0)$

$A(\tau)$	T^*					
	0	50	100	150	200	250
0	10 000	10 000	10 000	10 000	10 000	10 000
0·5	10 013	10 016	10 022	10 033	10 049	10 085
1·0	10 051	10 059	10 071	10 090	10 122	10 167
1·5	10 114	10 124	10 139	10 162	10 189	10 220

TABLE 4

Values of nondimensional frequency ratio $(\omega/\omega_0)10^4$ *for glass-epoxy (GE) triangular plates* $(r = 1{\cdot}5)$

$A(\tau)$	T^*					
	0	50	100	150	200	250
0	10 000	10 000	10 000	10 000	10 000	10 000
0·5	10 013	10 023	10 039	10 059	10 094	10 134
1·0	10 050	10 062	10 081	10 102	10 139	10 184
1·5	10 113	10 125	10 168	10 212	10 245	10 292

TABLE 5

Nondimensional frequency ratios $(\omega/\omega_0)10^4$ *for boron-epoxy (BE) triangular plates* $(r = 2{\cdot}0)$

$A(\tau)$	T^*					
	0	50	100	150	200	250
0	10 000	10 000	10 000	10 000	10 000	10 000
0·5	10 008	10 010	10 018	10 034	10 048	10 068
1·0	10 032	10 041	10 069	10 092	10 112	10 149
1·5	10 071	10 091	10 132	10 181	10 242	10 295

studied with several variations in the temperature parameter. For the particular type of temperature distribution considered in this paper, the frequency amplitude behavior is one of the hardening type, i.e. frequency increases with the amplitude of vibration. An increase in the temperature results in a slight increase in the frequency ratio. This behavior is observed at any moderately large amplitude of vibration.

REFERENCES

1. CHIA, C. Y., *Nonlinear Analysis of Plates*, New York, McGraw-Hill, 1980.
2. SATHYAMOORTHY, M., Recent research in nonlinear plate vibrations, *The Shock and Vibration Digest* (1987).
3. SUNAKAWA, M. and UEMURA, M., Deformation and thermal stresses in a rectangular plate subjected to aerodynamic heating, Report No. 359, Aeronautical Research Institute, University of Tokyo, 1960.
4. PAL, M. C., Large deflections of heated circular plates, *Acta Mechanica*, **8** (1969), 82–103.
5. PAL, M. C., Static and dynamic nonlinear behavior of heated orthotropic circular plates, *International Journal of Nonlinear Mechanics*, **8** (1973), 489–504.
6. BISWAS, P., Large thermal deflections of elastic plates by conformal transformations, *Proceedings of the International Conference on Numerical Methods in Thermal Problems*, Venice, Italy, 1981, pp. 462–472.
7. SUNAKAWA, M., Influences of temperature changes and large amplitude on free flexural vibration of rectangular plates, *Transactions of JSME*, **30** (1966), 558–576.
8. PAL, M. C., Large amplitude free vibration of rectangular plates subjected to aerodynamic heating, *Journal of Engineering Mathematics*, **4** (1970), 39–49.
9. JONES, R., MAZUMDAR, J. and CHEUNG, Y. K., Vibration and buckling of plates at elevated temperatures, *International Journal of Solids and Structures*, **16** (1980), 61–70.
10. NOWINSKI, J. L. and ISMAIL, I. A., Large oscillations of an anisotropic triangular plate, *Journal of the Franklin Institute*, **280** (1965), 417–424.
11. VENDHAN, C. P. and DHOOPAR, B. L., Nonlinear vibration of orthotropic triangular plates, *AIAA Journal*, **11** (1973), 704–709.
12. SATHYAMOORTHY, M., Nonlinear vibration of elliptical plates, *Journal of Sound and Vibration*, **70** (1980), 458–460.

32

Thermal Buckling of Symmetric Angle-ply Laminated Plates

T. R. TAUCHERT and N. N. HUANG

Department of Engineering Mechanics, University of Kentucky, Lexington, Kentucky, USA

ABSTRACT

An analysis of thermal buckling of symmetric angle-ply plates which are simply supported and subjected to a uniform temperature change is reported. The Rayleigh–Ritz technique is used to determine both the inplane stress resultants for the prebuckled state and the subsequent buckling behavior. Numerical results are presented for fiber-reinforced laminates, which show the effects of fiber orientation, number of layers, aspect ratio, and type of supports on the critical buckling temperature and the associated mode shape.

INTRODUCTION

Although buckling of laminated plates caused by applied edge loads has been studied by several investigators, thermal buckling of laminates has received little attention. One exception is the work of Stavsky,[1] which considers the stability of a circular laminated plate subjected to an axisymmetric temperature field. More recently, thermal buckling behaviors of thin[2] and thick[3] antisymmetric angle-ply rectangular plates have been investigated. It was demonstrated in Ref. 2 that when a simply-supported antisymmetric angle-ply laminate experiences a uniform temperature change, the plate middle surface remains undeformed until buckling occurs. An exact solution for the critical buckling temperature is given in Ref. 2.

In the case of a symmetric angle-ply laminated plate under a uniform temperature field, the prebuckled state is generally characterized by

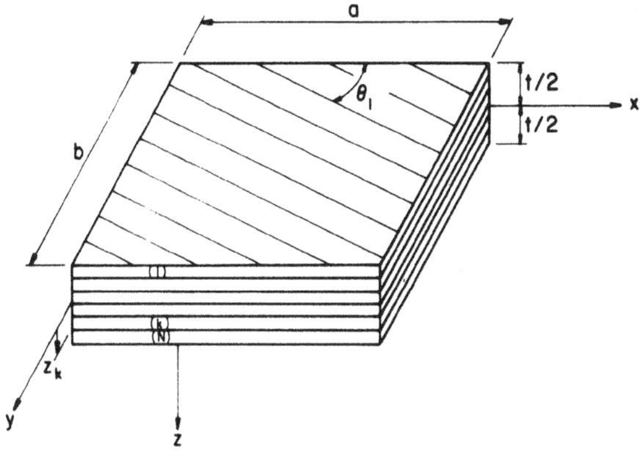

FIG. 1. Plate geometry.

nonzero inplane displacements, and subsequent buckling involves 'flexure–twist' coupling. As a result, obtaining closed form solutions does not appear feasible. In this paper the Rayleigh–Ritz method is employed in order to determine both the prebuckled deformations and the thermal buckling loads for simply-supported plates.

The laminate under consideration consists of N layers of a fiber-reinforced material (Fig. 1), with the layers taken to be macroscopically homogeneous and orthotropic. For each layer located above the middle surface $z = 0$ and having fiber orientation θ_k, there exists a corresponding layer located at an equal distance below the middle surface having the same fiber orientation.

GOVERNING EQUATIONS

Following the classical laminated plate theory, the force (N_x, N_y, N_{xy}) and moment (M_x, M_y, M_{xy}) resultants in a symmetric angle-ply laminate are related to the middle-surface displacement components (u, v, w), thermal forces $(N_x^{\mathrm{T}}, N_y^{\mathrm{T}}, N_{xy}^{\mathrm{T}})$ and thermal moments $(M_x^{\mathrm{T}}, M_y^{\mathrm{T}}, M_{xy}^{\mathrm{T}})$ through the constitutive equations:

$$\begin{Bmatrix} N_x \\ N_y \\ N_{xy} \end{Bmatrix} = \begin{bmatrix} A_{11} & A_{12} & A_{16} \\ A_{12} & A_{22} & A_{26} \\ A_{16} & A_{26} & A_{66} \end{bmatrix} \begin{Bmatrix} u_{,x} \\ v_{,y} \\ u_{,y} + v_{,x} \end{Bmatrix} - \begin{Bmatrix} N_x^{\mathrm{T}} \\ N_y^{\mathrm{T}} \\ N_{xy}^{\mathrm{T}} \end{Bmatrix} \tag{1}$$

$$\begin{Bmatrix} M_x \\ M_y \\ M_{xy} \end{Bmatrix} = \begin{bmatrix} D_{11} & D_{12} & D_{16} \\ D_{12} & D_{22} & D_{26} \\ D_{16} & D_{26} & D_{66} \end{bmatrix} \begin{Bmatrix} -w_{,xx} \\ -w_{,yy} \\ -2w_{,xy} \end{Bmatrix} - \begin{Bmatrix} M_x^{\mathrm{T}} \\ M_y^{\mathrm{T}} \\ M_{xy}^{\mathrm{T}} \end{Bmatrix} \tag{2}$$

in which a comma denotes differentiation with respect to the subscript. Here A_{ij} and D_{ij} are the extensional and bending stiffness coefficients for the laminate, defined in terms of the transformed reduced stiffness coefficients $(\bar{Q}_{ij})_k$ for the individual layers $k = 1, 2, \ldots, N$ by (see (4))

$$(A_{ij}, D_{ij}) = \sum_{k=1}^{N} \int_{z_{k-1}}^{z_k} (1, z^2)(\bar{Q}_{ij})_k \, \mathrm{d}z \qquad (i, j = 1, 2, 6) \tag{3}$$

The thermal forces and moments appearing in eqns (1–2) are defined by

$$\begin{Bmatrix} N_x^{\mathrm{T}} & M_x^{\mathrm{T}} \\ N_y^{\mathrm{T}} & M_y^{\mathrm{T}} \\ N_{xy}^{\mathrm{T}} & M_{xy}^{\mathrm{T}} \end{Bmatrix} = \sum_{k=1}^{N} \int_{z_{k-1}}^{z_k} \begin{bmatrix} \bar{Q}_{11} & \bar{Q}_{12} & \bar{Q}_{16} \\ \bar{Q}_{12} & \bar{Q}_{22} & \bar{Q}_{26} \\ \bar{Q}_{16} & \bar{Q}_{26} & \bar{Q}_{66} \end{bmatrix}_k \begin{Bmatrix} \alpha_x \\ \alpha_y \\ 2\alpha_{xy} \end{Bmatrix}_k (1, z) T \, \mathrm{d}z \tag{4}$$

For the present case of a symmetric laminate exposed to a uniform temperature distribution, say $T = T_1$, the thermal moments M_x^{T}, M_y^{T} and M_{xy}^{T} vanish, while the thermal forces N_x^{T}, N_y^{T} and N_{xy}^{T} are functions of T_1 alone.

In order to obtain a Rayleigh–Ritz solution to the problem, an expression for the plate's total thermoelastic potential energy is required. In the case of small deformation resulting from the temperature change $T = T_1$, the potential energy is given by[5]

$$\begin{aligned} \Pi_{\mathrm{S}} = \int_0^b \int_0^a \Bigg[& \frac{A_{11}}{2} u_{,x}^2 + A_{12} u_{,x} v_{,y} + \frac{A_{22}}{2} v_{,y}^2 \\ & + A_{16} u_{,x}(u_{,y} + v_{,x}) + A_{26} v_{,y}(u_{,y} + v_{,x}) \\ & + \frac{A_{66}}{2}(u_{,y} + v_{,x})^2 + \frac{D_{11}}{2} w_{,xx}^2 + D_{12} w_{,xx} w_{,yy} + \frac{D_{22}}{2} w_{,yy}^2 \\ & + 2D_{16} w_{,xx} w_{,xy} + 2D_{26} w_{,yy} w_{,xy} + 2D_{66} w_{,xy}^2 \\ & - N_x^{\mathrm{T}} u_{,x} - N_{xy}^{\mathrm{T}}(u_{,y} + v_{,x}) - N_y^{\mathrm{T}} v_{,y} \Bigg] \mathrm{d}x \, \mathrm{d}y \end{aligned} \tag{5}$$

For the buckling problem it is necessary to include the potential of the inplane forces due to the transverse deflection, in which case the total potential becomes

$$\Pi = \Pi_{\mathrm{S}} + \frac{1}{2} \int_0^b \int_0^a [N_x w_{,x}^2 + 2N_{xy} w_{,x} w_{,y} + N_y w_{,y}^2] \, \mathrm{d}x \, \mathrm{d}y \tag{6}$$

THERMAL BUCKLING SOLUTIONS

The laminate is presumed to be simply supported in such a manner that the edges are either (1) fixed against normal displacements but free to move in the tangential (inplane) direction, or (2) rigidly fixed against both normal and tangential displacements. The corresponding boundary conditions are given by:

Case I

$$\begin{aligned} x=0,a:\quad & w=0 \quad M_x=-D_{11}w_{,xx}-D_{12}w_{,yy}-2D_{16}w_{,xy}=0\\ & u=0 \quad N_{xy}=A_{16}u_{,x}+A_{26}v_{,y}+A_{66}(u_{,y}+v_{,x})-N_{xy}^{\mathrm{T}}=0\\ y=0,b:\quad & w=0 \quad M_y=-D_{12}w_{,xx}-D_{22}w_{,yy}-2D_{26}w_{,xy}=0\\ & v=0 \quad N_{xy}=A_{16}u_{,x}+A_{26}v_{,y}+A_{66}(u_{,y}+v_{,x})-N_{xy}^{\mathrm{T}}=0 \end{aligned} \tag{7}$$

Case II

$$\begin{aligned} x=0,a:\quad & w=0 \quad M_x=-D_{11}w_{,xx}-D_{12}w_{,yy}-2D_{16}w_{,xy}=0\\ & u=0, \quad v=0\\ y=0,b:\quad & w=0 \quad M_y=-D_{12}w_{,xx}-D_{22}w_{,yy}-2D_{26}w_{,xy}=0\\ & u=0, \quad v=0 \end{aligned} \tag{8}$$

For solutions via the Rayleigh–Ritz method, the displacement components u, v, w are assumed as follows:

Case I

$$\begin{aligned} u&=\sum_{i=1}^{m}\sum_{j=0}^{n} u_{ij}\sin\alpha_i x\cos\beta_j y\\ v&=\sum_{i=0}^{n}\sum_{j=1}^{m} v_{ij}\cos\alpha_i x\sin\beta_j y\\ w&=\sum_{i=1}^{p}\sum_{j=1}^{q} w_{ij}\sin\alpha_i x\sin\beta_j y \end{aligned} \tag{9}$$

Case II

$$u = \sum_{i=1}^{m} \sum_{j=1}^{n} u_{ij} \sin \alpha_i x \sin \beta_j y$$
$$v = \sum_{i=1}^{m} \sum_{j=1}^{n} v_{ij} \sin \alpha_i x \sin \beta_j y \qquad (10)$$
$$w = \sum_{i=1}^{p} \sum_{j=1}^{q} w_{ij} \sin \alpha_i x \sin \beta_j y$$

in which $\alpha_i = i\pi/a$ and $\beta_j = j\pi/b$. Note that the assumed solutions satisfy identically all displacement boundary conditions. However, for Case I the expressions (9) satisfy neither the force nor the moment boundary conditions (7), while for Case II the expressions (10) fail to satisfy the moment conditions.

Before investigating buckling of the laminate, the inplane forces in the prebuckled state must be determined. Substituting the assumed solutions (9) or (10) into eqn. (5), and minimizing Π_S with respect to u_{ij} and v_{ij}, yields a system of linear algebraic equations which can be solved for u_{ij} and v_{ij}. The stress resultants N_x, N_y and N_{xy} are then calculated using eqn. (1).

For Case II, in which the plate edges are fixed against both normal and tangential displacements, the coefficients u_{ij} and v_{ij} are found to be zero. In this case the inplane stress resultants are uniform within the plate and can be calculated exactly; in particular,

$$N_x = -N_x^{\mathrm{T}} \qquad N_y = -N_y^{\mathrm{T}} \qquad N_{xy} = -N_{xy}^{\mathrm{T}} \qquad (11)$$

On the other hand, when the plate edges are free to translate in the tangential direction, the inplane forces vary throughout the laminate. Comparing this case with that of an antisymmetric laminate under the same boundary conditions, it can be concluded that the variation of the inplane forces is due to the existence of nonzero stiffnesses A_{16}, A_{26} and thermal force N_{xy}^{T} (Ref. 6).

Assuming that the inplane stress resultants remain unchanged during buckling, the values N_x, N_y and N_{xy} for the prebuckled state are substituted into eqn. (6) for Π. Minimization of Π with respect to the Ritz coefficients w_{ij} leads to a set of simultaneous homogeneous equations in w_{ij}. Setting the determinant of the coefficient matrix equal to zero yields an approximate value for the critical buckling temperature $(T_1)_c$.

One special case for which the exact buckling temperature of a symmetric angle-ply plate is known is that of a laminate having an infinite number of layers. As $N \to \infty$, the stiffnesses $A_{16}, A_{26}, D_{16}, D_{26}$ and the thermal force N_{xy}^{T} approach zero.[4] In this case the equations governing thermal buckling reduce to those for an antisymmetric infinite-layer angle-ply laminate. An exact solution to the latter problem was derived in Ref. 3; the critical values of the thermal force N_x^{T} were shown to be

$$(N_x^{\mathrm{T}})_{\mathrm{c}} = [D_{11}\alpha_i^4 + 2(D_{12} + 2D_{66})\alpha_i^2\beta_j^2 + D_{22}\beta_j^4]/(\alpha_i^2 + R\beta_j^2) \qquad (12)$$

where $R \equiv N_y^{\mathrm{T}}/N_x^{\mathrm{T}}$ (a constant ratio). Finding the primary buckling load, i.e. the lowest critical value of $(N_x^{\mathrm{T}})_{\mathrm{c}}$ in eqn. (12), requires a searching procedure involving integer values of the mode numbers i and j.[3]

Finally, we note that eqn. (12) is also applicable in the case of unidirectional and cross-ply laminates, for which the ply angles are either 0° or 90°.

NUMERICAL RESULTS AND DISCUSSION

For illustrative purposes, the critical buckling temperature $(T_1)_{\mathrm{c}}$ has been calculated for various regular symmetric angle-ply laminates (with ply orientations $+\theta/-\theta/+\theta\cdots$). The Young's moduli, shear modulus, Poisson's ratio and thermal expansion coefficients for each lamina are arbitrarily taken as follows:

$$E_1/E_0 = 181{\cdot}0 \qquad E_2/E_0 = 10{\cdot}3 \qquad G_{12}/E_0 = 7{\cdot}17$$
$$\nu_{12} = 0{\cdot}28 \qquad \alpha_1/\alpha_0 = 0{\cdot}02 \qquad \alpha_2/\alpha_0 = 22{\cdot}5$$

in which E_0 and α_0 represent arbitrary reference values.

In determining the inplane stress resultants for the prebuckled state of the laminate, the upper limits of the series expressions (9) for u and v were taken to be $m = n = 16$. Results of the subsequent buckling analysis, based upon various values of the upper limits p and q of the series (9) and (10) for w, are given in Figs 2 and 3. Figure 2 shows a plot of the buckling temperature $(T_1)_{\mathrm{c}}$ versus the number of modes considered (p times q) for a square laminate ($b/a = 1$, $t/a = 0{\cdot}05$) having ply angle $\theta = 11{\cdot}25°$, edges free to move in the tangential direction (type I), and various numbers of layers. For the case of infinite N the Rayleigh–Ritz procedure yields the exact solution (given by eqn. (12)) providing $p = q = 2$. For the other values of N considered it is evident that reasonably good convergence of the Ritz solution is obtained when 64 modes are included ($p = q = 8$). Similar results

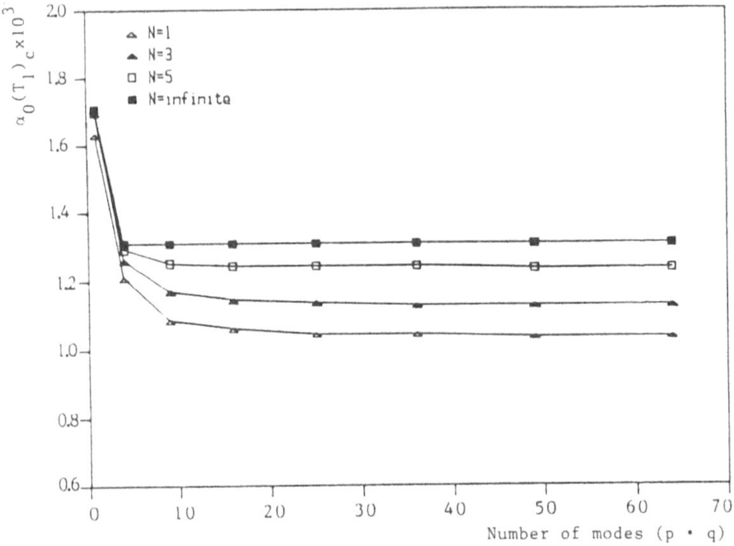

FIG. 2. Critical temperature for a square laminate ($b/a = 1$, $t/a = 0{\cdot}05$, $\theta = 11{\cdot}25°$) with type I edge conditions.

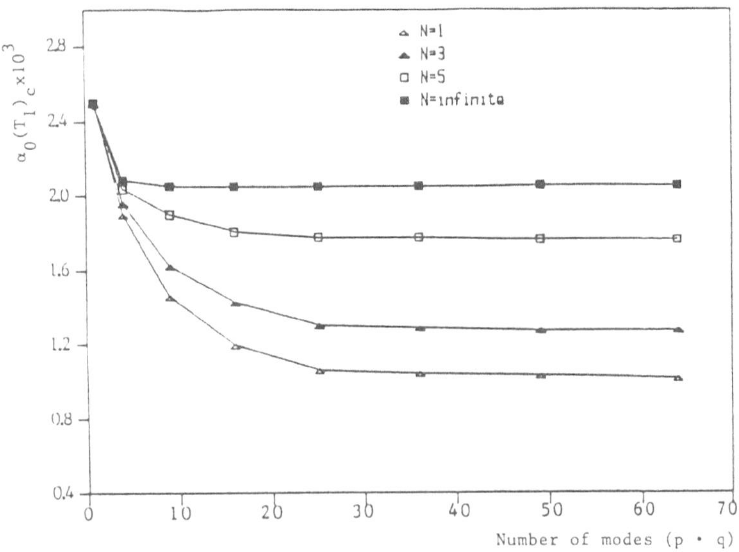

FIG. 3. Critical temperature for a rectangular laminate ($b/a = 2$, $t/a = 0{\cdot}05$, $\theta = 22{\cdot}5°$) with type II edge conditions.

are given in Fig. 3 for a rectangular plate ($b/a = 2$, $t/a = 0{\cdot}05$) having ply angle $\theta = 22{\cdot}5°$, edges fixed against translation (type II), and $N = 1, 3, 5$ and ∞. The Ritz solution for $N = \infty$ coincides in this case with the corresponding exact solutions when $p = q = 3$.

The influence of ply orientation θ on the critical buckling temperature for square laminates ($b/a = 1$, $t/a = 0{\cdot}05$) having 1, 3 and ∞ layers is illustrated in Fig. 4. It is seen that variations in θ may result in large changes in $(T_1)_c$. When $N \geq 3$ the resistance of the laminate to thermal buckling increases as the orientation of the reinforcement approaches 45°. It is also noted that the critical temperature for a given thickness laminate increases as the number of layers N increases. Since the bending stiffnesses D_{16} and D_{26} decrease with increasing N, it follows that the bending–twist coupling tends to decrease the resistance of the laminates against buckling. Furthermore, it is noted from Fig. 4 that, when the layer number N is large and/or the fiber direction is nearly parallel to the plate edges ($\theta = 0°$ or $\theta = 90°$), the difference in the buckling temperatures corresponding to the two types of edge conditions is small.

Buckled mode shapes for several different laminate configurations are presented in the form of normalized three-dimensional and contour plots in Figs. 5–7. Figures 5a and 5b show the buckling patterns of square ($b/a = 1$,

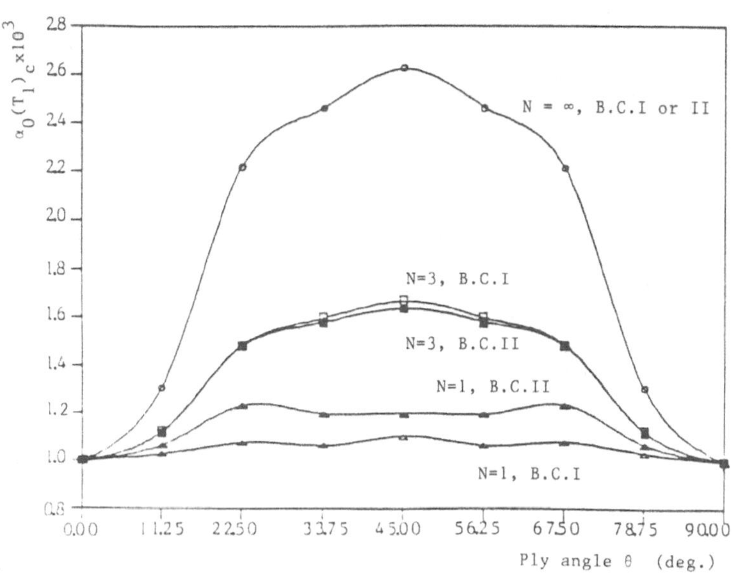

FIG. 4. Effect of ply angle on buckling temperature for square laminates.

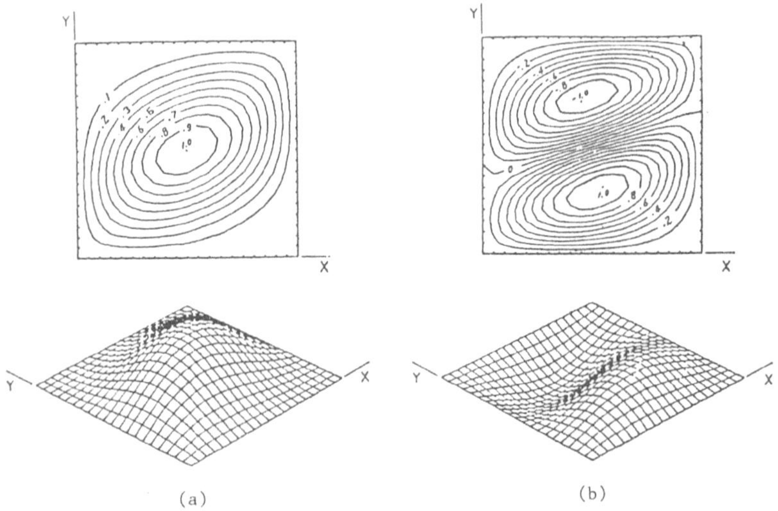

FIG. 5. Mode shape for a square laminate ($b/a = 1$, $t/a = 0{\cdot}05$, $\theta = 22{\cdot}5^\circ$, $N = 3$) with edge conditions: (a) type I; (b) type II.

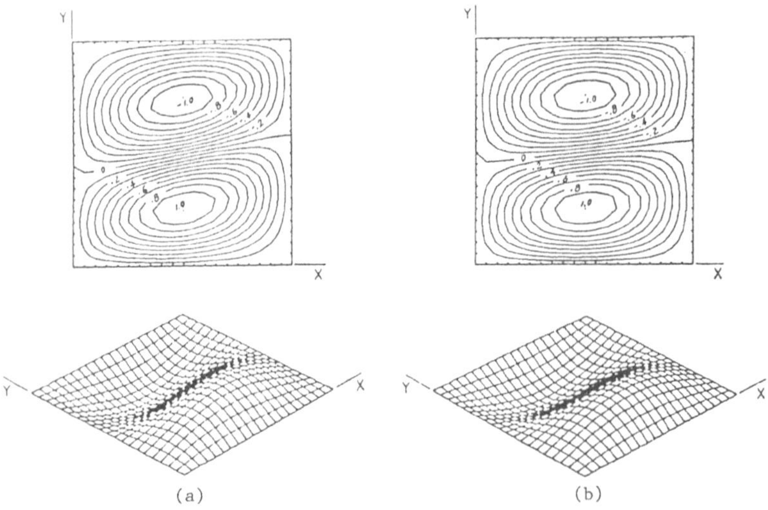

FIG. 6. Mode shape for a square laminate ($b/a = 1$, $t/a = 0{\cdot}05$, $\theta = 11{\cdot}25^\circ$) with type I edge conditions. (a) $N = 1$; (b) $N = 5$.

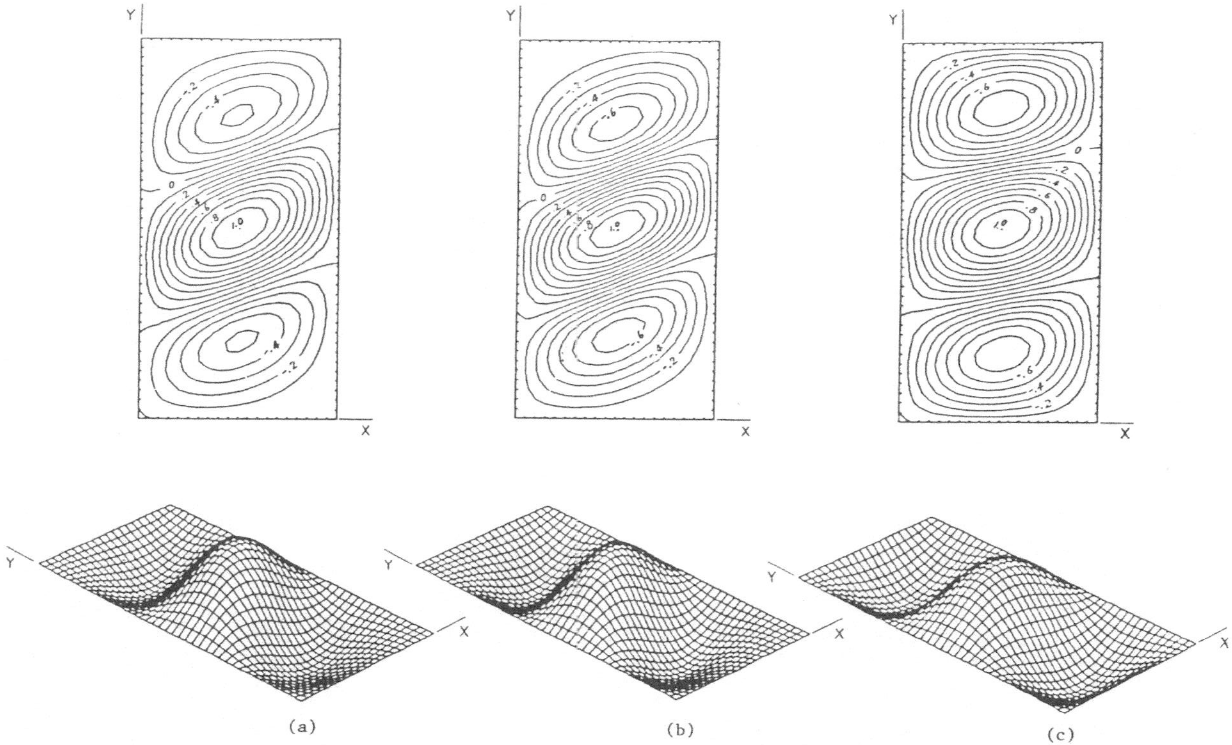

FIG. 7. Mode shape for a rectangular laminate ($b/a = 2$, $t/a = 0{\cdot}05$, $\theta = 22{\cdot}5°$). (a) $N = 1$, edge condition I; (b) $N = 1$, edge condition II; (c) $N = 5$, edge condition I.

$t/a = 0{\cdot}05$), three layer plates of ply angle $\theta = 22{\cdot}5°$, with edge conditions of the types I and II, respectively. Although the critical temperatures corresponding to the different boundary conditions are nearly equal ($\alpha_0(T_1)_c = 1{\cdot}487 \times 10^{-3}$ and $1{\cdot}483 \times 10^{-3}$, respectively), the associated mode shapes are seen to be totally different.

Buckling patterns for square plates, each having ply angle $\theta = 11{\cdot}25°$ and edges which are free to move tangentially but with different number of layers ($N = 1$ and $N = 5$), are shown in Figs 6a and 6b, respectively. Whereas the single-layer plate buckles at a significantly lower temperature ($\alpha_0(T_1)_c = 1{\cdot}032 \times 10^{-3}$ versus $1{\cdot}236 \times 10^{-3}$), the mode shapes for the two plates are seen to be quite similar.

Figures 7a and 7b show the buckled shapes of single-layer rectangular plates ($b/a = 2$, $t/a = 0{\cdot}05$) having identical fiber orientation ($\theta = 22{\cdot}5°$) but subject to different edge conditions (types I and II, respectively).

Unlike the earlier case of square plates (Fig. 5), the difference in support conditions results in a significant difference in the buckling temperature ($\alpha_0(T_1)_c = 0{\cdot}918 \times 10^{-3}$ for type I and $1{\cdot}019 \times 10^{-3}$ for type II). However, the corresponding mode shapes have the same general appearance, characterized by three half waves in the y-direction and one half wave in the x-direction.

Figure 7c relates to a plate having the same dimensions, ply angle and boundary conditions as the laminate considered in Fig. 7a, but having five layers rather than one. Note that the node lines (contour lines for $w = 0$) are more nearly parallel to the x-axis for the case $N = 5$ than for $N = 1$. Based upon an exact analysis[6] it can be shown that in the limit at $N \to \infty$, the node lines for an arbitrary angle-ply plate become parallel to the plates' shorter edges.

CONCLUDING REMARKS

Classical laminated plate theory has been employed to investigate thermal buckling of symmetric angle-ply laminates caused by a uniform temperature rise. It is found that inplane deformations generally occur prior to buckling, resulting in nonuniform stress resultants. Deflections transverse to the plate middle surface do not occur, however, until the temperature reaches a critical value $(T_1)_c$. The Rayleigh–Ritz technique provides a convenient means for obtaining approximate solutions to the inplane and buckling problems.

Through numerical examples it has been demonstrated that fiber orientation, number of layers, aspect ratio and edge conditions can have a significant influence upon the critical buckling temperature and the associated mode shape.

In the more general situation of an arbitrary laminate subject to a nonuniform temperature field, inplane and transverse deflections occur simultaneously, and transition from the regime of small deflections to the large-deformation regime progresses gradually. Research on large deformation and postbuckling behaviors of laminates under nonuniform thermal loading is under way, the results of which will be reported subsequently.

REFERENCES

1. STAVSKY, Y., Thermoelastic stability of laminated orthotropic circular plates, *Acta Mech.*, **22** (1975), 31–51.
2. TAUCHERT, T. R. and HUANG, N. N., Thermal buckling and postbuckling behavior of antisymmetric angle-ply laminates, *Proc. Int. Symp. Composite Materials and Structures*, Beijing, June 1986, pp. 357–362.
3. TAUCHERT, T. R., Thermal buckling of thick antisymmetric angle-ply laminates, *J. Thermal Stresses* (in press).
4. JONES, R. M., *Mechanics of Composite Materials*, Washington, Scripta, 1975.
5. TAUCHERT, T. R., Thermal stresses in plates—statical problems, in: *Thermal Stresses*, Vol. I (R. B. Hetnarski, ed.), Amsterdam, North-Holland, 1986, pp. 23–141.
6. HUANG, N. N., Thermal elastic buckling and large deformation behavior of rectangular angle-ply laminated plates, MS thesis, University of Kentucky, 1986.

33

Nonlinear Vibration and Bending of Laminated Anisotropic Circular Plates

K. SRINIVASAMURTHY and C. Y. CHIA

Department of Civil Engineering, University of Calgary, Calgary, Alberta T2N 1N4, Canada

ABSTRACT

This paper is analytically concerned with large amplitude free flexural vibration and static large deflection of a generally laminated anisotropic circular plate with clamped movable and immovable edges. The dynamic von Karman-type plate equations used in this study are expressed in terms of the transverse displacement and two in-plane displacements. A unified approximate solution is presented by use of the Galerkin procedure and the method of harmonic balance. Two in-plane equilibrium equations and the associated in-plane boundary conditions are satisfied exactly. Numerical results for nonlinear bending and vibration of laminated angle-ply and cross-ply circular plates are presented graphically for different high-modulus material properties, geometries of lamination and in-plane boundary conditions.

INTRODUCTION

Static large deflection and large amplitude free vibration of plates have been considered by a number of investigators based on the von Karman plate theory. Yamaki[1] discussed nonlinear free vibrations of an isotropic circular plate. Nonlinear forced vibration of the plate was considered by Kung and Pao.[2] Finite element formulation to study the large amplitude vibration of orthotropic circular plates was due to Venkateshwara Rao *et al.*[3] Alwar and Sekharreddy[4] employed the Chebyshev series to obtain nonlinear static and dynamic response of isotropic and orthotropic annular plates. Based on the Galerkin procedure, Satyamoorthy and Chia[5] investigated nonlinear static and dynamic behaviour of isotropic and orthotropic circular plates. Huang and Aurora[6] considered nonlinear oscillations of orthotropic annular plates of variable thickness. Using Chebyshev polynomials and the implicit Houbolt scheme, Nath and Alwar[7] discussed the static and dynamic responses of orthotropic circular plates. Based on Reissner's plate equations, nonlinear free vibration of thick orthotropic annular plates with varying thickness was studied by Reddy and Huang.[8] Satyamoorthy[9] reported a comparative study between two different approaches, namely stress function and displacement equations to the nonlinear vibration of isotropic and orthotropic circular plates. Venkateshwara Rao and Kanakaraju[10] investigated optimization of cylindrically orthotropic plates with a constraint on fundamental frequency. Ruei *et al.*[11] studied dynamic and static nonlinear analysis of cylindrically orthotropic circular plates with nonuniform edge constraints. Based on Berger's approximation and the related von Karman plate theory, several investigators[12–18] also studied the large amplitude vibration of homogeneous circular plates.

A comprehensive review of nonlinear vibrations of homogeneous and laminated rectangular plates can be found in Ref. 19 or elsewhere.

GOVERNING EQUATIONS

Consider an initially flat, circular, elastic plate of constant thickness h in the z-direction and radius a. The reference plane $z = 0$ is located at the undeformed middle plane. The type of plate under consideration is constructed of an arbitrary number of rectilinearly orthotropic homogeneous layers perfectly bonded together. Each layer is of arbitrary thickness, elastic properties and orientation of orthotropic axes with respect to the plate axes. The plate materials are assumed to be continuous everywhere and each layer obeys the generalized Hooke's law. The dynamic von Karman-type equations governing the geometrically nonlinear behaviour of the generally laminated plate are given in Ref. 19. As usual, the in-plane and rotatory inertias are neglected. These basic equations are written in terms of the transverse displacement, w, and two in-plane displacements, u and v, as

$$L_1u + L_2v - L_3w = -w_{,x}L_1w - w_{,y}L_2w \tag{1}$$

$$L_2u + L_4v - L_5w = -w_{,x}L_2w - w_{,y}L_4w \tag{2}$$

$$\begin{aligned} L_3u + L_5v - L_6w - \rho hw_{,tt} = &-q - (u_{,x} + \tfrac{1}{2}w_{,x}^2)L_7w - (v_{,y} + \tfrac{1}{2}w_{,y}^2)L_8w \\ &- (u_{,y} + v_{,x} + w_{,x}w_{,y})L_9w - w_{,x}L_3w - w_{,y}L_5w \\ &- 2(B_{12} - B_{66})[w_{,xy}^2 - w_{,xx}w_{,yy}] \end{aligned} \tag{3}$$

where L_i are the linear differential operators defined by

$$\begin{aligned} L_1 &= A_{11}(\ )_{,xx} + 2A_{16}(\ )_{,xy} + A_{66}(\ )_{,yy} \\ L_2 &= A_{16}(\ )_{,xx} + (A_{12} + A_{66})(\ )_{,xy} + A_{26}(\ )_{,yy} \\ L_3 &= B_{11}(\ )_{,xxx} + 3B_{16}(\ )_{,xxy} + (B_{12} + 2B_{66})(\ )_{,xyy} + B_{26}(\ )_{,yyy} \\ L_4 &= A_{66}(\ )_{,xx} + 2A_{26}(\ )_{,xy} + A_{22}(\ )_{,yy} \\ L_5 &= B_{16}(\ )_{,xxx} + (B_{12} + 2B_{66})(\ )_{,xxy} + 3B_{26}(\ )_{,xyy} + B_{22}(\ )_{,yyy} \\ L_6 &= D_{11}(\ )_{,xxxx} + 4D_{16}(\ )_{,xxxy} + 2(D_{12} + 2D_{66})(\ )_{,xxyy} \\ &\quad + 4D_{26}(\ )_{,xyyy} + D_{22}(\ )_{,yyyy} \\ L_7 &= A_{11}(\ )_{,xx} + 2A_{16}(\ )_{,xy} + A_{12}(\ )_{,yy} \\ L_8 &= A_{12}(\ )_{,xx} + 2A_{26}(\ )_{,xy} + A_{22}(\ )_{,yy} \\ L_9 &= A_{16}(\ )_{,xx} + 2A_{66}(\ )_{,xy} + A_{26}(\ )_{,yy} \end{aligned} \tag{4}$$

in which A_{ij}, B_{ij} and D_{ij} are the membrane, coupling and flexural rigidities of the plate[19] respectively.

In this work the composite plate is clamped along its edge and two types of in-plane boundary conditions are considered. The out-of-plane boundary conditions are

$$w = w_{,x} = w_{,y} = 0 \quad \text{along } x^2 + y^2 = a^2 \tag{5}$$

and the in-plane boundary conditions are the following cases:

1. Immovable (rigidly clamped)

$$u = v = 0 \quad \text{along } x^2 + y^2 = a^2 \tag{6}$$

2. Movable (loosely clamped)

$$\left.\begin{aligned} xN_x + yN_{xy} = 0 \\ yN_y + xN_{xy} = 0 \end{aligned}\right\} \quad \text{along } x^2 + y^2 = a^2 \tag{7}$$

METHOD OF SOLUTION

On the basis of a single-mode analysis a solution is formulated for the problem stated above. As usual, the transverse displacement is assumed in a separable form consisting of the product of a time function $f(t)$ and a spatial function

$$w = f(t)F(x, y) \tag{8}$$

where $f(t)$ is the function of time to be determined. To satisfy the out-of-plane boundary conditions the spatial function is taken to be

$$F(x, y) = \sum_{n=1}^{3} S_n(a^2 - x^2 - y^2)^{n+1} \tag{9}$$

A linear analysis is carried out in order to determine the coefficients S_n. The values of S_n are $S_1 = 0{\cdot}773\,13$, $S_2 = 0{\cdot}170\,98$ and $S_3 = 0{\cdot}055\,89$.

The two in-plane displacements u and v are assumed as follows:

$$u = \frac{h^2 f^2(t)}{a^{16}} \sum_{i=0}^{7} a^{2i} \sum_{n=0}^{(15-2i)} b_{in} x^{(15-2i-n)} y^n + \frac{h^2 f(t)}{a^8} \sum_{i=0}^{7} a^{2i} \sum_{n=0}^{(7-2i)} c_{in} x^{(5-2i-n)} y^n \tag{10}$$

$$v = \frac{h^2 f^2(t)}{a^{16}} \sum_{i=0}^{7} a^{2i} \sum_{n=0}^{(15-2i)} d_{in} x^{(15-2i-n)} y^n + \frac{h^2 f(t)}{a^8} \sum_{i=0}^{3} a^{2i} \sum_{n=0}^{(7-2i)} e_{in} x^{(15-2i-n)} y^n \tag{11}$$

in which b_{in}, c_{in}, d_{in} and e_{in} are constants to be determined so as to satisfy the two in-plane equilibrium equations as well as the appropriate boundary conditions.

The assumed displacements are substituted into two in-plane equilibrium equations (1) and (2) and boundary conditions (6) or (7). On comparing coefficients of like terms a system of linear algebraic equations in terms of 184 constants, b_{in}, c_{in}, d_{in} and e_{in}, are formulated. In order to solve for these constants 136 equations are obtained from in-plane equilibrium equations and the remaining 48 equations from in-plane boundary conditions. These equations are then solved to determine these constants.

The solution for three displacements, however, does not satisfy the equation of transverse motion exactly. The error caused by the trial function for w is minimized by the Galerkin procedure such that a nonlinear ordinary differential equation for the time function is obtained as follows:

$$p_1 \frac{\mathrm{d}^2 f}{\mathrm{d}t^2} + p_2 f + p_3 f^2 + p_4 f^3 = q \tag{12}$$

where p_i are constants not defined here. The time differential equation (13) is solved by the method of harmonic balance. In this method the time function $f(t)$ is expanded into a Fourier cosine series as

$$f(t) = \sum_{j=0}^{\infty} g_j \cos j\omega t \tag{13}$$

where ω is the nonlinear frequency of vibration and g_j are Fourier coefficients to be determined. This is substituted into the nonlinear time differential equation. Each term is converted into the first power of the cosine functions and then the coefficients of like cosine terms are equated. This results in a set of coupled nonlinear algebraic equations in the form

$$R_m g_m + R_{mjk} g_j g_k + R_{mjkl} g_j g_k g_l = 0 \qquad m = 0, 1, 2, \ldots \tag{14}$$

where

$$\begin{aligned} R_m &= -p_1 m^2 + p_2 \\ R_{mjk} &= \tfrac{1}{2} p_3 \qquad j + k = m \quad \text{or abs}\,(j-k) = m \\ R_{mjkl} &= \tfrac{1}{4} p_4 \qquad j + k + l = m \end{aligned} \tag{15}$$

or $\text{abs}\,(j+k+l) = m$ or $\text{abs}\,(j-k-l) = m$ or $\text{abs}\,(j-k+l) = m$.

The system of equations (14) are solved for a given set of parameters to obtain Fourier coefficients. Once these coefficients are found, transverse displacement w and in-plane displacements u and v can be determined.

NUMERICAL RESULTS AND DISCUSSION

Numerical results for symmetrically and unsymmetrically laminated angle-ply and cross-ply plates are presented. Calculations are performed for

TABLE 1
Numerical values of elastic constants

Material	E_L/E_T	G_{LT}/E_T	ν_{LT}
Glass–epoxy	3·0	0·50	0·25
Boron–epoxy	10·0	0·33	0·22
Graphite–epoxy	40·0	0·50	0·25

glass–epoxy (GL), boron–epoxy (BO) and graphite–epoxy (GR) composites. The elastic constants typical of these materials are listed in Table 1, where E_L and E_T are the tensile moduli in the filament and transverse directions, G_{LT} is the shear modulus and ν_{LT} is Poisson's ratio.

The static and dynamic results are compared with the existing data[5] for the case of orthotropic single-layer circular plates. Very good agreement was found. Details are not presented here.

The relation between transverse load q and central deflection w_0 of a two-layer boron–epoxy plate is shown in Fig. 1 for various values of the orientation angle. The curve for $\theta = 30°$ lies in between $\theta = 15°$ and 45°. This curve is very close to $\theta = 45°$ and not shown here. Figure 2 shows the load–deflection relation for a graphite–epoxy plate having different numbers of layers. For a given deflection the transverse load increases as the number of layers increases. The curve for a six-layer plate is close to that given by the uncoupled solution ($n = \infty$) for which $B_{ij} = 0$. The effect of the

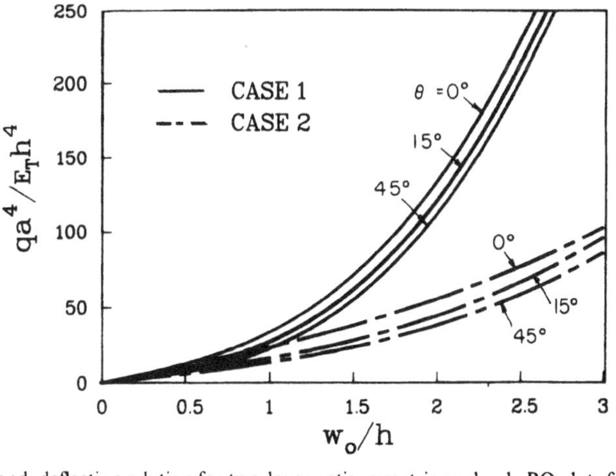

FIG. 1. Load–deflection relation for two-layer antisymmetric angle-ply BO plate for various values of orientation angle.

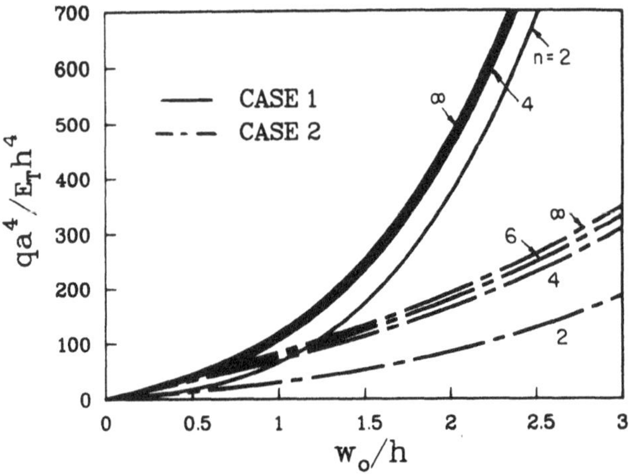

FIG. 2. Load–deflection relation for antisymmetric cross-ply GR plate for various number of layers.

coupling between bending and stretching, therefore, is to reduce the stiffness of the plate, and decreases with increasing the number of layers. It may be observed from this figure that the number of layers influences the magnitude of the load in case 2 more than in case 1.

The effect of number of terms in the time series (13) on a typical amplitude–frequency response of an unsymmetric angle-ply plate is given in Table 2, where ω is the nonlinear frequency and ω_0 the linear frequency of vibration. The result indicates that this series converges rapidly. The use of the first four terms in each time series should be adequate in engineering calculations but six terms are considered in the study. Figure 3 shows the frequency–amplitude response of a three-layer symmetric angle-ply boron–epoxy plate for different values of the fibre orientation angle. The

TABLE 2
Values of nondimensional amplitude, w_0/h, of a two-layer antisymmetric $\pm 45°$ angle-ply boron–epoxy plate for different number of terms in time series (13)

Number of terms	Case 1 $\omega/\omega_0 = 2{\cdot}146\,160$	Case 2 $\omega/\omega_0 = 1{\cdot}273\,359$
2	2·025 676 74	2·187 896 756
4	2·267 594 32	2·225 798 431
6	2·278 960 53	2·236 839 102

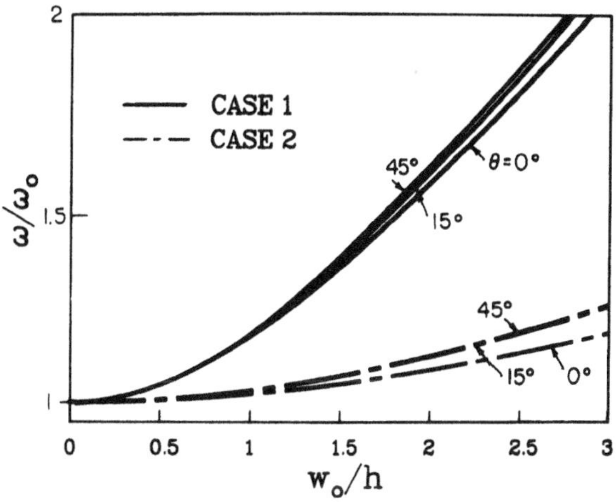

FIG. 3. Frequency–amplitude response for three-layer symmetric angle-ply BO plate for various values of orientation angle.

effect of the angle on the frequency–amplitude response of an anti-symmetric two-layer angle-ply plate is shown in Fig. 4. At small amplitudes the in-plane conditions have little effect on the frequency–amplitude response. The effect of the number of layers on dynamic behaviour of antisymmetric cross-ply graphite–epoxy plate is depicted in Fig. 5. The plate with a large number of layers mechanically behaves as a specially

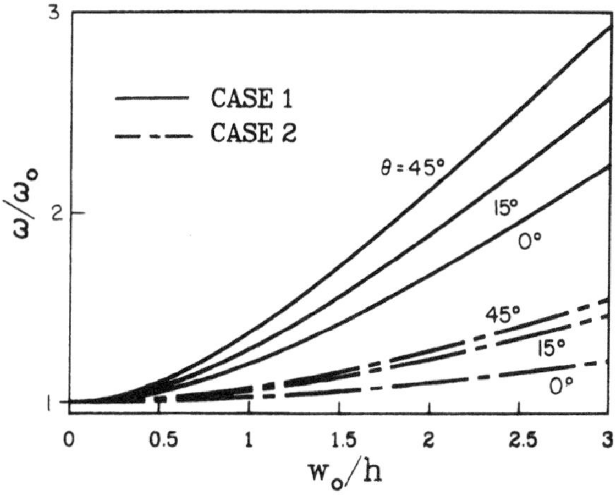

FIG. 4. Frequency–amplitude response for two-layer antisymmetric angle-ply BO plate for various values of orientation angle.

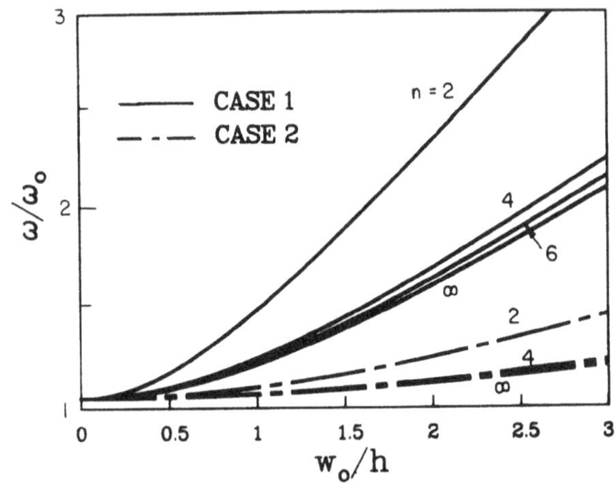

FIG. 5. Frequency–amplitude response for antisymmetric cross-ply GR plate for various number of layers.

orthotropic plate ($n = \infty$). Figure 6 shows the variation of the frequency ratio of a two-layer antisymmetric $\pm 45^\circ$ angle-ply plate with the amplitude for various material properties. High-modulus materials have low frequency ratios at large amplitudes.

In the cases considered above, the hardening type of nonlinearity is observed. That is, the frequency ratio increases with the amplitude of vibration.

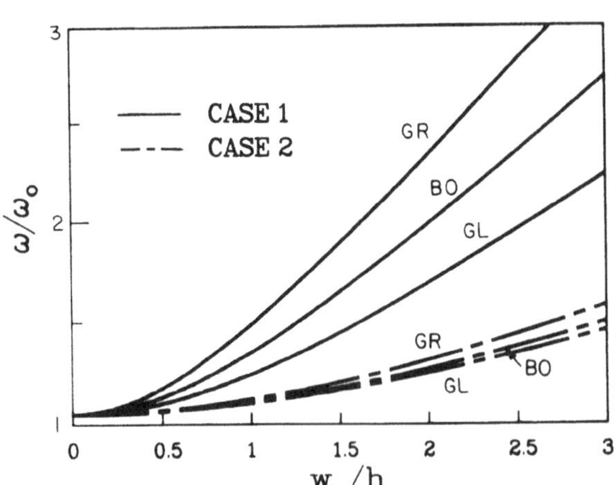

FIG. 6. Effect of material properties on frequency–amplitude response for two-layer antisymmetric $\pm 45^\circ$ angle-ply plate.

CONCLUSION

Based on the numerical results presented in this study, the following conclusions can be drawn regarding nonlinear static and dynamic behaviour of laminated circular plates.

The central deflection of an angle-ply plate increases as the fibre orientation angle increases for a given load. As the number of layers increases, the central deflection decreases. Thus, the coupling between bending and stretching reduces the stiffness of the plate.

As far as the dynamic behaviour of the plate is concerned, the hardening type of nonlinearity is observed. The frequency ratio increases substantially for all amplitudes in the immovable case but in the movable case the increase is not at all significant. Under a large number of layers the laminated plate behaves as a specially orthotropic plate.

REFERENCES

1. YAMAKI, N., Influences of large amplitudes on flexural vibrations of elastic plates, *Z. Angew. Math. Mech.*, **41** (1961), 501–510.
2. KUNG, G. C. and PAO, Y. U., Nonlinear flexural vibration of clamped plate, *ASME Trans. J. appl. Mech.*, **39** (1972), 1050–1054.
3. VENKATESHWARA RAO, G., KANAKARAJU, K. and RAJU, I. S., Finite element formulation for the large amplitude free vibrations of beams and orthotropic circular plates, *Comp. Struct.*, **6** (1976), 169–172.
4. ALWAR, R. S. and SEKHARREDDY, B., Large deflection static and dynamic analysis of isotropic and orthotropic annular plates, *Int. J. Nonlinear Mech.*, **14** (1979), 347–359.
5. SATYAMOORTHY, M. and CHIA, C. Y., Nonlinear vibration of orthotropic circular plates including transverse shear and rotatory inertia, *Proc. ASME Annual Winter Meeting on Modern Developments on Composite Materials and Structures*, 1979, pp. 357–372.
6. HUANG, C. L. and AURORA, P. R., Nonlinear oscillations of elastic orthotropic annular plates of variable thickness, *J. Sound Vib.*, **62** (1979), 443–453.
7. NATH, Y. and ALWAR, R. S., Nonlinear dynamic analysis of orthotropic circular plates, *Int. J. Solids Struct.*, **16** (1980), 433–443.
8. REDDY, J. N. and HUANG, C. L., Large amplitude free vibrations of annular plates of varying thickness, *J. Sound Vib.*, **79** (1981), 387–396.
9. SATYAMOORTHY, M., Nonlinear vibration of orthotropic circular plates—a comparison, *Fib. Sci. Technol.*, **16** (1982), 111–119.
10. VENKATESHWARA RAO, G. and KANAKARAJU, K., Optimization of cylindrically orthotropic circular plates, *J. appl. Math.*, **51** (1984), 354–360.
11. RUEI, K. H., JIANG, C. and CHIA, C. Y., Dynamic and static nonlinear analysis of cylindrically orthotropic circular plates with non-uniform edge constraints, *J. appl. Math. Phys.*, **35** (1984), 387–400.

12. VENKATESHWARA RAO, G. and KANAKARAJU, K., Large amplitude axisymmetric vibrations of orthotropic circular plates elastically restrained against rotation, *J. Sound Vib.*, **69** (1980), 175–180.
13. SATYAMOORTHY, M., Transverse shear and rotatory inertia effects on nonlinear vibrations of orthotropic circular plates, *Comp. Struct.*, **14** (1981), 129–134.
14. REDDY, J. N., HUANG, C. L. and SINGH, I. R., Large deflections and large amplitude vibrations of axisymmetric circular plates, *Int. J. Num. Meth. Engng*, **17** (1981), 527–541.
15. CHIA, C. Y. and SATYAMOORTHY, M., Nonlinear vibration of circular plates with transverse shear and rotatory inertia, *J. Sound Vib.*, **78** (1981), 131–137.
16. SATYAMOORTHY, M. and CHIA, C. Y., Nonlinear flexural vibration of moderately thick orthotropic circular plates, *Ingen. Archiv.*, **52** (1982), 237–243.
17. DUMIR, P. C., GANDHI, M. L. and NATH, Y., Nonlinear axisymmetric transient analysis of orthotropic thin annular plates, *Fib. Sci. Technol.*, **21** (1984), 23–40.
18. SHILKUST, D., Stability and vibrations of geometrically nonlinear cylindrically orthotropic circular plates, *J. appl. Mech.*, **51** (1984), 354–360.
19. CHIA, C. Y., *Nonlinear Analysis of Plates*, New York, McGraw-Hill, 1980.

34

The Use of Spline Functions in Calculating the Natural Frequencies of Anisotropic Rectangular Laminates

SHEMIN WANG

Department of Civil Engineering, Hebei Architecture and Engineering College, Handan City, People's Republic of China

and

D. J. DAWE

Department of Civil Engineering, University of Birmingham, PO Box 363, Birmingham B15 2TT, UK

ABSTRACT

A preliminary study is made of the effectiveness of cubic spline functions when used in potential-energy-based analyses of symmetrically-laminated, composite rectangular plates having orthotropic or anisotropic material properties. The problem under consideration is the determination of laminate natural frequencies and a first-order shear deformation plate theory is used. The analysis procedures employed are the Rayleigh–Ritz and finite strip methods. The numerical results presented show that the use of cubic spline functions is efficient for moderately-thick laminates and is particularly promising for anisotropic material. However, for truly thin laminates the efficiency is reduced considerably.

INTRODUCTION

When used as the basis for the analysis of laminated plates the classical plate theory (CPT) can give very misleading results for other than very thin laminates. The first-order shear deformation plate theory (SDPT)[1–4] removes the excessive constraint imposed by the Kirchhoff hypothesis by allowing initial normals to the middle surface to become non-normal, but straight, in the deformed configuration: this allows account to be taken of transverse, or through-thickness, shear deformation effects and of rotary inertia in dynamic applications. The resulting theoretical plate model yields results for the overall behaviour of moderately-thick plates which compare closely with the predictions of full three-dimensional elasticity theory and for this reason is adopted in the present study.

The concern here is with the calculation of the natural frequencies of lateral vibration of rectangular, fibre-reinforced composite laminates. Attention is restricted to symmetric lay-ups, so that out-of-plane and in-plane effects are uncoupled, but within this category of laminate allowance is made both for orthotropic properties with respect to the plate axes and for the more-difficult circumstance when the properties are anisotropic.

Analyses of the free vibration of rectangular homogeneous, isotropic plates and composite laminates within the context of SDPT, and incorporating a variety of edge conditions, have been described earlier by the second author and his colleagues.[5-11] Such analyses are based on the use of the Rayleigh–Ritz method (RRM) and one of its piecewise forms, the finite strip method (FSM). In using these potential energy methods in conjunction with SDPT, the assumed displacement fields comprise the definition of three independent reference quantities, namely the lateral deflection w and the two cross-sectional rotations ψ_x an ψ_y (along the x and y axes respectively). In the earlier works the definitions of w, ψ_x and ψ_y involve the use of series of smooth, continuous Timoshenko beam functions, with the specific definition of these functions changing with change in the plate edge conditions. The numerical results presented for natural frequencies demonstrate that convergence of results with use of an increasing number of series terms in the displacement field is very rapid for homogeneous plates[5-8] and for orthotropic laminates.[9-11] However, a limited number of results for an anisotropic square laminate show[9,10] a notably poorer rate of convergence, particularly for moderately-thick geometry.

In the present study the use of an alternative type of function in the definitions of the three references quantities is examined. These functions are spline functions which are a well-established mathematical tool, as evidenced by the existence of a number of books describing the functions, amongst which are those of Ahlberg *et al.*[12] and Prenter.[13] Their name arises from the drawing-office procedure of constructing a continuous smooth curve segment by segment using a flexible strip of material called a spline.

Spline functions have been used only to a limited extent in structural analysis but the literature does contain some applications to plates, based on the use of the CPT. These applications include the use of spline functions in Rayleigh–Ritz analyses by Antes[14] and by Mizusawa *et al.*[15] and, more recently, in a finite strip approach by Cheung and Fan.[16]

There are a number of types of spline function available but here attention is restricted to the cubic B_3-spline as used in the context of CPT in the earlier studies.[14–16] Attention is concentrated on the use of the spline functions in the RRM though some results are quoted for their use in the FSM as well. In this preliminary study only a few results can be presented but there is interest in seeing whether the use of spline functions, as compared to the use of beam functions, leads to an improvement in the analysis of anisotropic plates.

ANALYSIS

Laminate Equations

A typical rectangular plate is shown in Fig. 1. The components of translational displacement are u, v and w whilst ψ_x and ψ_y are the components of rotation along the x and y axes respectively. The plate is of uniform thickness h and, in general, is composed of a number nl, of layers of unidirectional fibre-reinforced composite material. The arrangement of layers is assumed to be such as to produce a symmetric laminate, with no bending–stretching coupling. Bending–twisting coupling is allowed,

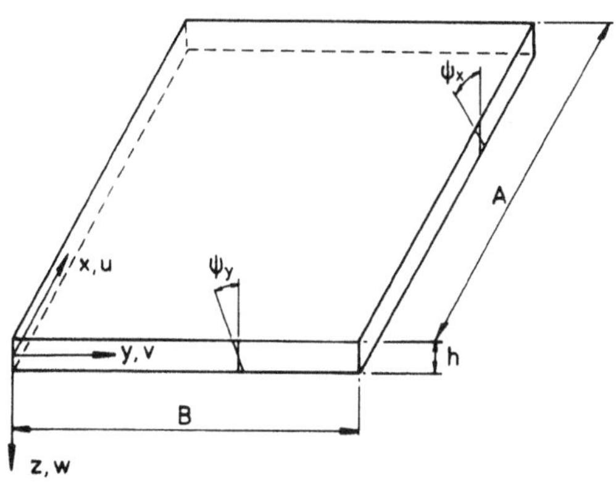

FIG. 1. Plate geometry.

though, and so the laminate can exhibit fully anisotropic bending properties through the presence of the D_{16} and D_{26} stiffness coefficients.

During free harmonic vibration with circular frequency p, the displacements u, v and w at a general point in the plate are assumed, within the context of SDPT,[2-4] to have the form

$$u = \exp(ipt)z\psi_x(x, y), v = \exp(ipt)z\psi_y(x, y), w = \exp(ipt)w(x, y) \qquad (1)$$

where $w(x, y)$, $\psi_x(x, y)$ and $\psi_y(x, y)$ are amplitudes of the motion at the point (x, y).

The maximum strain energy per unit middle-surface area of the vibrating laminate is

$$\Delta U = \frac{1}{2}\left[D_{11}\left(\frac{\partial\psi_x}{\partial x}\right)^2 + D_{22}\left(\frac{\partial\psi_y}{\partial y}\right)^2 + 2D_{12}\frac{\partial\psi_x}{\partial x}\frac{\partial\psi_y}{\partial y}\right.$$
$$+ D_{66}\left(\frac{\partial\psi_x}{\partial y} + \frac{\partial\psi_y}{\partial x}\right)^2 + 2D_{16}\frac{\partial\psi_x}{\partial x}\left(\frac{\partial\psi_x}{\partial y} + \frac{\partial\psi_y}{\partial x}\right)$$
$$+ 2D_{26}\frac{\partial\psi_y}{\partial y}\left(\frac{\partial\psi_x}{\partial y} + \frac{\partial\psi_y}{\partial x}\right)$$

$$+ A_{44}\left\{\psi_y^2 + 2\psi_y\frac{\partial w}{\partial y} + \left(\frac{\partial w}{\partial y}\right)^2\right\}$$
$$+ A_{55}\left\{\psi_x^2 + 2\psi_x\frac{\partial w}{\partial x} + \left(\frac{\partial w}{\partial x}\right)^2\right\}$$
$$\left. + 2A_{45}\left(\psi_x\psi_y + \psi_x\frac{\partial w}{\partial y} + \psi_y\frac{\partial w}{\partial x} + \frac{\partial w}{\partial x}\frac{\partial w}{\partial y}\right)\right] \qquad (2)$$

In this expression the stiffness coefficients are defined as

$$D_{ij} = \frac{1}{3}\sum_{l=1}^{nl}(Q_{ij})_l(h_l^3 - h_{l-1}^3) \qquad i,j = 1, 2, 6 \qquad (3)$$

$$A_{ij} = k_i k_j \sum_{l=1}^{nl}(Q_{ij})_l(h_l - h_{l-1}) \qquad i,j = 4, 5 \qquad (4)$$

where Q_{ij} for $i, j = 1, 2, 6$ are plane-stress reduced stiffnesses and Q_{ij} for $i, j = 4, 5$ are transverse shear stiffnesses.[4] The subscript l associated with Q_{ij}

refers to the lth layer and h_l is the distance from the laminate middle surface to the lower surface of the lth layer. The parameters $k_i k_j$ are the shear correction factors whose values must be prescribed: in this study these values are determined using the procedures of Chow[17] and Whitney.[18]

The maximum kinetic energy per unit middle-surface area of the vibrating laminate is

$$\Delta T = \frac{p^2}{2}\left[m_1 w^2 + m_2(\psi_x^2 + \psi_y^2)\right] \tag{5}$$

Here, with allowance made for variable material density,

$$m_1 = \sum_{l=1}^{nl} \rho_l(h_l - h_{l-1}) \quad \text{and} \quad m_2 = \frac{1}{3}\sum_{l=1}^{nl} \rho_l(h_l^3 - h_{l-1}^3) \tag{6}$$

where ρ_l is the material density of the lth layer.

To implement a Rayleigh–Ritz approach, whether of the continuous or piecewise type, we need to use assumed distributions of the reference quantities w, ψ_x and ψ_y in the above expressions for strain and kinetic energy densities. Such distributions must be admissible, of course, which means amongst other things that the kinematic boundary conditions of any particular plate must be satisfied precisely and that the natural boundary conditions must not be artificially prevented from being satisfied. In SDPT some standard boundary condition on, for example, an edge $x = \bar{x}$ (a constant) are:

$$\begin{array}{llll} \text{Simply supported} & w = 0, & \psi_y = 0, & M_x = 0 \\ \text{Clamped} & w = 0, & \psi_x = 0, & \psi_y = 0 \\ \text{Free} & M_x = 0, & M_{yx} = 0, & Q_x = 0 \end{array} \tag{7}$$

Here M_x, M_{xy} and Q_x are the bending moment (about the edge), twisting-moment and transverse shear force (all per unit length) respectively.

The Cubic B_3-spline Function[12,13]

The manner in which a reference quantity is represented by the B_3-spline function in one direction, chosen here to be the x-direction, is illustrated in Fig. 2. The plate dimension A in the x-direction is divided into q sections (Fig. 2a) which are taken to be of equal length, though this need not be so. Correspondingly there are $q + 3$ section knots, including one knot outside either end of the plate for the purposes of completing the definition of the function and of prescribing appropriate boundary conditions.

A function $f(x)$ (which represents each of the three reference quantities w,

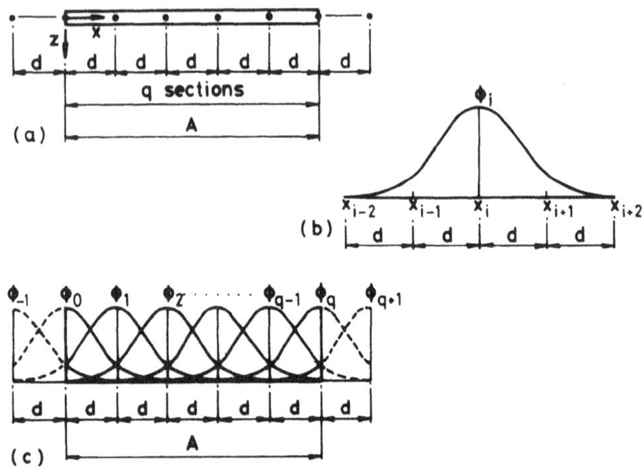

FIG. 2. Details of spline functions. (a) Division of plate length into sections; (b) local function ϕ_i, (c) combination of local functions.

ψ_x and ψ_y) is expressed as a summation of $(q+3)$ local cubic splines ϕ_i in the form

$$f(x) = \sum_{i=-1}^{q+1} \alpha_i \phi_i \tag{8}$$

where the α_i are coefficients to be determined. Each of the local spline functions ϕ_i has non-zero value over four consecutive sections, with its centre at a section knot located at $x = x_i$. Figure 2b shows a single-local function whilst Fig. 2c shows the combination of local functions contributing to the variation of $f(x)$ in eqn. (8).

The local spline function ϕ_i, with centre at $x = x_i$, is defined as

$$\phi_i(x) = \frac{1}{6d^3} \times \begin{cases} 0 & x < x_{i-2} \\ (x - x_{i-2})^3 & x_{i-2} \le x \le x_{i-1} \\ d^3 + 3d^2(x - x_{i-1}) + 3d(x - x_{i-1})^2 - 3(x - x_{i-1})^3 & x_{i-1} \le x \le x_i \\ d^3 + 3d^2(x_{i+1} - x) + 3d(x_{i+1} - x)^2 - 3(x_{i+1} - x)^3 & x_i \le x \le x_{i+1} \\ (x_{i+2} - x)^3 & x_{i+1} \le x \le x_{i+2} \\ 0 & x_{i+2} < x \end{cases} \tag{9}$$

The function is clearly a piecewise cubic polynomial function. The function itself and its first and second derivatives are continuous over the full range $x_{i-2} \leq x \leq x_{i+2}$ whilst at the extremities of this range the function and its first two derivatives have zero value.

Prescribed conditions at the boundaries at $x = 0$ or $x = A$ can be accommodated in a number of ways. One way is by modifying just three local spline functions, namely those functions at the end itself and immediately adjacent on either side of the end. For example, suppose that it is desired to prescribe a clamped boundary at $x = 0$. Then the conditions given by the second of eqns (7) can be met if, in the expressions for each of w, ψ_x and ψ_y, the local spline functions ϕ_{-1}, ϕ_0 and ϕ_1 are replaced with modified functions $\bar{\phi}_{-1}$, $\bar{\phi}_0$ and $\bar{\phi}_1$ such that

$$\bar{\phi}_{-1} = \phi_1, \bar{\phi}_0 = \phi_0 - 4\phi_{-1}, \bar{\phi}_1 = \phi_1 - \tfrac{1}{2}\phi_0 + \phi_{-1}$$

and $\bar{\phi}_{-1}$ is eliminated.

It is clear that the nature of the spline function is markedly different from that of the Timoshenko beam functions used in earlier work. The spline function uses the same basic function, whatever the end conditions, since such conditions are met by local modifications only. Furthermore, the spline function appears potentially to be more capable of representing rapid localised variations between the ends: the spline function approach is general and versatile, allowing for case of computer programming. On the other hand it would be expected that the Timoshenko beam functions, being individually tailored to match particular boundary conditions, would usually be more efficient than the spline function when used in predicting the natural frequencies of isotropic or orthotropic plates: certainly this would be the case where the end conditions are simple supports for then the sinusoidal beam functions represent the mode shapes exactly. For anisotropic laminates the orthogonal beam functions are less suitable than they are for orthotropic laminates and the splines function may, in prospect, allow better representation of the boundary conditions and of the skewed mode shapes.

Methods of Solution

The main numerical method used in generating the results presented in the next section is the Rayleigh–Ritz method. This method, when used in conjunction with series of Timoshenko beam modes, has been described quite fully in earlier reports.[6,8–11] In the present related report, concerned with the use of B_3-spline functions, only the barest details of the procedure are recorded.

In using the RRM the three fundamental quantities are assumed to vary spatially over the complete middle surface of the plate in the form

$$
\begin{aligned}
w(x, y) &= \sum_{n=-1}^{q+1} \sum_{m=-1}^{q+1} a_{mn} W_m(x) W_n(y) \\
\psi_x(x, y) &= \sum_{n=-1}^{q+1} \sum_{m=-1}^{q+1} b_{mn} \Psi_{xm}(x) \Psi_{xn}(y) \qquad (10) \\
\psi_y(x, y) &= \sum_{n=-1}^{q+1} \sum_{m=-1}^{q+1} c_{mn} \Psi_{ym}(x) \Psi_{yn}(y)
\end{aligned}
$$

The six unidirectional functions $W_m(x) \ldots \Psi_{yn}(y)$ are each B_3-spline functions, in the x- or y-directions, of the form of the function defined in eqns (8) and (9) and with appropriate modification to match the prescribed displacement conditions at the four plate edges. It is assumed that there are q sections across the plate in each coordinate direction. The a_{mn}, b_{mn} and c_{mn} are generalised displacements.

The RRM approach requires the substitution of the displacement field given by eqn. (10) into the expressions for strain and kinetic energies per unit plate area, eqns (2) and (5). Integration over the plate middle surface yields the whole-plate energies U and T and the usual minimisation procedure results in the linear eigenvalue form

$$(\mathbf{K} - p^2\mathbf{M})\mathbf{D} = \mathbf{0} \qquad (11)$$

Here $\mathbf{K}$ is a stiffness matrix (obtained from U), $\mathbf{M}$ is the corresponding consistent mass matrix (obtained from T), and $\mathbf{D}$ is the column matrix of all generalised displacements. It is noted that full integration is used in determining $\mathbf{K}$ and $\mathbf{D}$. The eigenvalues, p^2, of eqn. (11), and corresponding eigenvectors if desired, can be extracted using any of a number of standard procedures.

In the next section, for comparison purposes, some new results are also quoted which have been obtained using the finite strip method in conjunction with the use of the B_3-spline function. In the FSM the assumed spatial variation over a strip of each of the three fundamental quantities is a series of products of local B_3-spline functions in one direction (the longitudinal or x-direction) and polynomial functions of a selected order in the other (crosswise) direction. Details of the analytical procedure will not be given here but clearly the analysis has much in common with the earlier FSM approach based on the use of longitudinal Timoshenko beam functions and described at length elsewhere.[5,7,10,11]

NUMERICAL APPLICATIONS

In this preliminary study the results presented relate to the first four natural frequencies of vibration of only two types of composite plate.

The plates are of square planform, of side length A, and thicknesses, h, $0{\cdot}1A$ and $0{\cdot}01A$ are considered. The analysis of these plates has been considered in earlier work when utilising Timoshenko beam functions and this earlier work provides very useful comparative solutions. The main interest here is in examining the manner of convergence of RRM spline-function results and comparing with the manner of convergence of RRM beam-function results.[9,10]

In interpreting the results presented it should be noted that r is the number of Timoshenko beam functions used in representing each fundamental quantity in the x-direction (for the FSM) or in each direction (for the RRM), and similarly q is the number of spline sections along the plate in the x-direction or along each direction. The total number of degrees of freedom is N, the number of finite strips in a complete plate when using the FSM is N_s, and the type of finite strip in all cases corresponds to quartic crosswise polynomial interpolation. It should also be noted that the CPT values quoted apply to plates of both thin and moderately-thick geometry.

We first consider the vibration of a fibre-reinforced, five-layer, orthotropic cross-ply laminate, i.e. a 0°/90°/0°/90°/0° laminate where the fibre orientation angle is measured from the x-axis. All layers have identical material properties with

$$E_L/E_T = 30, \qquad G_{LT}/E_T = 0{\cdot}6, \qquad G_{TT}/E_T = 0{\cdot}5 \qquad \text{and} \qquad \nu_{LT} = 0{\cdot}25$$

where subscripts L and T refer to directions along and transverse to the fibres respectively. Each of the three 0° plies is of thickness $h/6$ and each of the two 90° plies is of thickness $h/4$. The shear correction factors are calculated[17,18] to be $k_4^2 = 0{\cdot}591\,39$ and $k_5^2 = 0{\cdot}873\,23$. (It is noted that values quoted in some earlier work[9,10] for this laminate are incorrect, having been interchanged inadvertently.) The laminate is simply supported at those of its edges which run parallel to the x-axis whilst being clamped at its other two edges, and is designated an SCSC laminate.

Results for the five-layer SCSC laminate are documented in Table 1. It is clear that for both moderately-thick ($h/A = 0{\cdot}1$) and thin ($h/A = 0{\cdot}01$) geometry the RRM beam-function approach gives excellent results for this orthotropic laminate, as has been established in a more general fashion in earlier work:[9–11] the Timoshenko beam functions are so suitable in this sort of application that very close comparison with exact, converged values (in the context of SDPT) is achieved using only $r = 2$. Turning to consideration of the results of the new RRM spline-function approach, it is seen from Table 1 that for the moderately-thick geometry the manner of convergence of frequency values is very satisfactory, given that in this sort of application it would not be expected that the approach would be more

TABLE 1

Calculated natural frequencies using beam and spline functions, for square, five-layer, orthotropic SCSC laminates

$$\Omega = p\frac{A^2}{h}\left[\frac{\rho}{(Q_{11})_t}\right]^{1/2} \quad \text{where } (Q_{11})_t = \sum_{l=1}^{nl}\frac{(Q_{11})_l(h_l - h_{l-1})}{h} = 0{\cdot}517\,745E_L$$

$\frac{h}{A}$	Type of function	Solution method		N	Values of Ω			
					Mode 1, 1	Mode 1, 2	Mode 2, 1	Mode 2, 2
0·1	Beam	RRM[10]:	$r = 2$	12	4·490	7·912	9·214	11·328
			$r = 3$	27	4·489	7·912	9·214	11·328
			$r = 4$	48	4·489	7·911	9·214	11·328
			$r = 5$	75	4·489	7·911	9·213	11·328
	Spline	RRM:	$q = 3$	56	4·491	7·966	9·252	11·393
			$q = 4$	89	4·489	7·925	9·222	11·344
			$q = 5$	120	4·489	7·912	9·215	11·330
			$q = 6$	161	4·489	7·910	9·213	11·327
		FSM:	$q = 6, N_s = 2$	161	4·489	7·910	9·213	11·324
0·01	Beam	RRM[10]:	$r = 2$	12	6·184	14·369	14·502	19·829
			$r = 3$	27	6·184	14·369	14·500	19·829
			$r = 4$	48	6·184	14·369	14·500	19·829
			$r = 5$	75	6·184	14·369	14·500	19·828
	Spline	RRM:	$q = 3$	56	6·403	22·054	15·981	26·720
			$q = 4$	89	6·232	16·060	14·890	21·322
			$q = 5$	120	6·196	14·871	14·612	20·258
			$q = 6$	161	6·188	14·532	14·534	19·964
		FSM:	$q = 6, N_s = 2$	161	6·184	14·382	14·533	19·846
	CPT Solution[10]			—	6·215	14·553	14·612	20·055

efficient than the RRM beam-function approach. There is also very close agreement between the RRM and FSM predictions. (In applying the FSM the spline functions are used in the x-direction, i.e. between simply-supported ends.) For the thin geometry, though, there is seen to be a quite dramatic loss of efficiency for the spline-function approach and this is an unexpected characteristic of the approach.

TABLE 2

Calculated natural frequencies, using beam and spline functions, for square anisotropic plates

$\frac{h}{A}$	Type of function	Solution method		N	Values of $p\frac{A^2}{h}\left(\frac{12(1-\nu_{LT}\nu_{TL})}{E_L}\right)^{1/2}$ Mode 1	Mode 2	Mode 3	Mode 4
0·1	Beam	RRM[9]:	$r=2$	12	15·17	23·86	29·43	36·85
			$r=3$	27	14·93	22·94	28·34	32·12
			$r=4$	48	14·78	22·76	27·86	31·73
			$r=5$	75	14·68	22·56	27·72	31·33
			$r=6$	108	14·63	22·50	27·55	31·24
		FSM[10]:	$r=6, N_s=3$	198	14·28	22·24	26·64	31·01
	Spline	RRM:	$q=3$	48	13·97	21·98	26·60	32·15
			$q=4$	75	13·94	21·59	26·30	31·07
			$q=5$	108	13·93	21·50	26·22	30·73
			$q=6$	147	13·93	21·47	26·19	30·62
			$q=7$	192	13·93	21·46	26·19	30·59
		FSM:	$q=6, N_s=2$	147	13·93	21·46	26·19	30·62
0·01	Beam	RRM[9]:	$r=2$	12	21·71	36·08	56·48	73·09
			$r=3$	27	21·45	33·44	52·84	53·16
			$r=4$	48	21·33	33·28	50·66	51·69
			$r=5$	75	21·27	33·05	50·54	51·45
			$r=6$	108	21·25	33·02	50·33	51·30
		FSM[10]:	$r=6, N_s=3$	198	21·22	33·00	50·43	51·21
	Spline	RRM.	$q=3$	48	23·83	67·90	79·84	106·66
			$q=4$	75	22·09	42·30	59·63	75·50
			$q=5$	108	21·52	36·06	54·55	64·24
			$q=6$	147	21·32	34·06	52·51	56·88
			$q=7$	192	21·24	33·34	51·63	53·07
		FSM:	$q=6, N_s=2$	147	21·23	33·00	52·12	51·01
	CPT Solution[19]			—	21·35	33·18	50·72	51·87

The second type of plate considered is a single-layer, anisotropic, square plate with all edges clamped[19] (i.e. a CCCC plate) and with material properties defined as

$$E_L/E_T = 10, \qquad G_{LT}/E_T = G_{TT}/E_T = 0{\cdot}25, \qquad \nu_{LT} = 0{\cdot}3$$

The fibre direction is inclined at 30° to the x-axis and hence bending–stretching coupling occurs since D_{16} and D_{26} are not zero. The shear correction factors used are $k_4^2 = k_5^2 = 5/6$.

Results for the anisotropic CCCC plate are given in Table 2. Considering first the results corresponding to the use of Timoshenko beam functions (presented originally in Ref. 9) it is clear that for both thickness ratios the manner of convergence of the RRM results is notably poorer for the anisotropic plate than it is for orthotropic plates. Further, for the moderately-thick plate the comparison between RRM and FSM values is not very close. It is apparent that the orthogonal Timoshenko beam functions are much less suitable for the analysis of anisotropic plates (where nodal lines are highly skewed[9]) than they are for orthotropic plates. When using the B_3-spline functions the results obtained for the moderately-thick plate are impressive: RRM results converge quickly with increase in q and there is close comparison between RRM and FSM results. For the thin plate, though, the spline function results are again disappointing, with poor rate of convergence exhibited in the RRM approach.

For both of the above types of plate it is clear that for moderately-thick geometry the CPT frequency predictions are unrealistically high.

DISCUSSION AND CONCLUSIONS

In this preliminary study a new approach to the determination of the natural frequencies of rectangular laminated plates has been presented in which cubic spline functions are incorporated in the assumed displacement fields used in Rayleigh–Ritz and finite strip method analyses, within the context of first-order shear deformation plate theory. Previous studies have shown that orthogonal Timoshenko beam functions are highly appropriate for orthotropic laminates but are markedly less so for anisotropic laminates. The main objective in introducing the cubic spline functions here was to discover whether their use would lead to improvement, vis à vis the use of Timoshenko beam functions, in predicting the frequencies of anisotropic plates.

It has been demonstrated that the use of the cubic spline functions gives results of good accuracy for moderately-thick plates and, in particular, does indeed yield superior results for an anisotropic plate. This suggests that the cubic spline functions may well be of considerable use in determining frequencies (and buckling stresses) of a range of moderately-thick anisotropic laminates.

Unfortunately the picture changes considerably for truly thin plates, for then the accuracy associated with the use of the cubic spline functions has been shown here to decrease very markedly, whether the plate has orthotropic or anisotropic properties. This effect appears to be a form of locking behaviour, of a similar nature to that associated with a number of kinds of shear-deformable, isoparametric plate finite elements. It may be possible to introduce remedial measures to nullify, or at least significantly reduce, the locking effect so that the cubic spline functions can be used successfully in analysing truly thin plates. Alternatively, it may be that other types of spline function may prove to be more satisfactory in this regard than the cubic spline function. It is planned to examine both these possibilities in a further, extended study of the use of spline functions in the analysis of laminates.

REFERENCES

1. Reissner, E., The effect of transverse shear deformation on the bending of elastic plates, *J. appl. Mech.*, **18** (1951).
2. Mindlin, R. D., Influence of rotary inertia and shear on flexural motions of isotropic elastic plastics, *J. appl. Mech.*, **18** (1951).
3. Yang, P. C., Norris, C. H. and Stavsky, Y., Elastic wave propagation in heterogeneous plates, *Int. J. Solids Struct.*, **2** (1966), 665–684.
4. Whitney, J. M. and Pagano, N. J., Shear deformation in heterogeneous anisotropic plates, *J. appl. Mech.*, **37** (1970), 1031–1036.
5. Dawe, D. J., Finite strip models for vibration of Mindlin plates, *J. Sound Vib.*, **59** (1978), 441–452.
6. Dawe, D. J. and Roufaeil, O. L., Rayleigh–Ritz vibration analysis of Mindlin plates, *J. Sound Vib.*, **69** (1980), 345–359.
7. Roufaeil, O. L. and Dawe, D. J., Vibration analysis of rectangular Mindlin plates by the finite strip method, *Computers and Structures*, **12** (1980), 833–842.
8. Roufaeil, O. L. and Dawe, D. J., Rayleigh–Ritz vibration analysis of rectangular Mindlin plates subjected to membrane stresses, *J. Sound Vib.*, **85** (1982), 263–275.
9. Dawe, D. J. and Craig, T. J., The influence of shear deformation on the natural frequencies of laminated rectangular plates, in: *Composite Structures—3*, (I. H. Marshall ed.). London, Elsevier Applied Science, 1985, pp. 660–676.
10. Craig, T. J. and Dawe, D. J., Flexural vibration of symmetrically-laminated composite, rectangular plates, including transverse shear effects, *Int. J. Solids Struct.*, **22** (1986), 155–169.

11. DAWE, D. J. and CRAIG, T. J., The vibration and stability of symmetrically-laminated composite rectangular plates subjected to in-plane stresses, *Composite Structures*, **5** (1986), 281–307.
12. AHLBERG, J. H., NILSON, E. M. and WALSH, J. L. *The Theory of Splines and their Applications*, New York, Academic Press, 1967.
13. PRENTER, P. M., *Splines and Variational Methods*, New York, Wiley, 1975.
14. ANTES, H., Bicubic fundamental splines in plate bending, *Int. J. Num. Meth. Engng*, **8** (1974).
15. MIZUSAWA, T., KAJITA, T. and NARUOKA, M., Vibration of stiffened skew plates by using B-spline functions, *Computers and Structures*, **10** (1979), 821–826.
16. CHEUNG, Y. K. and FAN, S. C., Static analysis of right box girder bridges by spline finite strip method, *Proc. Inst. Civ. Engrs*, **75** (Part 2) (1983), 311–323.
17. CHOW, T. S., On the propagation of flexural waves in an orthotropic laminated plate and its response to an impulsive load, *J. Comp. Mater.*, **5** (1971), 306–319.
18. WHITNEY, J. M., Shear correction factors for orthotropic laminates under static load, *J. appl. Mech.*, **40** (1973), 302–304.
19. ASHTON, J. E. and WHITNEY, J. M., *Theory of Laminated Plates* (Section 5.8), Stamford, Conn., Technomic Publishing Co. Inc., 1970.

35

Vibrations of Unsymmetrically Laminated Plates Subjected to Inplane Initial Stresses

C. A. Fisher and M. S. Ewing

Department of Engineering Mechanics, United States Air Force Academy, Colorado Springs, Colorado 80840, USA

and

A. W. Leissa

Department of Engineering Mechanics, Ohio State University, Columbus, Ohio 43210, USA

ABSTRACT

Exact, closed-form equations are presented for the free vibration frequencies of unsymmetrically laminated cross-ply and angle-ply plates having simply-supported edge conditions and subjected to uniaxial or biaxial inplane, static loading conditions. Parametric studies are made for both cross-ply and angle-ply plates showing the effects of changing inplane loading magnitude (N_x), biaxial stress ratio (N_y/N_x), aspect ratio (a/b), material stiffness ratios, number of plies and angle-ply orientation angle upon the frequencies.

INTRODUCTION

There are numerous structural applications where flat plates are subjected to static, inplane stresses and are subsequently caused to vibrate. It is generally known that the effect of such initially applied inplane stresses is to change the natural frequencies of a plate, increasing them with tensile stresses and decreasing them with compressive or shear stresses. This is important to know in a forced vibration situation, for the resonant peak amplitudes of the response will be shifted accordingly. It is also generally known that at sufficiently large values of compressive (or shear) stress the lowest natural frequency will approach zero, which corresponds to the critical buckling stress of the plate.

In the literature one can find on the order of a hundred published papers dealing with the vibrations of *isotropic, homogeneous* plates subjected to inplane initial stresses. Laminated composite plates are, of course, not homogeneous. However, if they are symmetrically laminated, they are representable by homogeneous, orthotropic or anisotropic plate theory. For such plates one can find approximately a dozen relevant references. Unsymmetrically laminated plates involve coupling between bending and stretching. This coupling raises the order of the governing differential equations of motion from four to eight, and the number of boundary conditions which must be satisfied along each edge from two to four. The resulting mathematical problem is considerably more difficult, and virtually no results may be found for vibrations in the presence of inplane stress.

Some data are available for the free vibrations of unsymmetrically laminated plates in the absence of inplane stresses. These were typically obtained by approximate methods such as finite elements or Ritz–Galerkin procedures. Exact, closed-form solutions for free vibration frequencies, as well as buckling stresses, for unsymmetrically laminated cross-ply and angle-ply plates having certain types of simply-supported edges were presented in a paper by Whitney and Leissa.[1] Subsequently, Jones[2] made corrections to the numerical results given in the previous paper,[1] and generated a formula for vibration frequencies of cross-ply plates which accounted for inplane stresses, although no numerical results for such problems were presented.

The purpose of the present chapter is threefold:

1. To derive a closed-form formula for the free vibration frequencies of unsymmetrically laminated, simply-supported, angle-ply plates subjected to inplane initial stresses.
2. To make a parameter study for both cross-ply and angle-ply plates and present useful numerical results showing how the frequencies are affected by the parameters.
3. To compare frequencies of orthotropic plates having inplane stresses with those of unsymmetrically laminated plates, to

determine the importance of using bending–stretching coupling theory for such problems.

For angle-ply plates having simply-supported edges with inplane normal displacement constraint, but no tangential constraint, the free vibration solution of Ref. 1 is generalized to include inplane initial stresses which may be either tensile or compressive, and which may act in either direction (uniaxial loading) or both (biaxial loading). The resulting solution is exact and takes the form of an explicit, closed-form formula for the frequencies.

ANALYSIS

The differential equations of motion for the free vibration of a laminated composite plate subjected to inplane static forces are given by (cf. Ref. 3, Appendix)

$$\begin{bmatrix} L_{11} & L_{12} & L_{13} \\ L_{12} & L_{22} & L_{23} \\ L_{13} & L_{23} & L_{33} \end{bmatrix} \begin{bmatrix} u \\ v \\ w \end{bmatrix} = \begin{bmatrix} 0 \\ 0 \\ N^{\mathrm{i}} \end{bmatrix} - \rho h \frac{\partial^2}{\partial t^2} \begin{bmatrix} -u \\ -v \\ w \end{bmatrix} \tag{1}$$

where u and v are inplane displacement components in the x and y directions, respectively; w is the transverse displacement; ρ is the mass per unit volume (averaged across the thickness in the case of a laminated plate); h is the total plate thickness; t is time; and N^{i} represents the static inplane stress effects:

$$N^{\mathrm{i}} = N_x^{\mathrm{i}} \frac{\partial^2 w}{\partial x^2} + 2N_{xy}^{\mathrm{i}} \frac{\partial^2 w}{\partial x\, \partial y} + N_y^{\mathrm{i}} \frac{\partial^2 w}{\partial y^2} \tag{2}$$

Here N_x^{i} $(=\sigma_x^{\mathrm{i}} h)$ and N_y^{i} $(=\sigma_y^{\mathrm{i}} h)$ represent the inplane normal stress resultants (force per unit length), taken to be positive in tension, and N_{xy}^{i} $(=\tau_{xy}^{\mathrm{i}} h)$ is the inplane shear stress resultant. It must be emphasized that in the present work $N_x^{\mathrm{i}}, N_y^{\mathrm{i}}, N_{xy}^{\mathrm{i}}$ are *initial* inplane stress resultants which are statically applied to the flat plate, and that the plate remains flat after their application. In an unsymmetrically laminated plate there is coupling between bending and stretching during its transverse vibratory motion, which causes additional inplane stresses to be generated within the plate, and which vary sinusoidally with time.

The L_{ij} are different operators given by Ref. 1:

$$L_{11} \equiv A_{11}\frac{\partial^2}{\partial x^2} + 2A_{16}\frac{\partial^2}{\partial x\,\partial y} + A_{66}\frac{\partial^2}{\partial y^2}$$

$$L_{22} \equiv A_{22}\frac{\partial^2}{\partial y^2} + 2A_{26}\frac{\partial^2}{\partial x\,\partial y} + A_{66}\frac{\partial^2}{\partial x^2}$$

$$L_{33} \equiv D_{11}\frac{\partial^4}{\partial x^4} + 4D_{16}\frac{\partial^4}{\partial x^3\,\partial y} + 2(D_{12}+2D_{66})\frac{\partial^4}{\partial x^2\,\partial y^2} + 4D_{26}\frac{\partial^4}{\partial x\,\partial y^3} + D_{22}\frac{\partial^4}{\partial y^4} \tag{3}$$

$$L_{12} = L_{21} \equiv A_{16}\frac{\partial^2}{\partial x^2} + (A_{12}+A_{66})\frac{\partial^2}{\partial x\,\partial y} + A_{26}\frac{\partial^2}{\partial y^2}$$

$$L_{13} = L_{31} \equiv -B_{11}\frac{\partial^3}{\partial x^3} - 3B_{16}\frac{\partial^3}{\partial x^2\,\partial y} - (B_{12}+2B_{66})\frac{\partial^3}{\partial x\,\partial y^2} - B_{26}\frac{\partial^3}{\partial y^3}$$

$$L_{23} = L_{32} \equiv -B_{16}\frac{\partial^3}{\partial x^3} - (B_{12}+2B_{66})\frac{\partial^3}{\partial x^2\,\partial y} - 3B_{26}\frac{\partial^3}{\partial x\,\partial y^2} - B_{22}\frac{\partial^3}{\partial y^3}$$

where the A_{ij}, B_{ij}, D_{ij} are the well-known stiffness coefficients which relate the inplane stress resultants (N_x, N_y, N_{xy}) and bending moments and twisting moment resultants (M_x, M_y, M_{xy}) generated within the plate *during* vibration to the midplane strains ($\varepsilon_x, \varepsilon_y, \gamma_{xy}$) and curvature changes ($\kappa_x, \kappa_y, \kappa_{xy}$) occurring during vibration:

$$\begin{bmatrix} N_x \\ N_y \\ N_{xy} \\ \hline M_x \\ M_y \\ M_{xy} \end{bmatrix} = \left[\begin{array}{ccc|ccc} A_{11} & A_{12} & A_{16} & B_{11} & B_{12} & B_{16} \\ A_{12} & A_{22} & A_{26} & B_{12} & B_{22} & B_{26} \\ A_{16} & A_{26} & A_{66} & B_{16} & B_{26} & B_{66} \\ \hline B_{11} & B_{12} & B_{16} & D_{11} & D_{12} & D_{16} \\ B_{12} & B_{22} & B_{26} & D_{12} & D_{22} & D_{26} \\ B_{16} & B_{26} & B_{66} & D_{16} & D_{26} & D_{66} \end{array}\right] \begin{bmatrix} \varepsilon_x^0 \\ \varepsilon_y^0 \\ \gamma_{xy}^0 \\ \hline -\kappa_x \\ -\kappa_y \\ -\kappa_{xy} \end{bmatrix} \tag{4}$$

Equations (1) have exact, closed-form solutions for two special, but important, cases for rectangular plates having uniform inplane normal stresses ($N_x^i =$ constant, $N_y^i =$ constant), no inplane shear stress ($N_{xy}^i = 0$) and simply-supported edge conditions: (1) unsymmetrical cross-ply and (2) antisymmetrical angle-ply.

In the case of unsymmetrically laminated cross-ply plates[1]

$$A_{16} = A_{26} = B_{16} = B_{26} = D_{16} = D_{26} = 0 \tag{5}$$

Solutions to eqns (1) may be taken in the form

$$u = a_{mn} \cos\frac{m\pi x}{a} \sin\frac{n\pi y}{b} \sin \omega t \tag{6a}$$

$$v = b_{mn} \sin\frac{m\pi x}{a} \cos\frac{n\pi y}{b} \sin \omega t \tag{6b}$$

$$w = c_{mn} \sin\frac{m\pi x}{a} \sin\frac{n\pi y}{b} \sin \omega t \tag{6c}$$

where m and n are integers; ω is the circular natural frequency; and a_{mn}, b_{mn}, c_{mn} are arbitrary coefficients. The displacement functions given by eqns (6) satisfy shear diaphragm (or S2) boundary conditions ($w = M_n = N_n = u_t = 0$) for a simply-supported plate having four edges $x = 0, a$ and $y = 0, b$. Substituting eqns (6) into (1) and neglecting inplane inertia terms [$\rho h(\partial^2 u/\partial t^2)$ and $\rho h(\partial v^2/\partial t^2)$], which are small for the flexural vibration modes, allows one to solve explicitly for the natural frequencies in the form

$$\rho h\omega^2 = N_x\alpha^2 + N_y\beta^2 + C_{33} + \frac{2C_{12}C_{13}C_{23} - C_{11}C_{23}^2 - C_{22}C_{13}^2}{C_{11}C_{22} - C_{12}^2} \tag{7}$$

where $\alpha = m\pi/a$, $\beta = n\pi/b$ and

$$\begin{aligned} C_{11} &= A_{11}\alpha^2 + A_{66}\beta^2 \\ C_{22} &= A_{22}\beta^2 + A_{66}\alpha^2 \\ C_{33} &= D_{11}\alpha^4 + 2(D_{12} + 2D_{66})\alpha^2\beta^2 + D_{22}\beta^4 \\ C_{12} &= C_{21} = (A_{12} + A_{66})\alpha\beta \\ C_{13} &= C_{31} = B_{11}\alpha^3 + (B_{12} + 2B_{66})\alpha\beta^2 \\ C_{23} &= C_{32} = (B_{12} + 2B_{66})\alpha^2\beta + B_{22}\beta^3 \end{aligned} \tag{8}$$

It is observed that eqn. (7) yields *all* the flexural vibration frequencies arising from the integers m and n. It is also seen that positive (i.e. tensile) values of N_x and N_y can only increase each frequency. Equation (7) was obtained in a similar form by Jones.[2] A similar form, restricted to antisymmetrically laminated cross-plies, was presented earlier by Whitney and Leissa.[1]

In the case of antisymmetrically laminated angle-ply plates[1]

$$A_{16} = A_{26} = D_{16} = D_{26} = B_{11} = B_{22} = B_{12} = B_{66} = 0 \tag{9}$$

Solutions to eqns (1) may be taken in the form

$$
\begin{aligned}
u &= a_{mn} \sin\frac{m\pi x}{a}\cos\frac{n\pi y}{b}\sin\omega t \\
v &= b_{mn} \cos\frac{m\pi x}{a}\sin\frac{n\pi y}{b}\sin\omega t \\
w &= c_{mn} \sin\frac{m\pi x}{a}\sin\frac{n\pi y}{b}\sin\omega t
\end{aligned}
\tag{10}
$$

These displacement functions satisfy hinge-free (or S3) boundary conditions ($w = M_n = u_n = N_{nt} = 0$) along the edges $x = 0, a$ and $y = 0, b$. Substituting eqns (10) into (1) (again neglecting tangential inertia) yields the frequency equation given previously as eqn. (7), except that the C_{ij} are now defined by

$$
\begin{aligned}
C_{11} &= -(A_{11}\alpha^2 + A_{66}\beta^2) \\
C_{22} &= -(A_{22}\beta^2 + A_{66}\alpha^2) \\
C_{33} &= D_{11}\alpha^4 + 2(D_{12} + 2D_{66})\alpha^2\beta^2 + D_{22}\beta^4 \\
C_{12} &= C_{21} = -(A_{12} + A_{66})\alpha\beta \\
C_{13} &= C_{31} = 3B_{16}\alpha^2\beta + B_{26}\beta^3 \\
C_{23} &= C_{32} = B_{16}\alpha^3 + 3B_{26}\alpha\beta^2
\end{aligned}
\tag{11}
$$

RESULTS

In conducting the various parametric studies for both angle ply and cross ply it was concluded that the most meaningful parameters to plot were

$$
\frac{\omega^2 \rho b^4}{E_2 h^3} \quad \text{versus} \quad \frac{N_x b^2}{E_2 h^3}
$$

For a given least plate dimension (b), transverse lamina modulus (E_2), density (ρ) and plate thickness (h), variations in ω and N_x may be studied. Substitution of the plate dimension (a) in the above two parameters was considered but dismissed due to a less favorable graphical presentation.

Angle Ply

The 'base problem' considered is a 45° angle-ply laminate with two layers where $N_y/N_x = 0$ and $E_1/E_2 = 15$. A family of frequency curves was generated for different integer values of aspect ratio, beginning with $a/b = 1$. Figure 1 illustrates the characteristic behavior of these angle-ply plates.

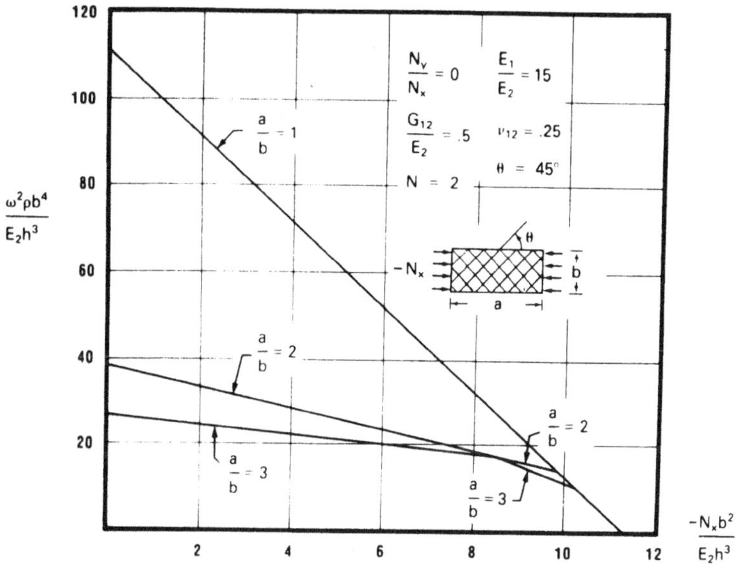

FIG. 1. Natural frequencies versus inplane loading for rectangular angle-ply laminated plates with various aspect ratios ($N_y/N_x = 0$).

All 'curves' shown are actually straight line segments, each corrc-sponding to a unique mode shape, i.e. a certain number of half-sine waves. Where two or more line segments make up a 'curve', the interpretation is that the mode shape of the fundamental mode changes as in-plane load is increased. For example, in Fig. 1 for $a/b = 2$, as $N_x b^2/E_2 h^3$ increases above approximately 9·7, the vibration mode shape shifts from one half-sine wave to two ($m = 2$ in eqn. (7)). When $N_x b^2/E_2 h^3$ equals 11·25, the critical buckling load, the plate buckles with the two half-sine wave mode shapes in the direction of loading.

In the next series of plots, the lay-up angle was varied between 0° and 90°. An inspection of Fig. 2 for an aspect ratio of one (square plate) reveals that lower ply angle plates have a higher critical buckling parameter and are therefore better suited for applications where buckling is a consideration. Also, for high ply angles (e.g. 75° and 90°), as N_x increases, the fundamental vibration mode shapes involve two or three half-sine waves. For $a/b = 1$, symmetry of buckling parameters exists for complementary angles. For aspect ratio plates higher than unity, however, this symmetry disappears. Interestingly, for integer values of plate aspect ratio, the critical buckling loads are constant only for 45° (11·25) and 90° (8·46) ply angles.

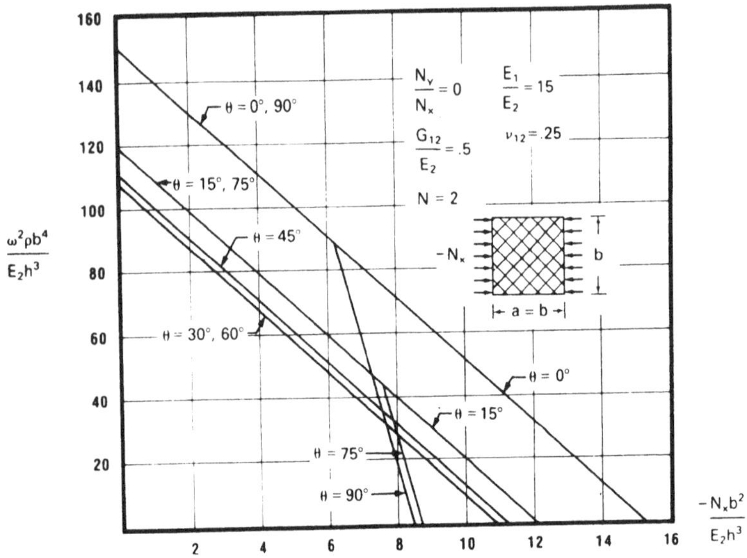

FIG. 2. Natural frequencies versus inplane loading for square angle-ply plates with various lay-up angles ($N_y/N_x = 0$).

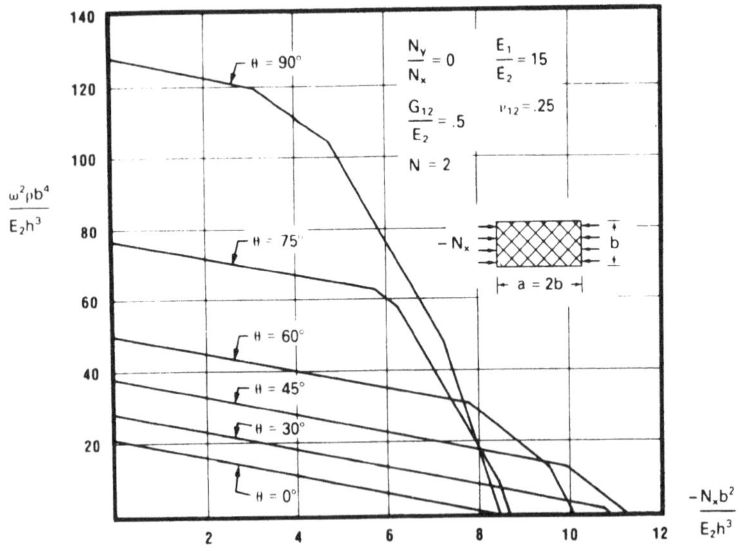

FIG. 3. Natural frequencies versus inplane loading for angle-ply plates with various lay-up angles ($a/b = 2$, $N_y/N_x = 0$).

TABLE 1
Vibration and buckling results for square angle-ply laminated plates
$(G_{12}/E_2=0{\cdot}5,\ \nu_{12}=0{\cdot}25,\ \theta=45°,\ N=2,\ N_y/N_x=0)$

E_1/E_2	Frequency parameter for $N_x=0$	Inplane loading parameter for buckling
1	34·634	3·509 1
10	89·482	9·006 4
15	111·06	11·253
20	132·03	13·377
30	173·31	17·560
40	214·26	21·708

Tables 1 and 2 tabulate the results for variations in E_1/E_2 and the number of plies, N. As the stiffness modulus ratio increases there is a corresponding increase in plate natural frequency and critical buckling load. Similarly, as the number of plies, N, is increased, the buckling and free vibration frequencies increase. However, for $N=2$ both values are less than half of the symmetric orthotropic ones (which occur for $N\rightarrow\infty$).

The next series of results consider the 45° angle-ply plate to be loaded in compression on all four edges ($N_y/N_x=1$). Figure 4 depicts the family of curves for different plate aspect ratios. As expected, the critical buckling load is dramatically reduced from the earlier loading where only two edges were placed in compression. In addition, up to an aspect ratio of 5, only double half-sine waves ($m=n=1$) for vibration and buckling mode shapes occur for 45° ply angles. Also, the critical buckling load is no longer a single value for integer values of aspect ratio, as had been observed earlier in Fig. 1. If the plate aspect ratio is held constant and the lay-up angle, θ, is

TABLE 2
Vibration and buckling results for square angle-ply laminated plates
$(E_1/E_2=25,\ G_{12}/E_2=0{\cdot}5,\ \nu_{12}=0{\cdot}25,\ \theta=45°,\ N_y/N_x=0)$

N	Frequency parameter for $N_x=0$	Inplane loading parameter for buckling
2	111·06	11·253
4	223·40	22·635
6	244·20	24·742
10	254·85	25·822
100	260·78	26·422

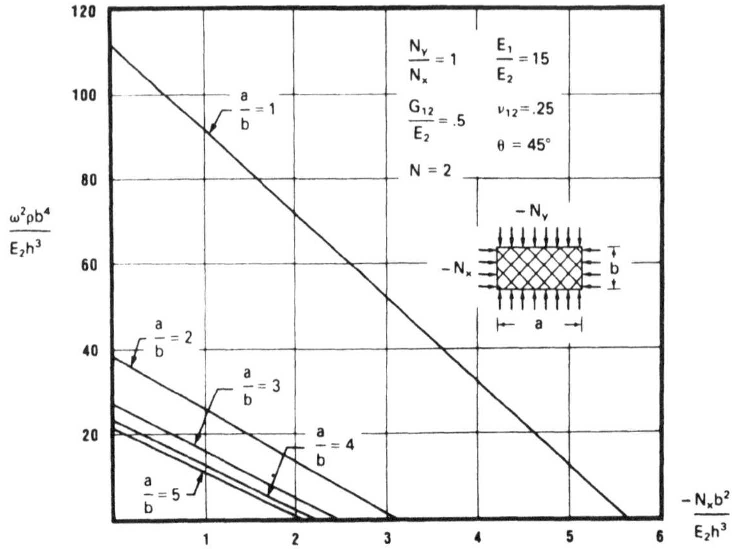

FIG. 4. Natural frequencies versus inplane loading for rectangular angle-ply plates with various aspect ratios ($N_y/N_x = 1$).

permitted to vary (Fig. 5) only for the $\theta = 0°$ and $\theta = 90°$ cases do the two half-sine wave vibration and buckling mode shapes occur.

Unusual results were observed for the loading represented by $N_y/N_x = -1$ (one side in compression, one in tension). From Fig. 6, for a square plate and a wide range of N_x the natural frequency is unaffected. As soon as the second and third half-sine wave mode shapes become dominant, little buckling margin remains. For plates of higher aspect ratio, increasing N_x actually causes an increase in natural frequency up to a maximum, namely the free vibration frequency of the square plate. This phenomena occurs because N_y provides a stiffening effect along a wider boundary than N_x for $a > b$. Figure 7 illustrates the effect of different lay-up angles, θ, on vibration and buckling for this 'tension–compression' loading. Unlike earlier loading cases shown in Fig. 2 ($N_y/N_x = 0$) and in Fig. 5 ($N_y/N_x = 1$), this loading situation results in a constant plate natural frequency for a large loading range, up to 90% of the initial buckling load. As a result, critical buckling occurs for small additional increases in N_x above the point where the natural frequency first decreases.

Cross Ply

Cross-ply parametric studies were conducted for completeness as well as for correlation with previously published results. Figures 8–10 and Tables 3

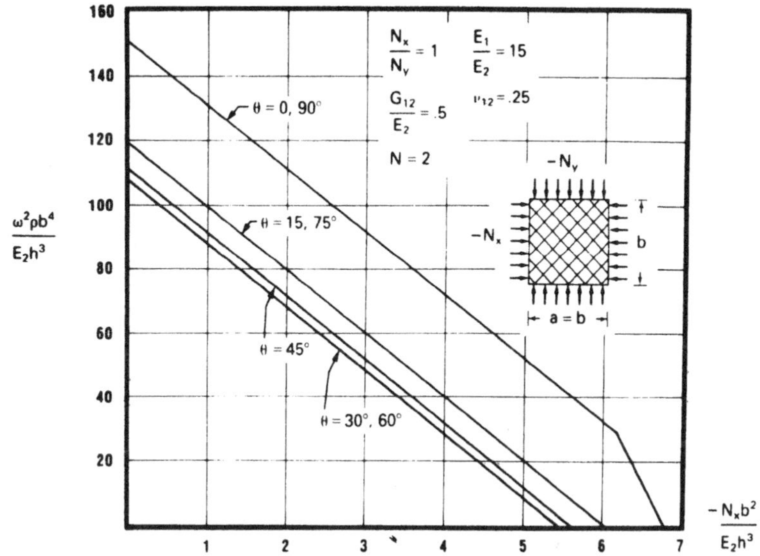

FIG. 5. Natural frequencies versus inplane loading for square angle-ply plates with various lay-up angles ($N_y/N_x = 1$).

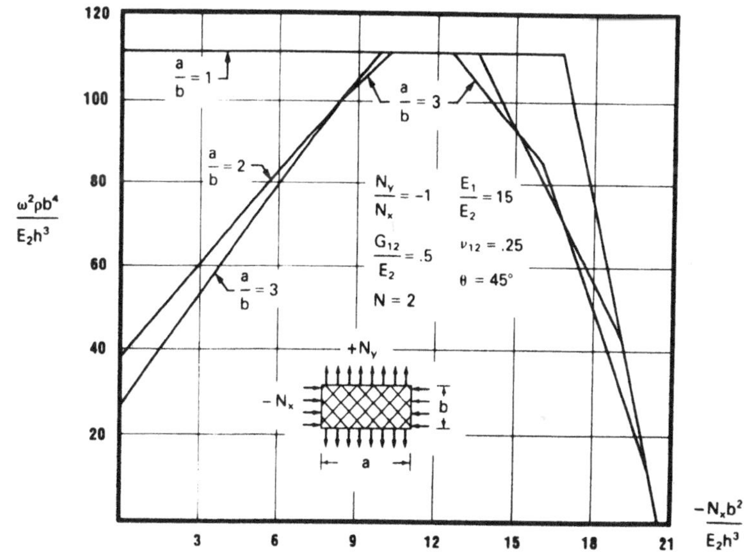

FIG. 6. Natural frequencies versus inplane loading for rectangular angle-ply plates with various aspect ratios ($N_y/N_x = -1$).

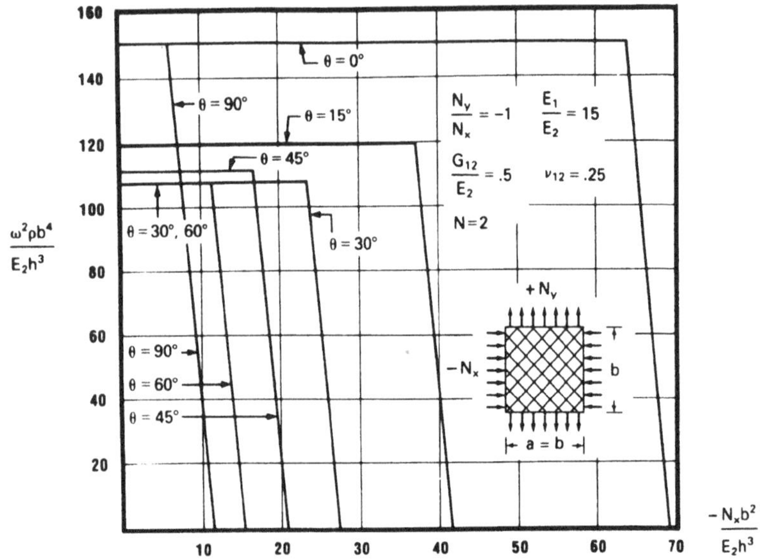

FIG. 7. Natural frequencies versus inplane loading for square angle-ply plates with various lay-up angles ($N_y/N_x = -1$).

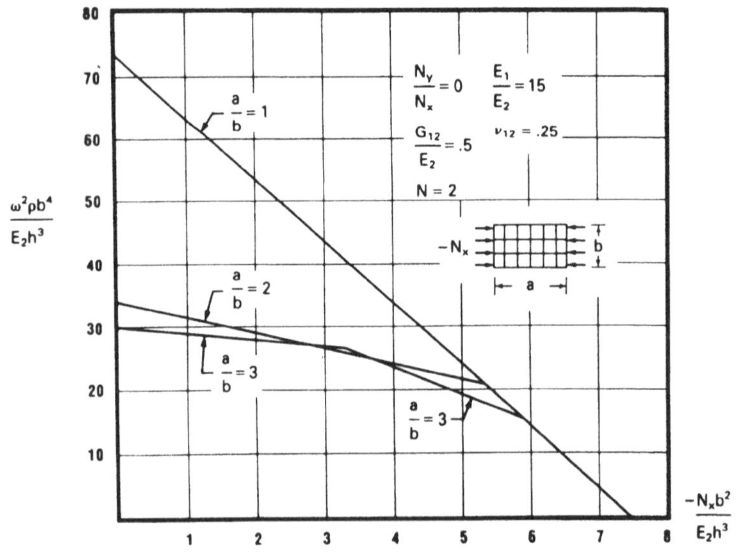

FIG. 8. Natural frequencies versus inplane loading for rectangular cross-ply laminated plates with various aspect ratios ($N_y/N_x = 0$).

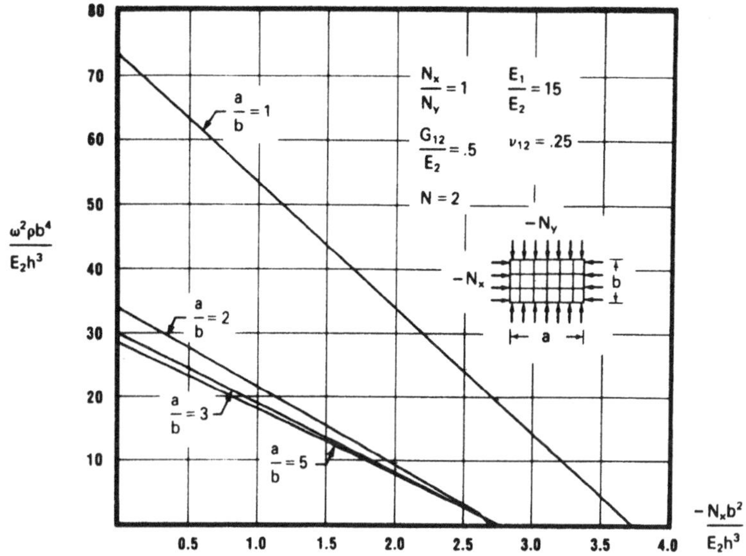

FIG. 9. Natural frequencies versus inplane loading for rectangular cross-ply plates with various aspect ratios ($N_y/N_x = 1$).

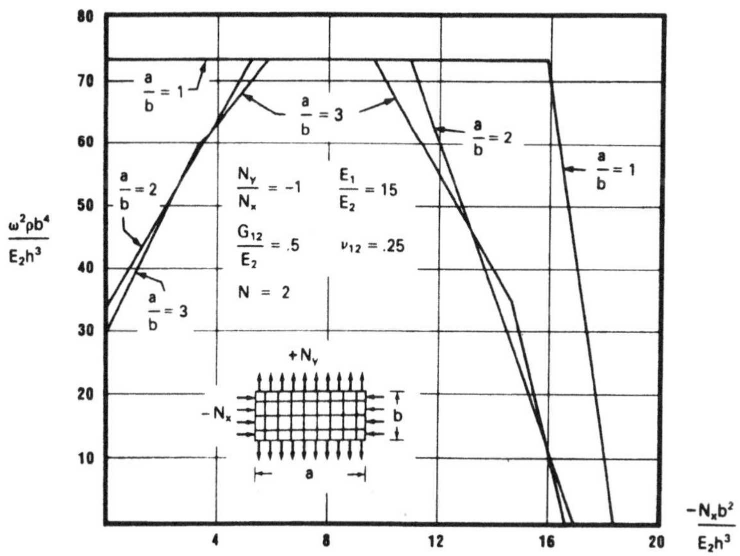

FIG. 10. Natural frequencies versus inplane loading for rectangular cross-ply plates with various aspect ratios ($N_y/N_x = -1$).

TABLE 3

Vibration and buckling results for square cross-ply laminated plates ($G_{12}/E_2 = 0{\cdot}5$, $\nu_{12} = 0{\cdot}25$, $\theta = 45°$, $N = 2$, $N_y/N_x = 0$)

E_1/E_2	Frequency parameter for $N_x = 0$	Inplane loading parameter for buckling
1	37·881	3·838 1
10	62·912	6·374 3
15	73·426	7·439 6
20	83·761	8·486 7
30	104·25	10·562
40	124·64	12·628

and 4 may be compared to the earlier similar graphs of angle-ply laminates. As expected, angle-ply and cross-ply laminates behave in much the same manner.

The 'base problem' is a two-layer plate with $N_y/N_x = 0$ and $E_1/E_2 = 15$. Figure 8 illustrates the effect of varying aspect ratios of the plate. These results are very similar to those of 45° angle-ply plates (Fig. 1). Tables 3 and 4 tabulate the results for variations in E_1/E_2 and the number of plies, N. Values for $N = 2$ are again less than half of the orthotropic ones ($N \rightarrow \infty$).

The case of hydrostatic loading ($N_y/N_x = 1$) is displayed in Fig. 9, where once again a/b is varied. The property of square plates having a significantly higher buckling load than plates with higher aspect ratios is well known for isotropic plates. It is seen here as well for the cross-ply plate. For 'tension–compression' buckling, Fig. 10 illustrates the same type of behaviors as for 45° angle-ply plates (Fig. 6). Interestingly, however, a square plate has approximately an 8% higher buckling load than other plates of integer aspect ratio. For 45° angle-ply plates, the buckling loads are essentially the same for all plates of integer aspect ratio.

TABLE 4

Vibration and buckling results for square cross-ply laminated plates ($E_1/E_2 = 25$, $G_{12}/E_2 = 0{\cdot}5$, $\nu_{12} = 0{\cdot}25$, $\theta = 45°$, $N_y/N_x = 0$)

N	Frequency parameter for $N_x = 0$	Inplane loading parameter for buckling
2	73·426	7·439 6
4	131·41	13·314
6	142·14	14·402
10	147·64	14·959
100	150·70	15·269

REFERENCES

1. Whitney, J. M. and Leissa, A. W., Analysis of heterogeneous anisotropic plates, *J. appl. Mech.* (ASME Transactions), **36**, No. 1 (1969), 261–266.
2. Jones, R. M., Buckling and vibration of unsymmetrically laminated cross-ply rectangular plates, *AIAA Journal*, **11**, No. 1 (1973), 1626–1632.
3. Leissa, A. W., Buckling of laminated composite plates and shell panels, Flight Dynamics Laboratory Report AFWAL-TR-85-3069, Wright-Patterson Air Force Base, 1985, 439 pp.

36

Maximum Frequency Design of Symmetric Angle-ply Laminates

ROBERT REISS and S. RAMACHANDRAN

Department of Mechanical Engineering, Large Space Structure Institute, Howard University, Washington, DC 20059, USA

ABSTRACT

This chapter considers the optimal design of classical laminates for maximum fundamental frequency. The constituent laminae are orthotropic and are stacked so that the geometry and material axis orientation are symmetric with respect to the plate's middle surface. The design variables are the orientations of the constituent laminae. Design derivatives of the fundamental frequency are used to determine conditions both necessary and sufficient to ensure at least a local maximum for the fundamental frequency. Specific results are presented for four- and eight-ply laminates subjected to various combinations of clamped, supported and free edge boundary conditions.

1. INTRODUCTION

1.1. Background

In comparison to the rather voluminous research studies, analytical and experimental, concerning laminated composites, the literature is noticeably lacking in studies on optimum design. In one of the earlier studies on this subject, Housner and Stein[1] determined, parametrically, the orientation of angle-ply laminates of prescribed thickness which maximized the shear buckling stress. Slightly more general but related optimization problems were solved using direct numerically based methods by Hirano.[2-3] Bert successfully determined the optimal orientation that maximized fundamental frequency for simply supported[4] and clamped[5] rectangular laminates, symmetric about their mid-plane. Bert's method, however, required *a priori* knowledge, exact or approximate, of the fundamental frequency in terms of the plate's flexural and torsional rigidities.

Tauchert and Adibhatla[6] and Cheng[7] consider the maximum stiffness design of laminated composites. The former study uses a Ritz method to represent the transverse laminate displacement and then they minimize the potential energy functional; Cheng, however, in his treatment of the same problem, differentiates the response equation with respect to the design variables and then invokes Gauss' theorem to obtain the design derivative of the compliance (work).

1.2. Review of the Fundamental Equations for Optimal Design

In a recent paper, Reiss[8] derived a general formula for the design derivatives (first and second) of eigenvalues for any self-adjoint distributed parameter structural system. Since that paper provides the underlying basis for the present study, the principal results of that paper will be reviewed here. In what follows, it is implicitly assumed that the fundamental frequency is not a repeated eigenvalue.

The class of eigenvalue problems to be considered are defined through the operator field equation

$$T^*E(S)Tw = \lambda M(S)w \tag{1.1}$$

Appropriate mixed boundary conditions must be adjoined to eqn. (1.1). Also, T and T^* are L_2 adjoint differential operators, S collectively denotes all of the design variables, M is an inertial operator, w is the eigenfunction and λ the corresponding eigenvalue.

Each eigenvalue λ of eqn. (1.1) is clearly a functional of the design S. Assuming the design changes by a small amount δS, the least eigenvalue λ_1 will change by $\delta\lambda_1$ where

$$\delta\lambda_1 = \frac{(Tw_1, \delta ETw_1)_\Omega - \lambda_1(w_1, \delta Mw_1)^2_\Omega}{(w_1, Mw_1)_\Omega} \tag{1.2}$$

In eqn. (1.2), δE and δM denote the design variations in the stiffness and inertial operators, respectively, the subscript 1 denotes the least eigenvalue and corresponding eigenfunction, and $(\cdot,\cdot)_\Omega$ denotes the L_2 bilinear form over the spatial domain Ω. If w_1 is normalized with respect to M, then the second variation $\delta^2\lambda_1$ is expressed by the relation

$$\delta^2\lambda_1 = (Tw_1, \delta^2 ETw_1)_\Omega - \lambda_1(w_1, \delta^2 Mw_1)_\Omega - 2\,\delta\lambda_1(w_1, \delta Mw_1)_\Omega$$
$$-2\sum_{n=2}^{\infty} \frac{[\lambda_1(w_n, \delta Mw_1)_\Omega - (Tw_n, \delta ETw_1)_\Omega]}{\lambda_n - \lambda_1} \tag{1.3}$$

For the class of problems under consideration in this paper, $\delta M = 0$. Thus the optimality criterion becomes

$$(Tw_1, \delta ETw_1)_\Omega = 0 \tag{1.4}$$

In light of eqn. (1.4), the second variation becomes

$$\delta^2 \lambda = (Tw_1, \delta^2 ETw_1)_\Omega - 2 \sum_{n=2}^{\infty} \frac{[(Tw_n, \delta ETw_1)_\Omega]^2}{\lambda_n - \lambda_1} \tag{1.5}$$

Equations (1.4) and (1.5) provide the starting point for this investigation.

2. PROBLEM FORMULATION

Consider a laminate consisting of orthotropic laminae which are perfectly bonded at common interfaces. If, for simplicity, the plate is symmetric with respect to its middle surface in both constituent laminae geometry and material axis orientation, then the stiffness coefficients coupling extensional strains to the bending moments vanish. In this case the plate equations of unforced motion are

$$D_{11}w_{,xxxx} + 4D_{16}w_{,xxxy} + 2(D_{12} + 2D_{66})w_{,xxyy} + D_{22}w_{,yyyy} + 4D_{26}w_{,xyyy} = \rho\omega^2 w \tag{2.1}$$

where ρ is the laminate density, ω is the natural flexural frequency and w is the amplitude of the transverse deflection of the laminate. The subscript following the comma denotes differentiation with respect to the indicated argument. The bending stiffnesses D_{ij} are given by

$$D_{ij} = \frac{1}{3} \sum_k \bar{Q}_{ij}(\theta_k)(z_k^3 - z_{k-1}^3) \tag{2.2}$$

where z_{k-1} and z_k are the respective coordinates of the top and bottom surfaces of the kth lamina, and $\bar{Q}_{ij}^k$, the transformed reduced stiffnesses of the kth lamina, depend only upon the lamina orientation θ_k through

$$\begin{aligned}
\bar{Q}_{11} &= U_1 + U_2 \cos 2\theta_k + U_3 \cos 4\theta_k \\
\bar{Q}_{12} &= U_4 - U_3 \cos 4\theta_k \\
\bar{Q}_{22} &= U_1 - U_2 \cos 2\theta_k + U_3 \cos 4\theta_k \\
\bar{Q}_{16} &= \tfrac{1}{2} U_2 \sin 2\theta_k + U_3 \sin 4\theta_k \\
\bar{Q}_{26} &= \tfrac{1}{2} U_2 \sin 2\theta_k - U_3 \sin 4\theta_k \\
\bar{Q}_{66} &= U_5 - U_3 \cos 4\theta_k
\end{aligned} \tag{2.3}$$

Here, $U_i (i = 1, \ldots, 5)$ are material constants[9] independent of material axes orientation.

Equations (2.1)–(2.3) may be cast in the form (1.1) by identifying the operators E, M, T and T^*, the design variables S and the eigenvalue λ. Thus

$$E = \begin{pmatrix} D_{11} & D_{12} & D_{16} \\ D_{12} & D_{22} & D_{26} \\ D_{16} & D_{26} & D_{66} \end{pmatrix} \tag{2.4}$$

$$T = \begin{bmatrix} \partial^2/\partial x^2 \\ \partial^2/\partial y^2 \\ 2\partial^2/\partial x \partial y \end{bmatrix}, \quad S = \begin{bmatrix} \theta_1 \\ \vdots \\ \theta_n \end{bmatrix} \tag{2.5}$$

$$M = \rho, \qquad \lambda = \omega^2, \qquad T^* = T^{\mathrm{T}} \tag{2.6}$$

After differentiating E with respect to θ_k and substituting that result and eqns (2.5) into eqn. (1.4), the optimality condition to determine the orientation of each lamina becomes

$$2U_2 \cos 2\theta_k I_3 + 8U_3 \cos 4\theta_k I_4 + U_2 \sin 2\theta_k I_1 - 2U_3 \sin 4\theta_k I_2 = 0 \tag{2.7}$$

where

$$I_3 = \iint\limits_{\Omega} w_{,xy} \nabla^2 w \, \mathrm{d}A \qquad I_4 = \iint\limits_{\Omega} w_{,xy}(w_{,xx} - w_{,yy}) \, \mathrm{d}A$$

$$I_1 = \iint\limits_{\Omega} (w_{,yy}^2 = w_{,xx}^2) \, \mathrm{d}A \qquad I_2 = \iint\limits_{\Omega} [(w_{,xx} - w_{,yy})^2 - 4w_{,xy}^2] \, \mathrm{d}A \tag{2.8}$$

The simplicity of eqn. (2.7) can be misleading. The integrals $I_1, \ldots, I_4$ are explicit scalar-valued functionals of the eigenvector w; they are also implicit functionals of the design set $\{\theta_1, \ldots, \theta_n\}$. In any event, they are scalars which are uniquely determined by the design set. By treating them as fixed scalars, it is obvious that each member θ_k of the design set must satisfy eqn. (2.7). It can easily be shown that regardless of the values of the scalars $I_1, \ldots, I_4$, there can be no more than four different solutions to eqn. (2.7). Thus, there can be no more than four different orientations in the design set $\{\theta_1, \ldots, \theta_n\}$. However, there may be many different sets of solutions $\{\theta_1, \ldots, \theta_n\}$. There will be as many different sets of solutions—each one containing no more than four different orientations—as there are stationary points for the fundamental frequency of the laminate.

The foregoing result was recently presented by Reiss.[10] The companion result for minimum compliance design was similarly and independently obtained by Cheng.[7] This result can be generalized to other cost functionals as well.

In many cases, the boundary conditions are symmetric in both the x and y directions. An example is the laminate with opposite edges clamped and the other pair of opposite edges supported. In this case, every specially orthotropic solution meeting the mid-plane symmetry requirements satisfies the optimality criterion eqn. (2.7). For such designs $Q_{16}^k = Q_{26}^k = 0$ for each k, and consequently the coupling stiffnesses D_{16} and D_{26} also vanish. Thus the eigenmode solutions to eqn. (2.1) are all either symmetric or skew-symmetric in each principal direction. As a result, both I_1 and I_2 vanish identically. And since both $\sin 2\theta$ and $\sin 4\theta$ vanish for $\theta = 0$ and $\theta = \pi/2$, eqn. (2.7) is identically satisfied by every specially orthotropic solution.

3. ANGLE-PLY LAMINATES

3.1. Optimality Criterion

We now consider a balanced four-ply symmetric laminate $[\theta/-\theta]_s$. For this laminate there is but one design variable θ. The flexural stiffnesses, determined from eqn. (2.2), are

$$\begin{aligned}
D_{11} &= \tfrac{16}{3}h^3(U_1 + U_2\cos 2\theta + U_3\cos 4\theta)\\
D_{12} &= \tfrac{16}{3}h^3(U_4 - U_3\cos 4\theta)\\
D_{22} &= \tfrac{16}{3}h^3(U_1 - U_2\cos 2\theta + U_3\cos 4\theta)\\
D_{16} &= \tfrac{12}{3}h^3(\tfrac{1}{2}U_2\sin 2\theta + U_3\sin 4\theta)\\
D_{26} &= \tfrac{12}{3}h^3(\tfrac{1}{2}U_2\sin 2\theta - U_3\sin 4\theta)\\
D_{66} &= \tfrac{16}{3}h^3(U_5 - U_3\cos 4\theta)
\end{aligned} \tag{3.1}$$

where h is the common thickness of the laminae.

Substitution of the stiffnesses (3.1) into eqn. (1.4) determines the optimality condition

$$2U_2\sin 2\theta I_1 - 4U_3\sin 4\theta I_2 + 3U_2\cos 2\theta I_3 + 12U_3\cos 4\theta I_4 = 0 \tag{3.2}$$

where $I_1, \ldots, I_4$ are defined through eqns (2.8).

A similar calculation for a symmetric balanced 8-ply laminate $[\theta/-\theta/-\theta/\theta]_s$ results in the optimality condition

$$8U_2 \sin 2\theta I_1 - 16U_3 \sin 4\theta I_2 + 3U_2 \cos 2\theta I_3 + 12U_3 \cos 4\theta I_4 = 0 \qquad (3.3)$$

3.2. Method of Solution

It now remains to solve the field equation (2.1) together with the optimality condition either (3.2) or (3.3) for ω_1, w_1 and θ. These non-linear equations can be solved iteratively. For any assumed design θ, eqn. (2.1) can be solved for ω_1 and w_1. The eigenfunction w_1 may now be used to compute the integrals $I_1, \ldots, I_4$, which are substituted back into the optimality condition to update θ.

In order to solve the eigenvalue equation (2.1), the lowest eigenmode w_1 is represented in a generalized double Fourier series

$$w_1(x, y) = \sum_{i=1}^{N} \sum_{j=1}^{N} \alpha_{ij} \Phi_i(x/a) \Theta_j(y/b) \qquad (3.4)$$

where a and b are the characteristic dimensions of the laminate in the x and y directions, respectively. A convenient choice for these shape functions are the eigenfunctions of a vibrating Bernoulli–Euler beam with corresponding opposite edge boundary conditions. Thus the representation (3.4) is necessarily complete and kinematically admissible. As N gets arbitrarily large, the partial sums will converge to the lowest eigenfunction.

The coefficients α_{ij} are determined by minimizing the appropriate Rayleigh quotient. This results in the well-known algebraic eigenvalue equation[11]

$$\begin{aligned}\sum_k \sum_l \{ & D_{11} R^{-3} A_{ki} b_{lj} + D_{12} R^{-1} (C_{ki} c_{jl} + C_{ik} c_{lj}) \\ & + D_{22} R B_{ki} a_{lj} + 4D_{66} R^{-1} E_{ki} e_{lj} + 2D_{16} R^{-2} (H_{ki} g_{jl} \\ & + H_{ik} g_{lj}) + 2D_{26} (G_{ki} h_{jl} + G_{ik} h_{lj}) - \rho\omega^2 R b^4 B_{ki} b_{lj} \} \alpha_{kl} = 0 \end{aligned} \qquad (3.5)$$

where R is the aspect ratio a/b.

There are a number of integrals involving the shape functions appearing in eqn. (3.5). Specifically these are

$$\begin{aligned} A_{ij} &= (\Phi_i'', \Phi_j'') & a_{ij} &= (\Theta_i'', \Theta_j'') \\ B_{ij} &= (\Phi_i, \Phi_j) & b_{ij} &= (\Theta_i, \Theta_j) \\ C_{ij} &= (\Phi_i'', \Phi_j) & c_{ij} &= (\Theta_i'', \Theta_j) \\ E_{ij} &= (\Phi_i', \Phi_j') & e_{ij} &= (\Theta_i', \Theta_j') \\ H_{ij} &= (\Phi_i'', \Phi_j') & h_{ij} &= (\Theta_i'', \Theta_j') \\ G_{ij} &= (\Phi_i', \Phi_j) & g_{ij} &= (\Theta_i', \Theta_j) \end{aligned} \tag{3.6}$$

where

$$(u, v) \equiv \int_0^1 u(x)v(x)\,\mathrm{d}x$$

The coefficients (3.6) are computed just once for each set of boundary conditions.

The eigenvalue equation (3.5) can be readily solved if the design variable θ is known. For, in this case, the only unknowns appearing in eqn. (3.5) are the eigenvalue ω and the eigenmatrix $\boldsymbol{\alpha}$ whose elements are α_{ij}.

It is convenient to introduce the matrices $\mathbf{A}$ with elements A_{ij}, $\mathbf{B}$ with elements B_{ij}, etc. The integrals appearing in eqn. (2.8) may now be conveniently expressed through

$$\begin{aligned} I_1 &= R^3\mathrm{tr}(\boldsymbol{\alpha}\mathbf{a}\boldsymbol{\alpha}^{\mathrm{T}}\mathbf{B}) - R^{-1}\mathrm{tr}(\boldsymbol{\alpha}\mathbf{b}\boldsymbol{\alpha}^{\mathrm{T}}\mathbf{A}) \\ I_2 &= R^3\mathrm{tr}(\boldsymbol{\alpha}\mathbf{a}\boldsymbol{\alpha}^{\mathrm{T}}\mathbf{B}) + R^{-1}\mathrm{tr}(\boldsymbol{\alpha}\mathbf{b}\boldsymbol{\alpha}^{\mathrm{T}}\mathbf{A}) \\ &\quad - 4R\mathrm{tr}(\boldsymbol{\alpha}\mathbf{c}\boldsymbol{\alpha}^{\mathrm{T}}\mathbf{C}^{\mathrm{T}}) - 2R\mathrm{tr}(\boldsymbol{\alpha}\mathbf{e}\boldsymbol{\alpha}^{\mathrm{T}}\mathbf{E}^{\mathrm{T}}) \\ I_3 &= \mathrm{tr}(\boldsymbol{\alpha}\mathbf{g}\boldsymbol{\alpha}^{\mathrm{T}}\mathbf{H}) + R^2\mathrm{tr}(\boldsymbol{\alpha}\mathbf{h}^{\mathrm{T}}\boldsymbol{\alpha}^{\mathrm{T}}\mathbf{G}^{\mathrm{T}}) \\ I_4 &= \mathrm{tr}(\boldsymbol{\alpha}\mathbf{g}\boldsymbol{\alpha}^{\mathrm{T}}\mathbf{H}) - R^2\mathrm{tr}(\boldsymbol{\alpha}\mathbf{h}^{\mathrm{T}}\boldsymbol{\alpha}^{\mathrm{T}}\mathbf{G}^{\mathrm{T}}) \end{aligned} \tag{3.7}$$

In eqns (3.7) the notation tr denotes the trace of the indicated matrix. The representation (3.7) facilitates the computation of $I_1, \ldots, I_4$, and consequently the solution of eqn. (2.7).

3.3. Sufficiency Condition

Every solution to eqns (2.1) and (3.2) or (3.3) provides a stationary value for the fundamental frequency. It will be necessary to examine the second design derivative of the frequency in order to determine whether or not the stationary value is indeed a local maximum. This second derivative is calculated from eqn. (1.5).

According to eqn. (3.5), there will be N^2 eigenfrequencies $\omega_i (i = 1, \ldots, N^2)$. It will be assumed that these eigenfrequencies are ordered so that

$$\omega_1 < \omega_2 \leq \omega_3 \ldots < \omega_{N^2}$$

Corresponding to each ω_k is an eigenmatrix $\boldsymbol{\alpha}^k$ calculated from eqn. (3.5). The second design derivative of the frequency is obtained by substituting eqn. (3.4) into eqn. (1.5). The result for a four-ply laminate is

$$\frac{d^2\omega^2}{d\theta^2} = \frac{32}{3} h^3 [2I_1 U_2 \cos 2\theta - 8I_2 U_3 \cos 4\theta - 3I_3 U_2 \sin 2\theta - 24 I_4 \sin 4\theta]$$
$$- \frac{16}{3} h^3 \sum_{k=2}^{N^2} \frac{1}{\omega_k^2 - \omega_1^2} [4J_1^k U_2 \sin 2\theta$$
$$+ 8J_2^k U_3 \sin 4\theta + 3J_3^k U_2 \cos 2\theta + 12J_4^k U_3 \cos 4\theta]^2 \tag{3.8}$$

where

$$J_1^k = R^3 \mathrm{tr}(\boldsymbol{\alpha}^k \mathbf{a} \boldsymbol{\alpha}^{\mathrm{T}} \mathbf{B}^{\mathrm{T}}) - R^{-1} \mathrm{tr}(\boldsymbol{\alpha}^k \mathbf{b} \boldsymbol{\alpha}^{\mathrm{T}} \mathbf{A}^{\mathrm{T}})$$
$$J_2^k = R \mathrm{tr}(\boldsymbol{\alpha}^k \mathbf{c}^{\mathrm{T}} \boldsymbol{\alpha}^{\mathrm{T}} \mathbf{C}^{\mathrm{T}}) - R^{-1} \mathrm{tr}(\boldsymbol{\alpha}^k \mathbf{b} \boldsymbol{\alpha}^{\mathrm{T}} \mathbf{A}^{\mathrm{T}}) \tag{3.9}$$
$$+ R \mathrm{tr}(\boldsymbol{\alpha}^k \mathbf{c} \boldsymbol{\alpha}^{\mathrm{T}} \mathbf{C}^{\mathrm{T}}) - R^3 \mathrm{tr}(\boldsymbol{\alpha}^k \mathbf{a} \boldsymbol{\alpha}^{\mathrm{T}} \mathbf{B}^{\mathrm{T}}) + 4R \mathrm{tr}(\boldsymbol{\alpha}^k \mathbf{e} \boldsymbol{\alpha}^{\mathrm{T}} \mathbf{E}^{\mathrm{T}})$$
$$J_{3,4}^k = \mathrm{tr}(\boldsymbol{\alpha}^{k\mathrm{T}} \mathbf{H} \boldsymbol{\alpha} \mathbf{g}) \pm R^2 \mathrm{tr}(\boldsymbol{\alpha}^k \mathbf{h} \boldsymbol{\alpha}^{\mathrm{T}} \mathbf{G}) + \mathrm{tr}(\boldsymbol{\alpha}^k \mathbf{g} \boldsymbol{\alpha}^{\mathrm{T}} \mathbf{H}) \pm R^2 \mathrm{tr}(\boldsymbol{\alpha}^k \mathbf{h}^{\mathrm{T}} \boldsymbol{\alpha}^{\mathrm{T}} \mathbf{G}^{\mathrm{T}})$$

With the aid of eqns (3.8) and (3.9), it is a straightforward matter to determine the sign of $d^2\omega^2/d\theta^2$ and therefore whether the stationary eigenfrequency is a local maximum, minimum or saddle point in the design space.

4. RESULTS

4.1. Approximate Closed-form Solutions

In the simplest case, the eigenfunction expansion (3.4) will contain but one term, i.e.

$$w_{11} \alpha_{11} \Phi_1(x/a) \Theta_1(y/b) \tag{4.1}$$

The eigenvalue problem (3.5) then admits the closed-form solution

$$\omega^2 = [D_{11} R^{-3} A_{11} b_{11} + 2D_{12} R^{-1} C_{11} c_{11} + R D_{22} B_{11} a_{11}$$
$$+ 4D_{66} R^{-1} E_{11} e_{11} + 4D_{16} R^{-2} H_{11} g_{11} + 4D_{26} G_{11} h_{11}] / \rho R b^4 b_{11} B_{11} \tag{4.2}$$

The matrices in eqn. (3.7) all consist of one element. Thus

$$
\begin{aligned}
I_1 &= (R^3 a_{11} B_{11} - R^{-1} b_{11} A_{11})\alpha_{11}^2 \\
I_2 &= (R^3 a_{11} B_{11} + R^{-1} b_{11} A_{11} - 6RC_{11}c_{11})\alpha_{11}^2 \\
I_3 &= (g_{11} H_{11} + R^2 h_{11} G_{11})\alpha_{11}^2 \\
I_4 &= (g_{11} H_{11} - R^2 h_{11} G_{11})\alpha_{11}^2
\end{aligned}
\tag{4.3}
$$

In the special cases in which all pairs of opposite edges are either clamped or supported, $H_{11} = G_{11} = h_{11} = g_{11} = 0$. In these cases, both I_3 and I_4 vanish identically and both of the optimality conditions (3.2) and (3.3) simplify to

$$U_2 \sin 2\theta I_1 - 2U_3 \sin 4\theta I_2 = 0 \tag{4.4}$$

for which the solutions are

$$\theta = 0, \pi/2, \frac{1}{2}\cos^{-1}\frac{U_2}{4U_3}\frac{I_1}{I_2} \tag{4.5}$$

Here

$$\frac{I_1}{I_2} = \frac{R^4 a_{11} B_{11} - b_{11} A_{11}}{R^4 a_{11} B_{11} + b_{11} A_{11} - 6R^2 C_{11} c_{11}} \tag{4.6}$$

For the special cases in which all four edges are simply supported, eqn. (4.6) becomes

$$\frac{I_1}{I_2} = \frac{R^4 - 1}{R^4 + 1 - 6R^2} \tag{4.7}$$

It is interesting to note that the solutions (4.5) and (4.7) are identical to Bert's[4] optimum design. Similarly, it can be shown that for the clamped plate, the ratio (4.6) becomes

$$\frac{I_1}{I_2} = \frac{R^4 - 1}{R^4 + 1 - 1{\cdot}814R^2} \tag{4·8}$$

Bert's[5] solution for this problem, again obtained quite differently, may be obtained from eqn. (4.7) by replacing the constant 1·814 by 1·785.

4.2. Exact Optimum Designs

When more than one term in the series (3.4) is retained, it is necessary to resort to numerical techniques to solve the eigenvalue equation (3.5) and the associated optimality condition. The following results were obtained using

TABLE 1
All four sides simply supported
($E_1/E_2 = 10$, $\nu_{12} = 0{\cdot}3$, $G_{12}/E_1 = 0{\cdot}025$)

R	$N=1$			$N=9$		
	θ°_{min}	θ°_{max}	θ°_{min}	θ°_{min}	θ°_{max}	θ°_{min}
1·0	0	45	90	0	45	90
1·2	0	50·1	90	0	49·4	90
1·3	0	52·9	90	0	54·7	90
1·7	0*	73·9*	90*	0	90	—

up to $N=9$ (81 terms in the double Fourier series), and successively iterating between the eigenvalue equation (3.5) and the optimality condition. The nature of the resulting stationary design is determined by evaluating the sign of eqn. (3.8).

Simply-supported plate. Table 1 summarizes typical results for a four-ply simply-supported plate for various aspect ratios R. For aspect ratios up to 1·3, the one-term approximation is reasonably close to the 81 term Fourier solution. However for larger values of R, the one-term approximation admits three stationary designs, whereas the correct solution has local extrema only for the orthotropic design. (The asterisks in Table 1 indicate the values are merely stationary.) The iterative method was found to be globally convergent to all stationary designs. For the most part, the solution was achieved in only a couple of iterations.

Clamped plate. Table 2 summarizes typical results for a four-ply plate clamped on all four edges. The orthotropic designs are local maxima for all

TABLE 2
All four sides clamped
($E_1/E_2 = 10$, $\nu_{12} = 0{\cdot}3$, $G_{12}/E_1 = 0{\cdot}025$)

R	$N=9$		
	θ°_{max}	θ°_{min}	θ°_{max}
1·0	0	45	90
1·2	0	21·5	90
1·27	0	7·2	90
2·0	—	0	90

aspect ratios. And for $R > 1$, the absolute maximum is $\theta_{max} = 90°$. For the data selected, the one-term approximation provides only two stationary solutions for $R > 1{\cdot}06$, namely 0° and 90°. For $1 < R < 1{\cdot}06$, three extrema were found.

The iterative method was found to be globally convergent only to the middle design for the smaller values of R, and globally convergent to both designs for the larger aspect ratios.

Two opposite edges clamped and two opposite edges supported. The boundaries x = constant are assumed to be the clamped edges. For $R = 1$, only the orthotropic designs provide extrema, and $\theta = 0°$ is the optimum. As R is increased from unity, there is a transition from two to three stationary designs. In this range, both orthotropic designs provide minima and the angle-ply is globally optimal. Another increase in R results in a second transition back to two stationary designs. For R less than the first transition value, the maximum value of the frequency is at $\theta = 0°$, but for R greater than the second transition value, the optimal orientation is $\theta = 90°$. Typical results are tabulated in Table 3.

Non-symmetric boundary conditions. It was pointed out earlier that when the boundary conditions are not symmetric the orthotropic designs are not stationary designs. A typical example is a laminate clamped at $x = 0$ and $y = 0$ but unsupported at the edges $x = a$ and $y = b$. For simplicity this laminate will be called CCFF. For a square CCFF plate, local maxima are found at $\pm 45°$, where $+45°$ provides an absolute maximum. As the plate departs from square, i.e. as R either increases or decreases, the number of extrema drops from four to two. For aspect ratios $R \geq 2$ or $R \leq 2$ the

TABLE 3
Opposite edges clamped and simply supported
($E_1/E_2 = 10$, $\nu_{12} = 0{\cdot}3$, $G_{12}/E_1 = 0{\cdot}025$)

R	$N = 5$		
	θ°_{min}	θ°_{max}	θ°_{min}
1·0	—	0	90
1·35	—	0	90
1·4	0	27·2	90
1·7	0	90	—
1·9	0	90	—

TABLE 4
CCFF laminate
($E_1/E_2 = 10$, $\nu_{12} = 0{\cdot}3$, $G_{12}/E_1 = 0{\cdot}025$)

R	$N = 6$			
	θ°_{min}	θ°_{max}	θ°_{min}	θ°_{max}
0·5	−87·5	2·9	—	—
0·7	−83·8	12·2	—	—
1·0	−78·9	−45	−11·1	45
1·2	—	—	−5·8	51·5
1·5	—	—	−3·4	86·3
2·0	—	—	−2·5	87·1

extremal solutions approach the orthotropic designs. Typical results are presented in Table 4.

REFERENCES

1. HOUSNER, J. M. and STEIN, M., Numerical analysis and parametric studies in the buckling of composite orthotropic compression and shear panels, NASA TN D-7996, 1975.
2. HIRANO, Y., Optimum design of laminated plates under shear, *J. Composite Materials*, **13** (1979), 329–334.
3. HIRANO, Y., Optimal design of laminated plates under axial compression, *AIAA Journal*, **17** (1979), 1017–1019.
4. BERT, C. W., Optimal design of a composite-material plate to maximize its fundamental frequency, *J. Sound and Vibration*, **50** (1977), 229–237.
5. BERT, C. W., Design of clamped composite-material plates to maximize fundamental frequency, *J. Mechanical Design*, **100** (1978), 274–278.
6. TAUCHERT, T. R. and ADIBHATLA, S., Design of laminated plates for maximum stiffness, *J. Composite Materials*, **18** (1984), 58–69.
7. CHENG, K.-T., Sensitivity analysis and a mixed approach to the optimization of symmetric layered composite plates, *Eng. Opt.*, **9** (1986), 233–247.
8. REISS, R., Design derivatives of eigenvalues and eigenfunctions for self-adjoint distributed parameter systems, *AIAA Journal*, **24** (1986), 1169–1172.
9. JONES, R. M., *Mechanics of Composite Materials*, Washington, D.C., Scripta Book Co., 1975.
10. REISS, R., Optimization of symmetric angle-ply laminates for maximum fundamental frequency, 7th DOD/NASA Conference on Fibrous Composites in Structural Design, Denver, June, 1985.
11. ASHTON, J. E. and WHITNEY, J. M., *Theory of Laminated Plates*, Westport, Ct., Technomic Publishing Co., 1970.

37

An Analytical Study of the Free Vibration Characteristics of Thin Orthotropic Cylinders

C. B. SHARMA

Department of Mathematics, University of Manchester Institute of Science and Technology, Manchester M60 1QD, UK

and

M. DARVIZEH

Department of Mechanical Engineering, Gilan University, PO Box 401, Rasht, Iran

ABSTRACT

In this paper a calculation of overall response of thin orthotropic cylindrical shells is presented. Transverse shear deformation and rotary inertia terms are included in the analysis because of their obvious importance in the case of composite shells. The modal forms are assumed to have the axial dependence in the form of a simple Fourier series. Various aspects studied here involve influence of boundary conditions, changes in shell geometrical parameters, change in the directions of orthotropy, etc., on the frequencies, mode shapes and modal forces. Analytical results are shown to be in good agreement with some available experimental and theoretical results.

1. INTRODUCTION

The study of mechanical behaviour of a thin orthotropic cylindrical shell (e.g. frequencies, mode shapes and modal forces, etc.) is now one of the established subjects. In fact there has been a growing use of composite shell-type structures in engineering in general and the aerospace industry (e.g. launch vehicles, re-entry vehicles, aircraft fuselages and spacecrafts, etc.) in particular. This is due to the versatility of such structures of very high modulus, high strength and low density, etc. A good overview of previous work has been given by Leissa.[1] There are also some good recent reviews on composite shell vibrations which can be found in the literature.[2] Some preliminary work on this problem is also available.[3,4]

In this paper a calculation of overall response of thin orthotropic cylindrical shells is presented. Transverse shear deformation and rotary inertia terms are included in the analysis because of their obvious importance in the case of composite shells. The modal forms are assumed to have an axial dependence in the form of a simple Fourier series (whose derivatives are legitimized by Stokes's transformation) instead of the usual exponential dependence. Various important aspects of the problem are studied. These include the effects of the inclusion of shear deformation and rotary inertia terms and changes in shell geometry on the frequencies, mode shapes and modal forces as a natural extension of the previous work.[5] Apart from this the pattern of behaviour of shell response and associated natural frequencies are examined as a function of changes in the directions of orthotropy. Included here is also the study of the influence of boundary conditions on the vibration characteristics of composite shells. It is shown that the analytical results of the present analysis are in good agreement with some available theoretical and experimental results.[6,7]

2. THEORETICAL CONSIDERATIONS

The analysis is formulated on the basis of Love's thin shell equations in terms of circumferential and axial coordinates θ and x respectively. The governing equations of motion for free vibration and the stress–strain relations including the transverse shear and rotary inertia terms are taken to be the same as given by Dong.[6]

2.1. Modal Forms

The displacement and rotation components for an orthotropic circular cylindrical shell can be approximated by the following relations:[7]

$$\begin{gathered} u = \psi_u(x)\cos p\theta \sin \omega t \qquad v = \psi_v(x)\sin p\theta \sin \omega t \\ w = \psi_w(x)\cos p\theta \sin \omega t \\ \beta_x = \psi_{\beta_x}\cos p\theta \sin \omega t \qquad \beta_\theta = \psi_{\beta_\theta}\sin p\theta \sin \omega t \end{gathered} \tag{1}$$

In (1) p is the circumferential wave number and ω is the circular frequency due to a harmonic time dependence. Functions $\psi_u, \psi_v, \psi_w, \psi_{\beta_x}, \psi_{\beta_\theta}$ represent the axial dependence of the respective modal forms.

2.1.1. Axial dependence of modal forms

The functions representing the axial dependence of modal forms are given in the following Fourier series forms:

$$\psi_u(x) = A_{op} + \sum_{m=1}^{\infty} A_{mp} \cos(m\pi x/L)$$

$$\psi_v(x) = \sum_{m=1}^{\infty} B_{mp} \sin(m\pi x/L)$$

$$\psi_w(x) = \sum_{m=1}^{\infty} C_{mp} \sin(m\pi x/L) \tag{2}$$

$$\psi_{\beta_x} = D_{op} + \sum_{m=1}^{\infty} D_{mp} \cos(m\pi x/L)$$

$$\psi_{\beta_\theta} = \sum_{m=1}^{\infty} E_{mp} \sin(m\pi x/L)$$

where $A_{mp}, B_{mp}, C_{mp}, D_{mp}, E_{mp}$ denote the Fourier coefficients. Clearly sine series always give zero values at the end points unless one specifies the affected terms, i.e.

$$\begin{array}{lll} \psi_v(0) = v_0 & \psi_w(0) = w_0 & \psi_{\beta_\theta}(0) = \beta_\theta^0 \\ \psi_v(L) = v_L & \psi_w(L) = w_L & \psi_{\beta_\theta}(L) = \beta_\theta^L \end{array} \tag{3}$$

These values are utilized when one uses Stoke's transformations in differentiating the sets of displacement functions.

2.2. Derivation of Eigenvalue Problem

The analysis presented here is capable of tackling the vibration characteristics of cylindrical shells with arbitrary end conditions. In this paper individual problems of a shell with simply supported, free-free and clamped-free boundary conditions are deduced as particular cases of ensuing general theoretical considerations. As a base problem for present investigations a shell with freely supported ends with no tangential constraint is used.[8] The boundary conditions for such a shell are given as

$$u = N_{x\theta} = Q_x = \beta_x = M_{x\theta} = 0 \quad \text{at } x = 0, L \tag{4}$$

It is observed that for this problem none of the ten boundary conditions are satisfied by the assumed modal displacement forms (2) on a term-by-term basis. Thus the constraint conditions to satisfy these boundary conditions will lead to the following eigenvalue problem:

$$[e_{ij}][\tilde{N}_x^0 \tilde{N}_x^L \tilde{M}_x^0 \tilde{M}_x^L v_0, v_L, w_0, w_L]^T = [0] \qquad (i, j = 1, 10) \tag{5}$$

2.2.1. *SNA–SNA end conditions*

The boundary conditions for the shell with simply supported ends with no axial constraint (SNA–SNA) are given as

$$N_x = v = w = M_x = \beta_\theta = 0 \quad \text{at } x = 0, L \tag{6}$$

It is clear that the modal displacement forms (2) satisfy the above boundary conditions on a term-by-term basis. The frequency equation for the SNA shell is directly obtained as a 5×5 determinantal equation from the equations of motion.

The elements of this frequency determinant are no longer infinite series but reduce to single algebraic terms. By solving the above frequency equation all the frequencies are obtained as eigenvalues. These eigenvalues are then used to calculate the corresponding eigenvectors to give modal displacements and modal forces for single and multilayered shells.

2.2.2. *Clamped-free boundary conditions*

Nonsymmetric sets of boundary conditions for a cantilever circular cylindrical shell are given as

$$u = v = w = \beta_x = \beta_\theta = 0 \quad \text{at } x = 0$$
$$N_x = N_{x\theta} = Q_x = M_x = M_{x\theta} = 0 \quad \text{at } x = L$$

The conditions $u = 0$ and $\beta_x = 0$ at $x = 0$ must be enforced by the release of the end force, N_x, and the end moment, M_x. The conditions $N_{x\theta} = Q_x = 0$ and $M_{x\theta} = 0$ at $x = L$ must be enforced with the release of end displacements v, w and rotation β_θ. The retaining of the rows and columns in (5) associated with $\tilde{N}_x^0$, $\tilde{M}_x^0$, v_L, w_L, β_θ^L leads to the following frequency equations, by the requirement that for a nontrivial solution of a homogeneous set the determinant of coefficients matrix is zero, i.e.

$$[e_{ij}][\tilde{N}_x^0, \tilde{M}_x^0, v_L, w_L, \beta_\theta^L]^T = [0] \qquad (i, j = 1, 3, 6, 8, 10)$$

2.2.3. Free-free end conditions

The boundary conditions for such a shell are given by

$$N_x = N_{x\theta} = Q_x = M_x = M_{x\theta} = 0 \quad \text{at } x = 0 \text{ and } x = L$$

From the properties of modal displacements (2) it is clear that the tangential displacement v, radial displacement w and rotation β_θ are identically zero at the ends $x = 0$ and $x = L$. Therefore, the releasing procedure is required to remove these unwanted geometric end conditions. By specifying v and w separately at the two ends and also releasing the rotations β_θ at both ends the corresponding eigenvalue problem can be derived from the enforcement of natural boundary conditions, $N_{x\theta} = Q_x = M_{x\theta} = 0$, at both ends. This is achieved by retaining the rows and columns associated with v_0, v_L, $w_0, w_L, \beta_\theta^0, \beta_\theta^L$ in the original eigenvalue equation (5), which now reduces to

$$[e_{ij}][v_0, v_L, w_0, w_L, \beta_\theta^0, \beta_\theta^L]^T = [0] \qquad (i, j = 5, 6, 7, 8, 9, 10)$$

3. DISCUSSION OF RESULTS

3.1. Simply Supported Shell with no Axial Constraint

3.1.1. Single-layer orthotropic shells

As an example, a single-layer orthotropic shell will be discussed where the elastic constants are given by

$$C_{11} = 2{\cdot}272\,237 \times 10^{11}\ \text{N/m}^2$$
$$C_{12} = 2{\cdot}751\,1657 \times 10^{9}\ \text{N/m}^2$$
$$C_{22} = 2{\cdot}086\,4268 \times 10^{9}\ \text{N/m}^2$$
$$C_{44} = C_{55} = C_{66} = 7{\cdot}941\,0837 \times 10^{9}\ \text{N/m}^2$$

and the geometrical parameters are given by $h/a = 0{\cdot}01$ and $L/a = 2$. All the figures are drawn for the axial wave number, $m = 1$. In Fig. 1 the effects of orthotropy on the natural frequencies are depicted. Frequency curves 1 to 5 correspond to the values of C_{11}/C_{22} given by 1, 3·3, 10·9, 0·3 and 0·092 respectively. Thus, curve 1 is for an isotropic case ($C_{11}/C_{22} = 1$), curve 2 is for the case of moderately increased axial stiffness ($C_{11}/C_{22} \approx 3{\cdot}3$) and curve 3 corresponds to highly increased axial stiffness ($C_{11}/C_{22} \simeq 10{\cdot}9$). It is clear that the sensitivity to axial stiffness dies down with increasing p. Curve 4 is for moderately increased meridional stiffness ($C_{22}/C_{11} \simeq 3{\cdot}3$) and curve 5 for a highly increased meridional stiffness ($C_{22}/C_{11} \simeq 10{\cdot}9$). It is obvious from curves 4 and 5 that the values of the frequency are highly exaggerated as p increases due to enhanced meridional stiffness.

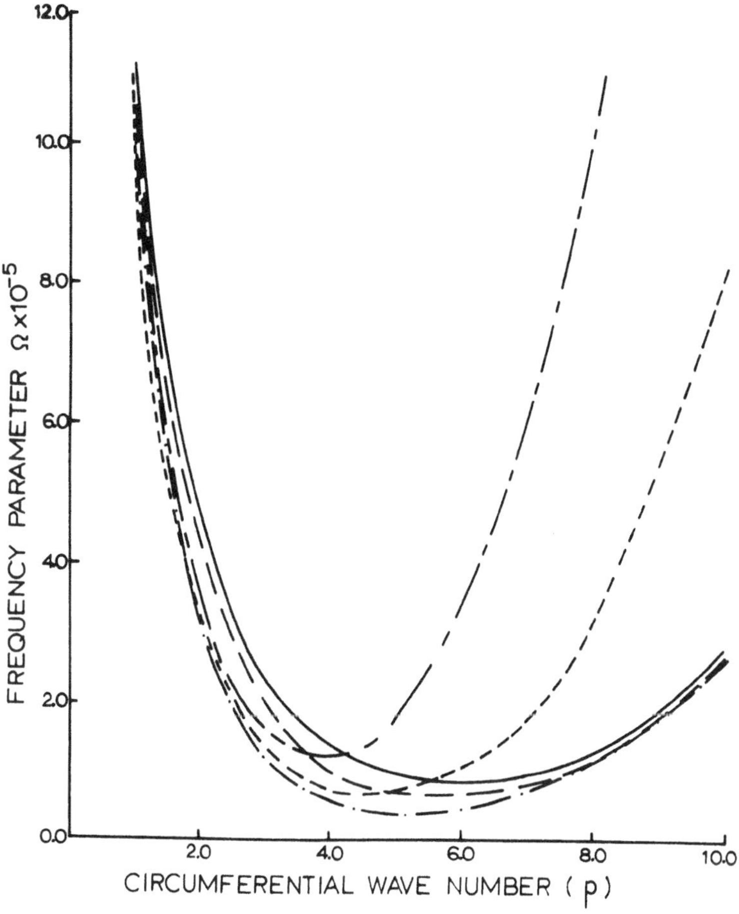

FIG. 1. Frequency variation with circumferential wave number for a simply supported cylindrical shell with variable orthotropy: —— $C_{11}/C_{22} \simeq 10{\cdot}9$; --- $C_{11}/C_{22} \simeq 3{\cdot}3$; -·-·- $C_{11}/C_{22} = 1$; ----- $C_{22}/C_{11} \simeq 3{\cdot}3$; —— —— $C_{22}/C_{11} \simeq 10{\cdot}9$; shell geometry: $h/a = 0{\cdot}01$, $L/a = 2$.

3.1.2. Mode shapes and modal forces and moments

Mode shapes, modal forces and moments of a single-layer orthotropic shell with given physical and geometrical properties in Section 3.1.1 are shown in Figs 2(a–j). Since there can be no unique basis for our purpose here we take up the maximum radial deflection amplitude as unity. It is found by comparing the cases where the previous results are available[9] (viz. u, v, w and β_θ) and there is good agreement. Figures 2(f–j) show the distribution of modal forces N_x, N_θ and moments M_x, M_θ respectively.

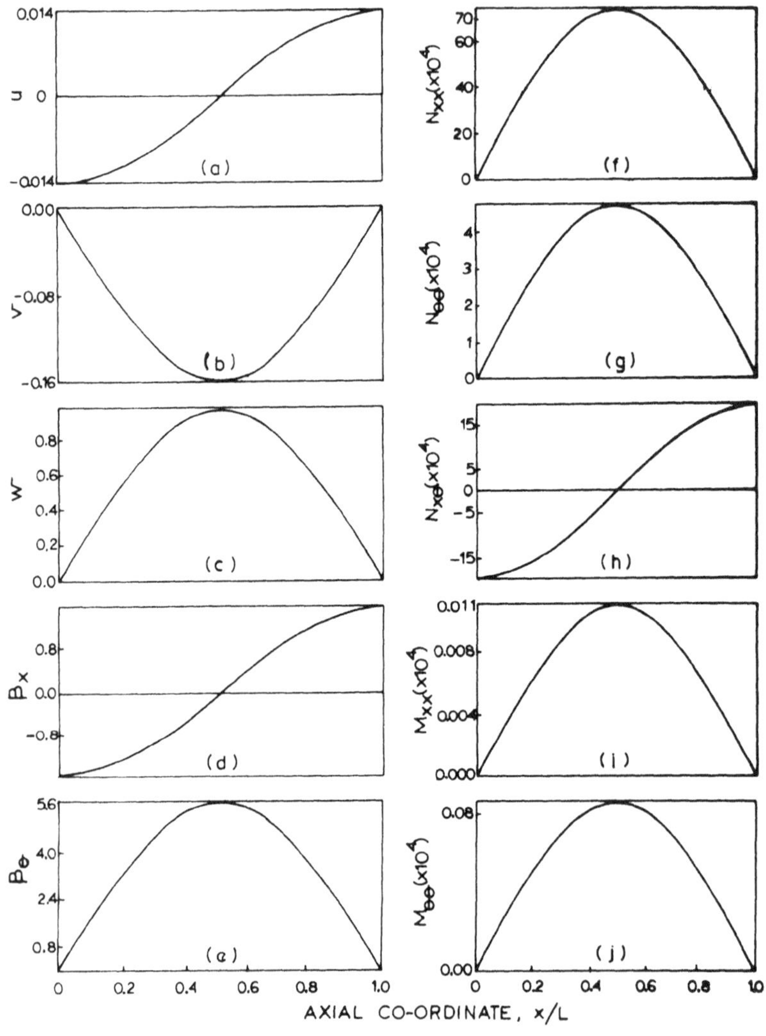

FIG. 2. Axial variation of mode shapes and modal forces and moments of an orthotropic simply supported cylindrical shell ($p = 6$, $\Omega \times 10^{-5} = 0{\cdot}859$, $C_{11}/C_{22} \simeq 10{\cdot}9$); shell geometry: $h/a = 0{\cdot}01$, $L/a = 2$.

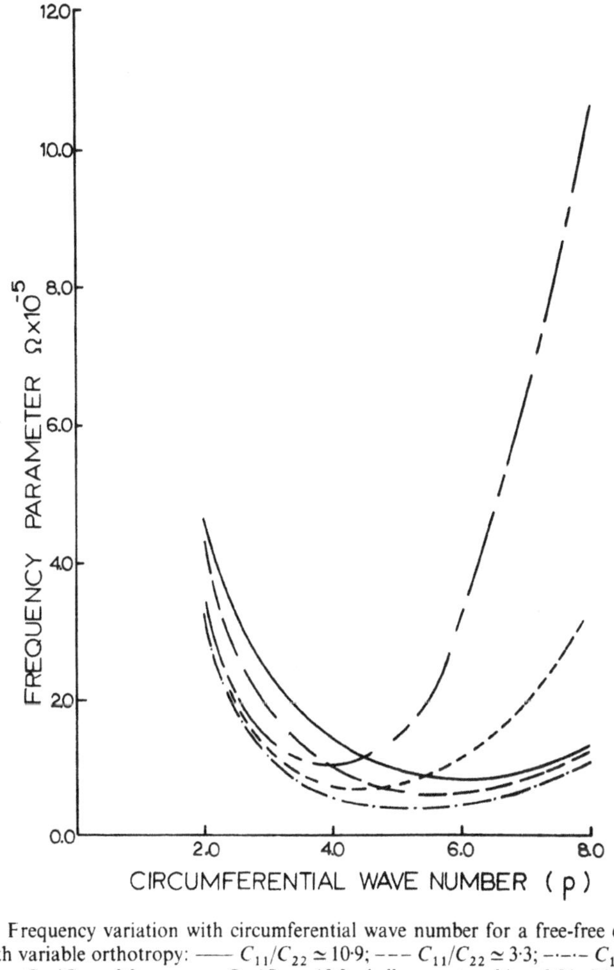

FIG. 3. Frequency variation with circumferential wave number for a free-free cylindrical shell with variable orthotropy: —— $C_{11}/C_{22} \simeq 10{\cdot}9$; --- $C_{11}/C_{22} \simeq 3{\cdot}3$; -·-·- $C_{11}/C_{22} = 1$; ----- $C_{22}/C_{11} \simeq 3{\cdot}3$; — — $C_{22}/C_{11} \simeq 10{\cdot}9$; shell geometry: $h/a = 0{\cdot}01$, $L/a = 2$.

Again the agreement with the corresponding results[9] (viz. N_x, M_θ) is found to be quite good.

3.2. Free-free Circular Cylindrical Shell

Single-layer orthotropic shells are considered with variable orthotropy (by giving different values to the ratio C_{11}/C_{22} of the elastic constants) and the shell geometry as for the 'SNA' shell. This will also help in revealing the influence of boundary conditions.

3.2.1. Frequency spectrum

Figure 3 shows the frequency variation with the circumferential wave number p. Five different curves correspond to various orthotropies as before, i.e. exaggerated axial stiffness ($C_{11}/C_{22} \simeq 10{\cdot}9$), moderate axial stiffness ($C_{11}/C_{22} \simeq 3{\cdot}3$), isotropic case ($C_{11}/C_{22} \simeq 1$), moderate meridional stiffness ($C_{22}/C_{11} \simeq 3{\cdot}3$) and exaggerated meridional stiffness ($C_{22}/C_{11} \simeq 10{\cdot}9$). It is observed that in each case frequency variations sensitivity to axial stiffness decreases with increasing circumferential wave number p. Whereas it is the other way round when one studies the influence of increased meridional stiffness, i.e. the values of frequency parameter are highly exaggerated as the value of circumferential wave number is increased. It is observed that in the case of highly exaggerated meridional stiffness ($C_{22}/C_{11} \simeq 10{\cdot}9$) the increments in frequency parameters as compared to other related cases are extremely high.

3.2.2. Mode shapes and modal stress resultants

Here, once again, since there can be no unique basis for discussion of the mode shapes and modal forces we have chosen for our purpose the maximum radial deflection amplitude as unity. Graphical presentation of mode shapes and modal stress resultants for free-free boundary conditions is given in Figs 4(a–j). The purpose of this investigation also includes the effect of boundary conditions on mode shapes and modal stress resultants. For this reason the physical and material properties of a free-free circular cylindrical shell are taken to be the same as for a SNA shell. It is also observed that the absolute values of deflections are significantly higher than the corresponding ones in the case of a SNA shell. Consequently the values of modal forces and moments are lower than the corresponding ones in the case of the SNA shell.

3.3. Clamped-free Circular Cylindrical Shell

The vibration characteristics of a clamped-free shell are discussed in this section. The various geometrical and material properties involved are the same as in the previous sections. Variation of natural analytical frequencies with respect to circumferential wave number is given in graphical form. Mode shapes, modal forces and moments are also presented in graphical forms.

3.3.1. Frequency spectrum

For this case variation of natural frequencies with respect to circumferential wave number for a clamped-free shell is given in Fig. 5. It is

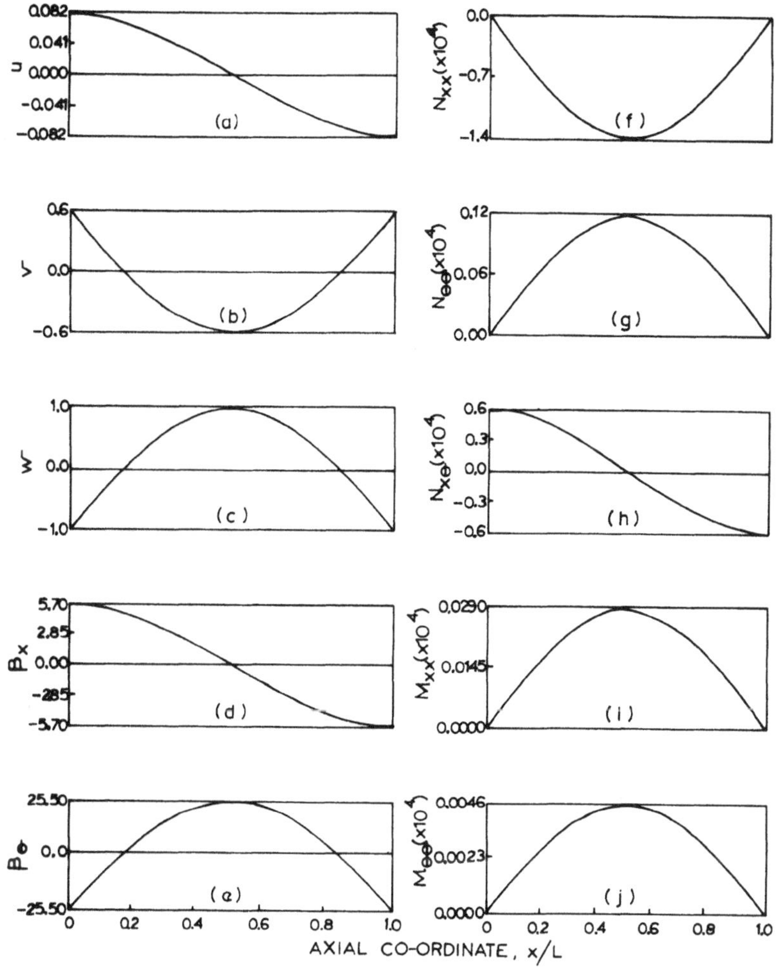

FIG. 4. Axial variation of mode shapes and modal forces and moments of an orthotropic free-free cylindrical shell ($p = 6, \Omega \times 10^{-5} = 0{\cdot}856, C_{11}/C_{22} \simeq 10{\cdot}9$); shell geometry: $h/a = 0{\cdot}01$, $L/a = 2$.

observed from Fig. 5 that the frequency values for the exaggerated axial stiffness ($C_{11}/C_{22} \simeq 10{\cdot}9$) are higher than the corresponding ones for moderate axial stiffness ($C_{11}/C_{22} \simeq 3{\cdot}3$) and the isotropic case ($C_{11}/C_{22} = 1$) for lower values of circumferential wave number p. As the values of p increase the natural frequencies for the three different cases converge towards each other. It can be inferred once again from this that the influence of axial stiffness is felt only for low p values but for high p values

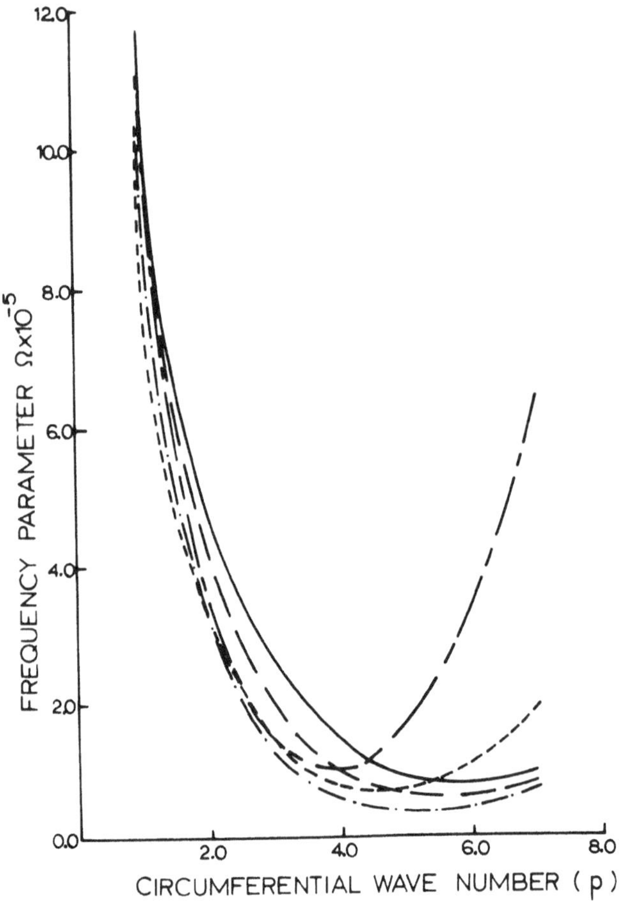

FIG. 5. Frequency variation with circumferential wave number for a clamped-free cylindrical shell with variable orthotropy: —— $C_{11}/C_{22} \simeq 10{\cdot}9$; --- $C_{11}/C_{22} \simeq 3{\cdot}3$; -·-·- $C_{11}/C_{22} = 1$; ----- $C_{22}/C_{11} \simeq 3{\cdot}3$; — — —- $C_{22}/C_{11} \simeq 10{\cdot}9$; shell geometry: $h/a \simeq 0{\cdot}01$, $L/a = 2$.

the influence of axial stiffness becomes minimal due to the increased membrane flexibility. For the two orthotropic cases ($C_{11}/C_{22} \simeq 10{\cdot}9$ and $C_{11}/C_{22} \simeq 3{\cdot}3$), meanwhile, minimum natural frequency corresponds to $p = 6$ whereas for the isotropic case ($C_{11}/C_{22} = 1$) it corresponds to $p = 5$. Effect of variation in meridional stiffness on frequency variation is observed by going through the moderate and exaggerated values of C_{22}. It is observed that the natural frequencies for these two cases are very close to those for the isotropic case for lower values of p (≤ 3). But as p increases, the

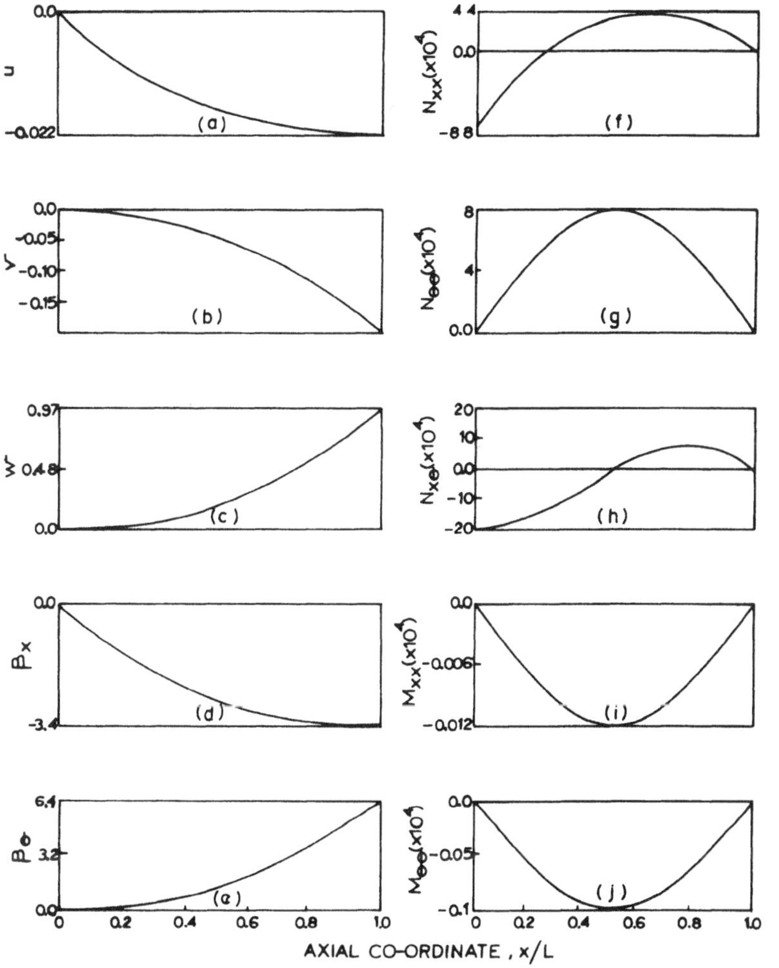

FIG. 6. Axial variation of mode shapes and modal forces and moments of an orthotropic clamped-free cylindrical shell ($p = 6$, $\Omega \times 10^{-5} \simeq 0{\cdot}853$, $C_{11}/C_{22} \simeq 10{\cdot}9$); shell geometry: $h/a = 0{\cdot}01$, $L/a = 2$.

values of the natural frequencies increase and become several times higher than the corresponding ones for the isotropic case and the first two orthotropic cases. This may be due to the fact that the effect of increased meridional stiffness is felt more and more strongly as the circumferential wave number p increases. It is also found that this change in material parameter brings about a change in p (which is now equal to 4) corresponding to the minimum natural frequency.

3.3.2. Mode shapes, modal forces and moments

Mode shapes, modal forces and moments for a single-layer orthotropic cantilever circular cylindrical shell are presented in Figs 6(a–j) and correspond to the minimum values of the frequency parameter. The geometrical and material properties are the same as those for the previous boundary conditions. Once again the modal forms are normalized with respect to the radial displacement w, whose normalized value is taken to be unity. It is observed that apart from the significant changes in the pattern of mode shapes, modal forces and moments, the absolute values of the modal forms $u, v, \beta_x, \beta_\theta$ are increased in comparison with the SNA shell. But these values are significantly lower than the corresponding ones for a free-free shell. It is also shown that the absolute value of N_x is lower than that for the SNA shell and significantly higher than that for the free-free shell. Meanwhile, the values of $N_\theta, N_{x\theta}, M_x$ and M_θ are higher than those for SNA and free-free circular cylindrical shells.

4. CONCLUDING REMARKS

In this paper a general solution procedure is presented to investigate the free vibration characteristics of single-layer orthotropic circular cylindrical shells. The analysis used here includes the effects of transverse shear deformation and rotary inertia terms. The study of the modal forms and the modal forces in a particular case reveals that for the cases where previous results are available the agreement is quite good. Frequency pattern is shown to be significantly influenced by the change in the direction of orthotropy and change in boundary conditions. By interchanging the axes of orthotropy of a cylindrical shell two different cases are obtained, one that exaggerates axial stiffness and the other that exaggerates meridional stiffness. Results here are given only for some particular cases but the approach can be utilized to calculate any of the required vibration characteristics for varying shell geometrical and material parameters and for different boundary conditions. The influence of boundary conditions is shown to lower the natural frequencies as we travel through the end conditions SNA, clamped-free and free-free respectively in that order. This effect is more significant for lower values of circumferential wave number p, but for higher values of p the effect of the change in the boundary conditions diminishes and the corresponding natural frequencies are found to be hardly distinct.

REFERENCES

1. LEISSA, A. W., *Vibration of Shells*, NASA SP-288, US Govt Printing Office, 1973.
2. VANDERPOOL, M. E. V. and BERT, C. W., Vibration of materially monoclinic, thick-wall circular cylindrical shells, *AIAA Journal* **19** (1981), 634–641.
3. SHARMA, C. B. and DARVIZEH, M., Free vibration characteristics of laminated orthotropic clamped-free cylindrical shells, Developments in Mechanics, *Proceedings of the 19th Midwestern Mechanics Conference*, Department of Engineering Mechanics, The Ohio State University, Columbus, Ohio, Vol. 13, Sept. 9–11, 1985.
4. SHARMA, C. B. and DARVIZEH, M., Free vibration characteristics of laminated orthotropic thin cylindrical shells with free-edges, 9th Polymodel Conference on Industrial Vibration Modelling, Newcastle upon Tyne, Armstrong Technology Centre, May 21–23, 1986.
5. DARVIZEH, M. and SHARMA, C. B., Natural frequencies of laminated orthotropic thin circular cylinders, *Thin Walled Structures*, **2** (1984), 207–217.
6. DONG, S. B., Free vibration of laminated orthotropic cylindrical shells, *J. Acoust. Soc. Amer.*, **44** (1968), 1628–1635.
7. CHUNG, H., Vibration of circular cylindrical shells, Technical Memorandum, Argonne National Laboratory, University of California, AUA-USERDA, 1975.
8. DARVIZEH, M., Free vibration characteristics of orthotropic thin circular cylindrical shells, Ph.D. thesis, submitted to the University of Manchester Institute of Science and Technology, 1986.
9. WARBURTON, G. B. and SONI, S. R., Resonant response of orthotropic cylindrical shells, *Journal of Sound and Vibration*, **53** (1977), 1–23.

38

Free Vibration Analysis of an Orthotropic Circular Cylindrical Shell of Laminated Composite

ANLIN YAO and FANGCHUN XIAO

Department of Mechanical Engineering, Southwestern Petroleum Institute, Nanchong, Sichuan, People's Republic of China

ABSTRACT

Based on the dynamic von Kármán-type equations governing the free flexural vibration of a shell, the weighted residual method is adopted. A new method for the free vibration analysis of an orthotropic circular cylindrical shell of laminated composite is obtained. The result is in good agreement with Werner Soedal's method. In this chapter, the method is applied to the numerical calculations and the validity of it is assessed by comparing with the natural frequencies in three types of pipe.

INTRODUCTION

The determination of the natural frequencies of an orthotropic circular cylindrical shell of laminated composite is an important engineering problem. This problem has been considered by Werner Soedal and also appears in the literature.[1]

The aim of this chapter is to introduce a new method for the free vibration analysis of an orthotropic circular cylindrical shell of laminated composite in order to compare with Werner Soedal's method. On the basis of the dynamic von Kármán-type equations governing the free flexural vibration of the shell, the weighted residual method is utilized. Therefore, a new method is presented and agrees with Werner Soedal's method. This method is simple and convenient.

THEORETICAL ANALYSIS

Consider a closed orthotropic circular cylindrical shell, simply supported at both ends. Let us assume that the shell is thin, even if it is composed of n laminations. Furthermore, we will assume again that displacements vary linearly through the shell thickness. The shell thickness is h, its length is l and its radius is a. As usual, the inplane and rotatory inertia effects are neglected. The dynamic von Kármán-type equations governing the free flexural vibration of the shell may be derived in the following form:[1]

$$D_{11}W_{,xxxx}+2(D_{12}+D_{33})\frac{1}{a^2}W_{,xx,\theta\theta}+\frac{D_{22}}{a^4}W_{,\theta\theta\theta\theta}+\rho hW_{,tt}=q_z-\frac{1}{a}\phi_{,xx} \quad (1)$$

$$\frac{A_{12}^2-A_{11}A_{22}}{a}W_{,xx}+A_{11}\phi_{,xxxx}+\frac{A_{22}}{a^4}\phi_{,\theta\theta\theta\theta}+\frac{A_{11}A_{22}-A_{12}^2-2A_{12}A_{33}}{A_{33}a^2}\phi_{,xx,\theta\theta}=0 \quad (2)$$

in which the subscripts preceded by a comma denote differentiation with respect to the corresponding coordinates, t is the time, ρ is the mass of the circular cylindrical shell per unit volume, W is the deflection of the shell, ϕ is the stress function, and A_{ij} and D_{ij} are the elements of $[A]$ and $[D]$, which are extensional and bending stiffness, respectively, and can be written as

$$A_{ij}=\sum_{k=1}^{n}(\bar{Q}_{ij})_k(h_{k+1}-h_k) \qquad (i,j=1,2,3) \quad (3)$$

$$D_{ij}=\sum_{k=1}^{n}(\bar{Q}_{ij})_k(h_{k+1}^3-h_k^3)/3 \qquad (i,j=1,2,3) \quad (4)$$

where the subscript k denotes the kth layer of laminated composite, h_k is the distance of the centroid of the kth layer of laminated composite from reference plane, and $(\bar{Q}_{ij})_k$ denotes the stiffness for the kth layer in the respective coordinate directions, which is a function of the state variables (i.e. stress, strain, etc.), but independent of the history of the stress–strain path. The values of the stiffness $(\bar{Q}_{ij})_k$ in the loading or progressive failure process can be determined as described in Refs 1 and 2.

The boundary conditions of the shell are

$$W(0, \theta, t) = W(l, \theta, t) = 0 \tag{5}$$

$$M_{xx}(0, \theta, t) = M_{xx}(l, \theta, t) = 0 \tag{6}$$

We are able to satisfy these boundary conditions by

$$W(x, \theta, t) = W(x, \theta)\, e^{i\omega t} \tag{7}$$

$$(x, \theta, t) = (x, \theta)\, e^{i\omega t} \tag{8}$$

Solving for the eigenvalues, we set $q_z = 0$; therefore, eqns (1) and (2) become

$$D_{11} W_{,xxxx}(x, \theta) + 2(D_{12} + D_{33})\frac{1}{a^2} W_{,xx,\theta\theta}(x, \theta)$$

$$+ \frac{D_{22}}{a^2} W_{,\theta\theta\theta\theta}(x, \theta) - \rho h \omega^2 W_{,tt}(x, \theta) - \frac{1}{a}\phi_{,xx}(x, \theta) = 0 \tag{9}$$

$$\frac{A_{12}^2 - A_{11}A_{22}}{a} W_{,xx}(x, \theta) + A_{11}\phi_{,xxxx}(x, \theta)\frac{A_{22}}{a^4}\phi_{,\theta\theta\theta\theta}(x, \theta)$$

$$+ \frac{A_{11}A_{22} - A_{12}^2 - 2A_{12}A_{33}}{A_{33}a^2}\phi_{,xx,\theta\theta}(x, \theta) = 0 \tag{10}$$

The boundary conditions of the shell become

$$W(0, \theta) = W(l, \theta) = 0 \tag{11}$$

$$M_{xx}(0, \theta) = M_{xx}(l, \theta) = 0 \tag{12}$$

To find the natural frequencies of the orthotropic circular cylindrical shell of laminated composite, the weighted residual method is adopted. Now we state as follows. The governing equations and the boundary conditions are[3,4]

$$\sum_{j=1}^{n} L_{ij}(W) - f_i = 0 \qquad (i = 1, 2, \ldots, n)\ (W \in V) \tag{13}$$

$$\sum_{j=1}^{n} G_{rj}(W) - g_r = 0 \qquad (r = 1, 2, \ldots, n)\ (W \in S) \tag{14}$$

where W is the deflection of the orthotropic circular cylindrical shell of laminated composite, or unknown function; L_{ij} and G_{rj} are differential operators; and f_i and g_r are known functions.

Let the trial functions be

$$W^*(x,\theta) = C_W \sin\frac{m\pi x}{l}\cos n(\theta-\psi) \tag{15}$$

$$\phi^*(x,\theta) = C_\phi \sin\frac{m\pi x}{l}\cos n(\theta-\psi) \tag{16}$$

where C_W and C_ϕ are parameters, and they are able to satisfy eqns (11) and (12) or eqn. (14). Hence substitution of eqns (15) and (16) into eqns (9) and (10) or eqn. (13) results in the internal residuals R_{Ii}, i.e.

$$R_{Ii} = \sum_{j=1}^{n} L_{ij}(W^*) - f_i = 0 \qquad (i = 1, 2, \ldots, n) \tag{17}$$

To eliminate the internal residuals R_{Ii}, we introduce the internal weighted functions W_{ki}, i.e.

$$W_{kiW} = W_{ki\phi} = \sin\frac{m\pi x}{l}\cos n(\theta-\psi) \tag{18}$$

Then the following residual equations are obtained:

$$\int_V R_{Ii} W_{ki}\, \mathrm{d}V = 0 \qquad (i = 1, 2, \ldots, n) \tag{19}$$

or

$$\int_V \left\{\left[D_{11}\left(\frac{m\pi}{l}\right)^4 + 2(D_{12}+D_{33})\left(\frac{n}{a}\right)^2\left(\frac{m\pi}{l}\right)^2 + D_{22}\left(\frac{n}{a}\right)^4 - \rho h\omega^2\right]\right.$$
$$\left.\times\, C_W - \frac{1}{a}\left(\frac{m\pi}{l}\right)^2 C_\phi\right\}\sin^2\frac{m\pi x}{l}\cos^2 n(\theta-\psi)\,\mathrm{d}V = 0 \tag{20}$$

$$\int_V \left\{(A_{11}A_{22} - A_{12}^2)\left(\frac{m\pi}{l}\right)^2 C_W \Big/ a + \left[A_{11}\left(\frac{m\pi}{l}\right)^4 + A_{22}\left(\frac{n}{a}\right)^4\right.\right.$$
$$\left.\left. + (A_{11}A_{22} - A_{12}^2 - 2A_{12}A_{33})\left(\frac{n}{a}\right)^2\left(\frac{m\pi}{l}\right)^2 \Big/ A_{33}\right]C_\phi\right\}$$
$$\times \sin^2\frac{m\pi x}{l}\cos n(\theta-\psi)\,\mathrm{d}V = 0 \tag{21}$$

Due to

$$C_W \neq 0 \qquad C_\phi \neq 0 \qquad \sin\frac{m\pi x}{l} \neq 0 \qquad \cos n(\theta - \psi) \neq 0$$

we obtain

$$\left[D_{11}\left(\frac{m\pi}{l}\right)^4 + 2(D_{12}+D_{33})\left(\frac{n}{a}\right)^2\left(\frac{m\pi}{l}\right)^2 + D_{22}\left(\frac{n}{a}\right)^4 - \rho h\omega^2\right]C_W - \frac{1}{a}\left(\frac{m\pi}{l}\right)^2 C_\phi = 0 \qquad (22)$$

$$\frac{A_{11}A_{22}-A_{12}^2}{a}\left(\frac{m\pi}{l}\right)^2 C_W + \left[A_{11}\left(\frac{m\pi}{l}\right)^4 + A_{22}\left(\frac{n}{a}\right)^4 + \frac{A_{11}A_{22}-A_{12}^2-2A_{12}A_{33}}{A_{33}}\left(\frac{n}{a}\right)^2\left(\frac{m\pi}{l}\right)^2\right]C_\phi = 0 \qquad (23)$$

For these two equations to be meaningfully satisfied, the determinant must equal zero. This gives us the natural frequencies of the orthotropic circular cylindrical shell of laminated composite for those modes where transverse deflection components dominate:

$$\omega^2 = \omega_{mn}^2 = \frac{1}{\rho h}\left\{\left[D_{11}\left(\frac{m\pi}{l}\right)^4 + 2(D_{12}+D_{33})\left(\frac{n}{a}\right)^2\left(\frac{m\pi}{l}\right)^2 + D_{22}\left(\frac{n}{a}\right)^4\right] + (A_{11}A_{22}-A_{12}^2)\left(\frac{m\pi}{l}\right)^4 \Big/ a^2\left[A_{11}\left(\frac{m\pi}{l}\right)^4 + A_{22}\left(\frac{n}{a}\right)^4 + (A_{11}A_{22}-A_{12}^2-2A_{12}A_{33})\left(\frac{n}{a}\right)^2\left(\frac{m\pi}{l}\right)^2 \Big/ A_{33}\right]\right\} \qquad (24)$$

This is a formula for calculating the natural frequencies of an orthotropic circular cylindrical shell of laminated composite. This formula is only applied to the shell which is simply supported at both ends. If the supports of the shell are changed, a different form of the formula is obtained.

NUMERICAL EXAMPLES

A composite pipe which is simply supported at both ends is composed of two layers; the internal layer material is GRP and the outer layer material is

TABLE 1

	Elastic moduli (MPa)		Shear modulus G_{LT} (MPa)	Poisson's ratio ν_{LT}	ρ (t/m³)
	E_L	E_T			
GRP	9 806·65	411·879 3	735·498 8	0·01	2·0
Steel	205 934·7		78 453·2	0·30	7·8

steel. The pipe thickness is 8 mm and the thickness of the GRP layer is 4 mm. The pipe diameter is 426 mm and its length is 2 m. The physical parameters of all materials are given in Table 1. Compute the natural frequency of the composite pipe.

At the same time, calculate the natural frequencies of the steel pipe and GRP pipe; their thickness is 8 mm, their diameter is 426 mm and their length is 2 m. They are simply supported at both ends. The results are given in Table 2 and Fig. 1.

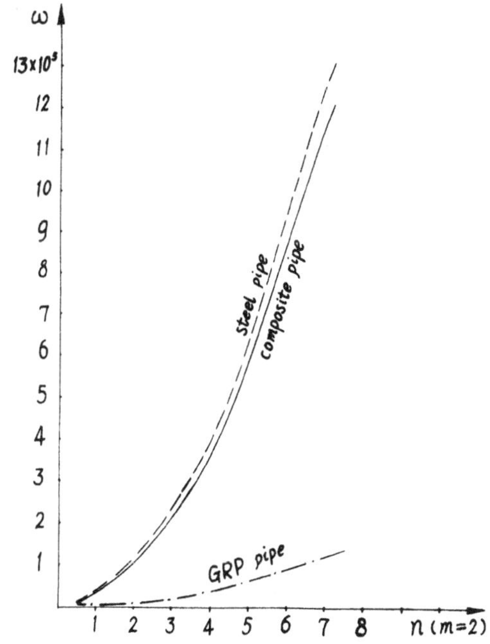

FIG. 1. n–ω curve for three types of pipe.

TABLE 2
The natural frequencies in three types of pipe

Pipe material	n ($m=2$)			
	1	3	5	7
Composite	33 954·99	219 107·8	590 833·5	114 843 3
Steel	37 567·13	243 009·6	655 389·5	127 396 5
GRP	6 481·536	23 108·36	57 722·8	109 798

CONCLUSION

This chapter analyses the natural frequencies of an orthotropic circular cylindrical shell of laminated composite and computes the natural frequencies of steel, GRP and composite pipe. As mentioned above, the concept of this chapter is clear and the method is convenient. We find that neither stringers nor rings alone can raise the natural frequencies for all m, n combinations. The result is in good agreement with Ref. 1.

REFERENCES

1. WERNER SOEDAL, *Vibration of Shells and Plates*, New York and Basel, Marcel Dekker, 1981, pp. 342–360.
2. LIU XILI and WANG BINGQUAN, *Basis of Composite Material Mechanics*, Beijing, Press of Arch. Industry of China, 1984, pp. 36–44.
3. YI XIANZHONG and XIAO FANGCHUN, Application of the weighted residual method in shear deformation theorem of circular cylindrical shells, Southwestern Petroleum Institute, 1986, pp. 1–4.
4. YI XIANZHONG, Analysis of precision of shear deformation theory of cylindrical shells, *Journal of Southwestern Petroleum Institute*, No. 3 (1986), 72–85.

39

Dynamic Buckling of Antisymmetrically Laminated Imperfect Rectangular Plates

VICTOR BIRMAN

School of Naval Architecture and Marine Engineering, University of New Orleans, PO Box 1098, New Orleans, Louisiana 70148, USA

ABSTRACT

This chapter deals with the dynamic buckling of antisymmetrically laminated angle-ply rectangular plates with initial imperfections. Three different problems are considered. First, nonlinear response of the plate subjected to time-dependent axial loading is studied. The problem is reduced to a single differential equation which is integrated numerically. In the particular case of a perfect plate the solution of the linear problem yields the condition of dynamic buckling. The third problem considered in the chapter is the behavior of an imperfect plate initially loaded by axial static stresses. The static response is determined first and the motion of the plate is superimposed on the static displacements in the second phase of the solution.

1. INTRODUCTION

The problems of dynamic buckling of structures are usually associated with their response to rapidly increasing in-surface compressive loads or time-dependent in-surface displacements of the boundaries. The first type of problem is typical for structures subject to impact, while the second type is particularly important in the studies of the response of structures in testing machines. The analytical approach to the solution of the linear dynamic buckling problem was proposed by Hoff,[1] who investigated the response of a slightly curved elastic column whose ends were axially displaced towards each other at a constant speed. The solution of the problem found in terms of Bessel functions appeared to be in good agreement with experimental results.[2] The problems of dynamic buckling of isotropic structures were treated numerically and experimentally by Vol'mir,[3,4] Ari-Gur *et al.*,[5] Babich and Khoroshun,[6] and Brigadirov and Tolokonnikov.[7] The recent paper of Saigal *et al.*[8] presents the finite element analysis of dynamic buckling of thin shells and the bibliography on dynamic buckling of spherical shells subjected to step or impulsive load.

The first work dealing with the behavior of nonisotropic structures subject to the compressive load increasing proportionally to time was published by Ekstrom.[9] He considered dynamic buckling of initially imperfect geometrically nonlinear simply-supported rectangular orthotropic plates. The problem, reduced to a single second-order nonhomogeneous differential equation with time-dependent coefficients, was solved numerically using a fourth-order Runge–Kutta method.

In this chapter we consider dynamic buckling of simply-supported imperfect angle-ply plates. The rate of loading is such that the time required for the plate to obtain considerable deformations is much longer than the time required for the pressure wave to travel through the plate. This allows us to neglect the effect of the axial inertia.[10]

Three different problems are considered. In the first problem we study the nonlinear dynamic deformations of initially imperfect plates subjected to a time-dependent axial load. The second problem is formulated as a particular case of the first one, i.e. initial imperfections are absent and the equations of motion are linear. This results in the question on the conditions of dynamic buckling of the perfect plate. Such a condition is found in the practical case where the plate is at rest at the moment of application of the load. The third problem deals with the dynamic response of an imperfect plate which was initially subject to the axial static load. The static deformations have to be determined in the first phase of the solution. Then the dynamic displacements due to the action of the time-dependent load are superimposed on the static solution.

2. DYNAMIC BUCKLING OF IMPERFECT PLATES: NONLINEAR PROBLEM

Consider an antisymmetrically laminated angle-ply rectangular plate subject to the axial compressive load of intensity $\bar{N}_x(t)$, which is a function of time (Fig. 1). The sides of the plate along the x and y axes have the lengths a and b respectively. The thickness of the plate is h.

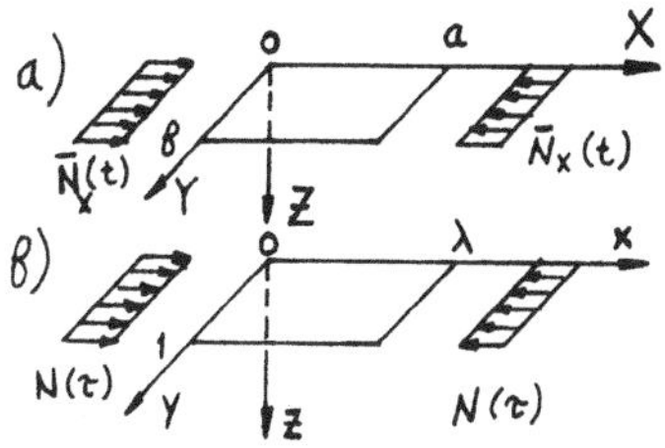

FIG. 1. Plate subjected to axial loading. (a) Dimensional scheme: $(x, y, z) = (X, Y, Z)/h$; (b) nondimensional scheme: $\lambda = a/b$.

The equations of nonlinear theory of laminated plates with initial imperfections representing the generalization of the equations of the von Kármán theory of isotropic plates were published by Stavsky and Hoff[11] and Tennyson *et al.*[12] In this chapter we adopt the nondimensional dynamic version of these equations used by Hui:[13,14]

$$L_d(w) + L_b(f) + \pi^4 w_{,\tau\tau} = f_{,yy}(w + w_0)_{,xx} + f_{,xx}(w + w_0)_{,yy} - 2f_{,xy}(w + w_0)_{,xy} \quad (1)$$

$$L_a(f) = L_b(w) + (w + 2w_0)_{,xy} w_{,xy} - (w + w_0)_{,xx} w_{,yy} - w_{0,yy} w_{,xx} \quad (2)$$

In eqns (1) and (2) w is the nondimensional out-of-plane displacement from the imperfect position, w_0 is the nondimensional initial imperfection and f is the nondimensional stress function:

$$w = \bar{w}/h \qquad w_0 = \bar{w}_0/h \qquad f = F/E_T h^3 \quad (3)$$

where $\bar{w}$, $\bar{w}_0$ and F are the out-of-plane displacement, initial imperfection and the stress function, and E_T is the modulus of elasticity in the transverse direction. The nondimensional time is defined as

$$\tau = \omega t \quad (4)$$

where

$$\omega = \pi^2 \sqrt{E_T h^3 / \rho b^4} \quad (5)$$

The nondimensional coordinates are

$$x = X/b \qquad y = Y/b \quad (6)$$

so that the lengths of the plate sides are $\lambda = a/b$ and 1 (see Fig. 1). The linear operators in (1) and (2) are given by

$$\begin{aligned} L_a(\ ) &= a_{22}(\ )_{,xxxx} + (2a_{12} + a_{66})(\ )_{,xxyy} + a_{11}(\ )_{,yyyy} \\ L_b(\ ) &= (2b_{26} - b_{61})(\ )_{,xxxy} + (2b_{16} - b_{62})(\ )_{,xyyy} \\ L_d(\ ) &= d_{11}(\ )_{,xxxx} + 2(d_{12} + 2d_{66})(\ )_{,xxyy} + d_{22}(\ )_{,yyyy} \end{aligned} \quad (7)$$

where a_{ij}, b_{ij} and d_{ij} are the elements of the nondimensional matrices $[\bar{A}_{ij}]$, $[\bar{B}_{ij}]$ and $[\bar{D}_{ij}]$ defined by

$$\begin{aligned}[\bar{A}_{ij}] &= E_{\mathrm{T}}h[A_{ij}]^{-1} \\ [\bar{B}_{ij}] &= -[A_{ij}]^{-1}[B_{ij}]/h \\ E_{\mathrm{T}}h^3[\bar{D}_{ij}] &= [D_{ij}] - [B_{ij}][A_{ij}]^{-1}[B_{ij}]\end{aligned} \tag{8}$$

The matrices of extensional, coupling and bending stiffnesses, $[A_{ij}]$, $[B_{ij}]$ and $[D_{ij}]$, are defined as usual in the theory of composite structures.

The initial imperfection and the transverse displacement of the simply-supported plate are given by

$$w_0 = W_0 \sin\frac{m\pi x}{\lambda}\sin n\pi y \qquad w = W(\tau)\sin\frac{m\pi x}{\lambda}\sin n\pi y \tag{9}$$

The substitution of (9) into (2) yields the nondimensional stress function:

$$f = c_0 W(\tau)\cos\frac{m\pi x}{\lambda}\cos n\pi y + [W^2(\tau) + 2W(\tau)W_0] \times\left(c_1\cos\frac{2m\pi x}{\lambda} + c_2\cos 2n\pi y\right) - \frac{N(\tau)y^2}{2} \tag{10}$$

where c_i are defined as in Ref. 12:

$$\begin{aligned}c_0 &= -C_{\mathrm{b}}\left(\frac{m}{\lambda}, n\right)\Big/ C_{\mathrm{a}}\left(\frac{m}{\lambda}, n\right) \\ c_1 &= n^2\Big/\left[32\left(\frac{m}{\lambda}\right)^2 a_{22}\right] \\ c_2 &= \left(\frac{m}{\lambda}\right)^2\Big/(32n^2a_{11})\end{aligned} \tag{11}$$

Here

$$\begin{aligned}C_{\mathrm{b}}\left(\frac{m}{\lambda}, n\right) &= (2b_{26} - b_{61})\left(\frac{m\pi}{\lambda}\right)^3 n\pi + (2b_{16} - b_{62})\frac{m\pi}{\lambda}(n\pi)^3 \\ C_{\mathrm{a}}\left(\frac{m}{\lambda}, n\right) &= a_{22}\left(\frac{m\pi}{\lambda}\right)^4 + (2a_{12} + a_{66})\left(\frac{m\pi}{\lambda}\right)^2(n\pi)^2 + a_{11}(n\pi)^4\end{aligned} \tag{12}$$

The nondimensional load is

$$N(\tau) = \frac{\bar{N}_x(t)b^2}{E_{\mathrm{T}}h^3} \tag{13}$$

The substitution of (9) and (10) into equation of motion (1) and the Galerkin procedure result in the following nonlinear differential equation:

$$W(\tau)_{,\tau\tau} + k_1 W(\tau) - k_\tau W(\tau) + k_2 W^2(\tau) + k_3 W^3(\tau) = k_\tau W_0 \tag{14}$$

The coefficients of this equation are

$$\begin{aligned} k_1 &= \frac{1}{\pi^4}\left[C_{\mathrm{d}}\left(\frac{m\pi}{\lambda}, n\right) + \frac{C_{\mathrm{b}}^2(m/\lambda, n)}{C_{\mathrm{a}}(m/\lambda, n)}\right] + 4\left(\frac{m}{\lambda}\right)^2 n^2(c_1 + c_2)W_0^2 \\ k_\tau &= \frac{1}{\pi^2}\left(\frac{m}{\lambda}\right)^2 N(\tau) \\ k_2 &= 6\left(\frac{m}{\lambda}\right)^2 n^2(c_1 + c_2)W_0 \\ k_3 &= 2\left(\frac{m}{\lambda}\right)^2 n^2(c_1 + c_2) \end{aligned} \tag{15}$$

where

$$C_{\mathrm{d}}\left(\frac{m}{\lambda}, n\right) = \left[d_{11}\left(\frac{m}{\lambda}\right)^4 + 2(d_{12} + 2d_{66})\left(\frac{m}{\lambda}\right)^2 n^2 + d_{22}n^4\right]\pi^4 \tag{16}$$

Equation (14) can be integrated numerically. The plate is usually at rest at the instant of application of the load, i.e. the initial conditions to be used are

$$W = W_{,\tau} = 0 \quad \text{at } \tau = 0 \tag{17}$$

3. DYNAMIC BUCKLING OF PERFECT PLATES: CONDITION OF BUCKLING

Consider a linear problem of dynamic buckling of the perfect plate subject to the load

$$N(\tau) = s\tau \tag{18}$$

where s is a coefficient representing the nondimensional rate of loading. The equation of motion is obtained from (14):

$$W(\tau)_{,\tau\tau} + (k_1 - c\tau)W(\tau) = 0 \tag{19}$$

with $c = k_\tau s$.

The solution of such equations was considered by Hoff,[1] Kamke[15] and Watson.[16] Introducing the new variables

$$\xi = k_1 - c\tau \qquad \eta = W(\tau) \tag{20}$$

one can transform (19) to

$$c^2(\mathrm{d}^2\eta/\mathrm{d}\xi^2) + \xi\eta = 0 \tag{21}$$

The solution of (20) is[15]

$$\eta = \xi^{0\cdot 5}[AJ_{1/3}(\xi^{1\cdot 5}/1\cdot 5c) + BY_{1/3}(\xi^{1\cdot 5}/1\cdot 5c)] \tag{22}$$

where A and B are constants of integration, and $J_{1/3}(\cdots)$ and $Y_{1/3}(\cdots)$ are the Bessel functions of the first and second kind respectively. Using the initial conditions (17) one obtains the set of two homogeneous algebraic equations:

$$AJ_{1/3}(\tilde{k}_1) + BY_{1/3}(\tilde{k}_1) = 0$$

$$A\left\{\frac{1}{2k_1^{0\cdot 5}} J_{1/3}(\tilde{k}_1) + \frac{k_1}{2c}[J_{-2/3}(\tilde{k}_1) - J_{4/3}(\tilde{k}_1)]\right\} \tag{23}$$

$$+ B\left\{\frac{1}{2k_1^{0\cdot 5}} Y_{1/3}(\tilde{k}_1) + \frac{k_1}{2c}[Y_{-2/3}(\tilde{k}_1) - Y_{4/3}(\tilde{k}_1)]\right\} = 0$$

where

$$\tilde{k}_1 = k_1^{1\cdot 5}/1\cdot 5c \tag{24}$$

The condition of dynamic buckling of perfect antisymmetrically laminated plates can be obtained from (23) if we require the existence of a nonzero solution. This condition can be written as

$$Y_{1/3}(\tilde{k}_1)[J_{-2/3}(\tilde{k}_1) - J_{4/3}(\tilde{k}_1)] = J_{1/3}(\tilde{k}_1)[Y_{-2/3}(\tilde{k}_1) - Y_{4/3}(\tilde{k}_1)] \tag{25}$$

The coefficient $\tilde{k}_1$ can be evaluated from (25). This coefficient represents the relationship between k_1 and c corresponding to dynamic buckling. If the analysis indicates that elastic dynamic buckling does not occur but the load increases as given by (18) the plastic effects have to be taken into account.

4. DYNAMIC BUCKLING OF IMPERFECT PLATES INITIALLY SUBJECTED TO CONSTANT AXIAL LOAD

Consider an imperfect plate subjected to axial loading $\bar{N}_{x0} = \text{const}$. The plate will experience a static deformation which can be found from the static version of eqns (1) and (2). The stress function is given by (10), where $N(\tau)$ must be replaced by

$$N_0 = \bar{N}_{x0}b^2/Eh^2 \tag{26}$$

and $W(\tau)$ must be replaced by $W_s = \text{const}$. The static out-of-plane displacement can be determined from

$$k_1 W_s - k_0 W_s + k_2 W_s^2 + k_3 W_s^3 = k_0 W_0 \tag{27}$$

where k_1, k_2 and k_3 are defined as in (15) and

$$k_0 = \frac{1}{\pi^2}\left(\frac{m}{\lambda}\right)^2 N_0 \tag{28}$$

Now the dynamic axial load $\bar{N}_{xt}(\tau)$ is applied to the plate. The motion of the plate due to this load will be superimposed on the basic static state. Equations of motion of the plate become

$$\begin{aligned} L_d(\tilde{w}) + L_b(\tilde{f}) + \pi^4 \tilde{w}_{,tt} = {} & \tilde{f}_{,yy}(w_s + w_0 + \tilde{w})_{,xx} + f_{,yy}\tilde{w}_{,xx} + \tilde{f}_{,xx}(w_s + w_0 + \tilde{w})_{,yy} \\ & + f_{,xx}\tilde{w}_{,yy} - 2\tilde{f}_{,xy}(w_s + w_0 + \tilde{w})_{,xy} - 2f_{,xy}\tilde{w}_{,xy} \end{aligned} \tag{29}$$

$$\begin{aligned} L_a(\tilde{f}) = {} & L_b(\tilde{w}) + (w_s + 2w_0 + \tilde{w})_{,xy}\tilde{w}_{,xy} + \tilde{w}_{,xy}w_{,xy} \\ & - (w_s + w_0 + \tilde{w})_{,xx}\tilde{w}_{,yy} - \tilde{w}_{,xx}w_{,yy} - w_{0,yy}\tilde{w}_{,xx} \end{aligned} \tag{30}$$

where $\tilde{w}$ and $\tilde{f}$ are dynamic fractions of the nondimensional out-of-plane displacement and the stress function defined by the equations similar to (3). The mode shape of the dynamic displacement is assumed to be the same as those of the static displacement and initial imperfection:

$$\tilde{w} = \tilde{W}(\tau)\sin\frac{m\pi x}{\lambda}\sin n\pi y \tag{31}$$

Then the substitution of the expressions for the static and dynamic displacements and the initial imperfection into (30) yields

$$\begin{aligned} \tilde{f} = {} & c_0 \tilde{W}\cos\frac{m\pi x}{\lambda}\cos n\pi y + (2W_s + 2W_0 + \tilde{W})\tilde{W} \\ & \times\left(c_1\cos\frac{2m\pi x}{\lambda} + c_2\cos 2n\pi y\right) - \frac{N_x(\tau)y^2}{2} \end{aligned} \tag{32}$$

where

$$N_x(\tau) = \bar{N}_{xt}(\tau)b^2/E_T h^3 \tag{33}$$

Finally, the equation of motion obtained from (29) after the substitution of w_0, w_s, $\tilde{w}$, f, $\tilde{f}$ and the application of the Galerkin procedure is

$$\tilde{W}(\tau)_{,\tau\tau} + p_1\tilde{W}(\tau) - p_{\tau 2}(\tau)\tilde{W}(\tau) + p_2\tilde{W}^2(\tau) + p_3\tilde{W}^3(\tau) = p_{\tau 1}(\tau) \tag{34}$$

where

$$
\begin{aligned}
p_1 &= \frac{1}{\pi^4}\left[C_d\left(\frac{m}{\lambda}, n\right) + \frac{C_b^2(m/\lambda, n)}{C_a(m/\lambda, n)}\right] \\
&\quad + 2\left(\frac{m}{\lambda}\right)^2 n^2(c_1 + c_2)(3W_s^2 + 6w_0 W_s + 2W_0^2) - \frac{1}{\pi^2}\left(\frac{m}{\lambda}\right)^2 N_0 \\
p_2 &= 6\left(\frac{m}{\lambda}\right)^2 n^2(c_1 + c_2)(W_s + W_0) \\
p_3 &= 2\left(\frac{m}{\lambda}\right)^2 n^2(c_1 + c_2) \\
p_{\tau 1}(\tau) &= \frac{1}{\pi^2}\left(\frac{m}{\lambda}\right)^2 (W_s + W_0)N_x(\tau) \\
p_{\tau 2}(\tau) &= \frac{1}{\pi^2}\left(\frac{m}{\lambda}\right)^2 N_x(\tau)
\end{aligned}
\tag{35}
$$

Equation (34) can be integrated numerically.

The initial conditions corresponding to the plate being at rest at the instant of load application are

$$\tilde{W} = \tilde{W}_{,\tau} = 0 \quad \text{at } \tau = 0 \tag{36}$$

5. NUMERICAL EXAMPLES

Numerical integration of eqn. (14) was carried out by the Runge–Kutta method. The material of the plate was graphite–epoxy with the following dimensionless characteristics:

$$E_L/E_T = 40 \qquad G_{LT}/E_T = 0{\cdot}5 \qquad \nu_{LT} = 0{\cdot}25$$

The lamination angle was $\pm 30°$ and the number of layers was assumed to be large so that the bending–stretching coupling was negligible.

The load was assumed to increase proportionally to time. The nondimensional rate of loading was defined when the dimensional load

$$\bar{N}_x(t) = \bar{s}t \tag{37}$$

was replaced by $N(\tau)$ given by (18). Then

$$s = \frac{\bar{s}b^2}{\pi^2 E_T h^3}\sqrt{\frac{\rho b^4}{E_T h^3}} \tag{38}$$

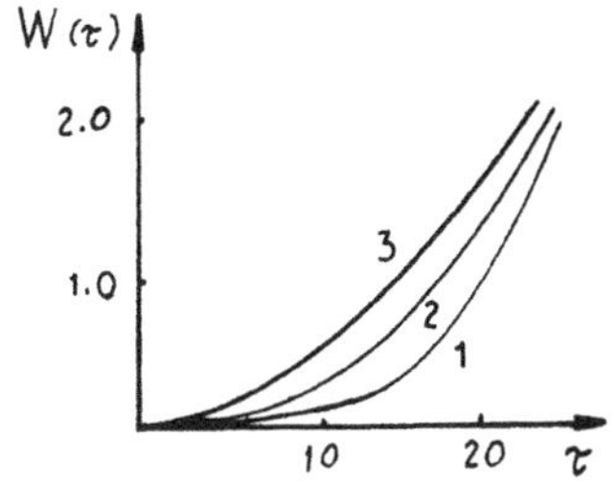

FIG. 2. Effect of initial imperfections on dynamic buckling of antisymmetrically laminated angle-ply plates. $\lambda = 1$, $m = n = 1$, $s = 3{\cdot}0$.

Curve	1	2	3
W_0	0·10	0·25	0·50

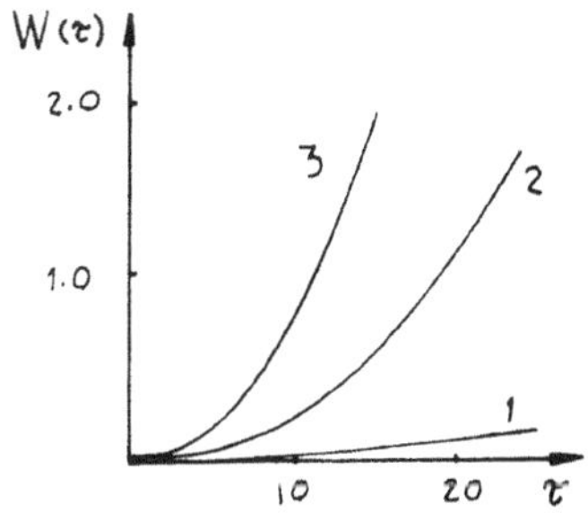

FIG. 3. Effect of rate of loading on dynamic buckling. $\lambda = 1$, $m = n = 1$, $W_0 = 0{\cdot}25$.

Curve	1	2	3
	1·0	3·0	5·0

The behavior of the square plate with different amplitudes of the initial imperfection is shown in Fig. 2. The displacements increase as a result of larger imperfections at the initial phase of motion. At the later phase the displacements exhibit a gradual growth combined with an oscillatory-type motion. This phase of motion is not shown since it corresponds to large deformations when plastic effects are unavoidable. The effect of the rate of loading is illustrated in Fig. 3.

REFERENCES

1. HOFF, N. J., The dynamics of the buckling of elastic columns, *ASME J. appl. Mech.*, **18** (1951), 68–74.

2. HOFF, N. J., Dynamic stability of structures, in: *Dynamic Stability of Structures* (Hermann, G. ed.), Oxford, Pergamon Press, 1967, pp. 7–64.
3. VOL'MIR, A. S., *Stability of Deformable Systems*, Moscow, Nauka, 1967 (in Russian).
4. VOL'MIR, A. S., *Nonlinear Dynamics of Plates and Shells*, Moscow, Nauka, 1972 (in Russian).
5. ARI-GUR, J., WELLER, T. and SINGER, J., Experimental and theoretical studies of columns under axial impact, *Int. J. Solids Struct.*, **18** (1982), 619–641.
6. BABICH, D. V. and KHOROSHUN, L. P., Dynamic loss of stability of an instantaneously compressed layered cylindrical shell, *Soviet appl. Mech.*, **16** (1981), 586–590.
7. BRIGADIROV, G. V. and TOLOKONNIKOV, L. A., Impact of a cylindrical shell against a rigid barrier, *Mech. Solids*, **18** (1983), 179–181.
8. SAIGAL, S., YANG, T. Y. and KAPANIA, R., Dynamic buckling of imperfection-sensitive shell structures, *Proc. 27th Structures, Structural Dynamics and Materials Conference*, Part 2, San Antonio, Texas, 1986, pp. 569–575.
9. EKSTROM, R. E., Dynamic buckling of a rectangular orthotropic plate, *AIAA J.*, **11** (1973), 1655–1659.
10. SEVIN, E., On the elastic bending of columns due to dynamic axial forces including effects of axial inertia, *ASME J. appl. Mech.*, **27** (1960), 125–131.
11. STAVSKY, Y. and HOFF, N. J., Mechanics of composite materials, in: *Composite Engineering Laminates* (Dietz, A. G. H. ed.), Cambridge, Mass., MIT Press, 1969, pp. 5–59.
12. TENNYSON, R. C., MUGGERIDGE, D. B., CHAN, K. H. and KNOT, N. S., Buckling of fiber-reinforced circular cylinders under axial compression, AFFDL-TR-72-102, 1972.
13. HUI, D., Soft-spring nonlinear vibrations of antisymmetrically laminated rectangular plates, *Int. J. Mech. Sci.*, **27** (1985), 397–408.
14. HUI, D., Effects of geometric imperfections on frequency–load interaction of biaxially compressed antisymmetric angle ply rectangular plates, *ASME J. appl. Mech.*, **52** (1985), 155–162.
15. KAMKE, E., *Differentialgleichungen*, 3rd edn, New York, Chelsea Publishing Company, 1959.
16. WATSON, G. N., *A Treatise on the Theory of Bessel Functions*, 2nd edn, Cambridge, The University Press, 1966.

40

Dynamic Response of CFRP Plates under the Action of Random Acoustic Loading

R. G. WHITE

Institute of Sound and Vibration Research, University of Southampton, Southampton SO9 5NH, UK

and

R. F. MOUSLEY

Royal Aircraft Establishment, M and S Department, R50 Building, Farnborough, Hants GU14 6TD, UK

ABSTRACT

Acoustic fatigue design procedures for metallic, stiffened skin and plate-type structures have been well established and validated in the past for aircraft structures. The advent of CFRP and its use in aircrcraft has necessitated re-appraisal of dynamic design techniques. This chapter concerns experimental and theoretical studies of CFRP plates under the action of random acoustic loading. Attention is given to the nature and levels of the dynamic strains induced in terms of statistical properties and relative modal contributions, the latter being important in consideration of using simple single mode formulae for dynamic response prediction. The effects of high levels of excitation, up to 160 dB, which can produce non-linear responses are discussed. The case of forced response of plates under the action of combined static in-plane compressive loading and acoustic excitation is also considered.

1. INTRODUCTION

Panel-type structures in aircraft are subjected to dynamic excitation by jet noise, turbulence, buffet, etc. There has been, in particular, considerable interest in the response of thin, stiffened-skin structures to jet noise, that is random acoustic loading. Originally, the problem confronting the designer was to design aluminium alloy structures composed of a thin skin attached, by rivetting, bonding or a combination of techniques, to stringers so that the built-up structure had adequate fatigue life. In order that dynamic response predictions can be made at the design stage it is obviously desirable that relatively simple design procedures be developed. A considerable amount of work has been carried out in the past on arrays of panels in what were then typical fuselage constructions although the design problem was simplified in two ways to yield reasonable dynamic response estimates for this type of structure under the action of random acoustic loading. The two simplifications were, first to consider a panel in an array to be a plate with boundary conditions which could be reasonably approximated by classical plate theory, and secondly to reduce the multi-modal response problem to that of fundamental mode response prediction only. These approximations generally form the basis of acoustic fatigue assessment procedures used successfully in the aircraft industry over about the past three decades. The methods are well established and proven. However, recent developments in aircraft construction, configurations and propulsion systems have necessitated their re-appraisal. In particular, composite constructions have been introduced and it is likely in the military application that structures may be exposed to higher levels of jet noise excitation than before.

In this chapter, the extension of established response prediction procedures to composite plates is examined and some experimental results are presented showing the dynamic strain response of panels under the action of random acoustic loading at high sound pressure levels.

2. PREDICTION OF THE RESPONSE OF PLATES TO RANDOM ACOUSTIC LOADING

The forced vibration of a plate by a sound field has been examined by many workers but a good statement of the problem is given by Clarkson[1] who simplified the multi-modal response prediction problem and developed the method which has been used for dynamic response prediction of metallic, panel-type structures.[2] Attention is drawn to this paper[2] because the approximations inherent in the single mode response prediction method are stated there, the method is validated for aircraft structures and extension of the procedure to composite panels is still based upon

Clarkson's work. Assuming that the response of a panel type structure can be represented by a plate which is vibrating predominantly in its fundamental mode, then the mean square bending stress at the point of interest is given by:

$$\overline{\sigma^2(t)} = \frac{\pi}{4\zeta} f_r G_p(f_r) \sigma_0^2 \tag{1}$$

where σ_0 = stress at the point of interest due to a uniform unit static pressure of unit magnitude;
ζ = equivalent viscous damping ratio associated with the fundamental resonance;
f_r = resonance frequency of the fundamental mode (Hz);
$G_p(f_r)$ = excitation pressure spectral density at frequency f_r.

This simple formula has been used as the basis of design methods for estimating the RMS stress in metallic, stiffened skin panels subjected to random acoustic loading.[3] The value ascribed to the equivalent viscous damping ratio, ζ, is obviously very important and for metallic structures without special damping treatments a value of $\zeta = 0.017$ has been successfully assumed to be typical.

3. COMPOSITE PLATE RESPONSE PREDICTION

With the introduction of composite structures and the use, in particular, of multi-layered CFRP constructions the initial problem was to consider whether or not, the well-established method outlined above could be reasonably applied to this type of structure. The secondary problem was that if the method could be applied, then how could the required parameters be estimated. The latter point arises because of the wide range of constituent materials which could be used in a variety of 'lay-ups', that is relative angles between fibres in each layer. Estimation of parameters is considered in this section, although as will become apparent, it is likely that in part some reliance will have to be placed upon experimentally determined values of viscous damping ratio (or loss factor). It must be noted at this stage, however, that most theoretical and experimental work on composite plates and structures involved strain prediction/measurement rather than stress, and fatigue work is usually based on surface strain criteria. Therefore, the RMS stresses estimated via eqn. (1) are converted to strains in the prediction process.

Natural frequencies of composite plates may be estimated by either the Rayleigh–Ritz method or a finite element (FE) technique. Examples of the Rayleigh–Ritz method applied to composite plates with free boundary conditions are given in Ref. 4. A comparison of the Rayleigh–Ritz and finite element methods applied to plates with static in-plane compression and a variety of boundary conditions is given in Ref. 5. For that application it was observed that although the FE method was very versatile it required much more computational effort than the Rayleigh–Ritz method. It must be noted, however, that the FE method may be the only method of analysis which could be applied with sufficient accuracy to an irregular structure such as a plate with a defect or hole in it. The lower frequencies of initially unstressed, specially orthotropic rectangular composite plates may be estimated via use of a data sheet.[6] Specially orthotropic plates are balanced laminates having layers of equal thickness, elastic properties and orientation which are symmetrically arranged about the plate central plane. It is not possible to use nomographs for plate natural frequency prediction, as have been used in the past, because as already stated above, the range of parameters for a composite plate is too wide. The current method is to use a programme[6] based on the Rayleigh–Ritz method which requires the elastic properties of the lamina to be known. These may either be deduced[7] from constituent fibre and matrix properties and fibre volume fraction by making a number of assumptions, or measured experimentally. If estimation of laminar properties is not possible, the required properties may be found by direct measurement on a multilayered composite having all fibres aligned in one direction. This is usually achieved by static tests of plates[8,9] and beams or via the dynamic testing of beam specimens in bending to determine flexural properties[10,11] or rods in torsion to determine shear properties.[12]

It can be seen from eqn. (1) that the damping of structures is important in controlling random response of panels and it would, of course, be the dominant factor in controlling resonant response if there were harmonic forcing at resonance. In the case of metallic, aluminium alloy structures it has already been observed in Section 2 above that a value of $\zeta = 0{\cdot}017$ sufficed for dynamic response prediction for that class of structure. This is a gross simplification, bravely made, which was adequate for skin-stringer structures. It must be realised that damping arises from a variety of sources in a built-up structure, such as material damping which is low for metals, damping at joints, friction, aerodynamic damping and acoustic radiation. In the case of composite structures the situation is more complicated. The same mechanisms prevail but material damping will in addition depend upon matrix properties and 'lay-up' to some extent. It has been proposed[13] that a typical equivalent viscous damping ratio for an individual plate within an array in a CFRP structure is 0·025 but it is advised that if this reference value is used, the estimated RMS strain may be factored by the square root of the ratio of damping values for other situations. Internal damping in a composite depends upon a range of parameters such as the effects of temperature on matrix energy dissipation; damage, etc. It is the authors' experience that composites have essentially linear damping and hence the material loss factor does not significantly depend upon strain amplitude. The damping properties of a plate may be estimated[14] with all edges simply supported or all edges clamped, the procedure having been validated for plates with free edges; it is probable that for the near future, experimentally determined values of ζ will often be used in RMS strain predictions.

The principal merit of CFRP as an aerospace structural material is the high stiffness-to-weight ratio of the composite compared with aluminium alloy. The internal damping of 'conventional' lay-up composites is not very high and it may be worthwhile to develop composites with enhanced internal dissipation. The damping properties of CFRPs have been well reviewed[15] and damping investigations carried out in Ref. 16. In the latter work,[16] it was shown that the internal damping of CFRP composites can be increased by careful material design, that is choice of form of construction and type of resin. A plate was manufactured which had a modal loss factor at the first resonance, when tested in the free–free condition, of approximately 0·01 at an acceptable fibre volume fraction without appreciable reduction in specific stiffness compared with the now usual type of construction. A loss factor of 0·01 represents about a tenfold increase on damping of plates with conventional construction, also tested in the freely-supported condition. (Note that these loss factors should not be compared with the value quoted above for a plate *in an array*.) The value of 0·01 is high for the modal loss factor of a freely suspended plate. Further effort should be devoted to development of the stiff, lightweight, heavily damped material!

The RMS strain prediction method now in use for rectangular composite plates[13] is based upon eqn. (1) which depends upon linear behaviour. It has already been established[17] that for very thin plates at high sound pressure levels the response to acoustic loading is non-linear and simple methods for response prediction are not applicable. For 'conventional' composite plates the linear prediction method can be applied reasonably to plates with six or more layers; this is essentially the 'thickness requirement' for linear behaviour, however non-linear plate vibration is well known to occur at large deflections relative to the thickness. In addition for plates with static, in-plane compressive loading, in the immediate pre-buckling and the whole of the post-buckling regimes, the vibration is wholly non-linear in nature with modal coupling, etc., and simple strain response prediction methods cannot be used;[5] this is discussed further in the next section. It is possible to use simple methods for estimating the lower resonance frequency of the plate but this should not be interpreted as an indicator that response prediction can be simplified in non-linear regimes.

4. EXPERIMENTAL STUDIES WITH CFRP PLATES INCLUDING THE EFFECTS OF STATIC, IN-PLANE COMPRESSION

The response of plates to random acoustic excitation has been examined at the ISVR in a progressive wave tube as shown in Fig. 1. Essentially, the apparatus consists of an electropneumatic driver (siren) connected via a horn to a hard-walled parallel duct with an absorbing termination. An aperture in the wall of the duct is used for mounting plate specimens in a supporting frame; the plate can then be exposed to acoustic excitation in the frequency range of approximately 40–800 Hz. A Wylie Laboratories

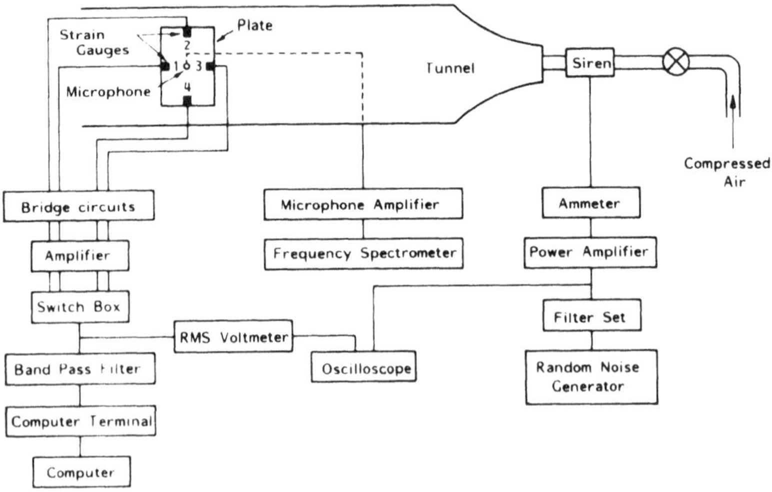

FIG. 1. Instrumentation for high intensity acoustic testing.

WAS 3000 siren of 30 000 W has been installed enabling a maximum SPL of 163 dB to be achieved. Signals from the strain measuring system were input to a digital signal processing system to enable the following strain parameters to be examined.

Strain Spectral Density

Modal contribution

This is obtained by dividing the running integral across the strain spectral density curves by the overall mean square strain. A large increase in the parameter at a given frequency is indicative of a large modal contribution at that frequency in the strain spectral density. Use of this parameter is very convenient for determining the quantitative contribution of individual resonances (modes) to the overall response.

Probability density

This should be of the form of a Gaussian distribution for linear strain signals. For non-linear responses the peak may be to one side of the mean (as also indicated by the cumulative probability) and can be indicated by evaluating the skewness of the probability density function. The sharpness of the peak of the function is indicated by the kurtosis value. Skewness and kurtosis are 0 and 3·0, respectively, for a Gaussian probability density.

Some experiments were carried out[18] on a CFRP plate at Sound Pressure Levels (SPL) up to 145 dB with a bandwidth of 90–700 Hz. The plate for these tests was of HT-S fibre and DX-210 resin at 0·6 fibre volume fraction, the 'lay-up' being of eight layers $(0^\circ, 90^\circ, 90^\circ, 0^\circ)_{SYMM}$. The rectangular plate, 410 mm × 280 mm, was clamped as well as possible in a steel frame and mounted in the tunnel wall. The response was measured via a strain gauge mounted along the longest side, measuring strain in the direction parallel to the shorter side which also meant that strain was being measured in a direction across the surface fibres (maximum strain). The fundamental natural frequency of the plate using the data sheet already referred to above,[6] assuming perfect edge clamping, was calculated to be 106 Hz and was subsequently measured to be 93 Hz; the difference was probably due to lack of rigidity in the clamping frame. A dynamic strain prediction, as already noted, of damping information is required. The viscous damping ratio of the first mode was measured to be $\zeta = 0{\cdot}0034$ (for modes 2, 3 and 4 respectively, values of 0·005, 0·0039 and 0·0045 were also measured). These values are as would be expected for a single plate in this mounting arrangement. The dynamic strain prediction programme[13] was used to estimate the RMS surface strains in the plate at the strain gauge position for the same values of SPL which were achieved in the experiments. The estimated and measured strain values are compared in Fig. 2 for overall SPL up to 145 dB. The single mode predictions using the data sheet were in very good agreement with the measured values. It can therefore be observed that for plates at moderate excitation levels, without static, in-plane loading, the prediction method is reliable. Although some non-linear effects were observed, the behaviour was sufficiently linear for the prediction method to work reasonably well.

Studies have also been carried out at much higher sound pressure levels and with static, compressive in-plane loading applied to a plate[5] via edge shortening. When static compressive, in-plane loading is applied the degree of initial imperfection in terms of initial 'flatness' of the plate is important in controlling the buckling behaviour. The effects of static, in-plane loading on natural frequencies of plates have been discussed[19] and the subject will

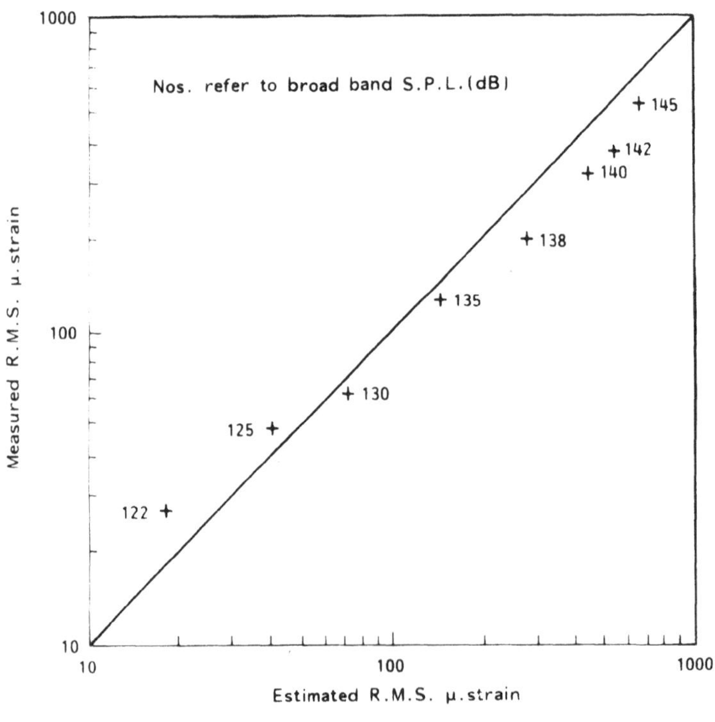

FIG. 2. Comparison of measured and estimated RMS strains in a CFRP plate. Broad band excitation.

not be considered in detail here. The buckling behaviour and natural frequencies of the test plate used in this part of the work reported upon here will, however, be briefly reported. The test plate, 300 mm × 400 mm × 2 mm was of CFRP at 0·6 fibre volume fraction with XAS fibres and 914C resin, the lay-up being $(45°, 0°, 0°, 90°, 0°, -45°, 0°)_{SYMM}$. The data presented here relate to clamped conditions on all sides. The plate mounted in its loading/clamping frame was subjected to static, in-plane compression via edge shortening u mm, the critical value of u to theoretically produce buckling being u_c. A Moiré fringe pattern [5,20] of the statically loaded plate is shown in Fig. 3 for $u/u_c = 2·8$, well above buckling, and it can be seen that static buckling occurred in an overall 2:1 type of mode. The Moiré method is most useful in this type of study; the results are much more informative in the laboratory than a figure in this chapter can indicate. Small head movements, for example, perpendicular to the plate axis show apparent 'movement' of the fringes which indicate the direction of bending in the static buckling mode shape relative to the plate axis (i.e. in the case shown, whether the areas near the top and bottom of the plate were deflected 'in' or 'out' relative to the initial plane of the plate). The inexpensive apparatus used to produce the Moiré fringe patterns simply consisted of a slide projector used as a light source and a piece of curtain net used as a grating. The method is described and better examples of fringe patterns given in Ref. 5.

Figure 4 shows theoretical prediction[5] of some resonance frequencies with various magnitudes of initial imperfection in the 1, 1 mode. That is the plate was considered to be initially deflected in the 1, 1 mode before the static compressive, in-plane load was applied. In Fig. 4, Q_i is the generalised transverse displacement in the 1, 1 mode; for $Q_i = 0$, the plate is flat. The buckling mode of this plate was predominantly 2:1 but there was some coupling with other modes, hence the rotation of the nodal pattern in Fig. 3. It can be seen from Fig. 4 that initial imperfection has a strong effect on natural frequencies and for the buckling mode, the effects are reversed in the pre- and postbuckle regimes. Strain responses to random acoustic excitation are shown in Figs. 5, 6 and 7 for the strain gauge position indicated; $u/u_c = 0$ in Fig. 5, the plate is just below buckling in Fig. 6 and is well into the postbuckle regime in Fig. 7. In each case the SPL was about 156 dB. The spectral and probability analyses show that even at this SPL the response is predominantly in the fundamental mode for $u/u_c = 0$, i.e. the simple case. This is also the situation in Fig. 6 for $u/u_c = 0·8$. In the postbuckle regime, Fig. 7, there are no strong modal contributions and the integral across the strain spectral density is indicative of a broadband strain

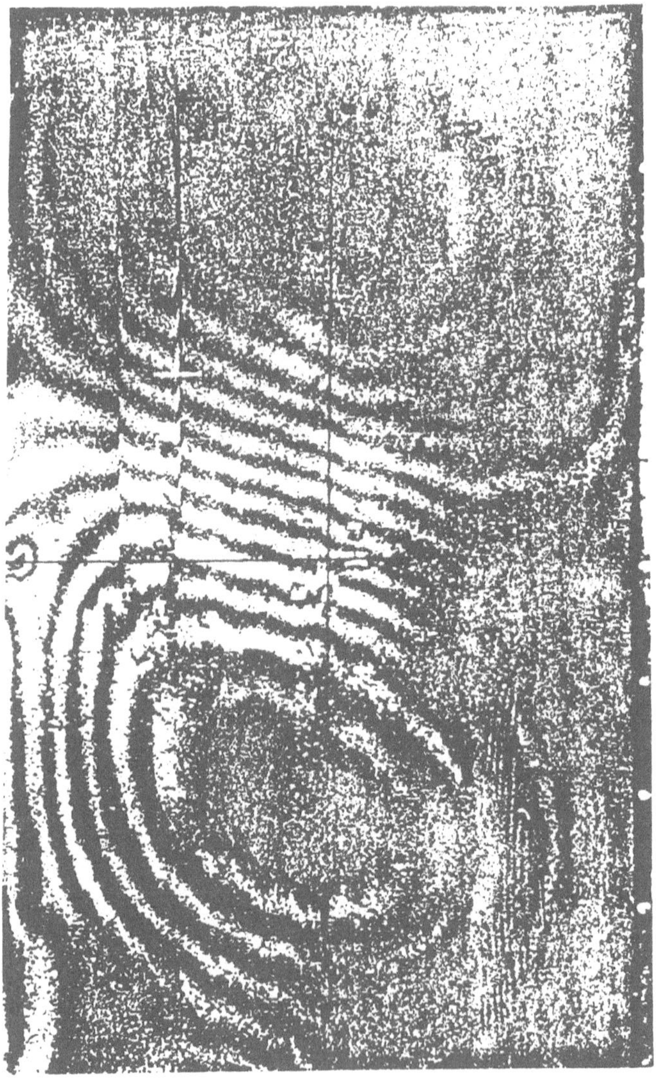

FIG. 3. The Moiré pattern of a composite plate under in-plane compression ($u/u_c = 2{\cdot}8$) (consecutive fringes represent a transverse displacement increment of 0·18 mm).

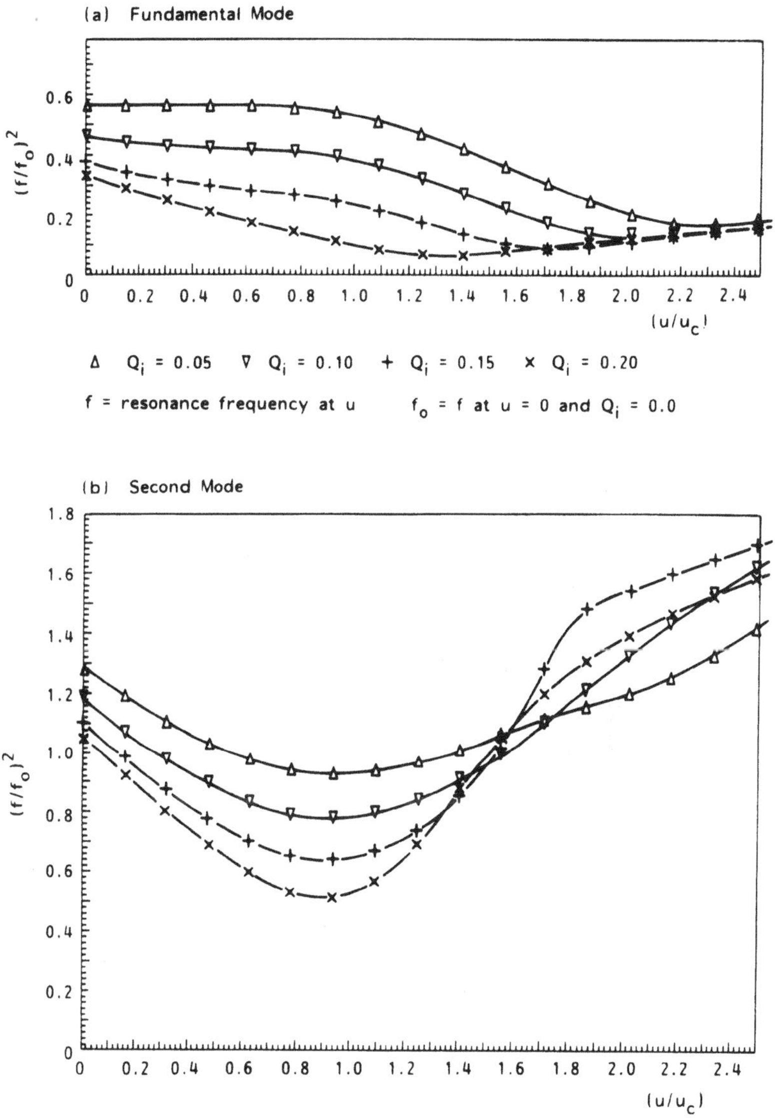

FIG. 4. Effects of (1, 1) mode imperfection on resonance frequencies of a composite plate (300 × 450 × 2 mm) all sides clamped; lay-up $|45, 0, 0, 90, 0, -45, 0|_{SYMM}$.

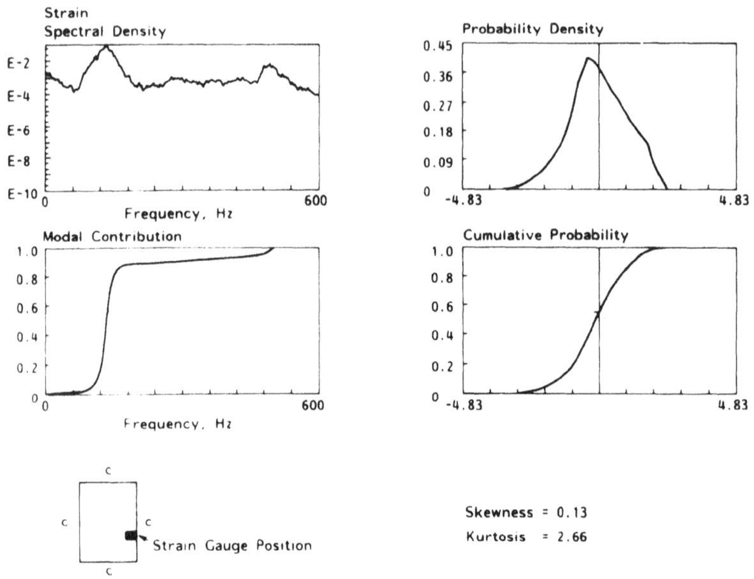

FIG. 5. Bending strain in a composite plate under acoustic loading. SPL = 155 dB, $u/u_c = 0{\cdot}0$.

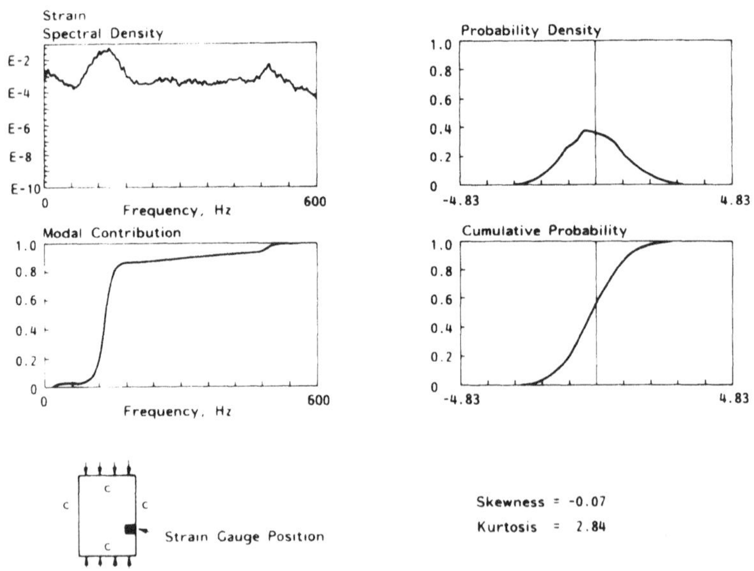

FIG. 6. Bending strain in a composite plate under combined acoustic loading and static, in-plane compression. SPL = 157 dB, $u/u_c = 0{\cdot}80$.

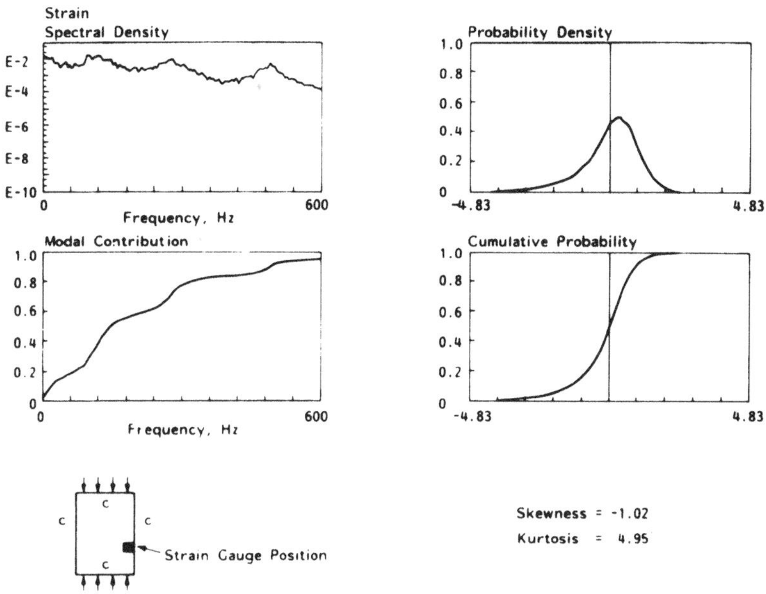

FIG. 7. Bending strain in a composite plate under combined acoustic loading and static, in-plane compression. SPL = 156 dB, $u/u_c = 2{\cdot}4$.

response. The probability plots indicate considerable non-linear behaviour in the postbuckle regime; this is also conveyed in the values of skewness and kurtosis noted on the figures. Highly skewed distributions were noted in a wide range of studies in Ref. 5 where kurtosis values of up to about 5 were measured. It is generally the case that for plates with static, compressive, in-plane loading combined with random acoustic loading, non-linear behaviour is only significantly observed in the prebuckle region at large vibration amplitudes (i.e. high excitation levels) whilst the postbuckle region is very non-linear with considerable non-linearity occurring even with small amplitude vibrations. Figure 8 shows the percentage contribution of the fundamental mode to the total RMS strain response at the strain gauge position already considered. It is clear that both bending and in-plane strain response contributions decrease with increasing compression until the mode cannot be identified in the response (see also the strain spectral density in Fig. 7). The fundamental mode is only found in the postbuckled in-plane strain response at low levels of excitation. It is clear that a single mode calculation of the type discussed above and used for the design of plate-type structures can only be used reasonably at static loads significantly below buckling. It is noteworthy, however, that the complete

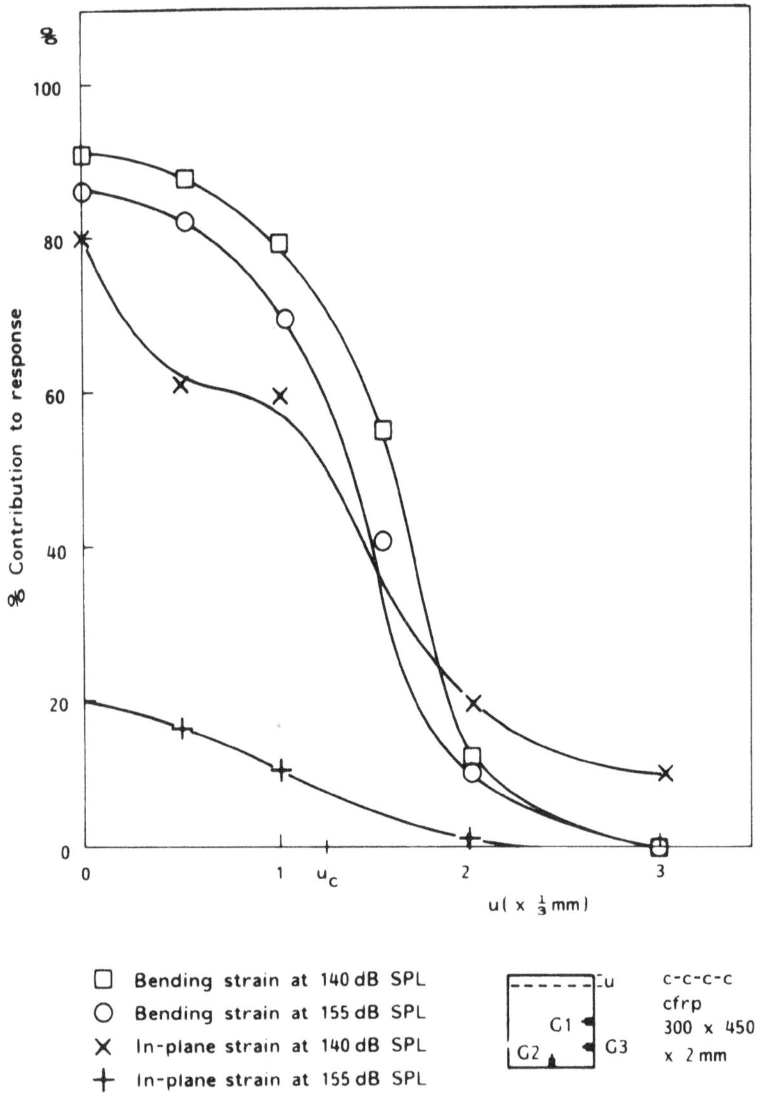

FIG. 8. Contribution of fundamental mode to response to random acoustic excitation.

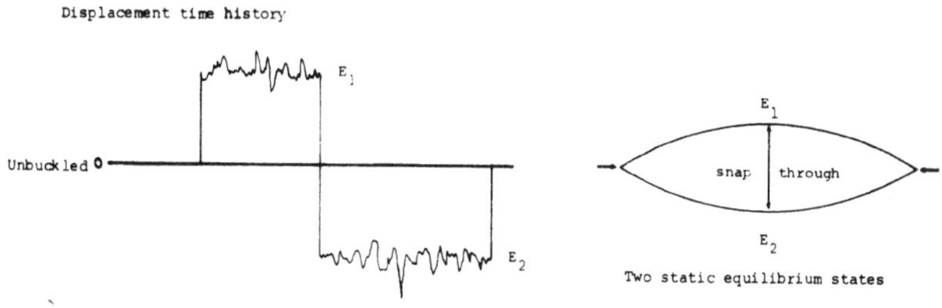

FIG. 9. Diagram to illustrate 'snap-through'.

range of experiments conducted,[5] lead to the following conclusions based upon analyses of the acoustically induced strains. It was found that the dynamic responses of postbuckled plates are not necessarily larger than the corresponding dynamic responses of prebuckled plates if no snap-through motion of large amplitude (compared with the oscillatory motion) occurs (snap-through is the term given to a transit from the environs of one static equilibrium to another, see Fig. 9). Snap-through motion of amplitude larger than the oscillatory component can occur when the compression parameter u/u_c is greater than 2·0 and the overall RMS bending strain value is then comparable to the static bending strain value. It is probably possible to take an RMS bending strain value equal to the static bending strain value for fatigue life estimation at high sound pressure levels, although account has to be taken of the frequency of snap-through.

5. CONCLUSIONS

Simple, single (fundamental) mode formulae for dynamic response prediction of panels in aircraft structures have been developed using plate models. These have been proved successful for metallic structures and are valid for composite structures of the multi-layered type. Modal damping is, of course, an important parameter in the estimation procedure and although a 'typical' value may be assumed it is probable that some knowledge of experimental data will also be required for the near future. It is hoped that the internal damping of composite structures will be eventually increased, in a predictable and controlled way by the use of more dissipative matrix materials and appropriate forms of construction.

Dynamic response prediction methods of the single mode type used for acoustic fatigue life estimation depend on the structure having linear dynamic response characteristics. It appears that in simple, statically unloaded plate-like constructions even at quite high sound pressure levels the degree of structural non-linearity is not great enough to preclude reasonable dynamic strain estimation. If compressive loading is also present, then the response can become non-linear at large deflections in the prebuckle regime and the postbuckle regime concerns wholly non-linear behaviour. It is clear that single mode calculations of the type discussed here can only be used reasonably at static, compressive loads significantly below buckling. In the postbuckle regime, at high sound pressure levels up to 160 dB, it is probably possible to take an RMS bending strain value equal to the static bending strain value for fatigue life estimation. It must be remembered, however, that in such a high level, combined strain state, the principal dynamic and static strains can act in perpendicular or parallel directions in different areas of a plate under uniaxial compression. As compression is critical in the fatigue behaviour of fibre reinforced materials, the effect of the direction of compression may be significant. Also, at high levels of compression, the fatigue failure mechanisms will be different from those experienced in pure bending. Thus although bending strain response predictions might be made, fatigue life estimation with high compressive loading will be difficult.

ACKNOWLEDGEMENT

The authors would like to thank the Royal Aircraft Establishment, Farnborough for the sponsorship, via an unclassified research agreement, of the programme of work on plates under combined loading.

REFERENCES

1. Clarkson, B. L., *Noise and Vibration*, Chap. 3. (R. G. White and J. G. Walker eds), Ellis Horwood, 1982.
2. Clarkson, B. L., Stresses in skin panels subjected to random acoustic loading, *Aeronautical Journal of the Royal Aeronautical Society*, **72** (1968), 1000–1010.
3. ESDU, The estimation of RMS stresses in stiffened skin panels subjected to random acoustic loading, Engineering Sciences Data Unit item number 72005, 1972 (Amended 1984).
4. Wright, G. C., Dynamic behaviour of fibre reinforced plastic beams and plates, Ph.D. thesis, University of Southampton, 1973.
5. Ng, C. F., Dynamic behaviour of postbuckled, composite plates under acoustic excitation, Ph.D. thesis, University of Southampton, 1986.
6. ESDU, Natural frequencies of rectangular, specially orthotropic laminated plates, Engineering Sciences Data Unit item number 83036, 1983.

7. ESDU, Estimation of the stiffnesses and apparent elastic properties of laminated flat plates, Engineering Sciences Data Unit item number 83035, 1983.
8. CURTIS, P. T., CRAG test methods for the measurement of the engineering properties of fibre reinforced plastics, RAE Technical Report 85099, 1985.
9. TSAI, S. W., MASSARD, T. N. and SUSUKI, I., Think composites, *Composites Design—1985*, Paris and Tokyo, Dayton, 1985.
10. WRIGHT, G. C., The dynamic properties of glass and carbon fibre reinforced plastic beams, *Journal of Sound and Vibration*, **21** (1972), 205–212.
11. WHITE, R. G., Some measurements of the dynamic properties of mixed, carbon fibre reinforced plastic beams and plates, *Aeronautical Journal of the Royal Aeronautical Society*, **79** (1975), 318–325.
12. WHITE, R. G. and PALMER, T. A., Control of the properties of carbon fibre reinforced plastics *AIAA Journal*, **22**(11) (1984), 1662–1669.
13. ESDU, Estimation of RMS strain in laminated skin panels subjected to random acoustic loading, Engineering Sciences Data Unit item number 84008, 1984.
14. ESDU, Estimation of damping in laminated and fibre reinforced plates, Engineering Sciences Data Unit item number 85012, 1985.
15. WILLWAY, T. A., Damping of composites, *Proceedings of the Workshop on Composites Design for Space Applications, ESTEC NOORDWIJK* ESA SP-743, 1986, pp. 47–52.
16. WILLWAY, T. A., Stiff, light, highly damped CFRPs and the effect of complex loads on damping, Ph.D. thesis, University of Southampton, 1986.
17. WHITE, R. G., A comparison of some statistical properties of the responses of Aluminium Alloy and CFRP plates to acoustic excitation, *Composites*, **9** (1978), 251–258.
18. KOUMBIS, N. B., Dynamic properties of carbon fibre reinforced plastics subjected to acoustic excitation, M.Sc. thesis, ISVR, University of Southampton, 1984.
19. WHITE, R. G., Some influences of material properties on the dynamic response of CFRP structures, *Proceedings of the Third International Conference on Composite Structures*, Paisley 1986, pp. 621–633.
20. MOUSLEY, R. F., A shadow Moiré method for the measurement of damage in composites, *Composite structures*, **4** (1985), 231–244.

41

Exact Finite Element Method for Sandwich Beams

M. T. HEINISUO, S. J. MALMI and A. I. J. MÖTTÖNEN

Tampere University of Technology, Box 527, 33101 Tampere, Finland

ABSTRACT

A finite element method for a Sandwich-type beam is presented. Linearly elastic theory is supposed to be valid both for the faces and the cores. The exact displacement functions which are the solutions of the governing differential equations are used as shape functions when deriving the stiffness equations. The stiffness matrices are presented for the beams with thin and thick faces. Static (including normal force) and dynamic stiffness matrixes are derived. Some numerical results are calculated with the computer program developed.

INTRODUCTION

Sandwich-type beams are widely used in the field of civil engineering (Fig. 1) because they are economical, light, tight and they have good insulation and strength capacity. The finite element method (FEM) is considered nowadays the most suitable way to perform the numerical calculations when solving the statics and dynamics of the structures.

Many FEM-formulations derived by using approximate shape functions for the displacements of the Sandwich beam are presented in the literature (e.g. Abel and Popov;[1] Thompson *et al.*;[2] Holt and Webber;[3] Murakami[4]). One FEM-formulation (Monforton[5]) using the exact shape functions was found. There the static stiffness matrix was derived for the laterally loaded Sandwich beam. In that paper the beam element had eight degrees of

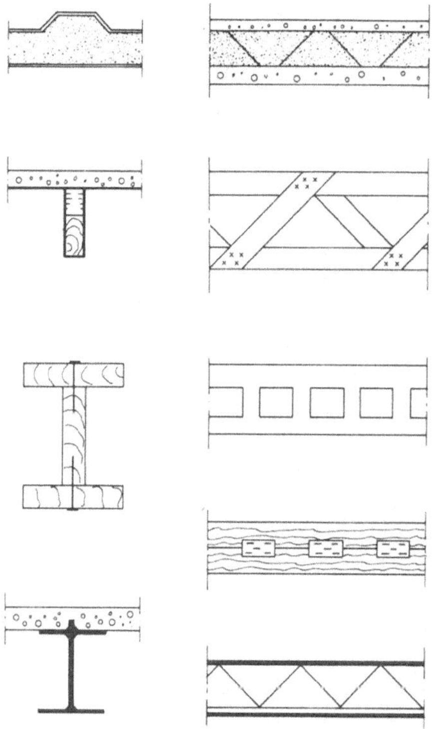

FIG. 1. Some sandwich beams.

freedom because the lateral displacement was divided into two parts: displacements due to the bending and the shear. In that formulation the displacement boundary conditions are difficult to handle. In the proposed method the degrees of freedom of the beam element are six and there are no difficulties in giving the displacement boundary conditions, because the total displacements are used when deriving the equations. Recently (Schwarze[6]) the exact force method for Sandwich beams was presented. More recently, Davies[7] has presented a similar matrix formulation for one of the cases considered in the present paper.

The static stiffness matrix for the Sandwich beam with thick faces was presented earlier by the first author (Heinisuo;[8] see also Davies[7]). In the present paper the method is generalized for the beams with thick and thin faces and for beams loaded both laterally and axially (buckling) and for a temperature gradient in the static case and also for dynamic lateral loading.

When comparing the proposed method with those using the approximate shape functions the benefit of the proposed method is obvious: the minimum amount of elements are needed in the solution procedure.

THE THEORY OF SANDWICH BEAMS

The following five limitations are valid for the theory.

1. The layers of the beam are connected with the continuous core. This condition is not fully satisfied when considering for example shear walls, nailed beams or composite steel concrete beams (see Fig. 1).
2. The layers are straight, parallel, elastic beams (shear deformation is neglected).
3. The slip-rule for the core material is linear and elastic and the core can resist only shear. Hence, the proposed method can be used incrementally in the non-linear case.
4. The lateral displacements of the layers are equal in every cross-section of the beam. This limitation excludes the local buckling of the beam.
5. The centroids of the layers remain on the line (Bernoulli's hypothesis). If the beam is made of two layers and a core or the beam is made of three symmetric layers and two equal cores then this limitation is useless. In a forthcoming study this hypothesis is tested and in the following only two-layer beams are considered.

A more detailed derivation of the theory of Sandwich beams is presented in the book of Stamm and Witte.[9] Only a short summary of the theory is presented here. Consider first a two-layer Sandwich beam element of length L and loaded laterally by the loading intensity q (Fig. 2). The bending

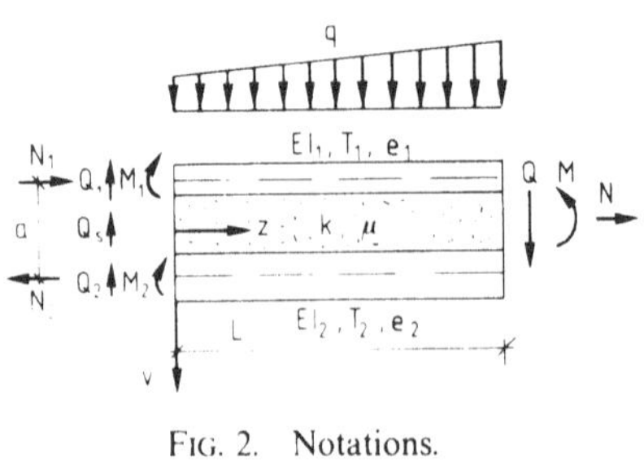

FIG. 2. Notations.

rigidity of the beam EI is divided into two parts $EI = EI_s + EI_o$ where EI_o is the sum of the rigidities of the faces, $EI_o = EI_1 + EI_2$ and EI_s is the Steiner term. The lateral displacement is v and γ is the rotation due to the slip of the core measured in the centroids of the faces. The shear rigidity of the core is k. Some values for the parameter k are presented in the Fig. 3. The equilibrium of the beam is governed by the differential equation (Stamm and Witte[9])

$$EI_o EI_s v^{(6)} - EIkv^{(4)} = -kq + EI_s q^{(2)} \tag{1}$$

The homogeneous boundary conditions can be expressed by the equation (Heinisuo[8])

$$\int_0^L (vQ - (v' - \gamma)M + \gamma M_o) = 0 \tag{2}$$

where Q is the total shear force and M is the total bending moment of the beam and $M_o = M - M_s$ is the sum of the moments of the faces ($M_o = M_1 + M_2$).

If the faces are thin ($EI \simeq EI_s$, $EI_o = 0$) the equations are simpler

$$-EIkv^{(4)} = -kq + EIq^{(2)} \tag{3}$$

$$\int_0^L (vQ - (v' - \gamma)M) = 0 \tag{4}$$

It is seen from the eqn. (2) that there are eight (four if the faces are thin) possibilities to give the boundary conditions. All these have a clear physical meaning as shown by Stamm and Witte.

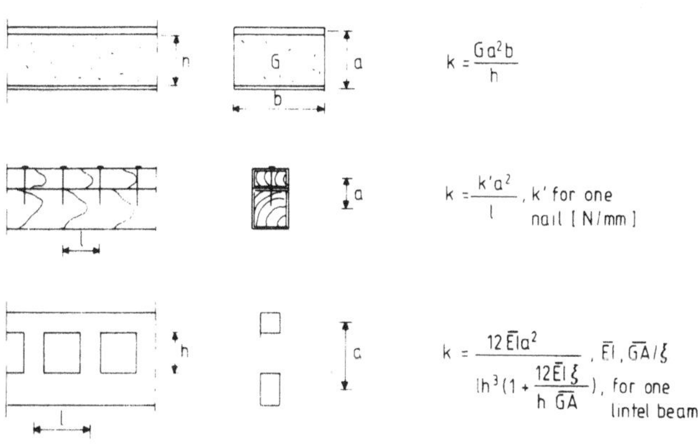

FIG. 3. Shear rigidities.

If the beam is also loaded with a normal force N (positive if tension) then the differential equations are (Stamm and Witte[9])

$$EI_o \neq 0: \; EI_o EI_s v^{(6)} - (EIk + EI_s N) v^{(4)} + kNv^{(2)} = -kq + EI_s q^{(2)} \quad (5)$$
$$EI_o = 0: \; -EI(k + N) v^{(4)} + kNv^{(2)} = -kq + EIq^{(2)} \quad (6)$$

If considering the lateral vibration of the beam with thick faces then, neglecting the damping and the rotational inertia moment, the differential equation of motion is (Stamm and Witte[9])

$$EI_o EI_s v^{(6)} - (EIk + EI_s N) v^{(4)} + kNv^{(2)} + EI_s \mu \ddot{v}^{(2)} - k\mu\ddot{v} = -kq + EI_s q^{(2)} \quad (7)$$

where the dots mean differentiation with respect to the time and μ is the mass per unit length of the beam. For the beam with thin faces the equation is

$$-(EIk + EIN) v^{(4)} + kNv^{(2)} + EI\mu\ddot{v}^{(2)} - k\mu\ddot{v} = -kq + EIq^{(2)} \quad (8)$$

If the motion is harmonic $\ddot{v} = -\omega^2 v$, where ω is the angular velocity, then the equations of the motion are

$$EI_o \neq 0: \; EI_o EI_s v^{(6)} - (EIk + EI_s N) v^{(4)} + (kN - EI_s \mu\omega^2) v^{(2)} + k\mu\omega^2 v = -kq + EI_s q^{(2)} \quad (9)$$

$$EI_o = 0: \; -EI(k + N) v^{(4)} + (kN - EI\mu\omega^2) v^{(2)} + k\mu\omega^2 v = -kq + EIq^{(2)} \quad (10)$$

The temperatures of the faces are: upper face T_1, lower face T_2. The coefficients of linear extension are accordingly e_1 and e_2. In the absence of bending stiffnesses the beam would bend into a curvature

$$\frac{e_1 T_1 - e_2 T_2}{a} = -\mathrm{N} \quad (11)$$

where a is the distance between the centroids of the faces. In this case the loading term $kEI_s\mathrm{N}''$ has to be added to the right hand sides of the aforementioned differential equations. In the following this approximation is used for the temperature loading.

In the following the exact FEM-formulations are given for the solution of the differential equations. It must be noted that in the present study the normal force is supposed to be known, i.e. the structure is statically determined when considering the normal forces. Also the normal force is supposed to be zero in the dynamic case and it is supposed that $EI \geq 3EI_o$ when the faces are thick. When deriving the FEM-formulations the exact solutions of the homogeneous differential equations are needed. The details for the calculations in the first case are given by the first author (Heinisuo[8]) and the other cases are studied in more detail in the diploma works of the two other authors (Malmi;[10] Möttönen[11]). The loading is given by the work equivalent nodal forces afterwards.

The following notations are used in the following

$$\xi = z/L,\ \alpha_1 = (EIk + NEI_s)/(EI_sEI_o),\ \alpha_2 = Nk/(EI_sEI_o)$$

$$\beta = EI_s/(kL^2),\ \delta = EI/(EI_o),\ v^4 = \mu\omega^2L^4/(EI),\ \alpha = EI_o/EI_s$$

$$\phi = \arccos\{[\delta^3 + \tfrac{9}{2}\beta^2v^4\delta(\delta - 3)]/(\delta^2 + 3\delta v^4\beta^2)^{3/2}\} \qquad -\pi/2 < \phi < \pi/2$$

$$\varepsilon = -\tfrac{2}{3}\sqrt{(\delta/\beta)^2 + 3\delta v^4}\cos(\pi/3 + \phi/3) + \tfrac{1}{3}\delta/\beta$$

$$\lambda_1^2 = EIk/(EI_sEI_o)$$

$$\lambda_2^2 = \tfrac{1}{2}(\sqrt{\alpha_1^2 - 4\alpha_2} - \alpha_1) \qquad -\infty < N < 0$$

$$\lambda_3^2 = \tfrac{1}{2}(\sqrt{\alpha_1^2 - 4\alpha_2} + \alpha_1) \qquad -\infty < N < \infty$$

$$\lambda_4^2 = \tfrac{1}{2}(\alpha_1 - \sqrt{\alpha_1^2 - 4\alpha_2}) \qquad 0 < N < \infty$$

$$\lambda_5^2 = -N/[EI_s(1 + N/k)] \qquad -k < N < 0$$

$$\lambda_6^2 = N/[EI_s(1 + N/k)] \qquad N < -k,\ N > 0$$

$$\lambda_7^2 = \tfrac{1}{3}[\delta/\beta + 2\sqrt{(\delta/\beta)^2 + 3\delta v^4}\cos\phi/3]$$

$$\lambda_8^2 = \tfrac{1}{3}[-\delta/\beta + 2\sqrt{(\delta/\beta)^2 + 3\delta v^4}\cos(\pi/3 - \phi/3)]$$

$$\lambda_9^2 = \tfrac{1}{3}[-\delta/\beta + 2\sqrt{(\delta/\beta)^2 + 3\delta v^4}\cos(\pi/3 - \phi/3)] \qquad \varepsilon < 0$$

$$\lambda_{10}^2 = \tfrac{1}{3}[\delta/\beta - 2\sqrt{(\delta/\beta)^2 + 3\delta v^4}\cos(\pi/3 + \phi/3)] \qquad \varepsilon > 0$$

$$\lambda_{11}^2 = \tfrac{1}{2}(\sqrt{(\beta v^4)^2 + 4v^4} - \beta v^4)$$

$$\lambda_{12}^2 = \tfrac{1}{2}(\sqrt{(\beta v^4)^2 + 4v^4} + \beta v^4) \tag{12}$$

The solutions of the corresponding homogeneous differential equations are

$$(2):\ v = C_1\sinh\lambda_1z + C_2\cosh\lambda_1z + C_3 + C_4z + C_5z^2 + C_6z^3 \tag{13}$$

$$(4):\ v = C_1 + C_2z + C_3z^2 + C_4z^3 \tag{14}$$

$$(6):\ v = C_3\cos\lambda_2z + C_4\sin\lambda_2z + C_1\cosh\lambda_3z + C_2\sinh\lambda_3z + C_5z + C_6 \qquad -\infty < N < 0 \tag{15}$$

$$v = C_1\cosh\lambda_3z + C_2\sinh\lambda_3z + C_3\cosh\lambda_4z + C_4\sinh\lambda_4z + C_5z + C_6 \qquad 0 < N < \infty \tag{16}$$

$$(7):\ v = C_1 \cos \lambda_5 z + C_2 \sin \lambda_5 z + C_3 z + C_4 \qquad -k < N < 0 \tag{17}$$

$$v = C_1 \cosh \lambda_6 z + C_2 \sinh \lambda_6 z + C_3 z + C_4 \qquad N < -k,\ N > 0 \tag{18}$$

$$(10):\ v/L = C_1 \cosh \lambda_7 \xi + C_2 \sinh \lambda_7 \xi + C_3 \cos \lambda_8 \xi + C_4 \sin \lambda_8 \xi + C_5 \cos \lambda_9 \xi + C_6 \sin \lambda_9 \xi \qquad \varepsilon < 0 \tag{19}$$

$$v/L = C_1 \cosh \lambda_7 \xi + C_2 \sinh \lambda_7 \xi + C_3 \cos \lambda_8 \xi + C_4 \sin \lambda_8 \xi + C_5 + C_6 \xi \qquad \varepsilon = 0 \tag{20}$$

$$v/L = C_1 \cosh \lambda_7 \xi + C_2 \sinh \lambda_7 \xi + C_3 \cos \lambda_8 \xi + C_4 \sin \lambda_8 \xi + C_5 \cosh \lambda_{10} \xi + C_6 \sinh \lambda_{10} \xi \qquad \varepsilon > 0 \tag{21}$$

$$(11):\ v/L = C_1 \cosh \lambda_{11} \xi + C_2 \sinh \lambda_{11} \xi + C_3 \cos \lambda_{12} \xi + C_4 \sin \lambda_{12} \xi \tag{22}$$

The equations needed in the solution procedure are (N = 0, thick faces)

$$\gamma = (EI_s EI_o/k^2) v^{(5)} - (EI_s(1 + N/k)/k) v^{(3)} - (\mu\omega^2 EI_s/k) v' - EI_s q'/k^2 \tag{23}$$

$$Q = Q_o + Q_s + Nv' = -EI_o v^{(3)} + k\gamma + Nv' \tag{24}$$

$$M = M_o + M_s = -EIv'' + EI_s \gamma' \tag{25}$$

$$M_o = -EI_o v'' \tag{26}$$

STIFFNESS MATRICES FOR SANDWICH BEAMS

The local coordinates are presented in Fig. 4. The components of the displacement vector $\{\mathbf{X}\}$ are

$$\begin{aligned} X_1 &= v(0), X_2 = v'(0) - \gamma(0), X_3 = \gamma(0) \\ X_4 &= v(L), X_5 = v'(L) - \gamma(L), X_6 = \gamma(L) \end{aligned} \tag{27}$$

if the faces are thick. If the faces are thin then

$$X_1 = v(0), X_2 = v'(0) - \gamma(0),\ X_3 = v(L), X_4 = v'(L) - \gamma(L) \tag{28}$$

The corresponding components of the force vector $\{\mathbf{P}\}$ are (thick faces)

$$\begin{aligned} P_1 &= -Q(0), P_2 = +M(0), P_3 = +M_o(0) \\ P_4 &= +Q(L), P_5 = -M(L), P_6 = -M_o(L) \end{aligned} \tag{29}$$

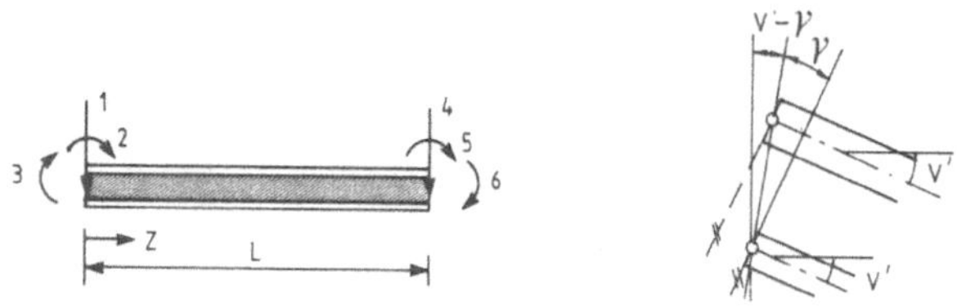

FIG. 4. Local coordinates.

and for the beam with thin faces

$$P_1 = -Q(0), P_2 = +M(0), P_3 = +Q(L), P_4 = -M(L) \qquad (30)$$

The relationship between displacements X_i and parameters C can be calculated. The result is

$$\{X\} = [\bar{K}]\{C\} \qquad (31)$$

and similarly

$$\{P\} = [\bar{\bar{K}}]\{C\} \qquad (32)$$

The final stiffness matrix is

$$[K] = [\bar{\bar{K}}][\bar{K}]^{-1} \qquad (33)$$

Table 1 presents the matrixes $[\bar{K}]$ and $[\bar{\bar{K}}]$ for the cases considered. The

TABLE 1

Matrices K and $\bar{\bar{K}}$

$$\bar{K} = \begin{pmatrix} 0 & 1 & 1 & 0 & 0 & 0 \\ -\lambda_1(EI/EI_s - 1) & 0 & 0 & 1 & 0 & 6EI_s/k \\ \lambda_1 EI/EI_s & 0 & 0 & 0 & 0 & -6EI_s/k \\ \mathrm{SH}\lambda_1 L & \mathrm{CH}\lambda_1 L & 1 & L & L^2 & L^3 \\ -\lambda_1(EI/EI_s - 1)\mathrm{CH}\lambda_1 L & -\lambda_1(EI/EI_s - 1)\mathrm{SH}\lambda_1 L & 0 & 1 & 2L & 6EI_s/k + 3L^2 \\ \lambda_1(EI/EI_s)\mathrm{CH}\lambda_1 L & \lambda_1(EI/EI_s)\mathrm{SH}\lambda_1 L & 0 & 0 & 0 & -6EI_s/k \end{pmatrix} \qquad (14)$$

$$\bar{\bar{K}} = \begin{pmatrix} 0 & 0 & 0 & 0 & 0 & 6EI \\ 0 & 0 & 0 & 0 & -2EI & 0 \\ 0 & -\lambda_1^2 EI_o & 0 & 0 & -2EI_o & 0 \\ 0 & 0 & 0 & 0 & 0 & -6EI \\ 0 & 0 & 0 & 0 & 2EI & 6EIL \\ \lambda_1^2 EI_o \mathrm{SH}\lambda_1 L & \lambda_1^2 EI_o \mathrm{CH}\lambda_1 L & 0 & 0 & 2EI_o & 6EI_o L \end{pmatrix} \qquad (14)$$

(continued)

TABLE 1—*contd.*

$$\bar{K}=\begin{pmatrix}1 & 0 & 0 & 0\\ 0 & 1 & 0 & 6EI_s/k\\ 1 & L & L^2 & L^3\\ 0 & 1 & 2L & 3L^2+6EI_s/k\end{pmatrix}\quad \bar{K}=\begin{pmatrix}0 & 0 & 0 & 6EI_s\\ 0 & 0 & -2EI_s & 0\\ 0 & 0 & 0 & -6EI_s\\ 0 & 0 & 2EI_s & 6EI_sL\end{pmatrix} \tag{15}$$

$$\bar{K}=\begin{pmatrix}1 & 0 & 1 & 0 & 0 & 1\\ 0 & \bar{K}_{22} & 0 & \bar{K}_{24} & 1 & 0\\ 0 & \lambda_3-\bar{K}_{22} & 0 & \lambda_2-\bar{K}_{24} & 0 & 0\\ \mathrm{CH}\lambda_3L & \mathrm{SH}\lambda_3L & \mathrm{C}\lambda_2L & \mathrm{S}\lambda_2L & L & 1\\ \bar{K}_{22}\mathrm{SH}\lambda_3L & \bar{K}_{22}\mathrm{CH}\lambda_3L & -\bar{K}_{24}\mathrm{S}\lambda_2L & \bar{K}_{24}\mathrm{C}\lambda_2L & 1 & 0\\ \bar{K}_{32}\mathrm{SH}\lambda_3L & \bar{K}_{32}\mathrm{CH}\lambda_3L & -\bar{K}_{34}\mathrm{S}\lambda_2L & \bar{K}_{34}\mathrm{C}\lambda_2L & 0 & 0\end{pmatrix} \tag{16}$$

where

$$\bar{K}_{22}=\lambda_3+EI_s\lambda_3^3(1+N/k-EI_o\lambda_3^2/k)/k$$

and

$$\bar{K}_{24}=\lambda_2-EI_s\lambda_2^3(1+N/k+EI_o\lambda_2^2/k)/k$$

$$\bar{K}=\begin{pmatrix}0 & \bar{K}_{12} & 0 & \bar{K}_{14} & -N & 0\\ EI_s\bar{K}_{32}\lambda_3-EI\lambda_3^2 & 0 & -EI_s\bar{K}_{34}\lambda_2+EI\lambda_2^2 & 0 & 0 & 0\\ -EI_o\lambda_3^2 & 0 & EI_o\lambda_2^2 & 0 & 0 & 0\\ -\bar{K}_{12}\mathrm{SH}\lambda_3L & -\bar{K}_{12}\mathrm{CH}\lambda_3L & \bar{K}_{14}\mathrm{S}\lambda_2L & -\bar{K}_{14}\mathrm{C}\lambda_2L & N & 0\\ -\bar{K}_{21}\mathrm{CH}\lambda_3L & -\bar{K}_{21}\mathrm{SH}\lambda_3L & -\bar{K}_{23}\mathrm{C}\lambda_2L & -\bar{K}_{23}\mathrm{S}\lambda_2L & 0 & 0\\ -\bar{K}_{31}\mathrm{CH}\lambda_3L & -\bar{K}_{31}\mathrm{SH}\lambda_3L & -\bar{K}_{33}\mathrm{C}\lambda_2L & -\bar{K}_{33}\mathrm{S}\lambda_2L & 0 & 0\end{pmatrix} \tag{16}$$

where

$$\bar{K}_{12}=-EI_s\lambda_3^3(EI_o\lambda_3^2/k-1-N/k)-N\lambda_3+EI_o\lambda_3^3$$

and

$$\bar{K}_{14}=-EI_s\lambda_2^3(EI_o\lambda_2^2/k+1+N/k)-N\lambda_2-EI_o\lambda_2^3$$

$$\bar{K}=\begin{pmatrix}1 & 0 & 1 & 0 & 0 & 1\\ 0 & \bar{K}_{22} & 0 & \bar{K}_{24} & 1 & 0\\ 0 & \lambda_3-\bar{K}_{22} & 0 & \lambda_2-\bar{K}_{24} & 0 & 0\\ \mathrm{CH}\lambda_3L & \mathrm{SH}\lambda_3L & \mathrm{CH}\lambda_4L & \mathrm{SH}\lambda_4L & L & 1\\ \bar{K}_{22}\mathrm{SH}\lambda_3L & \bar{K}_{22}\mathrm{CH}\lambda_3L & \bar{K}_{24}\mathrm{SH}\lambda_4L & \bar{K}_{24}\mathrm{CH}\lambda_4L & 1 & 0\\ \bar{K}_{32}\mathrm{SH}\lambda_3L & \bar{K}_{32}\mathrm{CH}\lambda_3L & \bar{K}_{34}\mathrm{SH}\lambda_4L & \bar{K}_{34}\mathrm{CH}\lambda_4L & 0 & 0\end{pmatrix} \tag{17}$$

TABLE 1—*contd.*

where

$$\bar{K}_{22} = \lambda_3 + EI_s\lambda_3^3(1 + N/k - EI_o\lambda_3^2/k)/k$$

and

$$\bar{K}_{24} = \lambda_4 + EI_s\lambda_4^3(1 + N/k - EI_o\lambda_4^2/k)/k$$

$$\bar{K} = \begin{pmatrix} 0 & \bar{K}_{12} & 0 & \bar{K}_{14} & -N & 0 \\ EI_s\bar{K}_{32}\lambda_3 - EI\lambda_3^2 & 0 & EI_s\bar{K}_{34}\lambda_4 - EI\lambda_4^2 & 0 & 0 & 0 \\ -EI_o\lambda_3^2 & 0 & -EI_o\lambda_4^2 & 0 & 0 & 0 \\ -\bar{K}_{12}\mathrm{SH}\lambda_3 L & -\bar{K}_{12}\mathrm{CH}\lambda_3 L & -\bar{K}_{14}\mathrm{SH}\lambda_4 L & -\bar{K}_{14}\mathrm{CH}\lambda_4 L & N & 0 \\ -\bar{K}_{21}\mathrm{CH}\lambda_3 L & -\bar{K}_{21}\mathrm{SH}\lambda_3 L & -\bar{K}_{23}\mathrm{CH}\lambda_4 L & -\bar{K}_{23}\mathrm{SH}\lambda_4 L & 0 & 0 \\ -\bar{K}_{31}\mathrm{CH}\lambda_3 L & -\bar{K}_{31}\mathrm{SH}\lambda_3 L & -\bar{K}_{33}\mathrm{CH}\lambda_4 L & -\bar{K}_{33}\mathrm{SH}\lambda_4 L & 0 & 0 \end{pmatrix} \quad (17)$$

where

$$\bar{K}_{12} = -EI_s\lambda_3^3(EI_o\lambda_3^2/k - 1 - N/k) - N\lambda_3 + EI_o\lambda_3^3$$

and

$$\bar{K}_{14} = -EI_s\lambda_4^3(EI_o\lambda_4^2/k - 1 - N/k) - N\lambda_4 + EI_o\lambda_4^3$$

$$\bar{K} = \begin{pmatrix} 1 & 0 & 0 & 1 \\ 0 & \lambda_5(1 + N/k) & 1 & 0 \\ \mathrm{C}\lambda_5 L & \mathrm{S}\lambda_5 L & L & 1 \\ -\bar{K}_{22}\mathrm{S}\lambda_3 L & \bar{K}_{22}\mathrm{C}\lambda_5 L & 1 & 0 \end{pmatrix} \quad (18)$$

$$\bar{K} = \begin{pmatrix} 1 & 0 & 0 & 1 \\ 0 & \lambda_6(1 + N/k) & 1 & 0 \\ \mathrm{CH}\lambda_6 L & \mathrm{SH}\lambda_6 L & L & 1 \\ \bar{K}_{22}\mathrm{SH}\lambda_6 L & \bar{K}_{22}\mathrm{CH}\lambda_6 L & 1 & 0 \end{pmatrix} \quad (19)$$

$$\bar{K} = \begin{pmatrix} 0 & 0 & -N & 0 \\ -N & 0 & 0 & 0 \\ 0 & 0 & N & 0 \\ N\mathrm{C}\lambda_5 L & N\mathrm{S}\lambda_5 L & 0 & 0 \end{pmatrix} \quad (18)$$

$$\bar{K} = \begin{pmatrix} 0 & 0 & -N & 0 \\ -N & 0 & 0 & 0 \\ 0 & 0 & N & 0 \\ N\mathrm{CH}\lambda_6 L & N\mathrm{SH}\lambda_6 L & 0 & 0 \end{pmatrix} \quad (19)$$

$$\bar{K} = \begin{pmatrix} L & 0 & L & 0 & L & 0 \\ 0 & (1+\varepsilon_1\beta)\lambda_7 & 0 & (1+\varepsilon_2\beta)\lambda_8 & 0 & (1+\varepsilon_3\beta)\lambda_9 \\ 0 & -\varepsilon_1\beta\lambda_7 & 0 & -\varepsilon_2\beta\lambda_8 & 0 & -\varepsilon_3\beta\lambda_9 \\ L\mathrm{CH}\lambda_7 & L\mathrm{SH}\lambda_7 & L\mathrm{C}\lambda_8 & L\mathrm{S}\lambda_8 & L\mathrm{C}\lambda_9 & L\mathrm{S}\lambda_9 \\ \bar{K}_{22}\mathrm{SH}\lambda_7 & \bar{K}_{22}\mathrm{CH}\lambda_7 & -\bar{K}_{24}\mathrm{S}\lambda_8 & \bar{K}_{24}\mathrm{C}\lambda_8 & -\bar{K}_{26}\mathrm{S}\lambda_9 & \bar{K}_{26}\mathrm{C}\lambda_9 \\ \bar{K}_{32}\mathrm{SH}\lambda_7 & \bar{K}_{32}\mathrm{CH}\lambda_7 & -\bar{K}_{34}\mathrm{S}\lambda_8 & \bar{K}_{34}\mathrm{C}\lambda_8 & -\bar{K}_{36}\mathrm{S}\lambda_9 & \bar{K}_{36}\mathrm{C}\lambda_9 \end{pmatrix}$$

where

$$\varepsilon_1 = \alpha\delta\beta v^4 + \lambda_7^2 - \alpha\beta\lambda_7^4 \qquad \varepsilon_2 = \alpha\delta\beta v^4 - \lambda_8^2 - \alpha\beta\lambda_8^4 \qquad \varepsilon_3 = \alpha\delta\beta v^4 - \lambda_9^2 - \alpha\beta\lambda_9^4 \quad (20)$$

(continued)

TABLE 1—*contd.*

$$\bar{K} = \frac{EI_s}{L^2}\begin{pmatrix} 0 & (\varepsilon_1 + \alpha\lambda_7^2)\lambda_7 & 0 & (\varepsilon_2 - \alpha\lambda_8^2)\lambda_8 & 0 & (\varepsilon_3 - \alpha\lambda_9^2)\lambda_9 \\ -L(\varepsilon_1 + \alpha\lambda_7^2) & 0 & -L(\varepsilon_2 - \alpha\lambda_8^2) & 0 & -L(\varepsilon_3 - \alpha\lambda_9^2) & 0 \\ -L\alpha\lambda_7^2 & 0 & L\alpha\lambda_8^2 & 0 & L\alpha\lambda_9^2 & 0 \\ -\bar{K}_{12}\mathrm{SH}\lambda_7 & -\bar{K}_{12}\mathrm{CH}\lambda_7 & \bar{K}_{14}\mathrm{S}\lambda_8 & -\bar{K}_{14}\mathrm{C}\lambda_8 & \bar{K}_{16}\mathrm{S}\lambda_9 & -\bar{K}_{16}\mathrm{C}\lambda_9 \\ -\bar{K}_{31}\mathrm{CH}\lambda_7 & -\bar{K}_{31}\mathrm{SH}\lambda_7 & -\bar{K}_{23}\mathrm{C}\lambda_8 & -\bar{K}_{23}\mathrm{S}\lambda_8 & -\bar{K}_{25}\mathrm{C}\lambda_9 & -\bar{K}_{25}\mathrm{S}\lambda_9 \\ -\bar{K}_{31}\mathrm{CH}\lambda_7 & -\bar{K}_{31}\mathrm{SH}\lambda_7 & -\bar{K}_{33}\mathrm{C}\lambda_8 & -\bar{K}_{33}\mathrm{S}\lambda_8 & -\bar{K}_{35}\mathrm{C}\lambda_9 & -\bar{K}_{35}\mathrm{S}\lambda_9 \end{pmatrix} \quad (20)$$

$$\bar{K} = \begin{pmatrix} L & 0 & L & 0 & L & 0 \\ 0 & (1 + \varepsilon_1\beta)\lambda_7 & 0 & (1 + \varepsilon_2\beta)\lambda_8 & 0 & 1 + \alpha\delta\beta^2 v^4 \\ 0 & -\varepsilon_1\beta\lambda_7 & 0 & -\varepsilon_2\beta\lambda_8 & 0 & -\alpha\delta\beta^2 v^4 \\ L\mathrm{CH}\lambda_7 & L\mathrm{SH}\lambda_7 & L\mathrm{C}\lambda_8 & L\mathrm{S}\lambda_8 & L & L \\ \bar{K}_{22}\mathrm{SH}\lambda_7 & \bar{K}_{22}\mathrm{CH}\lambda_7 & -\bar{K}_{24}\mathrm{S}\lambda_8 & \bar{K}_{24}\mathrm{C}\lambda_8 & 0 & \bar{K}_{26} \\ \bar{K}_{32}\mathrm{SH}\lambda_7 & \bar{K}_{32}\mathrm{CH}\lambda_7 & -\bar{K}_{34}\mathrm{S}\lambda_8 & \bar{K}_{34}\mathrm{C}\lambda_8 & 0 & \bar{K}_{36} \end{pmatrix} \quad (21)$$

$$\bar{K} = \frac{EI_s}{L^2}\begin{pmatrix} 0 & (\varepsilon_1 + \alpha\lambda_7^2)\lambda_7 & 0 & (\varepsilon_2 - \alpha\lambda_8^2)\lambda_8 & 0 & \alpha\delta\beta v^4 \\ -L(\varepsilon_1 + \alpha\lambda_7^2) & 0 & -L(\varepsilon_2 - \alpha\lambda_8^2) & 0 & -L\alpha\delta\beta v^4 & 0 \\ -L\alpha\lambda_7^2 & 0 & L\alpha\lambda_8^2 & 0 & 0 & 0 \\ -\bar{K}_{12}\mathrm{SH}\lambda_7 & -\bar{K}_{12}\mathrm{CH}\lambda_7 & \bar{K}_{14}\mathrm{S}\lambda_8 & -\bar{K}_{14}\mathrm{C}\lambda_8 & 0 & -K_{16} \\ -\bar{K}_{31}\mathrm{CH}\lambda_7 & -\bar{K}_{31}\mathrm{SH}\lambda_7 & -\bar{K}_{23}\mathrm{C}\lambda_8 & -\bar{K}_{23}\mathrm{S}\lambda_8 & -\bar{K}_{25} & -\bar{K}_{25} \\ -\bar{K}_{31}\mathrm{CH}\lambda_7 & -\bar{K}_{31}\mathrm{SH}\lambda_7 & -\bar{K}_{33}\mathrm{C}\lambda_8 & -\bar{K}_{33}\mathrm{S}\lambda_8 & 0 & 0 \end{pmatrix} \quad (21)$$

TABLE 1—*contd.*

$$\bar{K}=\begin{pmatrix} L & 0 & L & 0 & L & 0 \\ 0 & (1+\varepsilon_1\beta)\lambda_7 & 0 & (1+\varepsilon_2\beta)\lambda_8 & 0 & (1+\varepsilon_4\beta)\lambda_{10} \\ 0 & -\varepsilon_1\beta\lambda_7 & 0 & -\varepsilon_2\beta\lambda_8 & 0 & -\varepsilon_4\beta\lambda_{10} \\ LCH\lambda_7 & LSH\lambda_7 & LC\lambda_8 & LS\lambda_8 & LCH\lambda_{10} & LSH\lambda_{10} \\ \bar{K}_{22}SH\lambda_7 & \bar{K}_{22}CH\lambda_7 & -\bar{K}_{24}S\lambda_8 & \bar{K}_{24}C\lambda_8 & \bar{K}_{26}SH\lambda_{10} & \bar{K}_{26}CH\lambda_{10} \\ \bar{K}_{32}SH\lambda_7 & \bar{K}_{32}CH\lambda_7 & -\bar{K}_{34}S\lambda_8 & \bar{K}_{34}C\lambda_8 & \bar{K}_{36}SH\lambda_{10} & \bar{K}_{36}CH\lambda_{10} \end{pmatrix} \tag{22}$$

where

$$\varepsilon_4=\alpha\delta\beta v^{4}+\lambda_{10}^2-\alpha\beta\lambda_{10}$$

$$\bar{\bar{K}}=\frac{EI_s}{L^2}\begin{pmatrix} 0 & (\varepsilon_1+\alpha\lambda_7^2)\lambda_7 & 0 & (\varepsilon_2-\alpha\lambda_8^2)\lambda_8 & 0 & (\varepsilon_4+\alpha\lambda_{10}^2)\lambda_{10} \\ -L(\varepsilon_1+\alpha\lambda_7^2) & 0 & -L(\varepsilon_2-\alpha\lambda_8^2) & 0 & -L(\varepsilon_4+\alpha\lambda_{10}^2) & 0 \\ -L\alpha\lambda_7^2 & 0 & L\alpha\lambda_8^2 & 0 & -L\alpha\lambda_{10}^2 & 0 \\ -\bar{\bar{K}}_{12}SH\lambda_7 & -\bar{\bar{K}}_{12}CH\lambda_7 & \bar{\bar{K}}_{14}S\lambda_8 & -\bar{\bar{K}}_{14}C\lambda_8 & -\bar{\bar{K}}_{16}SH\lambda_{10} & -\bar{\bar{K}}_{16}CH\lambda_{10} \\ -\bar{\bar{K}}_{31}CH\lambda_7 & -\bar{\bar{K}}_{31}SH\lambda_7 & -\bar{\bar{K}}_{23}C\lambda_8 & -\bar{\bar{K}}_{23}S\lambda_8 & -\bar{\bar{K}}_{25}CH\lambda_{10} & -\bar{\bar{K}}_{25}SH\lambda_{10} \\ -\bar{\bar{K}}_{31}CH\lambda_7 & -\bar{\bar{K}}_{31}SH\lambda_7 & -\bar{\bar{K}}_{33}C\lambda_8 & -\bar{\bar{K}}_{33}S\lambda_8 & -\bar{\bar{K}}_{35}CH\lambda_{10} & -\bar{\bar{K}}_{35}SH\lambda_{10} \end{pmatrix} \tag{22}$$

$$\bar{K}=\begin{pmatrix} L & 0 & L & 0 \\ 0 & \lambda_{12}^2/\lambda_{11} & 0 & \lambda_{11}^2/\lambda_{12} \\ LCH\lambda_{11} & LSH\lambda_{11} & LC\lambda_{12} & LS\lambda_{12} \\ \bar{K}_{22}SH\lambda_{11} & \bar{K}_{22}CH\lambda_{11} & -\bar{K}_{24}S\lambda_{12} & \bar{K}_{24}C\lambda_{12} \end{pmatrix} \tag{23}$$

$$\bar{\bar{K}}=\frac{EI_s}{L^2}\begin{pmatrix} 0 & \lambda_{11}\lambda_{12}^2 & 0 & -\lambda_{11}^2\lambda_{12} & 0 \\ -L\lambda_{12}^2 & 0 & L\lambda_{11}^2 & 0 \\ -\bar{\bar{K}}_{12}SH\lambda_{11} & -\bar{\bar{K}}_{12}CH\lambda_{11} & \bar{\bar{K}}_{14}S\lambda_{12} & -\bar{\bar{K}}_{14}C\lambda_{12} \\ -\bar{\bar{K}}_{21}CH\lambda_{11} & -\bar{\bar{K}}_{21}SH\lambda_{11} & -\bar{\bar{K}}_{24}C\lambda_{12} & -\bar{\bar{K}}_{24}S\lambda_{12} \end{pmatrix} \tag{23}$$

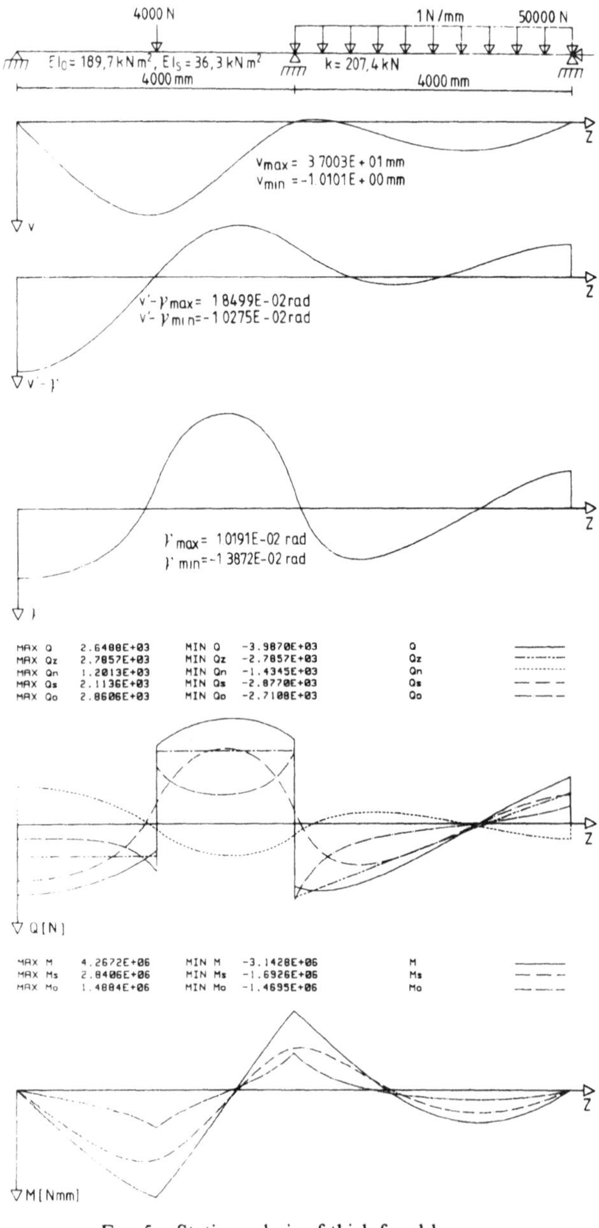

FIG. 5. Static analysis of thick faced beam.

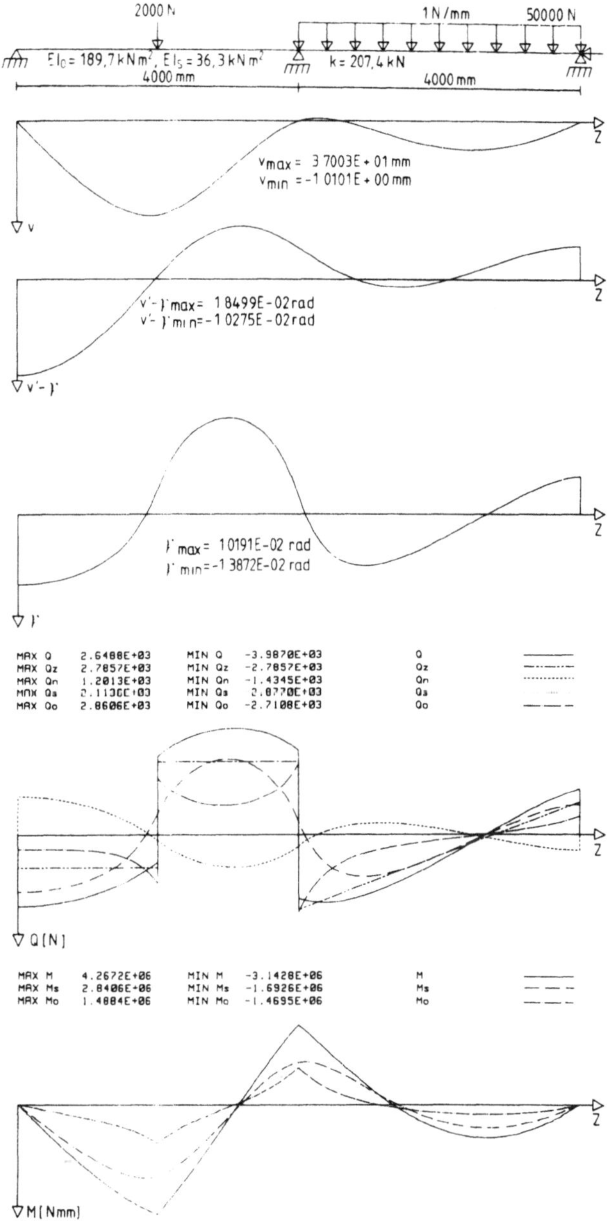
2000 N
1 N/mm
50000 N
$EI_0 = 189{,}7$ kNm², $EI_s = 36{,}3$ kNm²
k = 207,4 kN
4000 mm
4000 mm
Z
v
vmax = 3.7003E+01 mm
vmin = -1.0101E+00 mm
v'−γ'max = 1.8499E-02 rad
v'−γ'min = -1.0275E-02 rad
v'−γ'
γ'max = 1.0191E-02 rad
γ'min = -1.3872E-02 rad
γ'
MAX Q 2.6488E+03 MIN Q -3.9870E+03 Q
MAX Qz 2.7857E+03 MIN Qz -2.7857E+03 Qz
MAX Qn 1.2013E+03 MIN Qn -1.4345E+03 Qn
MAX Qs 2.1130E+03 MIN Qs 2.8770E+03 Qs
MAX Qo 2.8606E+03 MIN Qo -2.7108E+03 Qo
Q [N]
MAX M 4.2672E+06 MIN M -3.1428E+06 M
MAX Ms 2.8406E+06 MIN Ms -1.6926E+06 Ms
MAX Mo 1.4884E+06 MIN Mo -1.4695E+06 Mo
M [Nmm]

FIG. 5.—*contd.*

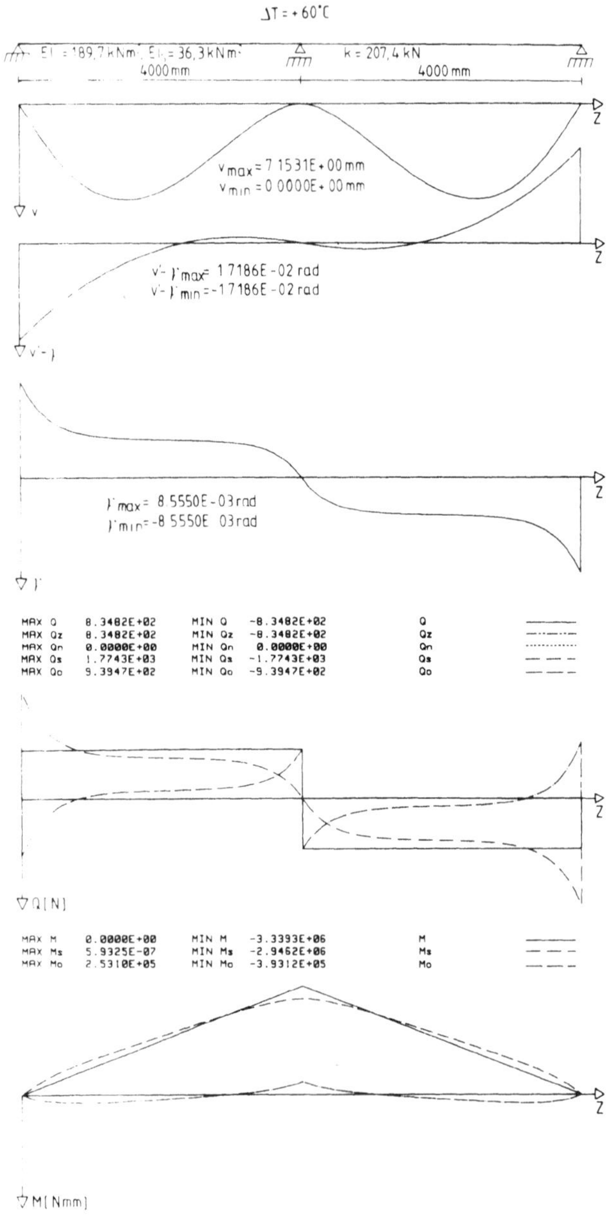
ΔT = + 60°C
k = 207,4 kN
4000 mm
4000 mm
z
v
v'-γ
γ
MAX Q 8.3482E+02 MIN Q -8.3482E+02 Q
MAX Qz 8.3482E+02 MIN Qz -8.3482E+02 Qz
MAX Qn 0.0000E+00 MIN Qn 0.0000E+00 Qn
MAX Qs 1.7743E+03 MIN Qs -1.7743E+03 Qs
MAX Qo 9.3947E+02 MIN Qo -9.3947E+02 Qo
Q [N]
MAX M 0.0000E+00 MIN M -3.3393E+06 M
MAX Ms 5.9325E-07 MIN Ms -2.9462E+06 Ms
MAX Mo 2.5310E+05 MIN Mo -3.9312E+05 Mo
M [Nmm]

FIG. 5.—*contd.*

following notation is used in the table: S = sin, C = cos, SH = sinh, CH = cosh.

The work equivalent nodal forces can be calculated according to the equations presented above. For the uniform temperature or moisture difference between upper and lower faces ($N = q = \omega = 0$) the result is

$$\{P\}^T = \{0\; EI_s\mathrm{N}\; 0\; 0\; -EI_s\mathrm{N}\; 0\}$$

NUMERICAL EXAMPLES

In the previous chapters the beams with thick and thin faces are handled separately. There are two reasons for this. Firstly, if a beam can be considered as a thin faced one then it is economical to use the theory of beams with thin faces, because its order is lower than the order of the theory of beams with thick faces. Secondly, it was found that the numerical stability was lost if the faces of the beam became thinner and when the theory of beams with thick faces was used. This happened due to the exponential functions in the equations. The last point is dependent on the capacity of

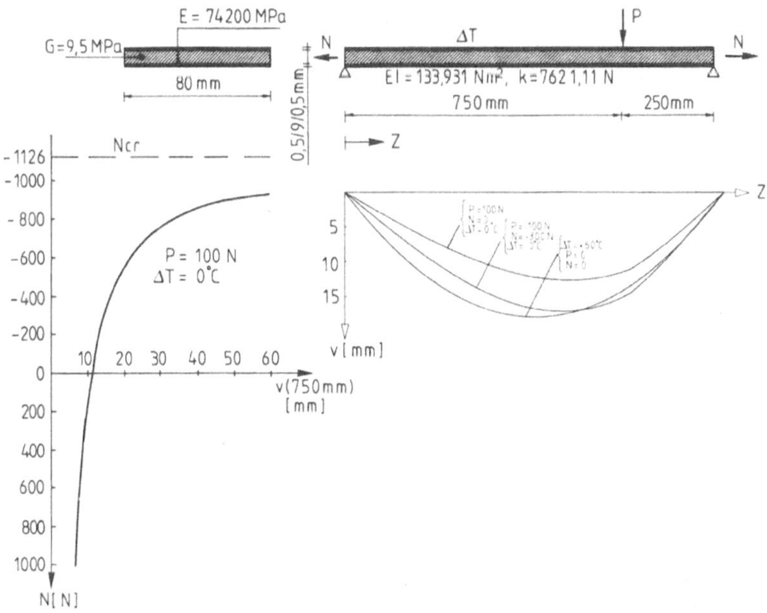

FIG. 6. Static analysis of thin faced beam.

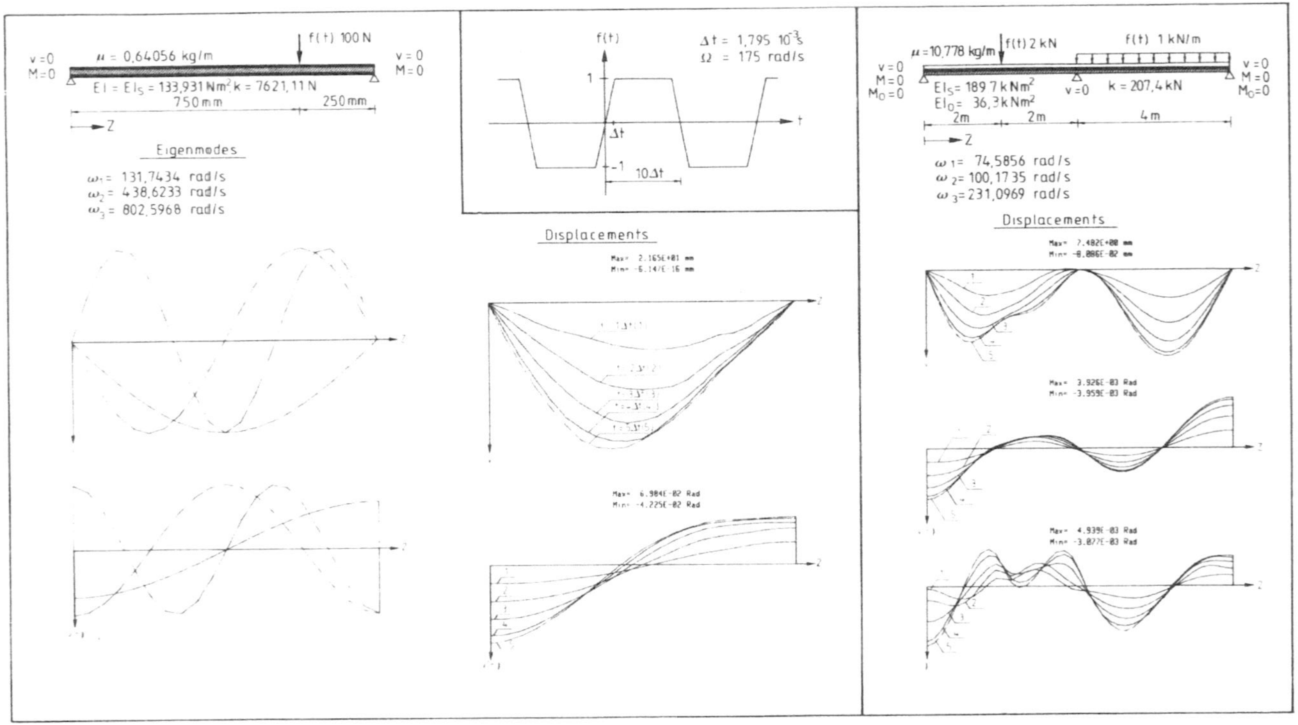

FIG. 7. Dynamic analysis.

the computer. The microcomputer HP 9836 was used in the present study and it was found that the stiffness matrix $[K]$ was correct with seven digits (in the case (14)) if the parameter $\lambda_1 L$ was in the range $\lambda_1 L \leq 20$. The beam must be divided into elements so that this condition is fulfilled in order to use eqn. (14).

The eigenvalue problems were solved by using the Sturm sequence property similar to Friberg.[12] As is well known the eigenmodes are needed, e.g. when designing the extra lateral supports: the optimum location of such supports are in the nodal points of the eigenmodes. It must be noted that if the theory of beams with thin faces is used and the higher natural frequencies are calculated then the beam may change from the range of beams with thin faces to the range of beams with thick faces. Also the shear deformations of the faces and the rotational inertia moments can change the natural frequencies in this case as in the cases of the well-known Timoshenko and Rayleigh beam.

Figure 5 presents some results for the beam with thick faces. The beam and the lateral loading are the same as in the example handled by Schwarze.[6] The effects of the temperature loading and the normal force are presented in the static case and the results for one dynamic loading case are presented in Fig. 7. The similar results for one beam with thin faces are presented in the Figs. 6 and 7. Some analytical results have been calculated by Mai for this case (Mai[13]).

More numerical examples concerning, for example, the analysis of layered wooden beams and the analysis of shear walls can be found from the Refs 8, 10 and 11.

ACKNOWLEDGEMENTS

The financial support of Finnish Academy is gratefully acknowledged.

REFERENCES

1. Abel, J. F. and Popov, E. P., Static and dynamic finite element analysis of Sandwich structures, *Proc. Conf. Matrix Methods Structural Mechanics*, AAFDL-TR-68-150, Ohio, Wright-Patterson, 1969, pp. 213–245.
2. Thompson, E. G., Goodman, J. R. and Vanderbilt, M. D., Finite element analysis of layered wood systems, *Journal of Structural Division, ASCE*, **101**, N ST12, Dec. (1975), 2659–2671.

3. Holt, P. J. and Webber, J. P. H., Finite elements for curved sandwich beams, *Aeronaut. Q.*, **28**, May (1977), 123–141.
4. Murakami, H., A laminated beam theory with interlayer slip, *J. appl. Mech.*, **51**, September (1984), 551–559.
5. Monforton, G. R., Stiffness matrix for sandwich beams with thick anisotropic laminated faces, *Computers Structures*, **10** (1979), 547–551.
6. Schwarze, K., Numerische Methoden zur Berechnung von Sandwich-elementen, *Stahlbau*, **12** (1984), 363–370.
7. Davies, J. M., The analysis of sandwich panels with profiled faces, *Proc. of Eight International Specialty Conference on Cold Formed Steel Structures*, University of Missouri-Rolta, St. Louis, Nov. 11–12, 1986.
8. Heinisuo, M., *Exact Stiffness Matrix for Sandwich Beam*, Tampere University of Technology, Department of Civil Engineering, Structural Mechanics, Report 9, Tampere, January 1986 (in Finnish).
9. Stamm, K. and Witte, H., *Sandwichkonstruktion*, Wien, Springer-Verlag, 1974.
10. Malmi, S., Dynamic analysis of sandwich beam by exact finite element method, Diploma work, Tampere University of Technology, Department of Civil Engineering, Structural Mechanics, Tampere, 1987 (in Finnish).
11. Möttönen, A., Exact finite element method for sandwich beam in combined bending and compression and in thermal loading, Diploma work, Tampere University of Technology, Department of Civil Engineering, Structural Mechanics, Tampere, 1987 (in Finnish).
12. Friberg, P. O., Beam element matrices derived from Vlasov's theory of open thin-walled elastic beams, *Int. J. Num. Meth. Engng*, **21** (1985), 1205–1228.
13. Mai, H. V., *Light weight structures, Kevryakennetekniikka*, 829**B**, Espoo, Otakustantamo, 1984 (in Finnish).

42

Degradation Models in Finite Element Analysis of Multilayered Composite Structures

D. GILLETTA and R. GIRARD

Office National d'Etudes et de Recherches Aérospatiales, BP 72, 92322 Châtillon Cedex, France

ABSTRACT

Non-linearities due to the material's degradation are taken into account in multilayered composite structures calculations. Two macroscopic models have been built in which the layer is supposed to be homogeneous.

In the first model, each layer behaves linearly until failure criteria are reached. Then some components of the layer's Hooke matrix are reduced to zero. The failure criteria used are those defined by Hashin.[1]

In the second model, internal damage variables are introduced in the layer's constitutive law. The failure criteria are then obtained by an instability condition. More global approaches of fracture may be achieved by translating the instability calculation from the layer to the whole stacking sequence, for example.

INTRODUCTION

The aim of this chapter is to take into account some non-linearities due to the material's degradation in the multilayered composite structures calculation.

Models have to be macroscopic, in order to be easily and efficiently inserted into a structure's calculation program. Each layer is supposed to be homogeneous. However, as the degradation mechanisms acting on fibres and matrix are very different, they are separately modelled. For fibres, the criterion involves stresses or strains in their direction, and for the matrix, stresses or strains in transversal directions.

Two kinds of model have been developed. The first one assumed a linear behaviour for the material until one of the criteria is reached. Then the layer's Hooke matrix is modified, by reducing to zero some components. This approach leads to a linear piecewise behaviour of the laminate. Criteria used are derived from those proposed by Hashin.[1]

This model seems to be quite reliable for rather small numbers of different layer orientations, but when this number increases, this kind of analysis is too drastic for the post failure behaviour conditions.

That is why a damage model with internal damage variables involved in the constitutive law of the layer has been implemented in a finite element program.

The failure criterion is defined by an instability condition. This instability is produced by a loss of the material's stiffness. The important point is that such instability conditions may be achieved at different levels in the structure: the layer, the whole stacking sequence, the entire structure. The use of more global failure criteria enables us to build up processes of fracture point determinations based on very few load increments.

1. Layer Modes of Degradation

This chapter deals with degradation models of layers presenting only one reinforcement direction. The layer is supposed to be a transversally isotropic and homogeneous body. The compliance of the layer, S, is

$$\left|\begin{matrix}\varepsilon_{11}\\ \varepsilon_{22}\\ \varepsilon_{12}\end{matrix}\right| = \left|\begin{matrix}\dfrac{1}{E_1} & -\dfrac{\nu_{12}}{E_1} & 0\\ -\dfrac{\nu_{12}}{E_1} & \dfrac{1}{E_2} & 0\\ 0 & 0 & \dfrac{1}{2G_{12}}\end{matrix}\right| \cdot \left|\begin{matrix}\sigma_{11}\\ \sigma_{22}\\ \sigma_{12}\end{matrix}\right| \cdot \left|\begin{matrix}\varepsilon_{23}\\ \varepsilon_{13}\end{matrix}\right| = \left|\begin{matrix}\dfrac{1}{2G_{23}} & 0\\ 0 & \dfrac{1}{2G_{13}}\end{matrix}\right| \cdot \left|\begin{matrix}\sigma_{23}\\ \sigma_{13}\end{matrix}\right|$$

where direction 1 is the fibre direction.

With respect to classical hypothesis of thick plates, the normal stress σ_{33} is supposed to have negligible effect.

Transverse cracking and fibre fracture are the main traits of the layer degradation. The visualization of the transverse cracking shows micro-cracks parallel to the fibre direction, whose distribution is very uniform on the whole layer. This fact justifies the use of macromodels for the layer. The previous kind of damage does not alter mechanical properties of the layer in the fibre direction except those in direction 2. Owing to the two different mechanisms by which the layer may break, the failure criteria defined for it, and the macrostresses or strains involved, have to be multiple.

For fibre breakage the criterion is based on macrostresses (strains) in direction 1. On the other hand, the matrix fracture is governed by macrovariables in directions 2 and 3.

2. Linear Model of Degradation[12]

(a) Failure criteria

In this model the layer has a linear behaviour up to the load factor for which failure criterion is reached. Hashin[1] criteria are used. Their form is (σ_{33} neglected):

Matrix modes

Tension: $\sigma_{22} \geq 0$

$$\frac{1}{Y^2}(\sigma_{12}^2 + \sigma_{13}^2) + \frac{1}{S_+^2}(\sigma_{22}^2) + \frac{1}{Z^2}(\sigma_{23}^2) = 1$$

Compression: $\sigma_{22} \leq 0$

$$\frac{1}{S_-}\left[\left(\frac{S_-}{2Z}\right)^2 - 1\right]\sigma_{22} + \frac{1}{4Z^2}\sigma_{22}^2 + \frac{1}{Z^2}\sigma_{23}^2 + \frac{1}{Y^2}(\sigma_{12}^2 + \sigma_{13}^2) = 1$$

Fibre modes

Tension: $\sigma_{11} \geq 0$

$$\frac{\sigma_{11}}{X_+^2} + \frac{1}{Y^2}(\sigma_{12}^2 + \sigma_{13}^2) = 1$$

Compression: $\sigma_{11} \leq 0$

$$\sigma_{11} = -(X_-)$$

Let $f(\sigma) = 1$ matrix criteria

$g(\sigma) = 1$ fibre criteria

The layer is undamaged until

$$f(\sigma) < 1 \quad \text{and} \quad g(\sigma) < 1$$

As soon as the matrix failure criterion is reached, the Hooke matrix of the layer becomes

$$\begin{vmatrix} K_{11} & 0 & 0 \\ 0 & 0 & 0 \\ 0 & 0 & 0 \end{vmatrix} \quad \begin{vmatrix} 0 & 0 \\ 0 & K_{55} \end{vmatrix}$$

When the fibre criterion is reached, in all the cases, we assumed that the layer is broken.

(b) The model's implementation in finite element analysis

The elements used are isoparametric and multilayered. The behaviour in the element's thickness is obtained by integration in this direction of each layer's behaviour.

The lowest load factor, r, for which the different layer degradations occur in each element is calculated for from a non-damaged state of the discrete structure by solving $f(r\sigma) = 1$ or $g(r\sigma) = 1$ for $r > 1$, when $g(r_{\min}\sigma) = 1$. Then the post-failure behaviour of the layers concerned will be $\sigma = 0$. The structure is then represented by a linear piecewise curve.

When the number of elements is large, some of them may be gathered together if $|r^i - r_{\min}| \leq \xi$, where r^i denotes the load factor defining the next layer's failure for the ith element and ξ is a precision parameter.

(c) Applications

Multilayered panels under uniaxial tensile load:

$$[0, 90_5]_s; \text{ glass–epoxy layers}$$

The knee point corresponds to the matrix failure in the 90° plies (Fig. 1):

$$\text{quasi-isotropic laminate } [0, \mp 45, 90]_s \text{ (carbon–epoxy layer)}$$

On the graph of Fig. 2 are sketched two calculations using almost the same failure criteria. But Chou *et al.*[2] use a discontinuous decrease in some

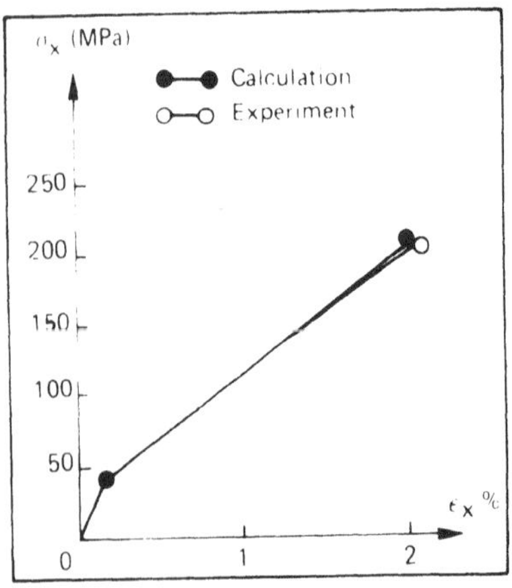

FIG. 1. $[0, 90_5]$.

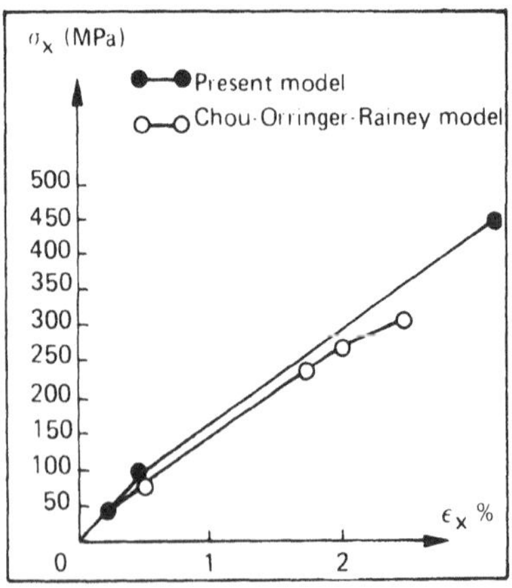

FIG. 2. $[0, 45, -45, 90]$.

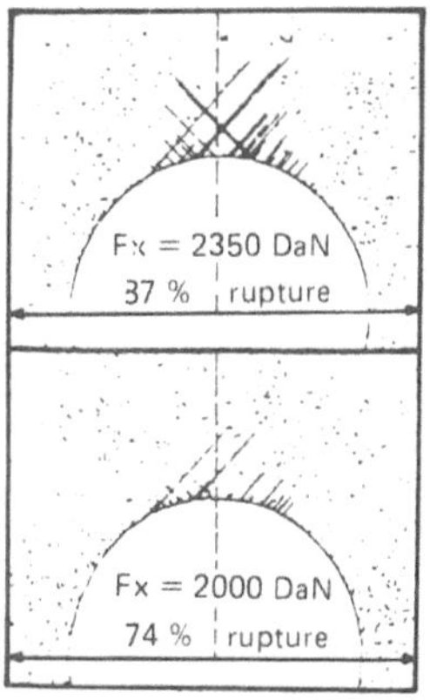

FIG. 3. Specimen X-rays.

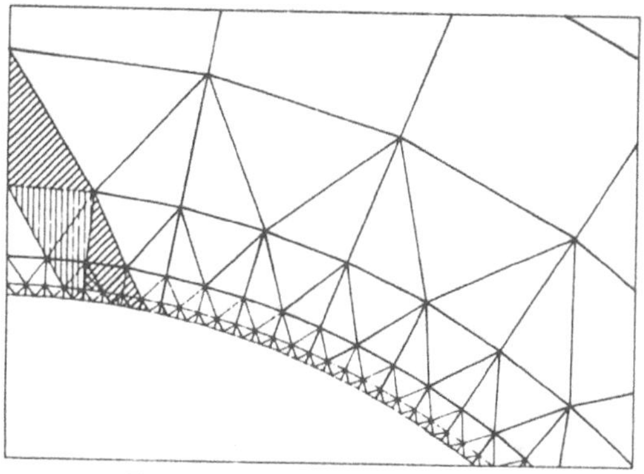

FIG. 4. Calculated damaged zone.

Young's moduli of the layer after the matrix criterion has been reached. This example proves that a progressive Young's moduli reduction is required when the number of fibre directions increases:

$[\mp 45]_s$ plate with a circular hole under uniaxial tensile loading

X-rays of the specimen[3] exhibit microcracks appearing first between 60° and 90° around the hole (see Fig. 3). This is well predicted by the model, as shown in Fig. 4. The growth of the damaged zone is quite well calculated up to failure.

Conclusion. The applications of the linear model point out its limits. This model is adapted for very few fibre directions involved in the stacking sequence. For example, $[\mp\theta]$; $[0, 90_n]$. Moreover, the ply-to-ply analysis may lead to a large number of load increments. That is why an intrinsic damage model has been developed. Continuous internal damage variables are then introduced in the layer's constitutive law.

3. Non-linear Damage Model[4–6]

(a) Layer damage model

A couple of damage variables (d, d') associated with the decrease in the loaded material stiffness[7–9] is introduced to model the layer's transverse cracking. The unilateral kind of transverse damage is also taken into account. This yields

$$G_{12} = G_{12}^{0}(1-d)$$

$$E_2 = E_2^{0}(1-d') \quad \text{Tension}$$

$$E_2 = E_2^{0} \quad \text{Compression}$$

E_2^o, G_{12}^o denote initial undamaged values of G_{12} and E_2. The layer's damage model is built up with the assumption of an in-plane stress state.

Let E be the elastic energy: $2E = \mathrm{Tr}\,[S^{-1}\varepsilon\,\bar{\varepsilon}]$.

The variables Y_d and $Y_{d'}$ are associated with (d, d') in the dispersion expression, and are defined by

$$2Y_d = -\left.\frac{\partial E}{\partial d}\right|_{\varepsilon} = 4G_{12}^o\varepsilon_{12}^2 \qquad 2Y_{d'} = -\left.\frac{\partial E}{\partial d'}\right|_{\varepsilon} = E_2^o\langle\varepsilon_{22}\rangle_+^2$$

with $\langle x\rangle_+ = x$ if $x > 0$ and $\langle x\rangle_+ = 0$ if $x \leq 0$.

The evolution of (d, d') is governed by

$$\mathbf{Y} = \operatorname*{Sup}_{t \leq \tau}\,[Y_{|t}] \quad \text{with } Y = \sqrt{2Y_d + b2Y_{d'}}$$

$\mathbf{Y}$ takes into account the irreversible status of the damage.

The constitutive law for (d, d') is

$$d = \frac{\langle \mathbf{Y} - Y_o\rangle_+}{Y_c} \qquad d' = \frac{\langle \mathbf{Y} - Y_o\rangle_+}{Y_c'}\delta \qquad \delta = \frac{\langle\varepsilon_{22}\rangle_+}{|\varepsilon_{22}|}$$

Y_0, Y_c, Y_c' are intrinsic parameters of the layer.

(b) Layer failure criteria

Matrix and fibre failure criteria are separately modelled. Owing to the fibre linear brittle behaviour, the failure criterion is $|\varepsilon_{11}| \leq \varepsilon_{11}^r$. On the other hand, the damageable elastic behaviour of the matrix leads to an unstable state. This instability point is chosen to define the layer's failure according to experiments in which fracture occurs just after the previous limit point.

The failure is then deduced from the variational problem:

$$\int_\Omega \mathrm{Tr}\,[\dot{\sigma}\dot{\varepsilon}^*]\,d\Omega = 0 \quad \forall\dot{\varepsilon}^* \tag{1}$$

where the dot denotes the derivative with respect to time.

Let the constitutive law of the layer be

$$\sigma = S^{-1}(d)\varepsilon \quad \text{and} \quad \dot{\sigma} = (S^{-1}(d) - B(\varepsilon))\dot{\varepsilon}$$

The solution of (1) is given by the condition $\det\,[S^{-1}(d) - B(\varepsilon)] = 0$; and the criterion is, when $\varepsilon_{22} \geq 0$,

$$\frac{4G_{12}^o\varepsilon_{12}^2}{\mathbf{Y}(Y_o + Y_c - Y)} + \frac{bE_2^o\varepsilon_{22}^2}{\mathbf{Y}(Y_o + Y_c - \mathbf{Y})} = 1$$

when $\varepsilon_{22} < 0$, the instability condition only gives a shear strain limit ε^r_{12}. So another limit ε^r_{22} is added[5] and gives the criterion

$$\frac{\varepsilon_{12}^2}{(\varepsilon^r_{12})^2} + \frac{\varepsilon_{22}^2}{(\varepsilon^r_{22})^2} = 1$$

(c) Fracture analysis of multilayered composite structures

According to experiment, the multilayered coupon fracture is well defined by the first ply failure. However, the use of internal damage variables enables us to translate the instability calculation from the layer to more global domains.

In the first step, instability has been calculated on the whole stacking sequence of each element. The following example (Fig. 5) shows that this approach leads to very close fracture load levels to those determined by the first ply failure. In the next step, the limit point of the structure's equilibrium path is determined. In the example described now, this point and first ply failure occurring in one of the elements of the discrete structure are reached for nearly the same load level (see Fig. 6).

A continuous method is used to achieve the calculation up to the limit point. A load parameter λ is added to the set of unknown displacements, $X = \{q_j, \lambda\}$.

The points of the equilibrium path are calculated for given values of the arc length parameter,[10] Δs. The solution is constrained to be in a plane

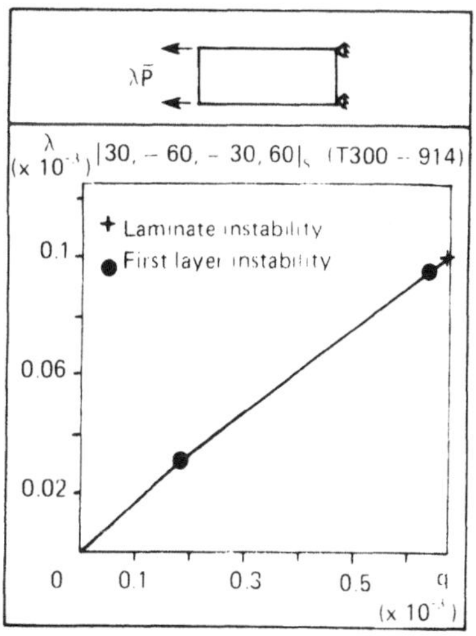

FIG. 5. Laminate and first ply instabilities.

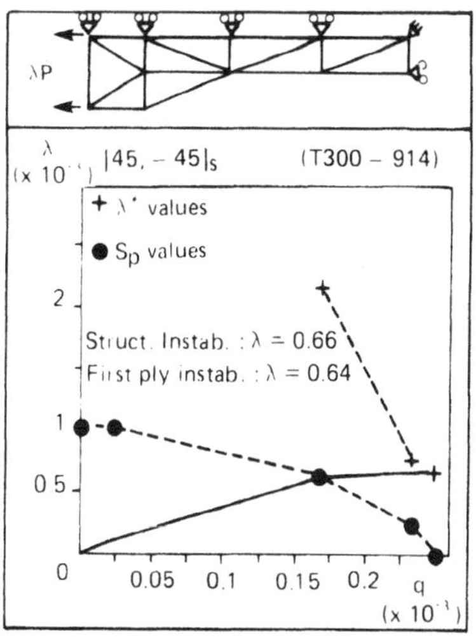

FIG. 6. Structure and first ply instabilities.

orthogonal to the known tangent vector, $\dot{X}(s_1)$, at a distance $\Delta s = s - s_1$ from the last calculated point, $X(s_1)$.

The non-linear system to solve is

$$\begin{cases} K(q_1)\Delta q - \bar{P}\Delta\lambda = (0_{I=1}\,|_{I>1}\, R(q_1)) \\ \dot{X}^{\mathrm{T}}(s_1)\Delta X = (\Delta s_{I=1}\,|\,0_{I>1}) \end{cases} \tag{2}$$

where K is the tangential stiffness matrix, R the internal forces and $\bar{P}$ the applied forces. The tangential vector, $\dot{X}(s)$, is defined by the derivative of X with respect to s.

The system (2) is solved by the Newton method; the subscript I indicates the Ith iteration.

The values of the arc length parameter, Δs, which set the different load increments, may be calculated automatically through extrapolation methods. The current stiffness parameter, S_p,[11] is one of the possible quantities on which extrapolations can be performed:

$$S_p = \frac{\dot{q}_1^t \bar{P}}{\dot{q}_m^{\mathrm{T}} \bar{P}} \quad \text{and} \quad \dot{q} = \frac{\Delta q}{\Delta \lambda}$$

q_1 and q_m are displacements for an initial state and the mth increment load state.

In the elastic range, S_p is equal to one and decreases to zero at the limit point (Fig. 6). The critical load step, in order to reach linearly an estimated value of the limit point, is given for each calculated point by

$$\Delta\lambda^* = -\frac{S_p^m}{S_p^m - S_p^{m-1}}(\lambda_m - \lambda_{m-1}) \quad \text{and} \quad \Delta s^* = \Delta\lambda^* \frac{\|X(s_m) - X(s_{m-1})\|}{\lambda(s_m) - \lambda(s_{m-1})}$$

The next point of the equilibrium path is determined by $\Delta s = \beta \Delta s^*$. β is chosen in such a manner that iteration convergence is ensured.

CONCLUSION

A linear model with a very simple analysis of fracture leads to ply-to-ply and large numbers of load increment calculations. On the other hand, a more sophisticated model requires non-linear analysis but enables us to build more global methods of fracture determination while getting local information through damage variables defined on each layer.

Just a few examples have been explored, so the reliability of the global approach must still be improved.

The next steps in the use of damage mechanics might concern the 'in-thickness properties' of the layer for out-of-plane loading or perhaps delamination growth problems.

Finally, the tensile and compressive layer behaviour are at the moment linearly modelled. But in compression non-linearities, probably due to fibre buckling or just bending, occur and then must be taken into account in the layer model.

REFERENCES

1. HASHIN, Z., Failure criteria for unidirectional fiber composites, *J. appl. Mech.*, **47** (1980), 329–334, June.
2. CHOU, S. C., ORRINGER, O. and RAINEY, J. H., Post failure behaviour of laminates. Part 1: No stress concentration, *J. Comp. Mat.*, **10** (1976), 371–381, Oct.
3. INIZAN, G., Endommagement d'une plaque composite trouée, en carbone–epoxy, sous chargement biaxial monotone et cyclique, *La Recherche Aérospatiale*, **1** (1986), 63–75.
4. GILLETTA, D., Composites 2D: Modélisation mécanique et identification de la couche élémentaire, Thèse Docteur-Ing., Université de Paris VI, E.N.S. Cachan, L.M.T., 1985.
5. COCHELIN, B. and VITTECOQ, E., Endommagement et rupture d'un composite 2D à fibres haute résistance, Mémoire DEA, Université Paris VI, E.N.S. Cachan, 1986.
6. GILETTA, D., GIRARD, H. and LADEVEZE, P., Composites 2D à fibres haute résistance: modélisation mécanique de la couche élémentaire, *Comptes rendus des 5ièmes Journées Nationales sur les Composites*, Paris, 1986, 685–699, Sept.
7. KACHANOV, L. M., Time on the rupture process under creep conditions, *T.V.Z., A.K.A.D., Nauk. SSR, O.T.D. Techn. Nauk*, **8** (1958).
8. CHABOCHE, J. L. and LEMAÎTRE, J., Aspect phénoménologique de la rupture par endommagement, *J. de Mécanique Appliquée*, **2**, n° 3 (1978), 317–365.
9. LADEVEZE, P., Sur une théorie de l'endommagement anisotrope, Rapport interne N° 34, L.M.T., E.N.S. Cachan, 1983.
10. RIKS, E., Some computational aspects of the stability analysis of non-linear structures, N.L.R., M.P., 82034 U, 1982.
11. BERGAN, P. G. and SOREIDE, T. H., Solution of large displacement and stability problems using the current stiffness parameter, *Int. Conf. Finite Element Nonlinear Solid. Struct. Mech.*, geilo, Vol. **1**, 1977, E02, 1–23.
12. GIRARD, R., Un modèle de dégradation pour les structures composites, Rapport tech. ONERA, N° 30/3043 RY 057 R, 1986.

43

Finite Element Analysis for Large Composite Structure

LIU XIA-SHI

Chinese Helicopter Research and Development Institute,
PO Box 825, Jingdezhen, Jiangxi, People's Republic of China

ABSTRACT

This report, aiming at the characteristics of laminated plate made of composite material, has developed a new method for finite element analysis for plate elements and the stiffness matrixes of isoparametric solid elements, super-parametric shell elements, solid-shell transition elements and elements with out-element nodes. These elements are available for analysing isotropic/orthotropic material, honeycomb sandwich and/or composite plates. They are able to distinctly represent the structure with complex shapes and solve the important subject in the research of composite material plates—layer stress analysis. From an engineering application standpoint, the maximum strength of various structures mentioned above has been studied and the method for solving non-linear problems of composite material has been recommended. Many analysis examples are also given at the end of the chapter. It is shown that the method reviewed is very effective.

THE STIFFNESS MATRIX OF PLATE STRUCTURE

There are 9 independent nodules of elasticity for orthotropic material. Because the plate is a transversely isotropic material, so the number of modules may decrease to 5. The following is a stress–strain relation of a typical laminated plate (with material symmetry axes 1, 2, 3)

$$\begin{bmatrix} \varepsilon_1 \\ \varepsilon_2 \\ \varepsilon_3 \\ \gamma_{12} \\ \gamma_{13} \\ \gamma_{23} \end{bmatrix} = \begin{bmatrix} S_{11} & S_{12} & S_{13} & & & \\ S_{21} & S_{22} & S_{23} & & & \\ S_{31} & S_{32} & S_{33} & & & \\ & & & S_{44} & & \\ & & & & S_{55} & \\ & & & & & S_{66} \end{bmatrix} \begin{bmatrix} \sigma_1 \\ \sigma_2 \\ \sigma_3 \\ \tau_{12} \\ \tau_{13} \\ \tau_{23} \end{bmatrix} \tag{1}$$

where:

$$\begin{aligned} S_{11} &= 1/El & S_{22} &= S_{33} = 1/E_t \\ S_{12} &= S_{13} = -\gamma_{lt}/El & S_{23} &= -\gamma_{tt}/E_t \\ S_{44} &= S_{55} = 1/G_{lt} & S_{66} &= 2(1+\nu_{tt})/E_t = 1/G_{tt} \end{aligned} \tag{2}$$

Here, E, G, r are Young's modulus, shear elastic modulus and Poisson's ratio respectively. l and t represent longitudinal and transverse directions respectively. γ_{lt} is a Poisson's ratio for transverse strain due to normal stress in l direction.

Using the inverse matrix, thus

$$\sigma = S^{-1}\varepsilon = C\varepsilon \tag{3}$$

where: C is stiffness matrix of K layers, S is stress–strain flexibility matrix.

The following is the stress–strain relation with respect to x, y, z plate axes.

$$\sigma = T^{-1}CT\varepsilon = D\varepsilon \tag{4}$$

where T is transformation matrix.

$$T = \begin{bmatrix} l_1^2 & l_2^2 & l_3^2 & 2l_1l_2 & 2l_1l_3 & 2l_2l_3 \\ m_1^2 & m_2^2 & m_3^2 & 2m_1m_2 & 2m_1m_3 & 2m_2m_3 \\ n_1^2 & n_2^2 & n_3^2 & 2n_1n_2 & 2n_1n_3 & 2n_2n_3 \\ l_1m_1 & l_2m_2 & l_3m_3 & l_2m_2 + l_2m_1 & l_1m_3 + l_3m_1 & l_2m_3 + l_3m_2 \\ l_1n_1 & l_2n_2 & l_3n_3 & l_1n_2 \quad l_2n_1 & l_1n_3 \quad l_3n_1 & l_2n_3 \quad l_3n_2 \\ m_1n_1 & m_2n_2 & m_3n_3 & m_1n_2 + m_2n_1 & m_1n_3 + m_3n_1 & m_2n_3 + m_3n_2 \end{bmatrix} \tag{5}$$

l_i, m_i, n_i $(i = 1, 2, 3)$ are direction cosines between the material's symmetric axes and plate axes x, y, z. (See Fig. 1.)

Thus, the elastic matrix D of K layers of plates can be determined by (4), although the matrix relates to fibre direction and is usually different for each ply.

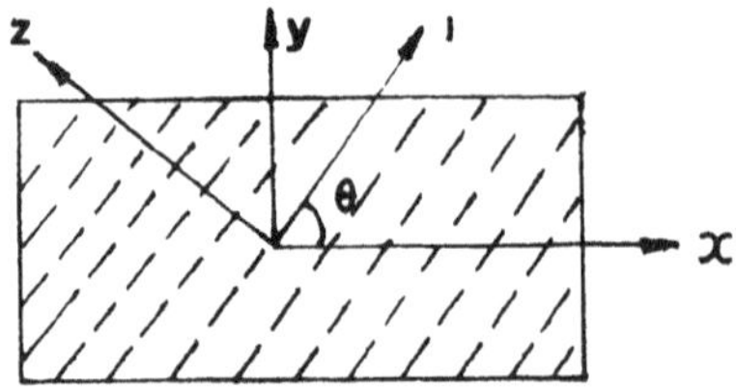

FIG. 1.

The element stiffness matrix K is expressed by following formula:

$$K = \int_V B^T D B \, dV = \int_{-1}^{1}\int_{-1}^{1}\int_{-1}^{1} B^T D B |J| \, d\xi \, d\eta \, d\zeta \tag{6}$$

where B is strain–displacement matrix, $|J|$ is Jacobian matrix determinant.

Formula (6) can be integrated by a Gauss integral, but the D matrix of each ply is different and not a continuous function of ζ, so in order to be able to use a Gauss integral formula with limits between 1 and -1, a variable ζ_K (the Kth layer) is induced, its limits between 1 and -1 at the Kth layer, thus the relation between ζ and ζ_K is as follows:

$$\zeta = -1 + \frac{1}{t}\left[-h_K(1-\zeta_K) + 2\sum_{j=1}^{k} h_j\right] \tag{7}$$

and

$$d\zeta = (h_k/t)\, d\zeta_k$$

where t is total plate thickness, h_k is the Kth ply thickness.

Substituting the above formula into (6), the plate element stiffness matrix is as follows:[1]

$$K = \sum_{k=1}^{m} \int_{-1}^{1}\int_{-1}^{1}\int_{-1}^{1} B^T D B |J| hk/t \, d\xi \, d\eta \, d\zeta_k \tag{8}$$

where m is the number of plies in the laminated plate.

Formulas (6) to (8) are mathematically tenable. Furthermore, when using values for the integral calculation, the results are much more accurate as the separate integration is better for approaching the integrand curve.

ELEMENT MODEL

In order to accurately represent the stress concentration area, it is commonly suggested to divide up the grid more finely. However, it is obviously unrealistic for the overall calculation owing to the complex structural shape.

Variable number-node isoparametric element series are adopted to ensure the accuracy, approach curve boundary and to minimize the number of elements.

1. The Hexahedron Isoparametric Elements with 8–20 Variable Number Nodes

The order of the element node numbers is shown in Fig. 2. The eight corners of the element are always used as nodes and others on the 12 edges

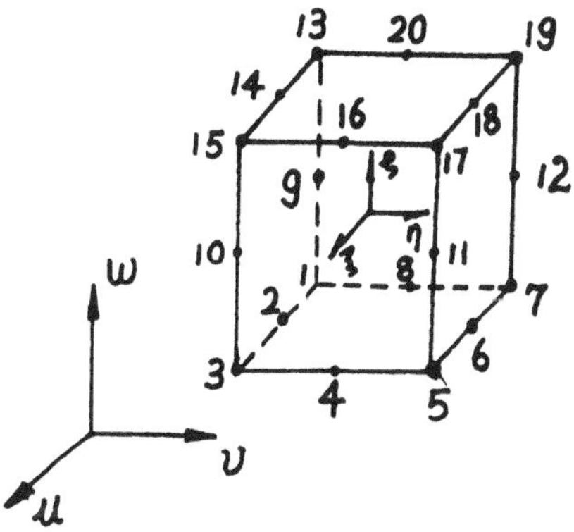

FIG. 2.

may be chosen if necessary. Thus, isoparametric elements with 8–20 variable number nodes is given. Its shape functions[9] are as follows:

$$\begin{cases} N_i(\xi,\eta,\zeta) = (1+\xi_i\xi)(1+\eta_i\eta)(1+\zeta_i\zeta)(1 - P_{i+1i} - P_{i-1i} \\ \qquad - P_{mi} + P_i + l_i\xi_i\xi + P_{i-1i}\eta_1\eta + P_{mi}\zeta_i\zeta)/8 \\ \qquad\qquad (i = 1,3,5,7,13,15,17,19) \qquad (9) \\ N_i(\xi,\eta,\zeta) = P_i(1-\xi^2)(1+\eta_i\eta)(1+\zeta_i\zeta)/4 \qquad (i = 2,6,14,18) \\ N_i(\xi,\eta,\zeta) = P_i(1+\xi_i\xi)(1-\eta^2)(1+\zeta_i\zeta)/4 \qquad (i = 4,8,16,20) \\ N_i(\xi,\eta,\zeta) = P_i(1+\xi_i\xi)(1+\eta_i\eta)(1-\zeta^2)/4 \qquad (i = 9,10,11,12) \end{cases}$$

where: (ξ_i, η_i, ζ_i) are the local coordinates of the ith node.

$$P_i = \begin{cases} 1 & \text{the } i\text{th point is a node} \\ 0 & \text{the } i\text{th point is not a node} \end{cases} \tag{10}$$

(i is the point number on the edges)

$$l_i = (-1)\frac{i-1}{2} \qquad (i = 1, 3, 5, 7, 13, 15, 17, 19) \tag{11}$$

$$m_i = 9, 10, 11, 12 \tag{12}$$

(with respect to $i = 1, 3, 5, 7, 13, 15, 17, 19$).

$i + l_i$, $i - l_i$, m_i are the middle point numbers of the three edges which cross the corner point i (given $i - l_i = 0$, then there are 8 nodes; given $i - l_i = 12$, there are 20 nodes). It can be proved that

$$\sum_{1}^{20} = N_i(\xi, \eta, \zeta) = 1$$

So the displacement interpolating functions meet the rigid body displacement and constant strain criterion.

2. The Triangular Prism Isoparametric Elements with 6–15 Variable Number Nodes (see Fig. 3)

In practical applications, a triangular prism isoparametric element with 6–15 variable number nodes is an ideal element from the standpoint of

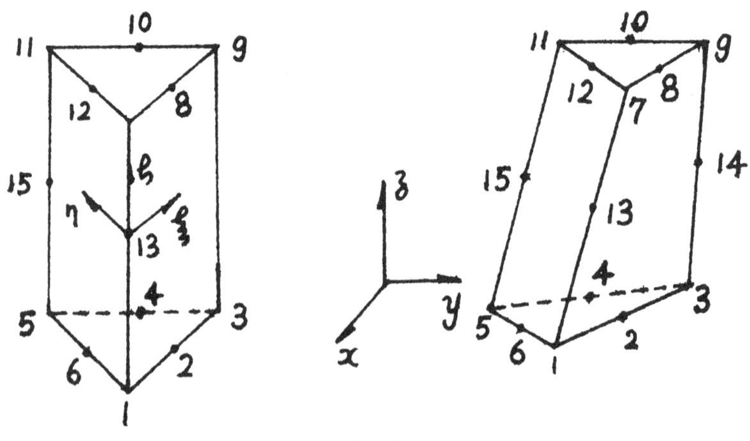

FIG. 3.

either calculating time or the accuracy. The local coordinate system and the order of the element node numbers are shown in Fig. 3a. The bottom of the triangular prism element is an isosceles right triangle ($0 \le \xi \le 1$, $0 \le \eta \le 1 - \xi$, $-1 \le \zeta \le 1$). The six corners are always regarded as six nodes, and the others on the middle points of each edge may be chosen based on the request.

Derive the shape functions using the same method as (9):

$$N_i(\xi,\eta,\zeta) = \begin{cases} 2P_iL_{j_i-1}L_{j_i}(1+\zeta_i\zeta) & i = 2,4,6,8,10,12 \\ P_iL_{i-12}(1-\zeta^2) & i = 13,14,15 \\ \frac{1}{2}L_{j_i}(1+\zeta_i\zeta) - \frac{1}{2}[N_{i+1}(\xi,\eta,\zeta) + N_{k_i}(\xi,\eta,\zeta) & \\ \quad + N_{12+j_i}(\xi,\eta+\zeta)] & i = 1,3,5,7,9,11 \end{cases} \tag{13}$$

where ζ_i are the local coordinates of the ith node in the ζ direction,

$$P_i = \begin{cases} 1\text{—the } i\text{th point is a node} \\ 0\text{—the } i\text{th point is not a node} \end{cases} \tag{14}$$

$$L_1 = 1 - \xi - \eta, \quad L_2 = \xi, \quad L_3 = \eta \tag{15}$$

$$k_i = \begin{cases} i-1 & i = 3,4,9,11 \\ 6 & i = 1 \\ 12 & i = 7 \end{cases} \tag{16}$$

$$j_i = 1 + \left[\frac{i}{2}\right] - 3\left[\frac{i}{6}\right] \tag{17}$$

Here (X) represents the integer part of X. The subscripts $i+1$, K_i and $12+j_i$ are the middle point numbers on three edges which cross the ith corner point. The following is the relationship between j_i and i:

i	1	2	3	4	5	6	7	8	9	10	11	12
j_i	1	2	2	3	3	1	1	2	2	3	3	1

Triangular prism elements are used on the structure boundaries, especially on a sharp edge corner, while the hexahedron elements are used on the other area. The triangular prism element is suitable for combining with a hexahedral element due to its quadrilateral sideplane.

3. 8-Node Superparametric Shells

Suppose the displacement functions are as follows:

$$\begin{bmatrix} u \\ v \\ w \end{bmatrix} = \sum_{i=1}^{8} N_i(\xi, \eta) \left(\begin{bmatrix} u_i \\ v_i \\ w_i \end{bmatrix} + \frac{\zeta_{ti}}{2} [\mathbf{V}_{1i} - \mathbf{V}_{2i}] \begin{bmatrix} \alpha_i \\ \beta_i \end{bmatrix} \right) \tag{18}$$

where

$$\mathbf{V}_{3i} = \begin{bmatrix} \Delta x_i \\ \Delta y_i \\ \Delta z_i \end{bmatrix}$$

$$\mathbf{V}_{1i} = \frac{i + \mathbf{V}_{3i}}{|i + \mathbf{V}_{3i}|} \tag{19}$$

$$\mathbf{V}_{2i} = \frac{\mathbf{V}_{3i} + \mathbf{V}_{1i}}{|\mathbf{V}_{3i} + \mathbf{V}_{1i}|}$$

i is unit vector along x axis (absolute coordinates), α_i, β_i are corner vectors.

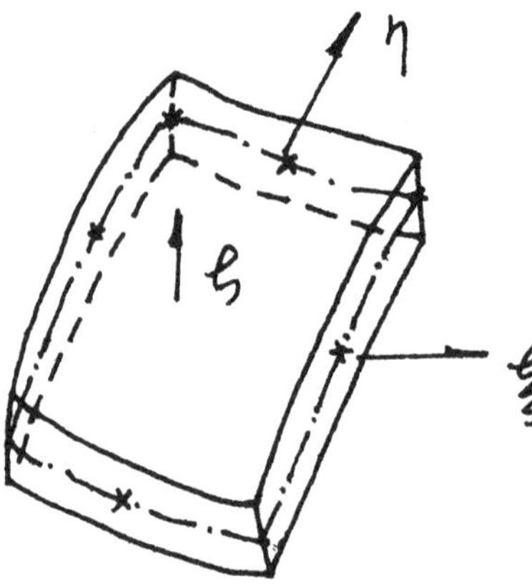

FIG. 4.

The shape functions are as follows:

corner point:

$$N_i = \tfrac{1}{4}(1 + \xi_0)(1 + \eta_0)(\xi_0 + \eta_0 - 1)$$

middle point:

$$\xi_i = 0 \qquad N_i = \tfrac{1}{2}(1 - \xi^2)(1 + \eta_0)$$
$$\eta_i = 0 \qquad N_i = \tfrac{1}{2}(1 + \xi_0)(1 - \eta^2) \tag{20}$$

The general equation of the stiffness matrix is still formula (8). The strain of the plate axis x, y, z is

$$\begin{bmatrix} \varepsilon_x \\ \varepsilon_y \\ \gamma_{xy} \\ \gamma_{xz} \\ \gamma_{yz} \end{bmatrix} = \begin{bmatrix} \dfrac{\partial u}{\partial x} \\ \dfrac{\partial V}{\partial y} \\ \dfrac{\partial u}{\partial y} + \dfrac{\partial V}{\partial x} \\ \dfrac{\partial w}{\partial x} + \dfrac{\partial u}{\partial z} \\ \dfrac{\partial w}{\partial y} + \dfrac{\partial V}{\partial z} \end{bmatrix} = B\delta \tag{21}$$

It is the same as isotropic thin sheet. Suppose $\sigma_z = 0$, then ε_z can be eliminated from the stress–strain function and $\sigma = D\varepsilon$ is obtained.

4. 15-node Solid-shell Transition Elements

There are many complex shape structures which consist of two parts—three-dimensional solid and shell elements. So a transition element is needed to connect a 20-node solid element with an 8-node shell element. The coordinates and displacement expressions of the transition element are as follows:

$$\begin{bmatrix} x \\ y \\ z \end{bmatrix} = \sum_{i=4}^{15} N_i(\xi, \eta, \zeta) \begin{bmatrix} x_i \\ y_i \\ z_i \end{bmatrix} + \sum_{j=1}^{3} N_j(\xi, \eta) \begin{bmatrix} x_j \\ y_j \\ z_j \end{bmatrix} + \sum_{j=1}^{3} N_j(\xi, \eta) \frac{\zeta}{2} \mathbf{V}_{ij} \tag{22}$$

$$\begin{bmatrix} u \\ v \\ w \end{bmatrix} = \sum_{i=4}^{15} N_i(\xi, \eta, \zeta) \begin{bmatrix} u_i \\ v_i \\ w_i \end{bmatrix} + \sum_{j=1}^{3} N_j(\xi, \eta) \begin{bmatrix} u_j \\ v_j \\ w_j \end{bmatrix}$$

$$+ \sum_{j=1}^{3} N_j(\xi, \eta) \zeta \frac{t_j}{2} [\mathbf{V}_{1j} - \mathbf{V}_{2j}] \begin{bmatrix} \alpha_j \\ \beta_j \end{bmatrix} \tag{23}$$

The shape functions $N_i(\xi, \eta, \zeta)$ of the 4–15 node element are the same as the 20-node isoparametric solid element. The shape functions $N_i(\xi, \eta)$ of 1–3

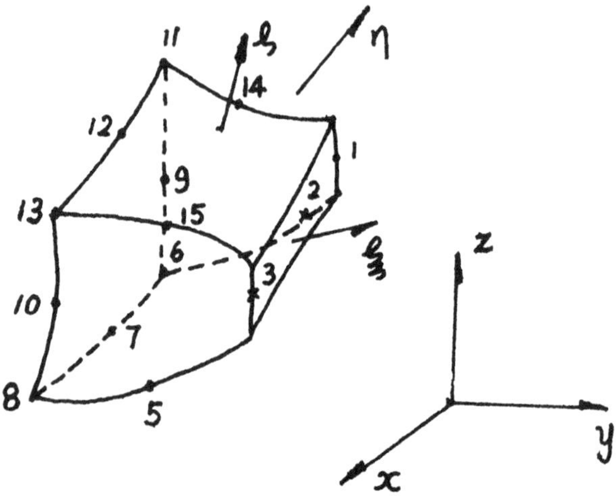

FIG. 5.

nodes are the same as corresponding shape functions of 8-node superparametric shell element.

5. Compatible Elements between Elements and Out-element Nodes[7]

Geometric shape and load distribution in the complex structure are rather complicated. The stress distribution varies greatly in some regions

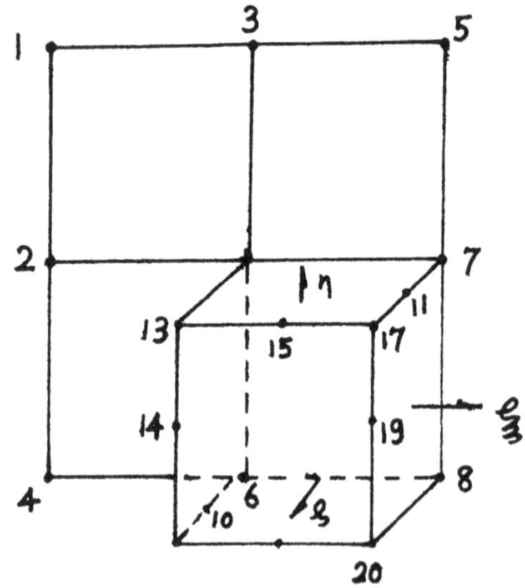

FIG. 6.

while less in others. Hence, different grid spacings should be adopted at various regions to minimize data preparation and numerous calculations. Variable number node or linear combinations of shape functions are commonly used at home and abroad, which affect the calculating accuracy, the author used a completely compatible element between element and out-element node to conform the boundary coordination. The shape functions are as follows:

$$N_1 = \tfrac{1}{64}(1+\xi)(1+\eta)(1-\zeta)(-\xi+\eta+2\zeta-2)$$

$$N_2 = \tfrac{1}{32}(1-\xi)(1+\eta)(1-\zeta)(1-\eta-2\zeta)$$

$$N_3 = \tfrac{1}{32}(1-\xi)(1+\eta)(1-\zeta)(1+\xi-2\zeta)$$

$$N_4 = \tfrac{1}{64}(1-\xi)(1+\zeta)[(-3+\eta)(\xi+\eta+2)+2(1+\eta)(1+\zeta)]$$

$$N_5 = \tfrac{1}{64}(1+\eta)(1-\zeta)[(-3-\xi)(-\xi-\eta+2)+2(1-\xi)(1+\zeta)]$$

$$N_6 = \tfrac{1}{32}(1-\xi)(3-\eta)(1-\zeta)(1+\xi-2\zeta)$$

$$N_7 = \tfrac{1}{32}(3+\xi)(1+\eta)(1-\zeta)(1-\eta-2\zeta)$$

$$N_8 = \tfrac{1}{64}(1-\zeta)\{(3+\xi)(3-\eta)(\xi-\eta)+2(1+\xi) \times [(1-\xi)(1+\eta)-4(1+\xi)(1-\eta)]\}$$

The other shape functions of N_9 to N_{20} are omitted here as they are the same as functions of the 20-node solid elements.

Now, let's discuss the convergency of the element with out-element nodes. It is clear that here there is no essential distinction between this element and common elements whose nodes are on the element boundaries. They are compatible with each other and include all the lienar terms, i.e. constant strain condition. Thus, they meet the general requirements of convergency criterion including the isoparametric element.

A compatible element between element and out-element nodes is also the hexahedron isoparametric element with 20-nodes. However, some nodes in this system are located neither on nor within the element boundaries, but beyond the element. Because they are intrinsically the same, it does not matter where the nodes are located, on, within or beyond the boundaries.

ELEMENT DIVISION AND SIMILAR ELEMENTS

It can be seen from formula (8) that the stiffness matrix of each ply needs to be calculated in the analysis. Obviously, this would increase the calculating time of the computer. Based on the formula of element stiffness matrix, several lemmata can be easily introduced as follows:

Lemma 1. If an element moves parallel, its stiffness matrix remains evidently unchanged.

Lemma 2. If a planar element is magnified (or reduced) in the same proportion as α, its stiffness matrix remains unchanged.

Lemma 3. If a three-dimensional element is magnified or reduced α times, its stiffness matrix is also magnified or reduced α times correspondingly.

Lemma 4. When an element rotates around a fixed point or axis, its stiffness matrixes are similar.

Using these lemmata, we can not only meet the analysis requirements for composite material structure but also ensure that the calculation time and dimension of the computer is not so influenced by the number of plies. In other words, neither extra computer storage nor extra calculating time are required with the increase of the number of plies. Furthermore, it can be divided into completely congruent elements. In brief, using these lemmata, a small computer can do a big job and gives satisfactory results.

THE METHOD FOR ANALYSIS OF LAYER STRESS

Generally, the laminated plate is subjected to both in-plane forces and moments, which are accompanied by transverse shear which may result in layer shearing stress. In addition, even under the in-plane forces, there is layer stress near the edges due to the different directions of the laminated plies and loading boundary condition of the free boundary. This is peculiar to a composite material and has never been seen in the condition of isotropic material. Because of the very low shearing and stripped strength, the existence of the layer shearing stress or normal stress will be liable to lead to a stripped failure between the layers, then causing the composite laminated plate to fail earlier. For this reason, research on the layer stress and strength of composite plates is one of the important subjects for developing composite materials at present.

Generally speaking, the layer stress analysis is settled by three-dimension elastic theory. In this report, each ply is regarded as a three-dimensional problem so as to present better the real load condition.

At present, there are still many workers using Puppo and Evensen's approximate theory which was presented in 1969, to analyse the layer stress. As to laminated plate under the generalized planar stress condition, a plate is simulated as a group of anisotropic plates separated by isotropic plates and the shearing stress is applied only at each ply of the isotropic plates, which play the role of 'adhesive agent' between the anisotropic layers. The elastic properties of the isotropic and anisotropic layers are determined by assuming the whole elastic properties of the plate model and the real plate to be the same. It is obvious that the analysis can be carried out successfully by creating the stiffness matrix of the element mentioned above.

THE COMBINED CALCULATION FOR LARGE STRUCTURES WITH BRANCH SYSTEM

As shown in Fig. 7, there are four branch systems (left and right planes; left and right tail planes) in a large aircraft structure. Although the corresponding branch systems are symmetrical, the fuselage (5) has no symmetry. Thus, the whole structure is not symmetrical. During free flying, the load in each plane of the aircraft is considered to be different.

This report describes the development of a combined calculation method, using branch system analysis, for settling the problem about storage and working time of the computer.

1. The Treatment for Branch Systems

As shown in Fig. 7, the structure is first divided into several parts, the fuselage is considered as a main system and the wing plane structures as branch systems. Thus, it forms five chain subsystems, and they are separated from each other by using four boundaries a, b, c and d. Owing to

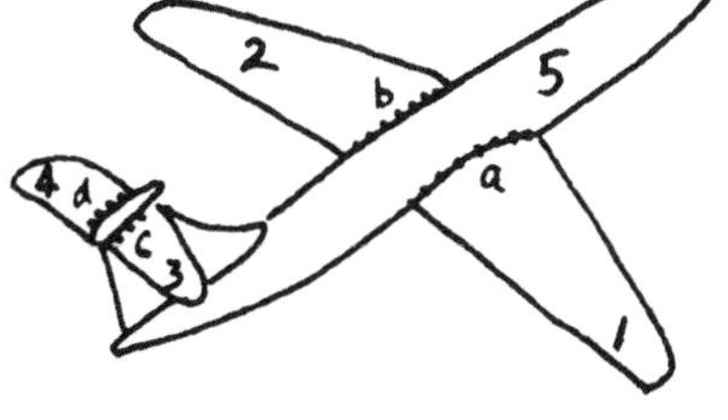

FIG. 7.

there being no coupled nodes between each subsystem, the stiffness matrix of the whole structure may be expressed as follows:

$$\begin{bmatrix} K_1 & & & & K_{1a} & & & & \\ & K_2 & & & & K_{2b} & & & \\ & & K_3 & & & & K_{3c} & & \\ & & & K_4 & & & & K_{4d} & \\ K_{a1} & & & & K_a & & & & K_{a5} \\ & K_{b2} & & & & K_b & & & K_{b5} \\ & & K_{c3} & & & & K_c & & K_{c5} \\ & & & K_{d4} & & & & K_d & K_{d5} \\ & & & & K_{5a} & K_{5b} & K_{5c} & K_{5d} & K_5 \end{bmatrix} \begin{bmatrix} \delta_1 \\ \delta_2 \\ \delta_3 \\ \delta_4 \\ \delta_a \\ \delta_b \\ \delta_c \\ \delta_d \\ \delta_5 \end{bmatrix} = \begin{bmatrix} \phi_1 \\ \phi_2 \\ \phi_3 \\ \phi_4 \\ \phi_a \\ \phi_b \\ \phi_c \\ \phi_d \\ \phi_5 \end{bmatrix} \qquad (24)$$

After condensation, the stiffness matrix may be as follows:

$$\begin{bmatrix} \bar{K}_a & & & & K_{a5} \\ & \bar{K}_b & & & K_{b5} \\ & & \bar{K}_c & & K_{c5} \\ & & & \bar{K}_d & K_{d5} \\ K_{5a} & K_{5b} & K_{5c} & K_{5d} & K_5 \end{bmatrix} \begin{bmatrix} \delta_a \\ \delta_b \\ \delta_c \\ \delta_d \\ \delta_5 \end{bmatrix} = \begin{bmatrix} \bar{\phi}_a \\ \bar{\phi}_b \\ \bar{\phi}_c \\ \bar{\phi}_d \\ \bar{\phi}_5 \end{bmatrix} \qquad (25)$$

$$\begin{aligned} \bar{K}_\beta &= K_\beta - K_{\beta i} K_i^{-1} K_{i\beta} & i &= 1, 2, 3, 4 \\ & & &\downarrow\ \downarrow\ \downarrow\ \downarrow \\ \bar{\phi}_\beta &= \phi_\beta - K_{\beta i} K_i^{-1} \phi_i & \beta &= a, b, c, d \end{aligned} \qquad (26)$$

After elimination of the branch system, the stiffness matrix obtained possesses spare symmetry.

2. The Method of Forming a Boundary Matrix

Now, let's discuss the calculation of $\bar{K}_\beta$ and $\bar{\phi}_\beta$, it can be seen that formula (26) contains $\bar{K}_i^{-1}$; it is very difficult to directly calculate K_i^{-1} when the rank number of K_i is high (usually above 1000). After careful analysis of the formula (26), it is found that the occurrence of K_i^{-1} is expressed in the form of $K_{\beta i} K_i^{-1} K_{i\beta}$, instead of the single form itself.

Hence, the inner nodes may be further divided into two groups—the pure inner nodes and edge inner nodes. The former are the inner nodes which the stiffness coefficients do not couple with the edge nodes. The latter are the inner nodes which the stiffness coefficients do couple with the edge nodes. From the point of view of structure, the edge inner nodes are virtually the inner nodes included in the boundary element. In addition, suppose the total number of degrees of freedom of the pure inner nodes in the *i*th subsystem are n_i; the degrees of freedom of the edge inner-nodes are n_b and of edge node, m.

It still possesses a commonality, when considering the characteristics of $K_{\beta i}K_{ii}^{-1}K_{i\beta}$, and according to the definition of the symbols, the matrix K_{ii}^{-1} may be changed as follows:

$$K_{ii}^{-1} = \left[\begin{array}{c|c} A & B \\ \hline C & D \end{array}\right]; \qquad K_{i\beta} = \left[\begin{array}{c} O \\ \hline E \end{array}\right]$$

The size for each divided part is:

$$A(n_i, n_i) \qquad B(n_i, n_b) \qquad C(n_b, n_i) \qquad D(n_b, n_b) \qquad E(n_b, m)$$

Thus, the required product of the matrix is

$$K_{\beta i}K_i^{-1}K_{i\beta} = [O \mid E^{\mathrm{T}}]\left[\begin{array}{c|c} A & B \\ \hline C & D \end{array}\right]\left[\begin{array}{c} O \\ \hline E \end{array}\right] = E^{\mathrm{T}}DE \tag{27}$$

In this, the rank number is taken to be $m \times m$. It is thus clear that it is only concerned with the submatrix D of K_i^{-1}. The subpart D may be calculated according to the definition of the converse matrix

$$K_iK_i^{-1} = I \tag{28}$$

When the different series of vectors in I are considered as many groups of load matrices, the calculation of the converse matrix is equal to that of combined equations. It may be settled according to the above mentioned procedure. Furthermore, only the subpart D is needed, i.e. only the last vector m in the unit matrix I, it may be written as

$$I_i = \left[\begin{array}{c} O \\ \hline I_{m \times m} \end{array}\right]$$

Thus, the work load for the inverse calculation has been further decreased. The time is greatly shortened compared with the calculation of the K_i division.

Also

$$K_{\beta i}K_i^{-1}\phi_i = [O \mid E^{\mathrm{T}}]\left[\begin{array}{c|c} A & B \\ \hline C & D \end{array}\right]\phi_i - E^{\mathrm{T}}[CD]\phi_i \tag{29}$$

It is clear that the required K_i^{-1} here is more than the subpart C. Because $B = C^{\mathrm{T}}$, and the subpart B has been incidentally calculated in the calculating process of subpart D, so the calculation for C does not increase the time.

3. The Utilization of Structural Character

The utilization of structural character may further simplify the calculation. Because the left wing plane part is symmetrical with the right wing plane part in the wing plane structure, in the part coordinate system, the stiffness matrices are the same in both left and right wing plane parts. Suppose the total coordinate system is identical with that of the right wing plane. The changed matrix of the part coordinate system to the total system is:

$$A = \begin{bmatrix} a & & \\ & a & \\ & & \ddots a \end{bmatrix}, \quad a = \begin{bmatrix} 1 & & \\ & 1 & \\ & & \ddots -1 \end{bmatrix} \tag{30}$$

Thus,

$$\bar{K}_b = (K_b - K_{b2}K_2^{-1}K_{2b}) = A_b^{\mathrm{T}}\bar{K}_a A_b \tag{31}$$

Similarly,

$$\bar{K}_d = A_d^{\mathrm{T}}\bar{K}_c Ad \tag{32}$$

Because the load in the right and left wing plane is different, $\bar{\phi}_\beta$ and $\bar{\phi}_d$ can not be obtained throughout the simple exchange of $\bar{\phi}_a$ and $\bar{\phi}_c$. However,

$$\begin{aligned} \bar{\phi}_b &= \phi_b - K_{b2}K_2^{-1}\phi_2 \\ &= A_b^{\mathrm{T}}\phi_a^1 - (A_b^{\mathrm{T}}K_{a1}A_2)(A_2^{\mathrm{T}}K_1^{-1}A_2)(A_2'\phi_1^1) \\ &= A_b^{\mathrm{T}}(\phi_a^1 - K_{a1}K_1^{-1}\phi_1^1) = A_b^{\mathrm{T}}\bar{\phi}_a^1 \end{aligned} \tag{33}$$

where ϕ_1^1 and ϕ_a^1 represent the loads in the left wing plane and are correspondingly added to the wing right plane.

Similarly,

$$\bar{\phi}_d = A_d^{\mathrm{T}}\bar{\phi}_c^1 \tag{34}$$

Thus, take the right wing-plane as a criterion, the right end-item of the left plane may be calculated; the storage and calculating time of the computer may be greatly decreased.

4. The Item Mergence

Considering there are a few nodes between each plane and fuselage, the step number of the matrix K_β ($\beta = a, b, c, d$) is not high, and usually does not exceed a hundred. In order to simplify the calculation, it is necessary to eliminate the boundary nodes. This can be done via the eliminating technique, i.e. item mergence:

$$\bar{K}_5\sigma_5 = \bar{\phi}_5 \tag{35}$$

where

$$\begin{aligned}\bar{K}_5 &= K_5 - \sum_{\beta=1}^{4} K_{5\beta}\bar{K}_\beta^{-1}K_{\beta 5}\\ \bar{\phi}_5 &= \phi_5 - \sum_{\beta=1}^{4} K_{5\beta}\bar{K}_\beta^{-1}\bar{\phi}_\beta\end{aligned} \tag{36}$$

Now, let's discuss the characteristics of the stiffness matrix $\bar{K}_5$: although K_β^{-1} is a full matrix, $K_{5\beta}$ contains many zero elements, the full matrix occurs only when the matrix product $K_{5\beta}K_\beta^{-1}K_{\beta 5}$ is located in the edge inner nodes; in other cases it is zero. Because it is not at all certain that the edge inner nodes in the main system are located at the front part, it is possible they are distributed inside, i.e. the coupled item is often located at the central part of the matrix. Thus, the matrix may be changed to the following form:

$$K_{5\beta}K_5^{-1}K_{\beta 5} = \begin{bmatrix} 0 \\ \hline \alpha^{\mathrm{T}} \\ \hline 0 \end{bmatrix} K_\beta^{-1}[0 \mid \alpha \mid\,] = \left[\begin{array}{c|c|c} 0 & 0 & 0 \\ \hline 0 & \alpha^{\mathrm{T}}K_\beta^{-1}\alpha & 0 \\ \hline 0 & 0 & 0 \end{array}\right]$$

where the rank number K_β^{-1} is very small, thus the matrix $\bar{K}_5$ still possesses the spare symmetry similar to the K_5. The only difference is that the band width concerning the edge inner nodes has been increased. By using the same method, it is not difficult to calculate the displacement σ_5 of the main system.

NUMERICAL CALCULATION EXAMPLES

1. Some Questions

The following examples are selected, because there are analytic solutions to be compared with. Using symmetry characteristics, a quarter of a plate is used for idealization (Fig. 8). In the following, the numbers of grids refer to a quarter of the plate unless otherwise specified.

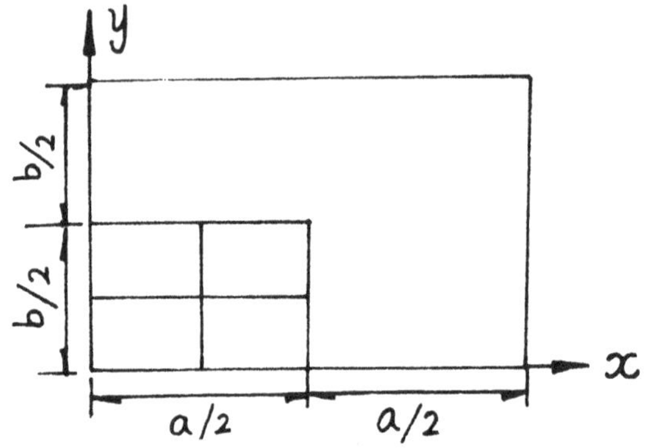

FIG. 8.

(1) A four-ply (0°/90°/90°/0°) simply-supported square plate with identical thickness is subjected to sine distribution load

$$q_0 \sin\frac{\pi x}{a} \sin\frac{\pi y}{b}$$

The mechanical properties:

$$G_{lt}/E_t = 0{\cdot}6 \qquad G_{tt}/E_t = 0{\cdot}5 \qquad \gamma_{lt} = 0{\cdot}25$$

Condition 1: Graphite–epoxy $E_l/E_t = 40$

Condition 2: Glass–epoxy $E_l/E_t = 3$

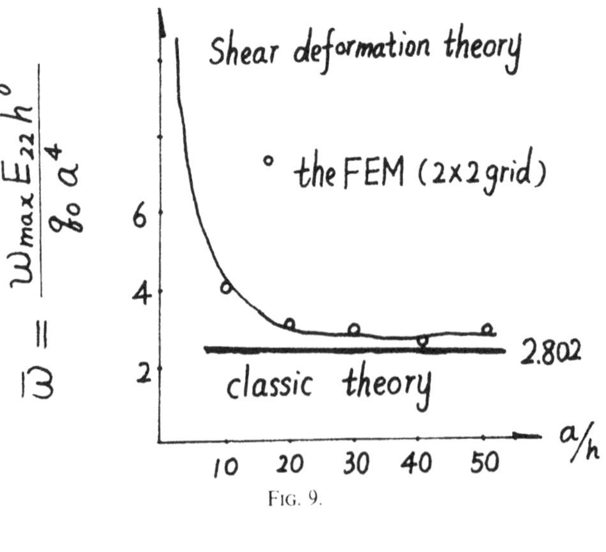

FIG. 9.

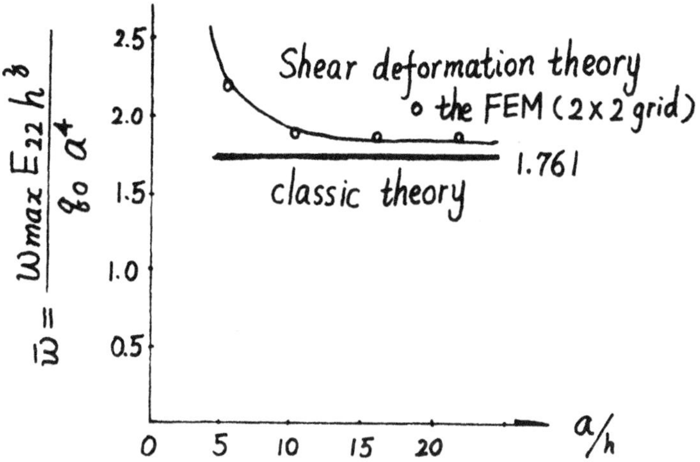

FIG. 10.

The calculation results are shown in Fig. 9 (graphite–epoxy) and Fig. 10 (glass–epoxy), by using a 4 × 4 grid spacing and equivalent load vector.

(2) A three-ply (0°/90°/0°) simply-supported rectangular plate ($b/a = 3$) with identical thickness is subjected to a sine distribution load

$$q_0 \sin\frac{\pi x}{a} \sin\frac{\pi y}{b}$$

The mechanical properties: $E_l/E_t = 25$, $G_{lt}/E_t = 0{\cdot}5$, $G_{tt}/E_t = 0{\cdot}2$, $\gamma_{lt} = 0{\cdot}25$. A 3 × 3 grid spacing and equivalent load vector are used in the calculation. The total degrees of freedom are less than that found in other references. Table 1 shows the results and comparison.

2. A Simulated Flexible Yoke of a Main Rotor Hub (see Fig. 11)

The material was glass–epoxy, type E-51 (618). In order to meet the request for further investigation, we adopted the variable node and

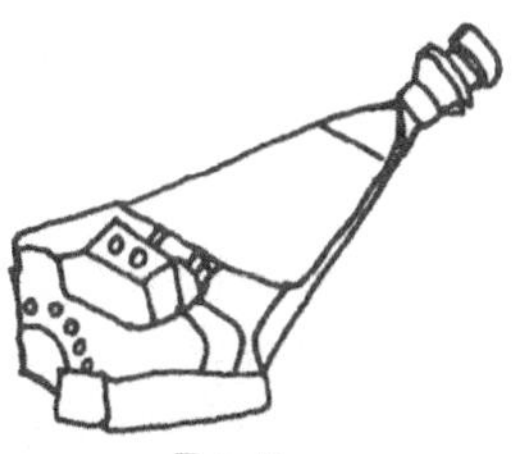

FIG. 11

TABLE 1

a/t	Centre deflection					
	A	B	C	Shell	Solid	Analytic solution
10	1·241	0·802	0·848	0·752	0·785	0·919
20	0·664	0·578	0·594	0·565	0·591	0·610
50	0·529	0·515	0·523	0·513	0·524	0·520
100	0·510	0·506	0·514	0·505	0·504	0·508
Classic theoretical solution 0·503						

Notes: 1. $W = W100E_t t^3/q_0a^4$.
2. A, Mawenga;[5] B, Reddy;[3] C, Wang Xing-wei *et al.*[13] Analytic solution, Pagano and Hatfield.[4]
3. The examples above were calculated by Mr Wand Shan-xin and Mr Zhang-jie.

out-element node in the transition between loose and dense grids, and made the following calculation

(1) Using variable node elements for the transition between different grids in the root area.
(2) Using variable node elements for the transition between different grids in the central area.
(3) Using out-element node elements for the transition between different grids in the root area.

The calculated results are as shown in Table 2.

TABLE 2

Condition	1	2	3
Max. displacement	48·55	56·32	81·54
Error	39·8%	30%	1·05%
Determining value	80·72	80·72	80·72

It can be seen from the results that we need to be careful to use the variable node elements and never use the non-even nodes; the out-element node elements lead to the same accuracy as the isoparametric elements. Thus the number of elements as well as the rank number have been decreased, and the working time and storage of computer have been minimized.

3. A Metal Main Rotor Hub of Some Types of Helicopter

Comparison of results from analysis and testing are in good agreement. Details are shown in Refs 11 and 12.

It can be seen from the above examples that the element created by using the described methods possesses the advantages of accuracy and convergency. Thus the finite element analysis method developed in this report is proved to be successful.

THE STRUCTURAL MAXIMUM STRENGTH

1. Equivalent Stress and Failure Criteria

Equivalent stress is determined by using Tsai–Hill's criteria of the planar stress conditions in accordance with the characteristics of the honeycomb sandwich and composite laminated plate. The equivalent stress of each ply is as follows:

$$\bar{\sigma}_i = \left[\sigma_l^2 - \sigma_l\sigma_t + \left(\frac{x}{y}\right)\sigma_t^2 + \left(\frac{x}{s}\right)\tau_{lt}^2\right]_i^{1/2}$$

and the failure criteria is $\bar{\sigma} \geq x$, where x, y and s represent, respectively, the main directional ultimate strengths of each ply. The signs of x and s are dependent on the signs of the working stress σ_l and σ_t.

When $\bar{\sigma} \geq x$, the following formula can be used to determine the failure mode:

$$\text{arc tg}(\sigma_t/\sigma_l) > \text{arc tg}(y/x) \tag{a}$$

$$\text{arc tg}(\tau_{lt}/\sigma_l) > \text{arc tg}(s/x) \tag{b}$$

Type 1. (a) is tenable, the matrix of the composite fails with a transverse tension–compression failure mode.

Type 2. If (b) is tenable, the matrix fails with a shearing failure mode.

Type 3. If both (a) and (b) are untenable, the fibre fails longitudinally with a tension–compression failure mode.

Using von Mises' criteria for metal structure, the equivalent stress is defined as:

$$\bar{\sigma} = \left\{\tfrac{1}{2}[(\sigma_x - \sigma_y)^2 + (\sigma_y - \sigma_z)^2 + (\sigma_z - \sigma_x)^2] + 3(\tau_{xy}^2 + \tau_{yz}^2 + \tau_{zx}^2)\right\}^{1/2}$$

When $\bar{\sigma} = \sigma_s$, the material begins to yield.

2. Variable Stiffness Method

The load causing the first layer to cease to be effective (or element yields) is defined as the critical load. When the load is greater than the critical load and increases continuously, the number of layers which fail will be increased gradually and thus will continuously change the system stiffness matrix. It is clear that calculation of the maximum structure is virtually a nonlinear analysis problem.

When the layer begins to fail (or element yields) a variable loading method is adopted. The corresponding displacement, strain and stress are δ_0, ε_0 and σ_0 respectively. Based on this, the active load is increased, and forms a corresponding stiffness matrix. For the stiffness matrix of the metal structure, it should be noted that the elasto-plastic matrix needs to be substituted for an elastic matrix in the yielded element, while the stresses adopted are the load-increased stresses σ_{i-1}. Thus the stiffness matrix is related to the former step stress and written as $K(\sigma_{i-1})$.

Solving the following basic equation

$$K(\sigma_{i-1})\,\Delta\delta_i = \Delta\phi_i$$

for $\Delta\delta_i$, $\Delta\varepsilon_i$ and $\Delta\sigma_i$, thus the displacement, strain and stress after the load increase are obtained, and the relationship between the stresses is $\sigma_i = \sigma_i + \Delta\sigma_i$.

The number of layers which fail increases with load and $\Delta\phi_i$ decreases with the load. This iteration should continue till the structure changes into a mechanism. The maximum structural strength is then determined, by considering the dispersion factor.

REFERENCES

1. PANDA, S. C. and NALAREJAN, R., Finite element analysis of laminated composite plates, *Int. J. Num. Meth. Engng*, **14**, No. I (1979), 69–79.
2. VINSON, J. R. and CHOU, T. W., *Composite Materials and their Use in Structures*, London, Elsevier Applied Science, 1975.
3. REDDY, J. N., A penalty plate-bending element for the analysis of laminated anisotropic composite plates, *Int. J. Num. Meth. Engng*, **15**, No. 8 (1980), 1187–1206.
4. PAGANO, N. J. and HATFIELD, S. J., Elastic behaviour of multilayered bidirectional composites, *AIAA* J., **10** (1972), 931–933.
5. MAWENGA, A. S. and DAVIES, J. D., Finite element bending analysis of multilayer plates, *Int. J. Num. Meth. Engng*, **8** (1974), 215–225.
6. LIU XIA-SHI, The finite element analysis for large dimension combination structure, *Aviation Journal*, **II** (1979).

7. WANG SHOU-MEI, The consistent element with out-element nodes, *Aviation Journal*, **II** (1980).
8. YIN ZE-YONG *et al.*, The finite element stress analysis for solid-shell combination components of engine, China Aviation Science Technique Literature, HJB840186.
9. LI RUI-XIA, The variable number nodes isoparametric elements and their application in axis symmetric problems, *Fudan Journal* (June 1982).
10. LI RUI-XIA, A 6-15 variable nodes triangular prism isoparametric element, *Computational Mathematics Journal of Universities* (Sept. 1984).
11. LIU XIA-SHI *et al.*, The overall stress analysis for helicopter main rotor hub yoke, *Aviation Science & Technique*, HK 78026.
12. LIU XIA-SHI *et al.*, The stress analysis for helicopter rotor hub, *Aviation Science & Technique*, HK 78027.
13. WANG XING-WEI *et al.*, Finite element analysis for composite laminated plates, *NAI Journal*, **III** (1983).

44

Large-scale Structural Analysis/Synthesis of Composite Structures by Finite Elements

STEVAN MAKSIMOVIĆ

Aeronautical Institute Niška b.b.,
11133 Žarkovo-Belgrade, Yugoslavia

ABSTRACT

This chapter presents work done toward the development of an efficient method for the structural analysis/synthesis of laminate fibre composite structures by using finite element techniques. An automated procedure is presented for designing minimum weight structures subject to strength and elastic stability constraints. In order to efficiently solve the optimization problem a variety of state-of-the-art techniques are used here. These include design variable linking, constraint deletion, the use of reciprocal variables and formal approximation techniques. Analytical expressions are employed for evaluating constraint gradients, which are calculated for active constraints. Eigenvalue buckling analysis and hybrid constraint approximations for laminate plates are introduced in this chapter. Using this procedure and efficient algorithm in process optimization, the number of actual finite element analyses is kept to a minimum. Several example problems are given to demonstrate that the method presented offers an efficient and practical optimum design procedure for different problems of aircraft structures treated.

1. INTRODUCTION

Optimal design of large-scale systems such as aircraft fibre composite structures is usually based on combining finite element analysis and optimization techniques into an integrated computer system. To achieve minimum mass, designs must be provided that simultaneously satisfy a multitude of local and global aircraft constraints, such as material strength, minimum gauge, buckling and flutter constraints. Structural optimization problems are generally characterized by (a) a large number of design variables, (b) a simple objective function and (c) indirect but well-behaved constraint functions.

The analysis of discretized structures is performed by finite element methods. The class of FE models considered in this chapter will be restricted to thin-walled structure assemblies of bars, membranes and shear panels. Each membrane or shear panel is made of one or several orthotropic layers. In this chapter an optimality criteria approach[1,2] for the optimum design of composite structures under multiple loading cases is used. The recurrence relationship for stress, displacement and buckling constraints, which result from Kuhn–Tucker necessity conditions for each type of constraint, are incorporated into a design algorithm. Herein the approximation concepts approach[3] is combined with an optimality criterion method to produce a very efficient algorithm for structural analysis/synthesis of large-scale structures.

2. FORMULATION OF THE OPTIMIZATION PROBLEM

The problem is to minimize the weight of a composite structures modelled with bar, membrane and shear panel finite elements. In its general form, the minimum weight problem is expressed as

$$\text{Minimize} \sum_{i=1}^{N} w_i X_i \tag{2.1}$$

Subject to behaviour and side constraints,

$$G_j(X) \equiv \bar{G}_j - G_j \geq 0 \qquad j \in Q_R \tag{2.2}$$

$$X_i^L \leq X_i \leq X_i^U \qquad i = 1, \ldots, N \tag{2.3}$$

where w_i is weight of ith element when X_i is equal to one, X_i is the design variable assigned to element i (e.g. bar area, thicknesses of each ply of the composite stack for each membrane and shear panel elements), N is the number of elements in the structural model, and X_i^L and X_i^U are lower and upper limits of X_i. The inequalities (2.2) represent behaviour constraints, which impose limitations on quantities describing the structural response (stresses, displacements and buckling loads). In order to establish the optimality conditions for the problem defined by eqns (2.1)–(2.3) we need the associated Lagrangian:

$$L(X,\gamma) = \sum_{i=1}^{N} w_i X_i + \sum_{j \in Q_R} \gamma_j (G_j - \bar{G}_j) \tag{2.4}$$

where γ_j is an undetermined Lagrangian multiplier. Differentiation of eqn. (2.4) with respect to the primary design variables once more will yield parts of the Kuhn–Tucker conditions, which must be satisfied by the optimum design.

3. CONSTRAINT APPROXIMATIONS

In order to reduce the number of detailed finite element structural analysis needed to obtain an optimum design, it is appropriate to construct explicit approximations for the constraints (2.2) retained during 0th stage of the design optimization process. The form of the constraint approximation is important, since it can have a significant effect on the computational efficiency of the search procedure. For some constraints, such as buckling, the use of constraint approximations based on first-order Taylor series may result in significant errors. Here a hybrid approximation is used for constraints which is given in the form

$$G_{j\mathrm{H}}(X) = G_j(X_j^0) + \sum_{i \in Q_R} A_i \left(\frac{\partial X_j}{\partial X_i}\right)^0 \tag{3.1}$$

where

$$A_i = \begin{cases} (X_i - X_i^0) & \text{if } \left(\dfrac{\partial X_j}{\partial X_i}\right)^0 < 0 \\ \dfrac{X_i^0}{X_i}(X_i - X_i^0) & \text{if } \left(\dfrac{\partial X_j}{\partial X_i}\right)^0 > 0 \end{cases} \tag{3.2}$$

where the superscript zero denotes quantities evaluated at the current design point X^0. With this approximation a given constraint may have a linear approximation with respect to one design variable and an inverse approximation with respect to another design variable.

3.1. The Buckling Constraints

The buckling constraints are constructed from the solutions of the eigenproblems associated with the finite plate elastic stability analysis and they are expressed in terms of buckling load factors λ_j. The critical load factors are defined through an eigenproblem:

$$Kq_j - \lambda_j K_G q_j = 0 \tag{3.1.1}$$

where K is the stiffness matrix, K_G is the geometric stiffness matrix and q_j is the eigenvector associated with the jth eigenvalue λ_j. The buckling load of the structure is given by $F_{CR} = \lambda_j F$, where λ_j is the lowest eigenvalue and F a set of applied loads. The constraint equation (2.2) for buckling can be written as

$$G_j \equiv \lambda_j - \bar{\lambda} \tag{3.1.2}$$

where $\bar{\lambda}$ is the lowest critical load factor. From eqn. (3.1.1) the eigenvalue λ_j is given as

$$\lambda_j = \frac{q_j^T K q_j}{q_j^T K_G q_j} \tag{3.1.3}$$

The optimization algorithm considered here requires the first partial derivatives of the constraints with respect to the design variables X_i. This gives

$$\frac{\partial \lambda_j}{\partial X_i} = \frac{1}{X_i} \frac{q_{ji}^T k_i q_{ji}}{q_j^T K_G q_j} \tag{3.1.4}$$

where k_i is the stiffness matrix and q_{ji} is the component of the buckling mode associated with the ith element. In order to construct explicit approximations of the buckling constraints it is necessary to obtain the first partial derivatives of λ_j with respect to flexural rigidities D_{rs} of the equivalent orthotropic plates. By using flexural rigidities D_{rs} eqn. (3.1.4) yields

$$\frac{\partial \lambda_j}{\partial D_{rs}} = \frac{q_{ji}^T \left| \frac{\partial k_i}{\partial D_{rs}} \right| q_{ji}}{q_j^T K_G q_j} \tag{3.1.5}$$

where (see Appendix)

$$\frac{\partial k_i}{\partial D_{rs}} = \frac{ab}{4} \left| \delta_{mn} \delta_{nq} \frac{\partial f_{mn}}{\partial D_{rs}} \right| \tag{3.1.6}$$

and

$$\frac{\partial f_{mn}}{\partial D_{11}} = \left(\frac{\pi m}{a}\right)^4 \qquad \frac{\partial f_{mn}}{\partial D_{12}} = 2\pi^4\left(\frac{mn}{ab}\right)^2$$
$$\frac{\partial f_{mn}}{\partial D_{22}} = \left(\frac{\pi n}{b}\right)^4 \qquad \frac{\partial f_{mn}}{\partial D_{66}} = 4\pi^4\left(\frac{mn}{ab}\right)^2 \tag{3.1.7}$$

The Lagrangian for the constraint of static stability can be written as

$$L(X,\gamma) = \sum_{i=1}^{N} w_i X_i + \sum_{j\in Q_R} \gamma_j(\lambda_j - \bar{\lambda}) \tag{3.1.8}$$

The optimality criterion is obtained by differentiating eqn. (3.1.8) with respect to X_i and setting the results equal to zero. This gives

$$w_i + \sum_{j\in Q_R} \gamma_j \frac{\partial \lambda_j}{\partial X_i} = 0 \tag{3.1.9}$$

Substituting eqn. (3.1.4) into this equation, the optimality criterion can be written as

$$1 = \sum_{j\in Q_R} \bar{\gamma}_j \frac{\bar{Q}_{ij}}{w_i X_i^2} \tag{3.1.10}$$

where

$$\bar{Q}_{ij} = X_i q_j^{\mathrm{T}} k_i q_j \tag{3.1.11}$$

and

$$\bar{\gamma}_j = \frac{\gamma_j}{q_j^{\mathrm{T}} K_{\mathrm{G}} q_j} \tag{3.1.12}$$

The redistribution of the material is carried out by using a recurrence relation. The recurrence relation modifies the design variables so that in the design space the initial design is moved towards a design that satisfies the optimality criterion. These relations for the stability constraints can be obtained from the optimality criterion equation (3.1.10). Here are given the three types of recurrence relation in the form

$$X_i^{n+1} = X_i^n \left[\sum_{j \in Q_R} \bar{\gamma}_j \frac{\bar{Q}_{ij}}{w_i X_i^2} \right]_n^{1/r} \tag{3.1.13}$$

$$X_i^{n+1} = X_i^n \left[1 + \frac{1}{r} \left(\sum_{j \in Q_R} \bar{\gamma}_j \frac{\bar{Q}_{ij}}{w_i X_i^2} - 1 \right) \right]_n \tag{3.1.14}$$

and

$$X_i^{n+1} = X_i^n \left[1 - \frac{1}{r} \left(\sum_{j \in Q_R} \bar{\gamma}_j \frac{\bar{Q}_{ij}}{w_i X_i^2} - 1 \right) \right]_n^{-1} \tag{3.1.15}$$

where $n + 1$ and n are the iteration numbers and r is the step size parameter.

Finally, by using eqns (3.1) and (3.2) a hybrid approximation of the buckling constraints can be written as

$$G_{jb}(X) \equiv \lambda_j(X_j^0) - 1 + \sum_{j \in Q_R} A_i \left(\frac{\partial \lambda_j}{\partial X_i} \right)^0 > 0 \tag{3.1.16}$$

where gradients $(\partial \lambda_j / \partial X_i)$ are given by eqns (3.1.4) and (3.1.5). It is known that buckling loads are very sensitive to changes in the values of the design variables. Using a constraint approximation as a linear or inverse function of the design variable may result in significant errors. By using a conservative constraint approximation (3.1.16) the effect of these errors can be minimized.

3.2. Stress Constraints

In this analysis the membrane is treated as a composite stack of orthotropic layers. Stresses in each layer i in the directions of orthotropy are given as

$$\sigma_i = E_i(\varepsilon_i - \alpha_i \Delta T) \tag{3.2.1}$$

with

$$\sigma_i = (\sigma_1, \sigma_2, \tau_{12})^{\mathrm{T}} \tag{3.2.2}$$

where E_i is the constitutive matrix of the orthotropic membrane, α_i are the thermal expansion coefficients and T is the temperature. Here the Hill–Tsai criterion is used in failure analysis of each layer. By using eqns (3.1) and (3.2) hybrid approximation of constraints may be written as

$$G_{jT}(X) \equiv 1 - T_j(X_j^0) - \sum_{i \in Q_R} A_i \frac{\partial T_j}{\partial X_i}(X^0) > 0 \tag{3.2.3}$$

with

$$T_j = \left| \left(\frac{\sigma_1}{F_1}\right)^2 + \left(\frac{\sigma_2}{F_2}\right)^2 - \frac{\sigma_1 \sigma_2}{RF_1F_2} + \left(\frac{\tau_{12}}{F_{12}}\right)^2 \right|^{1/2} \tag{3.2.4}$$

in which F_1, F_2 and F_{12} are the stresses of failure in uniaxial tension, compression and shear respectively.

4. THE SENSITIVITY ANALYSIS

The algorithm for structural optimization considered here needs information about the gradients of the structural weight and constraints. The gradients are computed from finite element matrices and their derivatives. The derivatives can be computed numerically or analytically. In this chapter an analytical approach is applied to four-node membrane and shear panel isoparametric elements. Here displacement constraints only are treated but the procedure is similar for all constraints. The finite element problem can be formulated as in Ref. 4:

$$ku = F \tag{4.1}$$

The displacement behaviour constraints and their derivatives with respect to design variables are given by

$$G_i(x) = 1 - \frac{u}{\bar{u}} > 0 \tag{4.2}$$

$$\frac{\partial G_i}{\partial \bar{x}} = -\frac{1}{\bar{u}} \frac{\partial u_i}{\partial \bar{x}} \tag{4.3}$$

where $\bar{u}$ denotes the allowable displacement. The derivatives of eqn. (4.1) with respect to any design variables x_i will be

$$K \frac{\partial u}{\partial x_i} + \frac{\partial K}{\partial x_i} u = \frac{\partial F}{\partial x_i} \tag{4.4}$$

The key to the sensitivity analysis is how to calculate the derivatives K and F with respect to x_i. The element stiffness matrix K_e is given in Ref. 4:

$$K_e = \int\int B^T DB|J| \, d\xi \, d\eta \tag{4.5}$$

The derivatives of the element stiffness matrix with respect to x_i are given by

$$\left|\frac{\partial K}{\partial x_i}\right|_e = \int\int \left(\frac{B^T}{x_i} DB|J| + B^T D \frac{\partial B}{\partial x_i} |J| + B^T DB \frac{\partial |J|}{\partial x_i} \right) d\xi \, d\eta \tag{4.6}$$

$$\left|\frac{\partial K}{\partial x_i}\right|_e = \sum_j^{NG} \sum_k^{NG} W_j W_k \left(\frac{\partial B^T}{\partial x_i} DB|J| + B^T D \frac{\partial B}{\partial x_i} |J| + B^T \frac{\partial |J|}{\partial x_i} \right)_{jk} \tag{4.7}$$

where W_j and W_k are the weighting factors of numerical integration, NG is the number of Gauss rules adopted, and $(\ )_{ik}$ is the integrand function of a Gauss sampling point with local coordinates ξ_j and η_k. In a similar manner all derivatives can be derived.

5. NUMERICAL RESULTS

In this section, numerical solutions of design problems formulated in previous sections are given, and the effect of various problem parameters on the efficiency and accuracy is investigated. Here is considered structural weight optimization of delta wing structures subject to displacements, strength and buckling constraints. This example is taken from Refs 5 and 6 but here is the coverplate modelled with fibre composite hat-stiffened panels, as shown in Figs 1, 2 and 3.

The skin panel is modelled using orthotropic isoparametric membrane elements and webs with shear panel elements.

Design variable linking is used to reduce the number of distinct fibre composite hat-stiffened panels to 16 (Fig. 3). The number of design variables involved in describing the skin is $16 \times 3 = 48$. Similarly, design variable linking is used for webs leading to 12 independent design variables for shear panels (Fig. 2). Lower limits are imposed on the design variables as follows: the lower limit is 0·02 on the thicknesses of the shear panel elements and 0·001 on the thicknesses of membrane elements. The wing was subjected to a single static load, uniformly distributed, $p = 144$. The geometric properties of hat-stiffened panels are given in Table 1. The local buckling is based on the assumption that each plate element (i.e. elements I

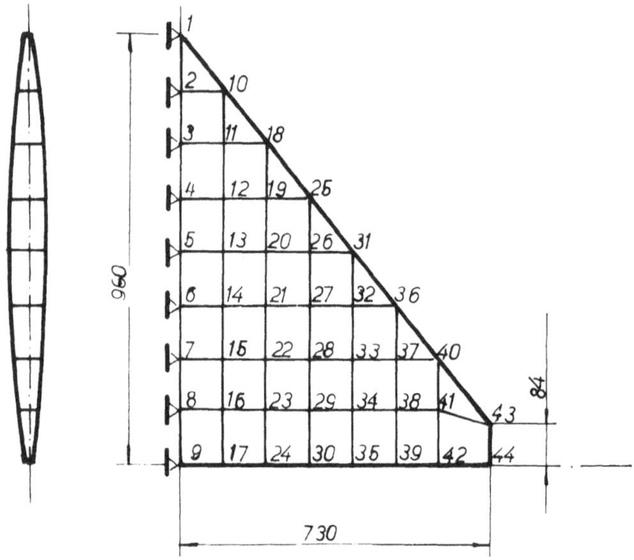

FIG. 1. Delta wing analysis model.

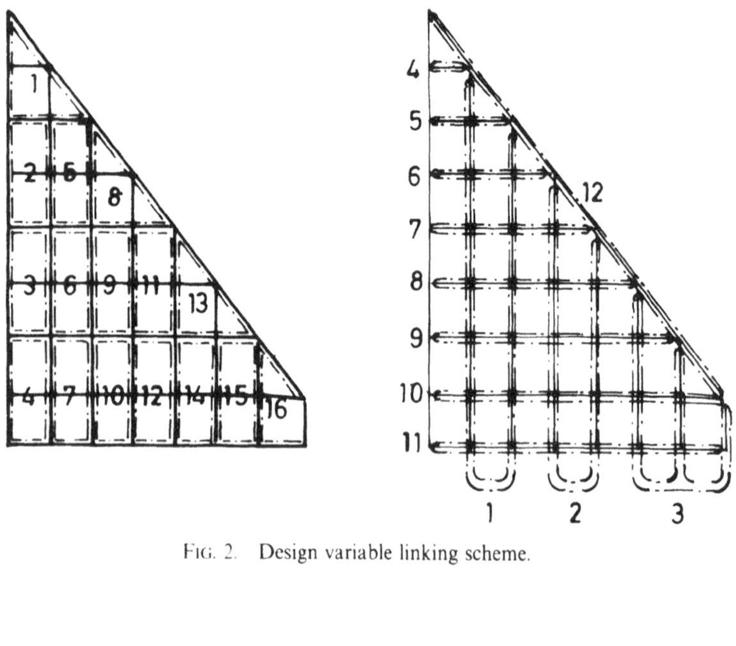

FIG. 2. Design variable linking scheme.

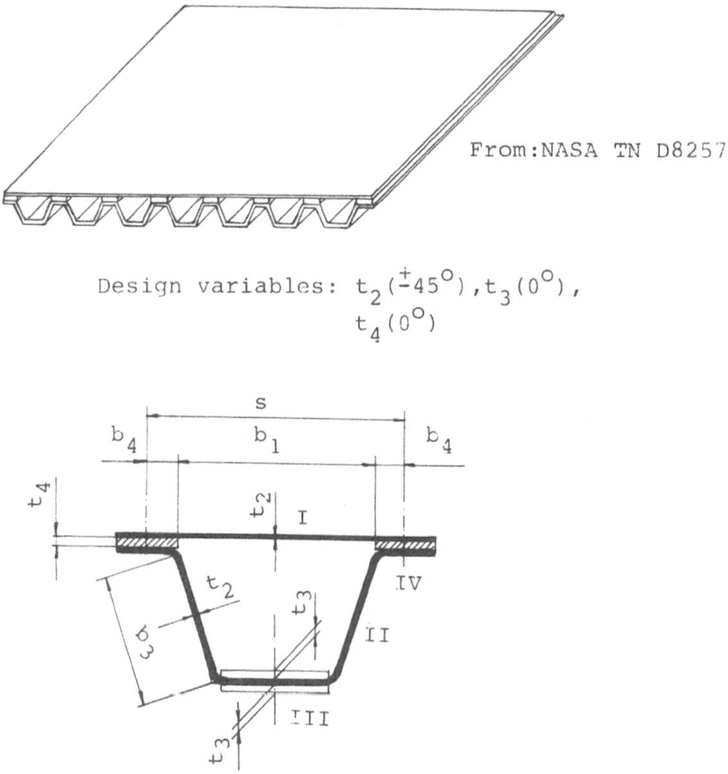

FIG. 3. Fibre-composite hat-stiffened panel.

TABLE 1
Geometric properties of hat-stiffened panels

Linked group number	b_1	b_2	b_3	b_4
1	1·0821	2·9830	0·9331	1·9672
2	1·5650	2·5271	1·1038	3·0533
3	1·2435	3·0320	0·9885	0·6545
4	1·3903	3·7200	1·0425	1·0033
5	3·3314	3·2082	2·9097	1·0074
6	1·5818	3·4545	1·1720	0·7639
7	1·2782	3·8752	1·0120	0·58114
8	4·8489	3·0544	2·0072	4·9034

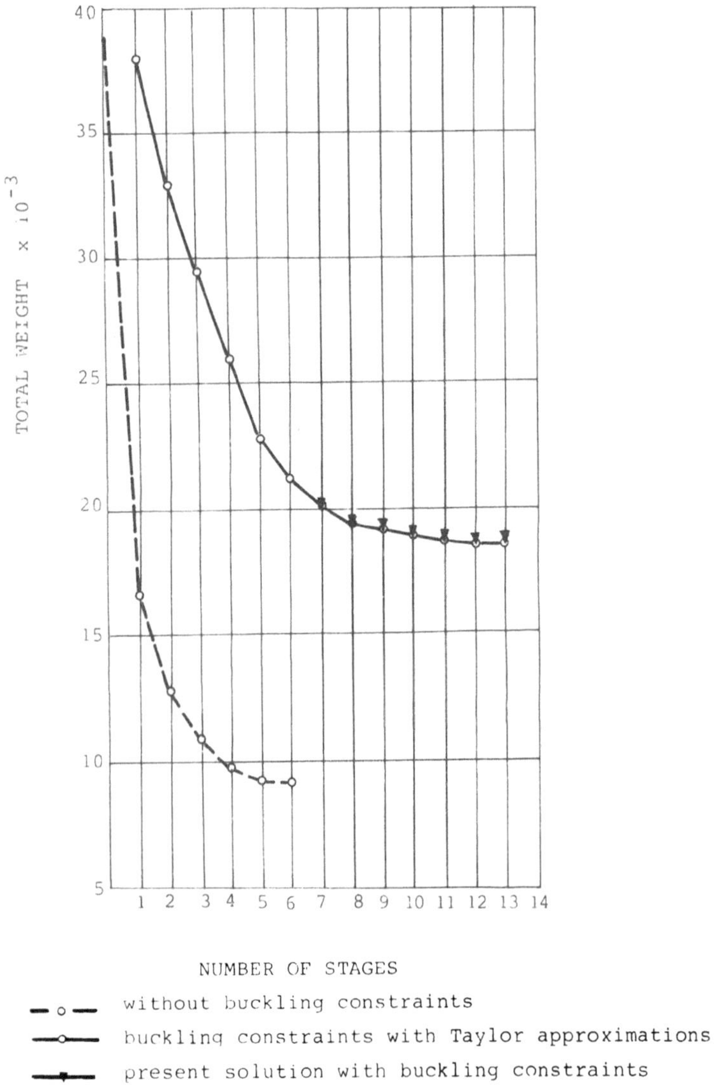

FIG. 4. Iteration histories for a delta wing.

TABLE 2
Material properties for delta wing

Composite material		Isotropic material
$E_L = 21 \times 10^6$	$F_T^t = 8 \times 10^3$	$E = 16{\cdot}4 \times 10^6$
$E_T = 1{\cdot}7 \times 10^6$	$F_T^c = 3 \times 10^4$	$\rho = 0{\cdot}16$
$G_{LT} = 0{\cdot}65 \times 10^6$	$F_{LT} = 1{\cdot}2 \times 10^4$	$\nu = 0{\cdot}3$
$F_L^t = 1{\cdot}8 \times 10^6$	$\rho = 0{\cdot}056$	$\sigma^U = \sigma^L = 125 \times 10^3$
$F_L^c = 1{\cdot}8 \times 10^5$	$\nu_{LT} = 0{\cdot}35$	

through IV of Fig. 3) is simply supported along the lines of the attachment plate element. The material properties of delta wing are given in Table 2.

The results of the structural weight optimization of delta wings are given in the form of total weight histories (Fig. 4). Here are considered three cases. In the first case the Hill–Tsai criterion is included as a constraint. In the second case Hill–Tsai and buckling constraints are used by using a Taylor series for constraint approximation.[7] Finally, in the third case Hill–Tsai and buckling constraints are used but here a hybrid approximation is introduced for buckling constraints. Figure 4 shows that inclusion of local buckling constraints drastically increases the total weight of the wing structure. Hybrid constraint approximations for buckling give conservative solutions.

6. CONCLUSIONS

The method in this chapter represents structural analysis/synthesis capability for complex, thin-walled composite structures modelled by finite elements. The practicability and its ability to deal efficiently with large-scale complex structures was clearly demonstrated by results of the weight optimization of composite wing structures with hat-stiffened coverplates. It has been shown that the combined use approximation concepts, optimality criteria and hybrid approximations for buckling constraints can be successfully applied to provide optimization of large-scale composite structures.

REFERENCES

1. FLEURY, C. and GERADIN, M., Optimality criteria and mathematical programming in structural weight optimization, *Comp. Struct.*, **8** (1978).

2. KHOT, N. S., BERKE, L. and VENKAYYA, V. B., Comparison of optimality criteria algorithms for minimum weight design of structures, *AIAA J.*, **17** (1978).
3. FLEURY, C. and SCHMIT, L. A., Dual method and approximation concepts in structural synthesis, NASA CR 3226, 1980.
4. ZIENKIEWICZ, O. C., *The Finite Element Method in Engineering Science*, London, McGraw-Hill, 1977.
5. SCHMIT, L. A. and MIURA, H., An advanced structural analysis/synthesis capability—ACCESS 2, *IJNME* **12** (1978), 353–377.
6. MAKSIMOVIĆ, S., Structural analysis/synthesis of laminated fibre composite structures, SAS—World Conference STRUCAD 86, October 1986, Paris.
7. MAKSIMOVIĆ, S., Optimization of composite structures by finite elements, First World Congress on Computational Mechanics, Sept. 22–26, 1986, Austin, Texas.
8. JONES, R. M., *Mechanics of Composite Materials*, London, McGraw-Hill, 1975.

APPENDIX

Here are given the basic parameters for buckling analysis of a rectangular homogeneous orthotropic plate subject to uniform inplane loading N_x, N_y, N_{xy}. The stiffness matrix is given by

$$K = \frac{ab}{4} |\delta_{mp} \quad \delta_{nq} \quad f_{mn}| \tag{A.1}$$

with

$$f_{mn} = \pi^4 \left| D_{11}\left(\frac{m}{a}\right)^4 + 2(D_{12} + 2D_{66})\left(\frac{m}{2}\right)^2\left(\frac{n}{2}\right)^2 + D_{22}\left(\frac{n}{6}\right)^4 \right| \tag{A.2}$$

$$\delta_{mp} = \begin{cases} 1 & \text{if } m = p \\ 0 & \text{if } m \neq p \end{cases} \qquad \delta_{nq} = \begin{cases} 1 & \text{if } m = q \\ 0 & \text{if } n \neq q \end{cases}$$

where a and b are dimensions of the plate. Given $(\partial\lambda_j/\partial D_{rs})$, the first partial derivatives of λ_j with respect to thicknesses as design variables (t_i) can be obtained from the expression

$$\frac{\partial \lambda_j}{\partial t_i} = \sum_{r,s} \frac{\partial \lambda_j}{\partial D_{rs}} \frac{\partial D_{rs}}{\partial t_i} \tag{A.3}$$

with

$$\begin{aligned} \frac{\partial D_{rs}}{\partial t_i} &= \frac{1}{12}\left|\left(\frac{\partial A_{rs}}{\partial t_i}\right)t^2 + 2A_{rs}t\right| \qquad && r, s = 1, 2, 6 \\ \frac{\partial A_{rs}}{\partial t_i} &= (C'_{rs})_i && r, s = 1, 2, 6 \end{aligned} \tag{A.4}$$

where constants C'_{rs} are defined in Ref. 8.

The geometric stiffness matrix is given by

$$K_G = \left| 8N_{xy}\beta_{mnpq} - \delta_{mp}\,\delta_{nq}\frac{\pi^2}{4}\left(\frac{b}{a}m^2N_x + \frac{a}{b}n^2N_y\right)\right|$$

where

$$\beta_{mnpq} = \left\{\begin{matrix} 0 \quad \text{if } p = m \text{ or } q = n \\ \dfrac{mnpq\,\delta^{pm}\,\delta^{qn}}{(p^2 - m^2)(q^2 - n^2)} \end{matrix}\right\}$$

$$\delta^{pm} = \left\{\begin{matrix} 1 \quad \text{if } (p + m) \text{ odd} \\ 0 \quad \text{if } (p + m) \text{ even} \end{matrix}\right\}$$

and

$$\delta^{nq} = \left\{\begin{matrix} 1 \quad \text{if } (q + n) \text{ odd} \\ 0 \quad \text{if } (q + n) \text{ even} \end{matrix}\right\}$$

In a similar manner K and K_G may be determined for nonuniform in-plane loading.

Index of Contributors

Subject Index

SPRINGER NATURE

GPSR Compliance

The European Union's (EU) General Product Safety Regulation (GPSR) is a set of rules that requires consumer products to be safe and our obligations to ensure this.

If you have any concerns about our products, you can contact us on ProductSafety@springernature.com

In case Publisher is established outside the EU, the EU authorized representative is:

Springer Nature Customer Service Center GmbH
Europaplatz 3
69115 Heidelberg, Germany

FSC
www.fsc.org
MIX
Papier | Fördert gute Waldnutzung
FSC® C083411

Kolibri 360 GmbH
Ferdinand-Jühlke-Straße 7
99095 Erfurt, Deutschland
produktsicherheit@kolibri360.de